Understanding and measuring the shelf-life of food

Related titles from Woodhead's food science, technology and nutrition list:

The stability and shelf-life of food (ISBN 1 85573 500 8)
The stability and shelf-life of a food product are critical to its success in the market place, yet companies experience considerable difficulties in defining and understanding the factors that influence stability over a desired storage period. This book is the most comprehensive guide to understanding and controlling the factors that determine the shelf-life of food products.

Taints and off-flavours in foods (ISBN 1 85573 449 4)
Taints and off-flavours are a major problem for the food industry. Part I of this important collection reviews the major causes of taints and off-flavours, from oxidative rancidity and microbiologically-derived off-flavours, to packaging materials as a source of taints. The second part of the book discusses the range of techniques for detecting taints and off-flavours, from sensory analysis to instrumental techniques, including the development of new rapid on-line sensors.

Colour in food – Improving quality (ISBN 1 85573 590 3)
The colour of a food is central to consumer perceptions of quality. This important new collection reviews key issues in controlling colour quality in food, from the chemistry of colour in food to measurement issues, improving natural colour and the use of colourings to improve colour quality.

Details of these books and a complete list of Woodhead's food science, technology and nutrition titles can be obtained by:

- visiting our web site at www.woodhead-publishing.com
- contacting Customer Services (email: sales@woodhead-publishing.com; fax: +44 (0) 1223 893694; tel.: +44 (0) 1223 891358 ext. 30; address: Woodhead Publishing Limited, Abington Hall, Abington, Cambridge CB1 6AH, UK)

Selected food science and technology titles are also available in electronic form. Visit our web site (www.woodhead-publishing.com) to find out more.

If you would like to receive information on forthcoming titles in this area, please send your address details to: Francis Dodds (address, tel. and fax as above; e-mail: francisd@woodhead-publishing.com). Please confirm which subject areas you are interested in.

Understanding and measuring the shelf-life of food

Edited by
R. Steele

CRC Press
Boca Raton Boston New York Washington, DC

WOODHEAD PUBLISHING LIMITED
Cambridge England

Published by Woodhead Publishing Limited
Abington Hall, Abington
Cambridge CB1 6AH
England
www.woodhead-publishing.com

Published in North America by CRC Press LLC
2000 Corporate Blvd, NW
Boca Raton FL 33431
USA

First published 2004, Woodhead Publishing Limited and CRC Press LLC

British Library Cataloguing in Publication Data
A catalogue record for this book is available from the British Library.

Library of Congress Cataloging-in-Publication Data
A catalog record for this book is available from the Library of Congress.

Woodhead Publishing Limited ISBN 1 85573 732 9 (book); 1 85573 902 X (e-book)
CRC Press ISBN 0-8493-2556-0
CRC Press order number: WP2556

The publisher's policy is to use permanent paper from mills that operate a sustainable forestry policy, and which have been manufactured from pulp which is processed using acid-free and elementary chlorine-free practices. Furthermore, the publisher ensures that the text paper and cover board used have met acceptable environmental accreditation standards.

Project managed by Macfarlane Production Services, Markyate, Hertfordshire
(e-mail: macfarl@aol.com)
Typeset by MHL Typesetting Limited, Coventry, Warwickshire
Printed by TJ International Limited, Padstow, Cornwall, England

Contents

Contributor contact details

Chapter 1

Professor R. P. Singh* and
B. A. Anderson
Department of Biological and
Agricultural Engineering
University of California, Davis
One Shields Avenue
Davis
CA 95616
USA

Tel: 530 752 0811
Fax: 530 752 5293
E-mail: rpsingh@ucdavis.edu
banderson@ucdavis.edu

Chapter 2

Mr R. Esse* and Mr A. Saari
Humidipak, Inc.
17613 Minnetonka Boulevard
Wayzata, MN 55391-3316
USA

E-mail: bob.esse@humidipak.com
al.saari@humidipak.com

Chapter 3

Dr P. S. Taoukis* and
Dr M. C. Giannakourou
Laboratory of Food Chemistry and
Technology
School of Chemical Engineering
National Technical University of
Athens
5 Iroon Polytechniou
Zografou Campus
15780 Athens
Greece

Tel: +30 210 7723171
Fax: +30 210 7723163
E-mail: taoukis@chemeng.ntua.gr

* = main contact.

Chapter 4

Dr R. E. Schouten* and
Dr O. van Kooten
Horticultural Production Chains
Wageningen University
Marijkeweg 22
6709 PG Wageningen
The Netherlands

Tel: +31 (0)317 482408
Fax: +31 (0)317 484709
E-mail: rob.schouten@wur.nl

Dr H. Jalink, Dr I. F. Kappers,
Dr J. F. H. Snel and Dr W. Jordi
Plant Research International
PO Box 16
6700 AA Wageningen
The Netherlands

Tel: +31 (0)317 475837
Fax: +31 (0)317 418094
E-mail: henk.jalink@wur.nl
Jan.snel@wur.nl
wilco.jordi@wur.nl

Chapter 5

Dr T. Deak
Budapest University of Economic Science
Faculty of Food Science
Budapest
Hungary

Tel: +36 1 372 6360
Fax: +36 1 372 6340
E-mail: tdeak@omega.kee.hu

Chapter 6

Professor A. Arnoldi
DISMA
University of Milan
Via Celoria 2
IT 20133
Milan
Italy

Tel: +39 02 50316814
Fax: +39 02 50316801
E-mail: anna.arnoldi@unimi.it

Chapter 7

Dr M. H. Gordon
School of Food Biosciences
The University of Reading
Whiteknights
P O Box 226
Reading RG6 6AP
UK

Tel: +44 (0) 118 3786723
Fax: +44 (0) 118 9310080
E-mail: m.h.gordon@reading.ac.uk

Chapter 8

Ms C. Davies
Unilever Colworth Research and Development
Sharnbrook
Bedfordshire MK44 1LQ
UK

Tel: +44 1234 222725
Fax: +44 1234 222881
E-mail:
Chrissie.Davies@Unilever.com

Chapter 9

Dr T. K. Singh and Professor K. R. Cadwallader*
Department of Food Science and Human Nutrition
University of Illinois at Urbana-Champaign
1302 W Pennsylvania Avenue
205 Agricultural Bioprocess Laboratory
Urbana
IL 61801
USA

Tel: 217 333 5803
Fax: 217 333 1875
E-mail: cadwlldr@uiuc.edu

Chapter 10

Dr G. D. Betts* and Dr S. J. Walker
Campden and Chorleywood Food Research Association
Chipping Campden
Gloucestershire GL55 6LD
UK

Tel: +44 (0) 1386 842071
Fax: +44 (0) 1386 842100
E-mail: g.betts@campden.co.uk
s.walker@campden.co.uk

Chapter 11

Dr I. A. Farhat
Division of Food Sciences
University of Nottingham
Sutton Bonington Campus
Loughborough
Leics LE12 5RD
UK

Tel: +44 (0) 115 9516134
Fax: +44 (0) 115 9516142
E-mail: imad.farhat@nottingham.ac.uk

Chapter 12

Professor V. Loureiro* and Dr M. Malfeito-Ferreira
Laboratório de Microbiologia
Instituto Superior de Agronomia
1349-017 Lisboa
Portugal

Tel: +35 1 21 365 34 48
Fax: +35 1 21 363 50 31
E-mail: vloureiro@isa.utl.pt

Dr A. Carreira
STAB Vida
Apartado 89, Sto António de Oeiras
2781-601 Oeiras
Portugal

Tel: +35 1 21 446 97 65
Fax: +35 1 21 446 97 40
E-mail: acarreira@stabvida.com

Chapter 13

Mr J. W. Irwin* and Dr N. Hedges
Unilever Colworth Research and Development
Sharnbrook
Bedfordshire
MK44 1LQ
UK

Tel: +44 (0) 1234 222453
Fax: +44 (0) 1234 222552
E-mail: john.irwin@unilever.com

Chapter 14

Professor S. Mizrahi
Department of Food Engineering and Biotechnology Technion
Israel Institute of Technology
Haifa 32000
Israel

Fax: +972 4 829 3399
E-mail: smizrahi@techunix.technion.ac.il

Chapter 15

Mr C. M. D. Man
Department of Applied Science
Faculty of Engineering, Science and the Built Environment
London South Bank University
London SE1 0AA
UK

Tel: +44 (0)207 815 7941
Fax: +44 (0)207 815 7999
E-mail: mandc@lsbn.ac.uk

Chapter 16

Professor P. A. Morrissey and Dr J. P. Kerry
Department of Food and Nutritional Sciences
University College Cork
Western Road
Cork
Ireland

Tel: 353 21 4902406
Fax: 353 21 4270244
E-mail: p.morrissey@ucc.ie

Part I

Factors affecting shelf-life and spoilage

1

The major types of food spoilage: an overview

R. P. Singh and B. A. Anderson, University of California, USA

1.1 Introduction

All food products are composed of biological raw materials. Biological products inherently spoil and deteriorate over time. This spoilage and deterioration cannot be completely stopped; however, it is the desire of food processors to slow this rate of deterioration as much as possible through formulation, processing, packaging, storage and handling. To properly address this problem of spoilage it is first important to understand what is meant by food spoilage and by what various modes can it take place.

Food spoilage can be defined in several different ways. Generally, a food is considered spoiled when it is no longer acceptable to the consumer. The worst case of spoilage is when it becomes a food safety issue, where the food product may cause the consumer illness or even death. However, less serious cases of food spoilage can simply be that the color, flavor, texture, or aroma of the food has deteriorated to the point that it is no longer acceptable. Another case of spoilage could be that the nutrients (e.g. vitamin content) in the food have deteriorated to the point that the food product no longer meets its declared nutritional value. The time it takes for a food product to reach one of these spoilage conditions is generally termed the product's shelf-life.

Many food products have some variation of open shelf-life dating marked on their containers. Open shelf-life dating varies widely by product, region, and manufacturer. Common open shelf-life dates include a 'sell-by-date', a 'best-if-used-by-date', or a 'better-if-used-by-date'. These dates help the consumer to decide how long the product may be stored prior to consumption and also help with stock rotation in grocery stores. Hopefully the food manufacturer conducted studies to properly determine the shelf-life of their product; however, often the date given assumes proper storage conditions of the product prior to consumption. It is important to understand what types of spoilage may occur in a product, how to best reduce the rate of deterioration of the product, and how to properly measure or detect its occurrence.

The three main categories of food spoilage that can occur are physical spoilage, chemical spoilage, and microbiological spoilage. There are some overlaps between the three categories, and often spoilage in one category can help to promote spoilage in another category. Table 1.1 lists the major mechanisms for spoilage or deterioration of various food categories. There are several main factors that influence most types of spoilage. These factors include temperature, pH, water activity, exposure to oxygen and light, and nutrients or chemicals available in the food product. Figure 1.1 shows graphically how one of these factors, water activity, affects various chemical and microbiological changes in typical food products. Each of the three main categories of spoilage is briefly reviewed in this chapter.

1.2 Physical instability

The first type of spoilage that can occur is due to physical changes or instability. This can include physical damage such as bruising of fresh fruits and vegetables or breaking of dry, brittle products such as potato chips and breakfast cereals. Bruising of fruits and vegetables can occur during transportation and distribution or by dropping the products. If the physical damage is severe, the product may become unacceptable to the consumer. There can be color changes due to enzymatic browning as cells are ruptured and there is often loss of water content at the bruise. In addition, bruising causes damage to cells and allows microbial growth to occur more readily. Breaking of dry, brittle products can make many products such as crackers, potato chips, ready-to-eat cereals, piecrusts, and many frozen foods unacceptable. The use of well-designed packaging systems that protect the products from vibrations and mechanical damage during distribution and handling can minimize the effects of bruising and breakage.

Most other physical changes or instabilities involve moisture or mass transfer of components in the food. A frequent cause of degradation of food products is a change in their water content (water loss, gain, or migration). The change in moisture alone may cause the product to become unacceptable, though frequently it also leads to other problems such as microbial or chemical degradation, which will be discussed in later sections. Moisture transfer occurs in foods due to gradients in chemical potential, which is directly related to the food's water activity (a_w). Water activity is defined as the equilibrium relative humidity for a product divided by 100. In bakery products such as bread, moisture migration can lead to staling. Staling involves moisture from the crumb (high a_w) migrating to the crust (low a_w). This causes the crumb to become dryer, firmer, and more crumbly, while the crust becomes tougher and less crisp (Piazza and Masi, 1995). Moisture transfer can occur between multi-component

Table 1.1 Deterioration/spoilage mechanisms and critical variables for various food products (Kilcast and Subramaniam, 2000; Labuza and Szybist, 2001)

Food product/ category	Type of spoilage/ deterioration	Critical storage variables
Milk	Oxidation, hydrolytic rancidity, bacterial growth	Oxygen, temperature
Milk powder	Oxidation, browning, caking	Oxygen, humidity, temperature
Dairy products	Oxidation, rancidity, lactose crystallization	Oxygen, temperature
Ice cream	Ice crystal or lactose crystal formation, oxidation	Temperature (freeze/thaw), oxygen
Fresh beef	Microbial (bacterial) growth, oxidation, moisture loss	Temperature, oxygen, light, humidity
Fresh poultry	Microbial growth	Temperature, oxygen
Fresh fish and seafood	Microbial growth, oxidation	Temperature, oxygen
Fruit	Enzymatic softening, microbial growth, bruising, moisture loss	Temperature, light, oxygen, handling, humidity
Leafy vegetables	Enzymatic activity, moisture loss/wilting, microbial growth	Temperature, light, oxygen, humidity
Crisp vegetables	Enzymatic softening, microbial growth, bruising, moisture loss	Temperature, light, oxygen, handling, humidity
Bread	Moisture migration (staling), starch retrogradation, microbial (mold) growth	Humidity, temperature, oxygen
Cereals	Moisture migration (softening), starch retrogradation, oxidation, breakage	Humidity, temperature, oxidation, handling
Soft bakery products	Moisture migration (staling), microbial (mold) growth, starch retrogradation	Humidity, temperature, oxygen
Crisp bakery/fried products (cracker, etc.)	Moisture migration (softening), oxidation, breakage	Humidity, temperature, oxygen, handling, light
Chocolate	Sugar crystallization (sugar bloom), fat crystallization (fat bloom), oxidation	Humidity, temperature, oxidation
Candy	Moisture migration (stickiness), sugar crystallization	Temperature, humidity
Beer	Oxidation, microbial growth	Oxygen, light, temperature
Coffee/tea	Oxidation, volatile loss	Oxygen, light, moisture
Frozen meats	Oxidation, freezer burn (drying)	Oxygen, temperature, humidity
Other frozen foods	Oxidation, ice crystal formation, texture change	Oxygen, temperature

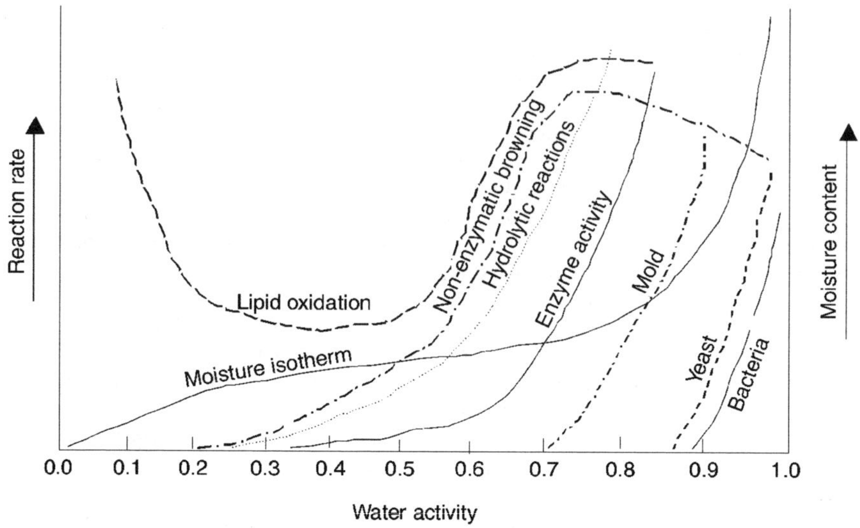

Fig. 1.1 Relative water activity stability map (from Labuza *et al.*, 1970).

foods that have components with different water activities. For example, moisture often transfers from fruit pieces to ready-to-eat cereal and from moist fillings to drier crusts of food products. Articles on moisture transfer in multi-domain foods have been published by Balasubrahmanyam and Datta (1994) and Labuza and Hyman (1998). Moisture changes can play a large role in the acceptability of many fruits and vegetables. Leafy vegetables will lose moisture to the environment and will wilt and have increased senescence if stored without proper packaging.

Temperature has a significant effect on the quality of fresh fruits and vegetables. Each crop has its own inherent rate of respiration and optimum temperature range to slow ripening and senescence and to maximize its postharvest life. In addition, climacteric fruits show a large increase in ethylene production during ripening. Ethylene is a strong plant growth regulator, and exposure to ethylene tends to accelerate the senescence of most crops. Therefore, eliminating or reducing exposure to ethylene will delay ripening and senescence in most crops, and will extend their postharvest life.

Temperature is also important in that most crops are susceptible to freeze damage when the temperature is lowered slowly and the product becomes partially frozen. This causes breakage of cells and damage to the product. Other crops, including most tropical fruits and vegetables, are sensitive to chilling injury, which occurs before the product begins to freeze (generally at temperatures of 5–15°C). Effects of chilling injury include pitting, water soaking, discoloration, development of off-flavors, accelerated senescence, or ripening/over-ripening. However, these effects are sometimes not apparent until weeks after the injury. Purvis (2002) discussed chill injury in green bell peppers. Other sources for information on postharvest handling of fruits and vegetables include Shewfelt and Prussia (1993), Willis (1998), Thompson *et al.* (1998), and Kader (2002).

Changes in glass transition temperature (T_g), which in turn affect the moisture mobility in a food product, can have a great impact on its shelf-life. The glass transition temperature for a product is the temperature at which it changes from a 'glassy' or brittle state to a 'rubbery' or soft state. The first known reference to these glassy and rubbery states in foods was from 1966 (White and Cakebread, 1966). Then in the 1980s, along with the pioneering work of Slade and Levine (1988, 1991) and Franks (1991), many more studies were conducted on the application of glass transition (polymer) theory to food products. The temperature at which the glass transition occurs depends on moisture content and composition.

Dry food products such as crackers are expected to be crisp. However, if they are stored in a high humidity environment, they will absorb water (lowering T_g) and undergo glass transition to become tough and soggy. Conversely, soft bakery products are expected to be moist and chewy; however, these products tend to lose moisture (raise T_g) to the point where they undergo glass transition and become glassy, hard, and brittle. Articles addressing these issues include Nikolaidis and Labuza (1996), Roudaut *et al.* (1998, 2002), and Li *et al.* (1998).

A change in temperature in relation to T_g also affects many different reaction rates. Above T_g, moisture is mobile and diffusion-limited reaction rates generally follow WLF (Williams–Landel–Ferry) kinetics (Williams *et al.*, 1955; Karel *et al.*, 1993), while below T_g there is much less mobility and reaction rates are generally much lower. This will be discussed further when chemical spoilage mechanisms are discussed.

Another effect of glass transition is caking of dry powders. As powder gains moisture, it undergoes glass transition and becomes amorphous. This causes the powder to stick together and cake. Recent work on the effect of glass transition on stickiness of food powders has been published by Jaya *et al.* (2002) and Ozmen and Langrish (2002).

Moisture loss can be a problem for even deeply frozen foods that can lose water, since the humidity (or water activity) in the environment at those temperatures is less than 100%. At −20°C, water activity is 0.82, and at −40°C, water activity is 0.68 (Fennema, 1985). Therefore moisture can evaporate or sublime from the surface to cause drying or freezer burn. To prevent moisture loss, frozen products should be completely sealed using moisture barrier packaging during storage.

Other physical changes that can occur in foods include crystal growth. This can include growth of ice crystals in frozen foods such as ice cream that cause its texture to become grainy. Large extracellular ice crystal growth occurs to a greater extent in foods that undergo slow freezing processes or multiple freeze/thaw cycles. Quick freezing processes lead to ice formation within cells that are smaller in size and more stable than are obtained during slow freezing processes; however, even rapid freezing can cause damage to the cell structures which can lead to enzymatic action in the product. The product does not need to completely thaw for moisture to move, but just to reach a temperature where liquid can migrate to form larger crystals. As temperature cycles in a freezer, temperature gradients will form in the product, and moisture will typically migrate along that temperature gradient (Reid, 1990). Emulsifiers and other water-binding agents can be added to help minimize the formation of large ice crystals during freeze/thaw cycles. In addition, by keeping the temperature of the product below its glass transition temperature throughout frozen storage, water in the food has much less mobility and will tend not to migrate (Levine and Slade, 1988). Recent studies on ice crystal growth have been published by Flores and Goff (1999), Russell *et al.* (1999), Bolliger *et al.* (2000), Ablett *et al.* (2002), and Downey (2002).

Similarly, crystallization of sugar can occur in some food products. This can occur in foods with high sugar content, where the amorphous or 'glassy' sugar undergoes glass transition either by uptake of moisture or by increase in temperature. Once in the rubbery state, sugar can crystallize and expel water. A good example of this is cotton candy, which becomes grainy when exposed to high humidity. Sugar bloom can occur in chocolate. When chocolate is stored in high humidity environments, moisture may condense on the surface; this brings sugar from the inside to the surface, giving a gray or white appearance. Sugar crystallization is also thought to be a component in the staling of sugar cookies and in graininess in candies and ice cream. Good reviews on sugar crystallization can be found in Hartel and Shastry (1991) and Hartel (1993).

Another well-studied type of crystallization is the migration and recrystallization of fat (cocoa butter) in chocolate, which is called fat bloom and appears as a whitish, greasy haze. Tempering of chocolate, which is the crystallization of cocoa butter into proper size, shape, and polymorphic structure, is an important step in minimizing fat bloom (Hartel, 1999). Improper tempering can lead to less stable forms of fat crystals and a greater likelihood of fat bloom. Other actions that can lead to fat bloom include partial melting and recooling of chocolate, abrasion to the surface, use of incompatible fats, or rapid cooling that causes stress cracks (Jana and Thakar, 1993). Recent studies on fat bloom in chocolate have been published by Tietz and Hartel (2000), Gao *et al.* (2001), Hodge and Rousseau (2002), and Walter and Cornillon (2002).

Another type of physical degradation that can occur is due to emulsion breakdown in products such as mayonnaise, margarine, and salad dressings. There are other mechanisms that cause product spoilage, such as lipid oxidation and microbial growth, which will be discussed later. Emulsions are thermodynamically unstable systems and consist of droplets of one phase dispersed in a second phase (often oil-in-water or water-in-oil). The long-term stability of the emulsion depends upon the emulsifier used and how well the phases are dispersed (droplet size). Emulsifiers such as egg yolk act at the surface of the droplets to lower surface tension, since they have ends that are both hydrophilic and hydrophobic. Then the emulsion becomes stable by the balancing of attractive forces, such as Van der Waals forces, and repulsive forces, such as electrostatic and steric interactions. These forces resist coalescence or creaming of droplets into one larger phase. Increasing the viscosity of the continuous phase can also be done to make the emulsion more stable. Destabilization of the emulsion can occur if it is not properly formed, by extreme vibration, or if the emulsifier is damaged, such as by partial freezing or extremely high temperatures. A review on the shelf-life of mayonnaise and salad dressings is given by Mistry and Min (1993) and a discussion on the stability of mayonnaise is given by Depree and Savage (2001).

1.3 Chemical spoilage (degradation)

Food spoilage can also occur in food products due to reaction or breakdown of the chemical components of the food, including its proteins, lipids, and carbohydrates. The rate at which the chemical reactions take place depends on many factors, which were mentioned earlier, including water activity, temperature, ($T - T_g$), pH, and exposure to light or oxygen. Each reaction has different optimum conditions. For example, enzyme activity is greatly reduced at low a_w, especially at moisture levels below the monolayer. Bell and White (2000) found that thiamin stability was more dependent on its relation to glass transition than to a_w. The occurrence of each chemical reaction has the potential of affecting color, flavor, aroma, and/or texture of the food product.

Protein degradation can involve reactions with proteins and other ingredients or enzymatic activity. Enzymes are complex proteins that act as catalysts that greatly increase chemical reaction rates. There are many types of enzymes that react with different chemical components in foods, many of which are formed by microorganisms. Ashie *et al.* (1996) gave a thorough review on enzymes and mechanisms to control their action.

Enzymes that act on other proteins include the protease plasmin. Plasmin can survive pasteurization temperatures and can cause degradation of dairy proteins in milk, and coagulation and gelation. Hayes and Nielsen (2000) studied the plasmin levels in fresh milk. Other proteases can attack proteins in meats and cause the meats to become mushy. These proteases can be more readily released from damaged cells, for example during multiple freeze/thaw cycles. Other proteases produced by microorganisms have the ability of breaking down meat and milk proteins. If pH is lowered (e.g. by microbial action) casein in milk will aggregate and precipitate from solution.

Oxidation of proteins can also occur. In meats, overexposure to oxygen can cause myoglobin and oxymyoglobin to oxidize into metmyoglobin, causing the meat color to turn from bright red to brown. This color change can be unappealing to consumers. Martinaud *et al.* (1997) studied oxidation processes of proteins from beef.

Enzymatic activities in fruits and vegetables can cause browning and softening of tissues. Typically these reactions are catalyzed by phenol oxidase enzymes, which react with phenol compounds and oxygen to form undesirable brown pigments. These reactions take place readily when cells are broken by bruising, cutting, or peeling. Nicolas *et al.* (1994) gave a review on enzymatic browning reactions in apples.

Nonenzymatic browning (Maillard browning) is a reaction that occurs between proteins (amino groups) and reducing sugars. The reaction scheme involves reaction to form an unstable Schiff's base, then transformation through the Amadori rearrangement. The reactions continue further through the Strecker degradation and polymerization reactions to form volatiles and dark pigments. This causes a browning of the color and sometimes changes in texture of the food product. In addition, Maillard browning is normally associated with a loss in nutritional value. The essential amino acid lysine, which readily reacts with reducing sugars, is quickly lost during nonenzymatic browning. Lievonen *et al.* (2002) recently looked at the effect of different reducing sugars on Maillard browning. Rate of Maillard browning has been found to vary with a_w, with a maximum rate typically occurring at a_w between 0.6 and 0.8, and lower reaction rates at both higher and lower a_w. Bell *et al.* (1998) measured the effects of a_w and T_g on Maillard browning rates. Maillard browning reactions are also very dependent on pH, and seldom occur at low pH. The reaction is also catalyzed with metal ions such as copper and iron.

Chemical spoilage of carbohydrates can include gelatinization/retrogradation reactions and browning reactions. Maillard browning, which was discussed earlier, is a major degradation reaction that can occur with reducing sugars. Other browning reactions such as caramelization can occur with carbohydrates, but require higher temperatures than products would typically be subjected to during distribution and storage.

Gelatinization is an important property of starches, which involves the swelling of starch granules with water and the subsequent loss of crystallinity and birefringence. The extent of gelatinization depends on many factors such as temperature, amount of shear, water activity, and presence of other components such as sugars and lipids. Retrogradation of starches is the reassociation or recrystallization process. Retrogradation occurs more readily with amylose than with amylopectin since it is a smaller unbranched molecule; therefore the use of waxy starches (low amylose content) can decrease the level of retrogradation. Often retrogradation will occur when the starch is exposed to freeze/thaw cycles and when moisture migration occurs, as in the staling of bread products. Hug-Iten *et al.* (1999) studied the retrogradation of starch using a light microscope during staling of bread, while Manzocco *et al.* (2002) studied it using X-ray diffraction analysis.

Oligosaccharides and polysaccharides in foods may also be broken down by hydrolytic reactions. In foods, one of the most common hydrolytic reactions is the production of starch into corn syrup using acid and enzymatic hydrolysis. Actions of enzymes on starch components are also important in food processes such as fermentation and brewing. Similar hydrolytic reactions can occur with pectin in fruits and vegetables by pectin lyase and polygalacturonase causing a softening of structure.

Lipid spoilage most often occurs due to oxidation reactions or action of lipolytic enzymes and other hydrolytic reactions. Lipid oxidation (oxidative rancidity) is the most important degradation method for fats and oils and occurs in many food products including fried foods, nuts, dried fruits and meats, milk powder, coffee, and margarine. During lipid oxidation, oxygen attacks unsaturated fats to form color changes, off-flavors, and sometimes toxic substances. The reacting oxygen can be dissolved within the oil or food product, may be in the headspace of the package, or may permeate through the package during storage. To minimize oxidation of lipids, all attempts should be made to keep oxygen from contacting the food product and to use a proper package design that acts as a barrier to oxygen transmission. The number and location of double bonds on the fatty acids or triglycerides is one factor that affects the rate of oxidation. Light and heat are other important considerations since they can catalyze the oxidation reaction. Other trace elements in the product can affect the rate and extent of oxidation. Trace metals such as copper and iron can greatly catalyze the oxidation reaction. On the other hand, tocopherols and other antioxidants have the ability to slow or prevent oxidation by reacting with radical oxygen in the product. Antioxidants such as citric acid, BHA, BHT, EDTA, TBHQ, and rosemary extract are sometimes added to help reduce oxidation. A review of lipid oxidation reactions and its effects was published by St Angelo (1996).

Hydrolytic rancidity caused by chemical reactions or action of lipolytic enzymes is another mechanism of lipid degradation. This reaction involves the cleaving of free fatty acids from triglyceride molecules in the presence of water. The free fatty acids that are released have shorter chain lengths, lower flavor thresholds, and sometimes off or rancid flavors and odors. Most lipolytic enzymes can be inactivated by heating above 60°C, and the reactions can be reduced by minimizing moisture (Kristott, 2000).

1.4 Microbial spoilage

Action by microorganisms is a common means of food spoilage and the most common cause of foodborne illness. Microbial spoilage is a major concern for so-called perishable foods such as fresh fruits, vegetables, meats, poultry, fish, bakery products, milk, and juices. Potential food spoilage microorganisms include bacteria, fungi (mold and yeast), viruses, and parasites. The growth of most microorganisms can be prevented or slowed by adjusting initial microbial load, adjusting temperature of storage, reducing water activity, lowering pH, use of preservatives, and using proper packaging. Some microorganisms simply cause spoilage of foods, while others can cause illness or even death if consumed. Table 1.2 provides a listing of selected microorganisms typically associated with food spoilage or foodborne illness as well as their minimum growth conditions and typical food vehicles (Banwart, 1989; Cousin, 1996; Robinson *et al.*, 2000; Jay, 2000). However, not all microbial growth is undesirable. Microorganisms are used and desired in the production of many fermented food products such as cheese, beer, wine, soy sauce, and sauerkraut.

Table 1.2 Selected microorganisms typically associated with food spoilage or foodborne illness (Banwart, 1989; Cousin, 1996; Robinson, 2000; Jay, 2000)

Sample microorganisms	Minimum conditions for growth			Food systems where commonly found
	Temperature (°C)	a_w	pH	
Spoilage microorganisms				
Most molds	<0	0.80	<2.0	
Most yeasts	−5	0.88	1–5	
Halophilic bacteria		0.75	4.5 (most)	salted fish
Xerophilic molds		0.61	1.5–3.5	
Osmophilic yeasts		0.61	1.5–3.5	
Lactic acid bacteria	4	0.94	3.5	fruits and vegetables, beer, milk, vacuum packed meats
Micrococci	4	0.90	5.0	fresh meats, cured meats
Acetobacter spp.	5	0.95	2.6	fruits, beer, wine
Acinetobacter spp.	1	0.96	5.5	fresh meats, poultry, and milk
Alternaria spp.	1	0.75	2.7	fruit and vegetable black rot, grains
Aspergillus niger	0	0.80	1.2	fruit and vegetable black rot, processed meats
Aspergillus spp.	0	0.64	2.0	grains, fruits and vegetables, nuts
Bacillus subtilis	5	0.95	4.2–5	vegetables, fresh meats and poultry, bread, milk
Botrytis cinerea	−2	0.93	2.5	fruits and vegetable gray rot, processed meats

Candida spp.	0	0.70	1.3	meats, poultry, seafood, dairy
Enterobacter aerogenes	2	0.95	4.4	fresh meats and poultry
Fusarium spp.	−3	0.87	2.2	vegetable dry rot, fruits, and grains
Mucor spp.	0	0.80	3.0	fruits, vegetables, cheese
Penicillium spp.	−6	0.78–0.90	1.9	meats, fruits, vegetables, grains
Pseudomonas spp.	<0	0.97	5.5	vegetable rot, meats, poultry, eggs
Rhizopus stolonifer	5	0.93	2.5	breads, vegetable soft rot, fresh meats
Trichosporon spp.	0	0.87	2.0	seafood, meat, dairy, fruits
Human pathogens				
Bacillus cereus	10	0.92	4.9	fish, fresh meats, water
Campylobacter jejuni	25	0.95	4.9	meats, milk
Clostridium botulinum	3.3	0.93	4.6	improperly canned foods
Escherichia coli O157:H7	15	0.95	4.0	vegetables, meats, poultry, milk
Listeria monocytogenes	0	0.92	4.3	poultry, dairy products, meats, vegetables
Salmonella spp.	7	0.94	4.0	poultry, meats, dairy products
Staphylococcus aureus	6 (10 for toxin)	0.86 (0.9 for toxin)	4.0 (4.5 for toxin)	meats, poultry
Vibrio parahaemolyticus	5	0.94	4.8	fish and seafood
Yersinia enterocolitica	−2	0.96	4.2	fresh meats, milk, seafood

There are numerous types of bacteria that can grow and cause spoilage in many different foods. Bacteria are single-celled organisms that are one to five microns in size. They can be round, rod, or spiral in shape, and reproduce by binary fission. Bacteria that cause foodborne illness include *Escherichia coli* O157:H7, *Bacillus cereus*, *Salmonella* spp., *Campylobacter jejuni*, *Clostridium* spp., *Listeria monocytogenes*, and *Vibrio* spp., among others. Many other bacteria species cause food spoilage without causing illness. Some bacteria are able to form spores as a protection mechanism when subjected to harsh conditions.

Yeasts are another type of microorganism that can cause food spoilage, but are also used in many fermentation operations. Yeasts are single-celled fungi that are round or cylindrical in shape and three to five microns in size. They multiply by budding or binary fission. Important yeasts in food include *Candida* spp., *Dekkera* spp., *Saccharomyces* spp., and *Zygosaccharomyces* spp.

Molds are another type of fungi with a larger cell size (30 to 100 microns) that form chains and branches. Molds come in many different shapes, sizes, and colors, and they can be seen with the naked eye when they form their branched structure. Molds reproduce by producing spores either sexually or asexually. Important molds that often cause spoilage of food products include *Aspergillus* spp., *Fusarium* spp., *Penicillium* spp., and *Rhizopus* spp. Some species of *Aspergillus* are able to produce secondary metabolites called aflatoxins, which can also cause illness (Ellis *et al.*, 1991).

Viruses are much smaller microorganisms that can only grow and reproduce inside living cells. They are only 0.02 to 0.25 microns in size and contain either DNA or RNA and a protein to reproduce. Viruses are not able to grow in food products, but may survive once in the food. Viruses may be present in the food because it came from an infected plant or animal, or they may enter the food product by contamination from insects, rodents, infected water or food handlers. If a food handler does not wash their hands well after using the restroom, there is a chance of fecal contamination. Viruses that have been known to cause foodborne illness include Hepatitis A, Norwalk virus, rotaviruses, and BSE (Bovine Spongiform Encephalopathy) or mad-cow disease.

In order for microorganisms to cause food spoilage, they must first be present on the food. Food products may become contaminated at the farm (e.g. milk from infected cows), by the end user, or at any point in between (e.g. further processing, packaging, distribution, etc.). Once present on the food, whether or not the microorganism grows depends upon the conditions of itself, the food, and the environment. Each microorganism has specific requirements and conditions both of the food itself and of the environment that must be met in order to grow, including nutrient requirements, water activity, pH, temperature, oxygen availability, and presence of antimicrobial agents.

Each of the factors mentioned affects which microorganisms are able to grow. Scott (1957) was one of the first to suggest that the equilibrium thermodynamics (or water activity) was important in determining the growth of microbes rather than water content. Since then, water activity has been adopted as one of the most important factors governing microbial growth. Slade and Levine (1991) suggested that glass transition was a more important factor in microbial growth, though many of their arguments were disputed by Chirife and Buera (1995, 1996). Most likely both are important, though water activity is the most generally used and accepted as an extremely important factor that affects all microorganisms. Virtually no microbe can grow at water activities less than 0.6, but almost all can grow at water activities above 0.9. Most fresh foods have water activities above 0.95, so are susceptible to microbial growth. Most bacteria cannot grow below an a_w of 0.91, though there are halophilic bacteria that can grow down to an a_w of 0.75. Most yeasts cannot grow below an a_w of 0.88, though some osmophilic yeasts can grow down to an a_w of 0.6. Most molds cannot grow below an a_w of 0.8, though some xerophilic molds can grow at 0.65. These are just general guidelines and depend on the components of the system; for example Chirife and Buera (1996) studied a semi-moist dog food stored for 20 days at 34°C. They found that if it was formulated with glucose and glycerol, there was mold growth, but if formulated with fructose and propylene glycol, there was no growth even though both were at a water activity of 0.89. Sautour *et al.* (2002) and Gock *et al.* (2003) recently published studies on the effect of temperature and a_w on several molds.

Other factors that greatly affect the growth of microorganisms include pH and oxidation–reduction potential. Most microorganisms grow best at a pH near 7.0. In products that have very low pH (<3.7) such as most citrus fruit, only lactic acid bacteria and certain yeasts and molds can grow. The oxidation–reduction potential is normally expressed as an Eh value. The value of Eh at which microorganisms grow determines whether they are aerobic or anaerobic. Aerobic microbes require positive Eh values, while anaerobe microbes require negative Eh values. Facultative aerobes can grow under both positive and negative Eh values (Jay, 2000).

Another factor that affects the growth of organisms is temperature. Depending on the temperature of storage, microorganisms may grow rapidly, grow slowly, stop growing, or even die. There are three main classes of microorganisms depending on the temperature at which they grow best. Mesophiles are organisms that grow best between 30°C and 40°C, but also grow between approximately 10°C and 45°C. Psychrotrophs like colder temperatures. They grow best between 20°C and 30°C, but can grow at cold temperatures below 7°C (Champagne *et al.*, 1994). Thermophiles prefer higher temperatures; they grow best between 55°C and 65°C, but can also grow well at 45°C to 55°C. As temperature is increased above 60°C, some microorganisms begin to die. At even higher temperatures, microorganisms will die in less time.

The nutrient content of the food also affects microbial growth. In order to grow, microorganisms require water, a carbon source for energy, a nitrogen source, and certain vitamins and minerals. Bacteria have the highest nutrient requirement, while molds have the lowest requirement, and yeasts have a requirement somewhere in the middle. Certain components of foods can actually have a negative impact on microbes and act as antimicrobial agents. Some of these antimicrobial agents are found naturally in foods, while others are added as preservatives. Examples of these agents include lysozyme, riboflavins, anthocyanins, and thymol. Islam *et al.* (2002), Savard *et al.* (2002), and Gill and Holley (2003) recently discussed the use of preservatives on meats and vegetables.

The composition of gases in the environment surrounding the food can also have an effect in slowing microbial growth. Elevated concentrations of carbon dioxide (above about 10%) tend to slow the growth of molds and other microorganisms. Often modified atmosphere packaging (MAP) or controlled atmosphere packaging (CAP) are used to change the storage environment of a food product and slow microbial growth.

Food preservation processes extend the shelf-life of foods by killing microorganisms or by changing the food and/or environmental conditions to stop or slow their growth. Food preservation processes designed to kill certain microorganisms include thermal processing and more recently nonthermal processes such as ultra-high-pressure processing, pulsed electric field (PEF) processing, and irradiation. In addition, antimicrobial agents such as hydrogen peroxide, chlorine, and ozone can be used to kill microorganisms, but the agents cannot be present in the finished product. Andrews *et al.* (2002) discuss the use of chlorine dioxide and aqueous chlorine on shrimp. Other preservation techniques that slow or prevent growth include drying, refrigeration, freezing, modified atmosphere packaging, and adjusting pH and alcohol content.

Thermal processing is one of the most common methods for killing microorganisms in foods. Pasteurization is a thermal treatment that reduces microbial spoilage populations to prolong shelf-life. There is a time–temperature relationship for thermal processing; lower temperatures require longer times to achieve the same degree of killing. Therefore, thermal treatments typically used for pasteurization are LTLT (low temperature – long time), HTST (high temperature – short time) or UHT (ultra high temperature – very short time). Sterilization is a thermal process where all viable microorganisms are killed (there is only a probability that one cell survives). Each microbe has its own susceptibility to heat. Often thermal death kinetics for a certain microbe are quantified using D- and z-values. The D-value is the time required at a given temperature to achieve a one-log reduction in population, while the z-value gives the temperature change required to achieve a 90% change in the D-value.

Microorganisms can cause food spoilage by breaking down the chemical components, or certain pathogenic microorganisms can cause foodborne illness by being present on the food or by producing toxins. The easiest chemical components for microbes to degrade are carbohydrates since they generally use the carbon as a source of energy. Simple sugars and small molecule carbohydrates are normally degraded more quickly than complex carbohydrates, such as cellulose and lignin. However, degradation of cellulose can occur by microorganisms that produce cellulolytic enzymes (Cousin, 1996).

The degradation of pectin by microorganisms is a major cause of spoilage in fruits and vegetables and is called soft rot. Some bacteria and molds can produce pectinolytic enzymes, such as polygalacturonase, pectinesterases, and pectin or pectate lyases. In order to minimize pectin degradation by microorganisms, fruits and vegetables can be treated with chlorine wash water, kept dry after washing and kept at refrigerated temperatures (Cousin, 1996).

Microorganisms that produce proteases can degrade proteins in foods. Deamination reactions that can occur typically produce ammonia, while decarboxylation reactions produce carbon dioxide. Other products from protein degradation reactions can include organic acids, hydrogen sulfide, mercaptans, and other undesired compounds. Bacteria are the microorganisms typically found to spoil meats and other products with high protein concentrations including *Pseudomonas* spp., *Enterobacter* spp., and *Flavobacterium* spp. Some enzymes produced are rather heat resistant and can remain active even after pasteurization to cause coagulation of milk.

Lipids are more difficult to break down by microorganisms. There must be some moisture present in foods for microorganisms to grow. Some microorganisms do produce lipases, which cause hydrolytic rancidity in lipids, including *Pseudomonas* spp., *Lactobacillus* spp., *Aspergillus* spp., *Rhizopus* spp., and *Saccharomyces* spp. A recent study on lipase from *Pseudomonas fragi* was published by Alquati *et al.* (2002), and Thompson *et al.* (1999) gave a review on microbial lipase activity. Typically spoilage by hydrolytic rancidity only occurs in raw meats, fish, and milk and dairy products.

Other organisms that can cause spoilage of foods are parasites. Parasites are not able to live on their own, but infect host animals and plants and live off them. Parasite infections are typically found in meats, fish, and shellfish. Cestodes (tapeworms), such as *Diphyllobothrium latum*, are found in beef, pork and fish. Nematodes (roundworms), such as *Trichinella spiralis*, are sometimes found in pork, fish, and shellfish. Trematodes, such as *Clonorchis sinensis*, are sometimes found in fish, shellfish, and even certain vegetables such as water chestnuts and bamboo. Protozoa, such as *Cryptosporidium parvum*, are sometimes found in drinking water, fruits and vegetables, and have become an increasingly prevalent health risk. Recent studies on *Cryptosporidium* spp. have been published by Teunis *et al.* (2002), Hanes *et al.* (2002), Limor *et al.* (2002), Robertson and Gjerde (2001a,b), and Walker *et al.* (2001). Most infection by parasites can be prevented by enhanced food sanitation and proper cooking of foods and possibly with the use of irradiation (Orlandi *et al.*, 2002).

1.5 Future trends

In the future, it is believed that more studies will be done to understand the relationship of the various spoilage mechanisms and the glass transition temperature. Glass transition theory has been applied to food only for approximately the last 20 years. In addition, there are a significant number of articles in the recently published literature on the stability of intermediate moisture foods and their spoilage mechanisms. Also, with each new case of foodborne illness, there is increasing research in the study of foodborne pathogens. For example, the parasite *Cryptosporidium parvum* was recently found to cause illness in drinking water. As a result there has been a great increase in the number of studies focused on this pathogen. It is also expected that there will be considerable future research on methods for preservation of foods and for destroying microorganisms, such as nonthermal methods of ultra-high-pressure and pulsed electric field technology. Such technologies will be aimed at making our food supply more safe, healthy and appealing to the consumer.

1.6 Sources of further information

There are several reference books and review articles that cover the various modes of food spoilage. A sampling of these sources (in addition to the chapter's references) is given below:

ASHIE, I.N.A., SMITH, J.P., SIMPSON, B.K. (1996), 'Spoilage and shelf-life extension of fresh fish and shellfish', *Critical Reviews in Food Science and Nutrition* 36 (1/2), 87–121.

CHARALAMBOUS, G. (1986), *The Shelf Life of Foods and Beverages*, Proceedings of the 4th International Flavor Conference, Rhodes, Greece, 23–26 July 1985, New York, Elsevier.

CHARALAMBOUS, G. (1986), *Handbook of Food and Beverage Stability: Chemical, Biochemical, Microbiological, and Nutritional Aspects*, Orlando, FL, Academic Press.

CLIVER, D.D., RIEMANN, H.P. (2000), *Foodborne Diseases*, 2nd edn, Boston, Academic Press.

ESKIN, N.A.M., ROBINSON, D.S. (2001), *Food Shelf Life Stability: Chemical, Biochemical, and Microbiological Changes*, Boca Raton, FL, CRC Press.

FENNEMA, O.R. (1996), *Food Chemistry*, New York, Marcel Dekker.

GOFF, H.D. (1994), 'Measuring and interpreting the glass transition in frozen foods and model systems', *Food Research International* 27 (2), 187–189.

HALLBERG, L.M., CHINACHOTI, P. (2002), 'A fresh perspective on staling: the significance of starch recrystallization on the firming of bread', *Journal of Food Science* 67 (3), 1092–1096.

HANSEN, J.N. (1994), 'Nisin as a model food preservative', *Critical Reviews in Food Science and Nutrition* 34 (1), 69–93.

JAY, J.M. (2000), *Modern Food Microbiology*, Gaithersburg, MD, Aspen Publishers.

LABUZA, T.P. (1982), *Shelf-life Dating of Foods*, Westport, CT., Food & Nutrition Press.

MAN, C.M.D., JONES, A.A. (2000), *Shelf-life Evaluation of Foods*, Gaithersburg, MD, Aspen Publishers.

NGUYEN-THE, C., CARLIN, F. (1994), 'The microbiology of minimally processed fresh fruits and vegetables', *Critical Reviews in Food Science and Nutrition* 34 (4), 371–401.

PARK, S., WOROBO, R.W., DURST, R.A. (1999), '*Escherichia coli* O157:H7 as an emerging foodborne pathogen: a literature review', *Critical Reviews in Food Science and Nutrition* 39 (6), 481–502.

REINECCIUS, G. (1991), 'Off-flavors in foods', *Critical Reviews in Food Science and Nutrition* 29 (6), 381–402.

SINGH, K. (1999), 'Spectrophotometric determination of iron in sugars using benzyltriethyl ammonium chloride', *International Sugar Journal* 101 (1211), 554–558.

TAUB, I.A., SINGH, R.P. (1998), *Food Storage Stability*, New York, CRC Press.

TROLLER, J.A., CHRISTIAN, J.H.B. (1978), *Water Activity and Food*, New York, Academic Press.

1.7 References

ABLETT, S., CLARKE, C.J., IZZARD, M.J., MARTIN, D.R. (2002), 'Relationship between ice recrystallization rates and the glass transition in frozen sugar solutions', *Journal of Science Food & Agric*ulture 82 (15), 1855–1859.

ALQUATI, C., GIOIA, L., SANTAROSSA, G., ALBERGHINA, L., FANTUCCI, P., LOTTI, M. (2002), 'The cold-active lipase of *Pseudomonas fragi*: heterologous expression, biochemical characterization and molecular modeling', *European Journal of Biochemistry* 269 (13), 3321–3328.

ANDREWS, L.S., KEY, A.M., MARTIN, R.L., GRODNER, R., PARK, D.L. (2002), 'Chlorine dioxide wash of shrimp and crawfish as an alternative to aqueous chlorine', *Food Microbiology* 19 (4), 261–267.

ASHIE, I.N.A., SIMPSON, B.K., SMITH, J.P. (1996), 'Mechanisms for controlling enzymatic reactions in foods', *Critical Reviews in Food Science and Nutrition* 36 (1/2), 1–30.

BALASUBRAHMANYAM, G., DATTA, A.K. (1994), 'Prevention of moisture migration in fondant coated biscuit', *Journal of Food Engineering* 21 (2), 235–244.

BANWART, G.J. (1989), *Basic Food Microbiology*, New York, Van Nostrand Reinhold.

BELL, L.N., WHITE, K.L. (2000), 'Thiamin stability in solids as affected by the glass transition', *Journal of Food Science* 65 (3), 498–501.

BELL, L.N., TOUMA, D.E., WHITE, K.L., YIN, H.C. (1998), 'Glycine loss and Maillard browning as related to the glass transition in a model food system', *Journal of Food Science* 63 (4), 625–628.

BOLLIGER, S., WILDMOSER, H., GOFF, H.D., THARP, B.W. (2000), 'Relationships between ice cream mix viscoelasticity and ice crystal growth in ice cream', *International Dairy Journal* 10 (11), 791–797.

CHAMPAGNE, C.P., LIAING, R.R., ROY, D., MAFU, A.A. (1994), 'Psychrotrophs in dairy products: their effects and their control', *Critical Reviews in Food Science and Nutrition* 34 (1), 1–30.

CHIRIFE, J., BUERA, M.P. (1995), 'A critical review of some non-equilibrium situations and glass transitions on water activity values of foods in the microbiological growth range', *Journal of Food Engineering* 25 (4), 531–552.

CHIRIFE, J., BUERA, M.P. (1996), 'Water activity, water glass dynamics, and the control of microbial growth in foods', *Critical Reviews in Food Science and Nutrition* 36 (5), 465–513.

COUSIN, M.A. (1996), Food Microbiology Course Notes. Purdue University, West Lafayette, IN.

DEPREE, J.A., SAVAGE, G.P. (2001), 'Physical and flavour stability of mayonnaise', *Trends in Food Science and Technology* 12, 157–163.

DOWNEY, G. (2002), 'Quality changes in frozen and thawed, cooked puréed vegetables containing hydrocolloids, gums and dairy powders', *International Journal of Food Science and Technology* 37 (8), 869–877.

ELLIS, W.O., SMITH, J.P., SIMPSON, B.K. (1991), 'Aflatoxins in food: occurrence, biosynthesis, effects of organisms, detection, and methods of control', *Critical Reviews in Food Science and Nutrition* 30 (3), 403–439.

FENNEMA, O.R. (1985), *Food Chemistry*, New York, Marcel Dekker.

FLORES, A.A., GOFF, H.D. (1999), 'Recrystallization in ice cream after constant and cycling temperature storage conditions as affected by stabilizers', *Journal of Dairy Science* 82 (7), 1408–1415.

FRANKS, F. (1991), 'Hydration phenomena: an update and implications for the food processing industry', in: Levine, H., Slade, L. (eds), *Water Relationships in Foods*, New York, Plenum Press, 1–19.

GAO, Y.Y., XIONG, C.H., CHEN, C.S. (2001), 'Study on the correlation of polymorphic transition and fat-blooming of three kinds of chocolates', *Food Science, China* 22 (5), 11–13.

GILL, A.O., HOLLEY, R.A. (2003), 'Interactive inhibition of meat spoilage and pathogenic bacteria by lysozyme, nisin, and EDTA in the presence of nitrite and sodium chloride at 24°C', *International Journal of Food Microbiology* 80 (3), 251–259.

GOCK, M.A., HOCKING, A.D., PITT, J.I., POULOS, P.G. (2003), 'Influence of temperature, water activity, and pH on growth of some xerophilic fungi', *International Journal of Food Microbiology* 81 (1), 11–19.

HANES, D.E., WOROBO, R.W., ORLANDI, P.A., BURR, D.H., MILIOTIS, M.D., ROBL, M.G., BIER, J.W., ARROWOOD, M.J. (2002), 'Inactivation of *Cryptosporidium parvum* oocysts in fresh apple cider by UV irradiation', *Applied Environmental Microbiology* 68 (8), 4168–4172.

HARTEL, R.W. (1993), 'Controlling sugar crystallization in food products', *Food Technology* 47 (11), 99–104.

HARTEL, R.W. (1999), 'Chocolate: fat bloom during storage', *The Manufacturing Confectioner* 79 (5), 89–99.

HARTEL, R.W., SHASTRY, A.V. (1991), 'Sugar crystallization in food products', *Critical Reviews in Food Science and Nutrition* 30 (1), 49–112.

HAYES, K.D., NIELSEN, S.S. (2000), 'Plasmin levels in fresh milk whey and commercial whey protein products', *Journal of Dairy Science* 83 (3), 387–394.

HODGE, S.M., ROUSSEAU, D. (2002), 'Fat bloom formation and characterization in milk chocolate observed by atomic force microscopy', *Journal of the American Oil Chemical Society* 79 (11), 1115–1121.

HUG-ITEN, S., HANDSCHIN, S., CONDE-PETIT, B., ESCHER, F. (1999), 'Changes in starch microstructure on baking and staling of wheat bread', *Lebensmittel-Wissenschaft und Technologie* 32 (5), 255–260.

ISLAM, M., JINRU, C., DOYLE, M.P., CHINNAN, M. (2002), 'Effect of selected generally recognized as safe preservative sprays on growth of *Listeria monocytogenes* on chicken luncheon meat', *Journal of Food Protection* 65 (5), 794–798.

JANA, A.H., THAKAR, P.N. (1993), 'Fat bloom in chocolate and confectionery coatings – a review', *Indian Food Industry* 12 (4), 33–39.

JAY, J.M. (2000), *Modern Food Microbiology*, Gaithersburg, MD, Aspen Publishers.

JAYA, S., SUGHAGAR, M., DAS, H. (2002), 'Stickiness of food powders and related physico-chemical properties of food components', *Journal of Food Science and Technology* 39 (1), 1–7.

KADER, A.A. (2002), *Postharvest Technology of Horticultural Crops*, Oakland, CA,

University of California, Agriculture and Natural Resources Communication Services.

KAREL, M., ROOS, Y., BUERA, M.P. (1993), 'Effects of glass transitions on processing and storage', in: Blanshard, J., Lillford, P. (eds), *Glassy States in Food*. Loughborough, Nottingham University Press.

KILCAST, D., SUBRAMANIAM, P. (2000), *The Stability and Shelf-life of Food*, Boca Raton, FL, CRC Press, Woodhead Publishing.

KRISTOTT, J. (2000), 'Fats and oils,' in: Kilcast D., Subramaniam, P. (eds), *The Stability and Shelf-life of Food*, Boca Raton, FL, CRC Press, Woodhead Publishing, 279–309.

LABUZA, T.P., HYMAN, C.R. (1998), 'Moisture migration and control in multi-domain foods', *Trends in Food Science and Technology* 9 (2), 47–55.

LABUZA, T.P., SZYBIST, L.M. (2001), *Open Dating of Foods*, Trumbull, CT., Food & Nutrition Press.

LABUZA, T.P., TANNENBAUM, S.R., KAREL, M. (1970), 'Water content and stability of low moisture and intermediate moisture foods', *Food Technology* 24, 543–550.

LEVINE, H., SLADE, L. (1988), 'Principles of "cryostabilization" technology from structure/property relationships of carbohydrate/water systems – a review', *Cryoletters* 9, 21.

LI, Y., KLOEPPEL, K.M., HSIEH, F. (1998), 'Texture of glassy corn cakes as a function of moisture content', *Journal of Food Science* 63 (5), 869–872.

LIEVONEN, S.M., LAAKSONEN, T.J., ROOS, Y.H. (2002), 'Nonenzymatic browning in food models in the vicinity of the glass transition: effects of fructose, glucose and xylose as reducing sugar', *Journal of Agricultural and Food Chemistry* 50 (24), 7034–7041.

LIMOR, J.R., LAL, A.A., LIHUA, X. (2002), 'Detection and differentiation of *Cryptosporidium* parasites that are pathogenic for humans by real-time PCR', *Journal of Clinical Microbiology* 40 (7), 2335–2338.

MANZOCCO, L., NICOLI, M.C., LABUZA, T.P. (2002), 'Study of bread staling by X-ray diffraction analysis', *Italian Journal of Food Science* 14 (3), 235–245.

MARTINAUD, A., MERCIER, Y., MARINOVA, P., TASSY, C., GATELLIER, P., RENERRE, M. (1997), 'Comparison of oxidative processes on myofibrillar proteins from beef during maturation and by different model oxidation systems', *Journal of Agricultural and Food Chemistry* 45 (7), 2481–2487.

MISTRY, B.S., MIN, D.B. (1993), 'Shelf life of mayonnaise and salad dressings', in: Charalambous, G. (ed.), *Shelf Life Studies of Foods and Beverages: Chemical, Biological and Nutritional Aspects,* Amsterdam, Elsevier Science Publishers.

NICOLAS, J.J., FLORENCE, C., RICHARD-FORGET, F.C., GOUPY, P.M. (1994), 'Enzymatic browning reactions in apple and apple products', *Critical Reviews in Food Science and Nutrition* 34 (2), 95–108.

NIKOLAIDIS, A., LABUZA, T.P. (1996), 'Glass transition state diagram of a baked cracker and its relationship to gluten', *Journal of Food Science* 61 (4), 803–806.

ORLANDI, P.A., CHU, D.M.T., BIER, J.W., JACKSON, G.J. (2002), 'Parasites and the food supply', *Food Technology* 56 (4), 72–81.

OZMEN, L., LANGRISH, T.A.G. (2002), 'Comparison of glass transition temperature and sticky point temperature for skim milk powder', *Drying Technology* 20 (6), 1177–1192.

PIAZZA, L., MASI, P. (1995), 'Moisture redistribution throughout the bread loaf during staling and its effect on mechanical properties', *Cereal Chemistry* 72 (3), 320–325.

PURVIS, A.C. (2002), 'Diphenylamine reduces chilling injury of green bell pepper fruit',

Postharvest Biology and Technology 25, 41–48.

REID, D.S. (1990), 'Optimizing the quality of frozen foods', *Food Technology* 44 (7), 78–82.

ROBERTSON, L.J., GJERDE, B. (2001a), 'Factors affecting recovery efficiency in isolation of *Cryptosporidium* oocysts and *Giardia* cysts from vegetables for standard method development', *Journal of Food Protection* 64 (11), 1799–1805.

ROBERTSON, L.J., GJERDE, B. (2001b), 'Occurrence of parasites on fruits and vegetables in Norway', *Journal of Food Protection* 64 (11), 1793–1798.

ROBINSON, R.K, BATT, C.A., PATEL, P.D. (2000), *Encyclopaedia of Food Microbiology*, New York, Academic Press.

ROUDAUT, G., DACREMONT, C., LEMESTE, M. (1998), 'Influence of water on the crispness of cereal-based foods – acoustic, mechanical, and sensory studies', *Journal of Texture Studies* 29 (2), 199–213.

ROUDAUT, G., DACREMONT, C., PAMIES, B.V., COLAS, B., LE MESTE, M. (2002), 'Crispness: a critical review on sensory and material science approaches', *Trends in Food Science and Technology* 13 (6–7), 217–227.

RUSSELL, A.B., CHENEY, P.E., WANTLING, S.D. (1999), 'Influence of freezing conditions on ice crystallisation in ice cream', *Journal of Food Engineering* 39 (2), 179–191.

ST ANGELO, A.J. (1996), 'Lipid oxidation in foods', *Critical Reviews in Food Science and Nutrition* 36 (3), 175–224.

SAUTOUR, M., SOARES-MANSUR, C., DIVIES, C., BENSOUSSAN, M., DANTIGNY, P. (2002), 'Comparison of the effects of temperature and water activity on growth rate of food spoilage moulds', *Journal of Industrial Microbiology and Biotech*nology 28 (6), 311–315.

SAVARD, T., BEAULIEU, C., BOUCHER, I., CHAMPAGNE, C.P. (2002), 'Antimicrobial action of hydrolyzed chitosan against spoilage yeasts and lactic acid bacteria of fermented vegetables', *Journal of Food Protection* 65 (5), 828–833.

SCOTT, W.J. (1957), 'Water relations of food spoilage microorganisms' *Advances in Food Research*, 7, 83–127.

SHEWFELT, R.L., PRUSSIA, S.E. (1993), *Postharvest Handling: A Systems Approach*, San Diego, Academic Press.

SLADE, L., LEVINE, H. (1988), 'Structural stability of intermediate moisture foods – a new understanding', in: Mitchell, J., Blanshard, J. (eds), *Food Structure – its Creation and Evaluation*. London, Butterworth, 115–147.

SLADE, L., LEVINE, H. (1991), 'Beyond water activity: recent advances based on an alternative approach to the assessment of food quality and safety', *Critical Reviews in Food Science and Nutrition* 30, 115–360.

TEUNIS, P.F.M., CHAPPELL, C.L., OKHUYSEN, P.C. (2002), '*Cryptosporidium* dose response studies: variation between isolates', *Risk Analysis* 22 (1), 175–183.

THOMPSON, C.A., DELAQUIS, P.J., MAZZA, G. (1999), 'Detection and measurement of microbial lipase activity: a review', *Critical Reviews in Food Science and Nutrition* 39 (2), 165–187.

THOMPSON, J.F., MITCHELL, F.G., RUMSEY, T.R., KASMIRE, R.F., CRISOSTO, C.H. (1998), *Commercial Cooling of Fruits, Vegetables, and Flowers*, Davis, CA, University of California.

TIETZ, R.A., HARTEL, R.W. (2000), 'Effects of minor lipids on crystallization of milk fat-cocoa butler blends and bloom formation in chocolate', *Journal of the American Oil Chemists' Society* 77 (7), 763–771.

WALKER, M., LEDDY, K., HAGAR, E. (2001), 'Effects of combined water potential and temperature stresses on *Cryptosporidium parvum* oocysts', *Applied Environmental*

Microbiology 67 (12), 5526–5529.

WALTER, P., CORNILLON, P. (2002), 'Lipid migration in two-phase chocolate systems investigated by NMR and DSC', *Food Research International* 35, 761–767.

WHITE, G.W., CAKEBREAD, S.H. (1966), 'The glassy state in certain sugar-containing food products', *Journal of Food Technology* 1, 73–82.

WILLIAMS, M.L., LANDEL, R.F., FERRY, J.D. (1955), 'The temperature dependence of relaxation mechanisms in amorphous polymers and other glass forming liquids', *Journal of the American Chemical Society* 77, 3701.

WILLIS, R.B.H. (1998), *Postharvest: An introduction to the physiology and handling of fruit, vegetables, and ornamentals*, Sydney, Australia, UNSW Press.

2

Shelf-life and moisture management

R. Esse and A. Saari, Humidipak Inc., USA

2.1 Introduction: moisture activity and shelf-life

Many manufactured food products are adversely affected by moisture changes which directly impact their shelf-life and quality when they are consumed. These foods will lose desirable texture characteristics if allowed to lose or gain too much moisture. Brown sugar becomes hard and lumpy; raisins become hard. Ready-to-eat cereals lose their favored crisp textures if they gain moisture. Jerky becomes tough and dry.

In addition, numerous other changes are affected by variations in moisture level. Some dry grain-based products can become rancid more rapidly through free radical oxidation at low humidities and thus become unacceptable. Labile nutrients such as vitamins and natural colors such as chlorophyll are oxidized more rapidly if stored at low moisture levels. On the other hand, if the moisture level is elevated, enzyme-mediated hydrolysis rates are increased significantly and the Maillard type of non-enzymatic browning is enhanced.

Even small variations in storage temperatures will lead to localized high moisture conditions in an intermediate moisture food. These areas can be prime locations for microbial spoilage such as by loci of bacteria causing food infections or toxins of various types. Active moisture management systems should include humidity regulation.

Packaging materials are used to control the ingress or egress of moisture vapor. Even if the packaging film has excellent moisture barrier properties, it cannot preserve the product in its optimal condition. The product, as produced, may have a moisture content slightly different from the optimum to achieve the longest shelf-life because of variability in the ingredients or the processing/ manufacturing system. A package may have minute leaks because of flex cracking of the material or flaws in the heat seal. The body of the package itself may have some measurable permeability to moisture vapor. These factors affect the changing moisture level of a food, significantly impacting the shelf-life and quality of a food product.

The optimal approach is to have an active moisture regulation system that can react to and manage the changing conditions that take place over the life of a product. This chapter will help the reader understand such principles as water activity, moisture isotherm and moisture management/regulation systems. Understanding these principles is vital for development and distribution of products intended to be distributed nationally requiring a shelf-life of six months or more. They may be helpful also in the preparation of 'fresh' products which are distributed and consumed within a week or two.

2.2 Water activity and moisture management

To understand and apply a moisture management system, we must first have a basic understanding of water activity. This term, abbreviated to A_w, is a measurable value for all food products. It is a ratio and is expressed as a decimal fraction of 1.00 to two or three significant figures. Water activity (A_w) is defined as:

$$A_w = P_s/P_w$$

where P_s is the vapor pressure of a product or solution and P_w is the vapor pressure of pure water.* While the value is not precisely according to Raoult's Law, it is an adequate estimate for almost all situations. Further, since it is empirically measured, it reliably serves the purpose and is satisfactory for food product applications.

Values of A_w range from 0.00 (absolutely dry) to 1.00 (pure water). Thus one obtains values such as 0.33 or 0.62 for water activities of specific products, a ready-to-eat cereal or dried fruit, respectively. This is a well-understood measurement by practitioners of food research. Instruments to directly measure A_w are readily available. Among the most reliable, moderately priced instruments are the dewpoint measuring meters which yield a numerical readout of the A_w for a sample within a few minutes. A careful experimenter can attain a repeatability of 0.002.

When reporting results to a non-scientist, it may be conceptually preferable to convert the A_w value to relative humidity. We know the term relative humidity

* A simplified derivative of Raoult's Law which describes the effect of changes in number (molal) of ions in true solution on the vapor pressure, boiling point and freezing point of the solvent. Raoult's Law ($P_A = X_A P_S$ where P_A is the partial pressure of the solvent P_s is the vapor pressure of pure solvent at the same environmental temperature and pressure, and X_A is the mole fraction of the solute) applies to ideal solutions. This formula is adequate for normal purposes, as instruments available today quickly determine the A_w of a material. It is obvious that a food product high in mono- or disaccharides will have a much lower A_w at a particular value for total moisture content than a product low in soluble compounds of low molecular weight.

from weather reports and from being exposed to environments which can be expected to be comfortable, or uncomfortable. By definition:

$$\text{Relative humidity (RH)} = A_w \times 100$$

For the purposes of this discussion, the experimental determination of A_w represents a degree of accuracy adequate and rigorous enough for the reader to better understand and identify applications that may require moisture management. Thus in the examples above, the food products can be said to have a 33% or 62% relative humidity, respectively, in the headspace of closed containers of the products.

When the water vapor pressure of the food and the air surrounding it are equal, they are in equilibrium. This is not a static system, but a dynamic system where the loss of water molecules from the product equals the gain of water molecules from the environment. When this food product is exposed to an environment above or below this equilibrium point, the protective package and its barrier level will determine how much the food will be impacted. The second factor is the environment to which the package and its product are exposed. In drier climates the product may lose moisture and in more humid areas it will gain. If by chance it is exposed to a 33% relative humidity, no net moisture change will take place because the interior of the package is in equilibrium with its environment.

Formulated or natural food products each have a unique A_w at which their texture is optimal. Changing the formulation can also change the A_w value, particularly if there is a change in a solute. For example, adding sucrose as a sweetener will reduce the A_w of a product. If a monosaccharide such as glucose or fructose is added, the A_w reduction will be almost double since a unit weight of glucose will add approximately 1.9 times as many molecules in solution as the same unit weight of sucrose. (See footnote on page 25.)

If two or more products with different values of A_w are placed in a package, they will tend to converge to an intermediate A_w. Consequently, none of the components will be at their optimal moisture content. If it is necessary to combine such components in a single package such as a cookie with a fruit preserve filling, all of the components need to be reformulated to a common A_w, otherwise the cookie portion will seem 'soggy', lacking crispness, and the filling will be firm and hard to chew.

It is a great challenge to produce a succulent fruit filling and a crisp cookie at an intermediate A_w. By selecting a mixture of sugars, flours, fats, emulsifiers, etc., a reasonably acceptable product with a long shelf stability (6–12 months) can be produced. However, such a product has a relatively narrow tolerance to changes in moisture or A_w.

Natural products such as fruits, vegetables and cereal grains move through an A_w range as they grow from small green specimens to fully ripened edible products. Ripening often involves conversion of biopolymers such as starch to glucose or fructose as part of the process, thus reducing the A_w. Apples develop reduced A_w during ripening at a given total moisture content. During storage, the apples will consume some of the glucose to provide energy to sustain life. In time, the apples will become wrinkled and much less crisp as they lose moisture. Lettuce and other leafy vegetables wilt, losing turgidity and thereby the desirable crisp bite. This moisture loss can be slowed markedly by placing the food item in a more humid environment where it will achieve its longest shelf-life. It will be in acceptable flavor or eating quality for the longest period of time. In some products this can be days, in others it might be months.

However, there is a downside to high humidity environments. Relatively small fluctuations in temperature may lead to condensation of water on the package or the product. This localized A_w of essentially 1.0 will encourage all microorganisms to grow. Cycling of temperature may tend to draw moisture out of a product. The rate of moisture loss is usually more rapid than take-up, so a product subjected to frequent temperature cycles will have a net loss of moisture. Since such environments are usually at a much lower humidity, the product suffers a net loss with each temperature cycle.

It should also be noted that, contrary to perception, refrigerated spaces have a relatively low relative humidity, generally in the 30–40% RH range. The dewpoint of the air in a refrigerator is a function of the temperature of the cooling refrigeration coils in the chamber, be it a home refrigerator or a cold storage facility.

All food products have an optimal A_w. In some cases, a slight change in the moisture content can make the product unacceptable. Examples might be freeze-dried mushrooms or powdered tomato base which each become unacceptable with only a slight increase in moisture. Beef jerky can have a considerable change in texture between an A_w of 0.74 and 0.76, a tolerance of less than 0.01. When this tolerance is very tight, it may mean that the product will need to be protected from the exposed environment with a high moisture barrier package.

Other products have a high tolerance to fluctuation in their moisture/A_w values. In fruit/berry preserves, a range of A_w within ± 0.1 is hardly noticeable. Pasta products can be exposed to widely fluctuating moisture levels and not be significantly harmed. We thus see various pastas available in supermarkets in non-barrier bags or even in unlined paperboard cartons.

2.2.1 Moisture sorption isotherm

A moisture sorption isotherm is the result of a plot of the A_w vs moisture content of the product equilibrated against a broad range of A_w. Typically, seven to nine values of A_w are employed from about 0.05 up to 0.85. In practice, the product that is exposed to these various humidity levels is also examined for changes that have been triggered by the increase or decrease in initial moisture level. This will then suggest how much moisture the product can gain or lose before its quality is affected. Figure 2.1 shows an example of typical moisture isotherms of a ready-to-eat cereal and of paper (cellulose).

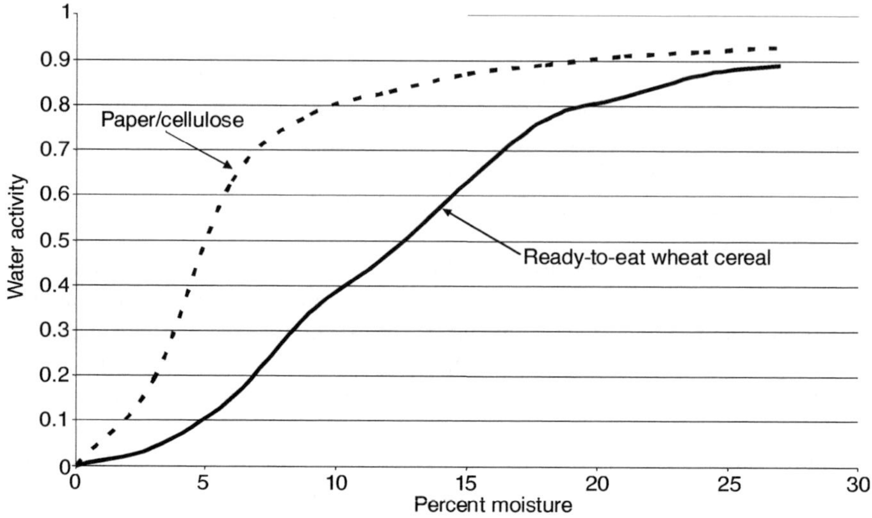

Fig. 2.1 Moisture sorption isotherms for paper and wheat cereal.

The interpretation of the moisture sorption isotherm curve is the source of basic information for the product and packaging development people to determine how much moisture barrier is required in a package for a particular product. The comparison of the isotherms for cellulose (paper) and a ready-to-eat wheat cereal in Fig. 2.1 shows that products are very different. The moisture content of the cellulose changes little between A_w values of 0.2 and 0.6. The difference in properties will probably be relatively minor at any A_w between those points. However, above A_w 0.7, a small change in A_w results in a large change in moisture content, leading to progressively weaker paper as the A_w increases. This tells us that the strength of paper is compromised as its A_w increases above 0.7. So, if paper were packaged for a tropical, humid climate, it requires material with significant moisture barrier properties.

The cereal has a minimal change in A_w until the water content reaches about 7%. Above 10% water, the A_w changes rather rapidly with small increases in moisture. This indicates that there are considerable changes in the cereal's structure after the A_w reaches about 0.3. Observations on the cereal show that it exhibits much less crispness, becoming rather tough and gummy at an A_w above 0.5. Because the ready-to-eat wheat cereal has a narrow tolerance to moisture loss or gain, it must be protected in a package with a material that is a very good barrier to moisture vapor.

However, the choice of formula or package is not always simple. During storage, changes in the product such as crystallization of sugars will release water and cause an increase in A_w of the product while its moisture content remains essentially constant. For example, the baked product chocolate brownies prepared with sucrose can have an A_w of 0.65 when fresh, but after a few months in a high barrier package, the product can have an A_w of up to 0.80

and be moldy as well as very crumbly. A significant portion of the sucrose has crystallized so a formulator must add a 'doctor', a substance that inhibits crystallization of sucrose. Monosaccharides such as glucose, or in its crude form, corn syrup, will serve this purpose well.

2.3 The effects of moisture on the storage stability of food

There are at least three important categories of the food sold in grocery stores:

- Manufactured products developed for convenient at-home preparation. These include bakery mixes (high-ratio cakes, angel food cakes, cookies, brownies, icings, muffins, biscuits), skillet dinners, beverages, sliced and shredded cheeses, and sausages, among others.
- Manufactured ready-to-eat products such as cookies, yogurt, butter/margarine, cereals, snack foods, fruit leather, nuts, etc.
- Food products including fresh fruits and vegetables and ingredients for home-prepared meals.

A large number of food products, called Intermediate Moisture Foods (IMF), are best at or near water activities at which microorganisms grow. Because of their contribution to the threat of foodborne illness, the relationship between moisture content and microbiological spoilage of food is of utmost importance. Food infections by *Shigella* spp., *Klebsiella* spp., *Escherichia* spp., *Vibrio* spp., *Salmonella* spp., among others, bring about much human misery through gastrointestinal distress. Food intoxications brought about by secretions of organisms such as *Clostridium botulinum, Staphylococcus* spp., *and Bacillus cereus* are serious matters, even fatal to the victims. Many molds produce very toxic substances with insidious effects such as carcinogenicity, mutagenicity, neurotoxicity, estrogenic and allergic consequences.

As noted in Table 2.1, the A_w of the environment of the organism is a crucial factor in reproduction of these organisms in food and food products. Some of the toxin-producing organisms are affected by the A_w of the product. The product development team must be aware of these detrimental effects and take appropriate action in moisture management, processing and packaging. From this table, it is obvious that microbiological problems do not occur during storage at an A_w less than 0.6.

Under both home and commercial storage and distribution, products are subjected to changes in temperature. The air in a package can hold considerably more moisture at elevated temperatures (46 g/m^3 at 100°F/37.8°C) than at refrigerated temperatures (6.5 g/m^3 at 40°F/4.4°C). Thus, at the higher temperature, the A_w (% relative humidity) of the air is reduced substantially, while at the lower temperature the A_w (% relative humidity) of the air is increased. As the temperature increases, the contents of a package will give off moisture, attempting to restore A_w equilibrium between the air and the contents of the package. Conversely, on cooling, the air will pump some moisture back

Table 2.1 Examples of the effect of water activity (A_w) on the growth of microorganisms

A_w (water activity)	Typical food items	Genera of microorganisms which grow (very slowly at lowest A_w of range) under this A_w
0.95–1.00	Fresh foods and meats, breads, approximately 40% sucrose, 8% NaCl	*Pseudomonas*, *Escherichia*, *Proteus*, *Bacillus*, *Clostridium*, *Shigella*, *Klebsiella*
0.91–0.95	Medium cheeses, cured meat (ham), retail fruit juice concentrate, 55% sucrose, 7% NaCl	*Salmonella*, *Vibrio*, *Serratia*, *Lactobacillus*, yeast – *Rhodotorula*
0.87–0.91	Fermented hard sausage, dry cheese, margarine, 65% sucrose, 15% NaCl	Most yeasts – *Candida*, *Torulopsis*, *Hansenula*, *Micrococcus*
0.80–0.87	Commercial fruit juice concentrate, chocolate syrup, maple and fruit syrup, flour, fruit cake, fondants, high-ratio cake	*Saccharomyces*, mycotoxigenic penicillia, *Staphylococcus aureus*
0.75–0.80	Fruit and berry preserves, marmalade, marshmallows, meat jerky	Halophilic bacteria, mycotoxigenic *Aspergillus* sp.
0.65–0.75	Rolled oats, fudge, marshmallow, raisins, fruit preserves, molasses, nuts, soft prunes	Xerophilic molds (*Aspergillus candidus*, *A. chevalieri*)
0.60–0.65	Dried fruit (<20% water), toffee, caramels, honey	Osmophilic yeasts, molds *Aspergillus echinulatus*, *Monascus bisporus*
0.50–0.60	Pasta (12% water), spices	No microbial growth
0.40–0.50	Whole egg powder (5% water)	No microbial growth
0.30–0.40	Cookies, crackers, bread crusts (5% water)	No microbial growth
0.20–0.30	Whole milk powder, dried vegetables, ready-to-eat cereals, hard cookies	No microbial growth

into the product, but more likely form some condensate. The product is likely to establish a gradient of A_w from very high on the surface to the original A_w of the bulk product, because in most cases the diffusion process is comparatively slow in intermediate moisture foods. This thin layer of high A_w is a fertile layer for microbial growth, be it bacterial, yeast or mold. A complete moisture management system will help to control this potential microbial problem.

Some products such as meat jerky are significantly above the minimum A_w for mold growth. In such cases, the products are normally packaged in a substantially oxygen-free environment, as molds are aerobic organisms. Usually, a packet of an oxygen scavenging system is included in the package to react and remove oxygen that diffuses from the air through the package, assuring that anaerobic conditions are maintained.

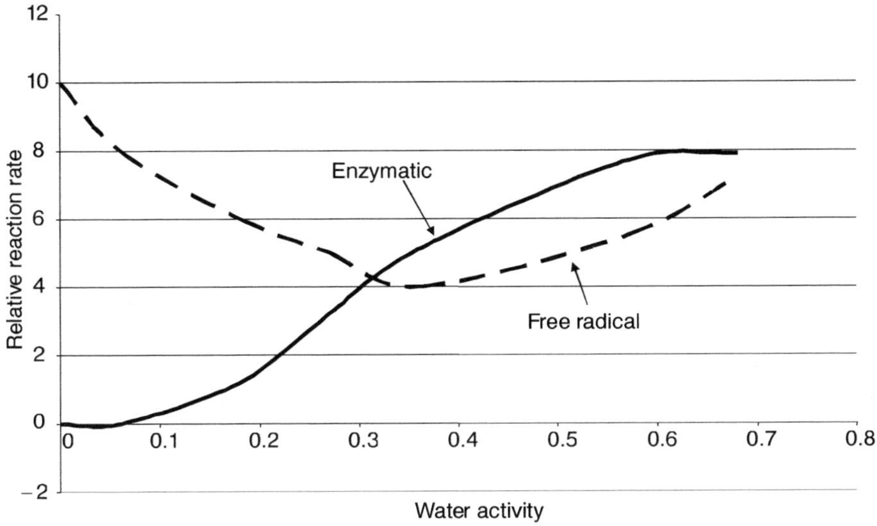

Fig. 2.2 Effect of A_w on oxidation – enzymatic and free radical.

Several other detrimental effects that are inhibited by an optimal A_w of a food are pointed out in Figs 2.2 and 2.3. These curves are generic, but typical of the reactions covered. Detailed discussion is beyond the scope of this chapter.

Figure 2.2 points out that lipidoxidases begin to catalyze oxidation of unsaturated fats at a significant rate above an A_w of about 0.3. The rate of oxidation increases rapidly as the A_w increases. On the other hand, the rate of non-enzymatic, free radical oxidation of unsaturated lipids decreases from A_w =

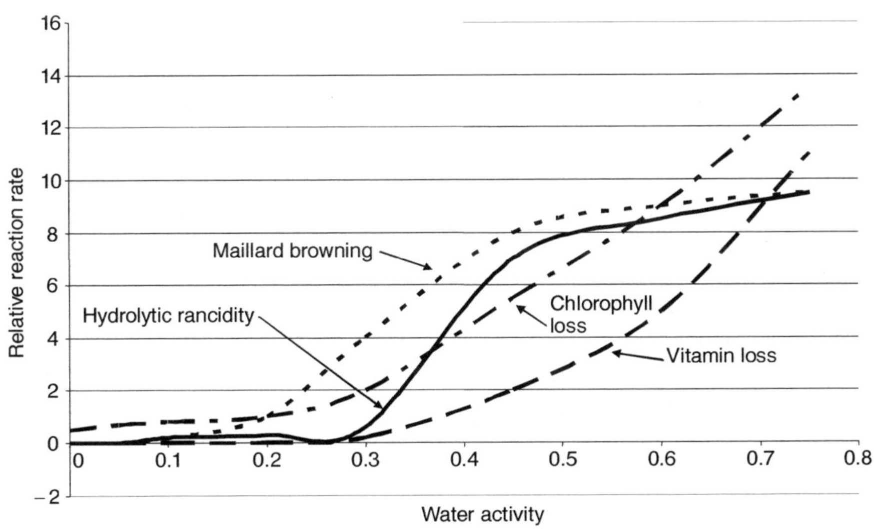

Fig. 2.3 Effect of A_w on various changes during storage.

0.0 to about $A_w = 0.35$ upon which the rate gradually increases with A_w. This is explained by the observation that in most dry cereal products, the A_w of 0.35 corresponds to a moisture level of 8–10% which is the amount of water necessary to form a protective monolayer over the surfaces of polysaccharides and proteins that are associated with unsaturated lipids. This monolayer acts as a barrier to free radical oxygen attack on the carbon-to-carbon double bond system.

Figure 2.3 shows the relationship between four other modes of deterioration of a food and the A_w of the product. *Hydrolytic rancidity* is the enzymatic hydrolysis of fatty acids from glycerol. Normally this is of little consequence except when the fatty acids are largely short or intermediate chain-length from butyric to lauric. Butyric and the 'goat acids', capric, caproic and caprylic, are volatile enough to be recognized as off-flavors at very low concentrations, except in cheeses, where they are accepted. Myristic and lauric acids impart an undesirable 'soapy' taste to a product in which they develop.

Maillard browning reaction is undesirable because of the color change in a product and production of bitter components in a food product. This complex series of reactions begins with an amine (protein or amino acid) combining with a carboyl moiety (reducing sugar) and progresses to a broad range of compounds including dark-colored polymers. The rate accelerates as the A_w increases above 0.25–0.3.

Color losses occur as the A_w increases. This process is illustrated by the curve for chlorophyll which begins to deteriorate above $A_w = 0.35$. Similar curves can be drawn for carotenoids, anthocyanins, and others.

Vitamins such as ascorbic acid are susceptible to oxidative processes but are less reactive at very low A_w conditions. Other vitamins such as Vitamin A are subject to free radical autooxidation processes as shown in Fig. 2.1.

2.4 How moisture management systems work: the case of meat jerky

In many cases, meat jerky has the best, moist, most tender texture at an A_w near the A_w at which microorganisms grow. A system that controls the A_w in a narrow range near this optimal point will enable the manufacturer to formulate and assure the best product possible with such a system with assurance of microbiological safety.

There are a number of factors as well as moisture that impact the shelf-life of a food product. They can include oxygen, ultraviolet light, heat and freezing conditions as well as other factors that may be unique to certain products. Means of controlling these factors that accelerate a product's deterioration have been commercially available, with the exception of moisture.

Now, for the first time, a practical moisture management system has been developed and is one more tool for the food scientist or academic researcher to apply from their tool chest. They can now manage the A_w of a product to keep it at its optimum level. An inexpensive, safe, effective moisture management system as described below has not been available previously. As discussed above, moisture management means that the product's optimal water activity is maintained over time as the product and package are exposed to varying environmental conditions. It will provide moisture to the product when required or remove it by absorbing it when conditions demand, so that the A_w remains constant in the package. Both scenarios are available at any time during the life of the moisture management system and are designed to keep the product at the optimum A_w.

Chemists, and even the alchemists, understood many years ago that addition of a solute to water or other solvent will raise the boiling point and reduce the freezing point of the solvent. Later it was observed that a saturated salt solution will create a headspace environment of a specific relative humidity as a function of which salts are selected. These phenomena are collectively known as colligative properties and are all a function of particles in solution in a unit weight of solvent (Raoult's Law).

Food chemists have employed this technique to help develop stable intermediate moisture food products such as fruit leather, confections and sweet baked goods. Known constant humidity environments have been generated in small chambers such as desiccators with saturated salt solutions. Experimental products are equilibrated against these known environments, establishing optimized formulae for such products.

These saturated salt solutions are not practical in a food package subjected to the typical distribution system because of the potential for spillage. Electronic control systems are not economically feasible for individual packages. Desiccants such as silica gel, molecular sieves or hydration devices employing a wick from a reservoir of water are acceptable only if the objective is to keep the product very dry or very moist, respectively. These systems cannot maintain a narrow range of water activity in the package.

A two-way moisture management technology that has been patented by Humidipak, Inc.* basically consists of a saturated salt solution with excess crystalline solute thickened with a gum system. The thickening, to a viscosity of about 5000 or more centipoises (cps), serves three important functions:

- Excess solute is suspended to assure uniform capability to absorb moisture from unit to unit.
- Loss of the contents of the pack is minimized if the pack is punctured, suffers stress cracks or a compromised seal.
- It permits use of more permeable films.

This filling is packaged in a material that is highly permeable to water vapor (high WVTR). This technology delivers the desired relative humidity in a range of less than $A_w = 0.02$ within a closed environment over an extended period of time under normal commercial distribution temperatures. Two patents have been issued and three more are currently pending.

* Humidipak, Inc., Minnetonka, Minnesota, USA.

This two-way moisture management system is currently available as packets, tubes and tubs (shallow plastic cups or trays). Food-compatible water activities available include 0.95, 0.84, 0.80, 0.78, 0.75, 0.73, 0.70, 0.65, 0.62 and 0.32, all made with food-grade materials. Other A_w values are available such as 0.69, 0.58, 0.52, 0.45 and 0.13 but not of food-grade materials.

An oxygen scavenger system can be added to many of these two-way humidity regulators. While these scavengers do not respond as rapidly as the dry, commercially available oxygen scavenger packets, the combined regulator/scavenger system has sufficient reaction rate to be practical in nitrogen-flushed applications such as meat jerky. Further, packets can be designed to meter some flavors into the package so that the product can seem to be fresher than a food product in a package without the moisture management system. It has also been found that mold growth can be further reduced by adding a mold inhibitor to the regulator packet, or applying some inhibitor onto the surface of the pouch.

2.4.1 Meat jerky

Meat jerky has been a fast-growing product in the food industry. Jerky is marinated beef, or turkey, or game or mixtures that has been dried to a USDA standard of identity level of moisture. The maximum moisture allowed is on the order of 30% with a water activity on the order of 0.8. According to an expert taste panel, at A_w of 0.78 to 0.8, 'natural style' jerky is subjectively relatively easy to bite and has a moist mouthfeel. The texture of jerky becomes less succulent and tougher as the water activity decreases from 0.75 and it becomes decidedly less palatable as the value drops below 0.73 to a hard, tough product giving a noticeably dry mouthfeel.

Analysis of numerous samples of jerky from numerous manufacturers obtained by diverse sellers such as grocery and convenience stores, sports stores and warehouse markets shows a wide variety in the A_w of product available to the consumer. Some of these results are summarized in Fig. 2.4.

Table 2.2 shows that a Humidipak moisture regulator maintains the A_w of jerky at room temperature and reasonably well at a storage temperature of 100°F (37.8°C) for 90 days, a condition that predicts the storage deterioration equivalent to about one year at room temperature (70°F or 21°C).

In the experiments summarized above, according to a panel, the jerky without Humidipak was definitely inferior to the jerky with humidity regulators at an A_w of 0.78.

2.4.2 Humidity restoration in beef jerky

Since the regulator is two-way, it can restore a product that has been dehydrated. However, the process is slow in jerky, thus it is far more important to maintain the proper moisture/water activity of jerky during storage and distribution.

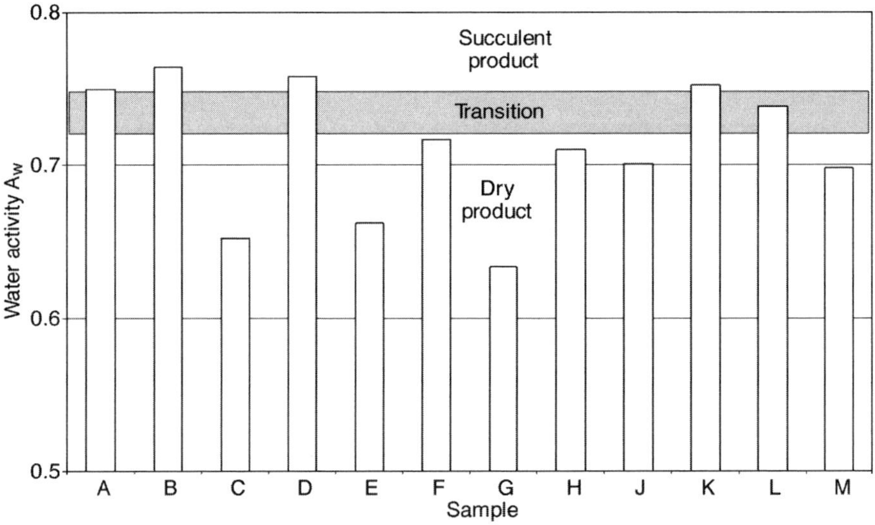

Fig. 2.4 The A_w of jerky by various manufacturers obtained for several types of retail outlets. The shaded area between about A_w 0.73 and 0.75 is the transition area between a moist, soft, succulent texture above 0.75 and the firm/hard, dry texture exhibited by jerky below $A_w = 0.73$.

Figure 2.5 shows how a package of jerky manufactured at A_w of 0.78 lost enough moisture after 11 months in distribution to fall to a very poor quality product with $A_w = 0.67$. When a Humidipak regulator at $A_w = 0.75$ was placed into the jerky package, the A_w of the jerky increased to 0.73 in three months. This rehydrated jerky had an acceptable texture. Had a higher humidity Humidipak been added, the jerky would probably have attained a value of over 0.75. However, mold could probably have developed since nitrogen flush equipment was unavailable.

It is difficult to produce and hold the A_w of a manufactured product to within ±0.02. By introducing a two-way regulator into the package, the initial product will be adjusted to the target specification by either absorbing or giving off water to the product. This is especially important for food items that are near the A_w where microbial growth can occur.

Table 2.2 A_w valves for storage of jerky with or without a humidity regulator

	With		Without	
Time (days)	Room temp.	100°F	Room temp.	100°F
0	0.79–0.80	0.79–0.80	0.79–0.80	0.79–0.80
90	0.78–0.79	0.75–0.77	0.76–0.74	0.73–0.74

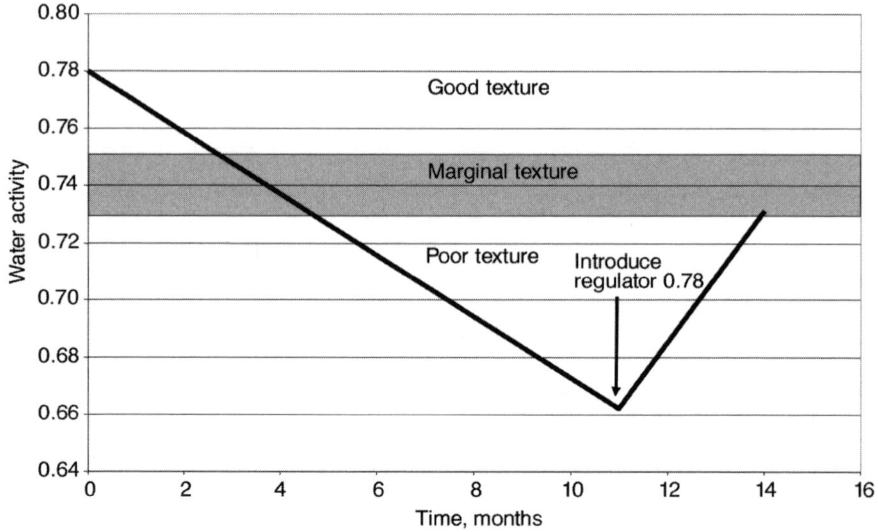

Fig. 2.5 Restoration of moisture to dehydrated jerky with an A_w 0.75 humidity regulator.

While restoration of moisture to a dehydrated product is possible by many methods, it does not return most products to the same organoleptic condition as the original. Therefore it is very important to maintain a constant humidity in the package so that the water activity of the contents remains the same.

2.4.3 Mold inhibition

Standard practice in the jerky industry employs commercially available oxygen scavenger packets to maintain a low level of oxygen in the headspace. Such an oxygen scavenging system has been incorporated into the humidity regulator. Table 2.2 shows that the Humidipak with scavenger packet is just as effective as commercially available oxygen scavenger packets. These values are a composite of field tests conducted with two manufacturers and represent some 30 samples of each test scheme.

The control consisted of the product as manufactured with nitrogen flushing of the filled jerky pouch with a commercial 50 ml oxygen scavenger packet. The other two test samples had Humidipak humidity regulators at 78% relative humidity. The data from the regulator with scavenger system is reported in the middle column; the data with plain regulator and the same oxygen scavenger packet as the control is given in the right column. These samples were stored for 90 days at 100°F (37.8°C) and 25% relative humidity, conditions that generally predict storage of one year at 70°F (21°C). Samples held for 90 days at 70°F had values approximately 25% of these headspace analyses as well as weight loss of the jerky through the package.

Inhibitors work, but less effectively, if incorporated into the filling of the humidity regulator. A more effective method is to 'print' a pattern of concentrated potassium sorbate solution at the same time that graphics are printed on the film. In any case, because of the low volatility of sorbic acid, the regulator must be slightly acidic and be in very near proximity with the product. It will add several weeks to the life of a properly hydrated jerky after opening a nitrogen flushed, resealable package.

2.5 Application of moisture management systems to food and other products

The current commercial uses of moisture management systems are in the tobacco markets which are very similar to food products in their deterioration modes and the potential impact of moisture on shelf-life of these products. The primary initial use of a moisture management system has been in the cigar area. Here it is used to keep high quality cigars that were purchased from a retailer with a walk-in humidor in a pristine state until consumed by the user (see Fig. 2.6). These cigars may sell for \$5 to \$35 each. It is also being used commercially for such items as the 25 count boxes of cigars that are packaged in very humid Central American countries and protects the products as they travel via various carriers and through unknown environments to the end point.

A further application is in the retail packaging of expensive cigars being merchandised in gasoline stations or convenience stores. Cigars of high quality that would normally only be sold from a walk-in humidor at a tobacco shop can now be packaged and fully protected over their shelf-life under ambient conditions in any geographic location. For cigars, this means keeping them exposed to the optimum level of 70% relative humidity which the moisture management system is fully capable of doing.

Laboratory testing is underway for a wide variety of applications, only a few of which are itemized below. Because of security and confidential disclosure agreements that are in place, other products under test can not be revealed at this time.

Beef jerky is a product that can become very tough and chewy when moisture levels drop and are not at their optimum. If the moisture is too high, the product

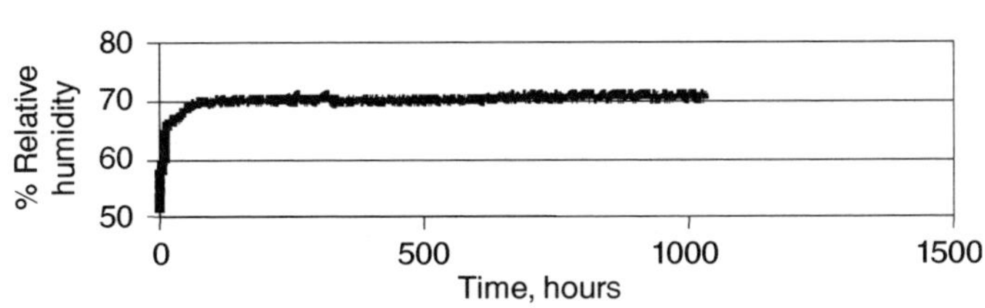

Fig. 2.6 Performance of the Humidipak moisture regulator (new 70% Humidipak in Humidor Bag, ambient conditions) in a portable premium cigar humidor (zippered high barrier film pouch).

can become moldy or fail to meet the legal definition of jerky. The moisture management system can target about 0.02 A_w below this optimum point and provide high quality performance for the product over its 12 to 18 month shelf-life.

The unique flavor nodes and free-flowing characteristics of many *dry spices* are significantly harmed when moisture levels are too high or too low. The consumer expects these products to last indefinitely, but it is apparent to the gourmet using them that the standard packaging used is often inadequate for protecting a spice for more than a very few months. This is especially true if the reclosable container is opened repeatedly under dry conditions.

Dried fruit products can be a delicacy when available in the proper state. But again this is a critical and relatively narrow moisture range where the product is in its pristine and delectable state. A moisture management system has proven to yield long-term stability and can even be used to salvage dried-out products back to nearly their original condition.

When a *stringed instrument* made of wood, such as a violin or guitar, is exposed to a low or high humidity, the wood will lose or gain moisture. This will result in the wood shrinking or expanding, which brings about warping that changes the sound characteristics of the instrument. Or worse, cracks may appear and even cause the adhesive joints to break apart. The appearance of cracks will reduce the value of the instrument by as much as one half, which can be several hundred thousands of dollars (or priceless in the case of a Stradivarius). Owners of these instruments will readily relate to such issues when holding guitars, violins, etc., under widely varying conditions that change with the seasons. It is almost farcical to explain the methods being used today to provide moisture protection. These instruments should be in the 50% relative humidity range. Wet sponges are the most common element in many of the systems used today. Naturally they try to reach 100% RH but because of leakage in the packaging may average out at something in the middle of the range. They also require daily monitoring and most likely the addition of distilled water for them to provide any impact on the instrument. A suitably designed moisture management system could deliver several months of optimum conditions for an instrument without any required daily effort.

Museums around the world expend considerable time and effort to maintain the optimum temperature and humidity conditions for a wide variety of historical and extremely valuable *works of art*. This may include timeframes for the artworks while on display, in warehouses or while being transported over short and long distances. Today, one of the systems in use requires the conditioning and monitoring of silica gel and, if not properly handled, can cause significant harm. A moisture management system as described herein using salt solutions can provide an ongoing and specifically targeted relative humidity atmosphere for each application over an extended period of time.

Lastly, *nylon connectors* have an optimal humidity at which they do not crack when compressed or blister when heat is applied. *Wood veneers* can be formed over small radii only over a narrow range of moisture content. *Powdered materials* such as laser printer cartridges will tend to produce fuzzy images if too dry where static charges develop, or form lumps if too moist.

2.5.1 Packaging materials

The salt solution required for a specific application can vary from 5 grams to 60 grams or more as a function of what the requirements are. It is often packaged in a pouch or other container and sealed shut to prevent leakage. This pouch or container is then placed, along with the food product being protected, in an outer package that has a moisture barrier to the outside environmental conditions.

A few examples of materials that have been commercialized or thoroughly laboratory tested for holding the salt solution are as follows:

- Printing/35# Kraft Paper/1.5 mil extruded Hytrel* film
- Thermoformed polypropylene cup/heat seal coating/Tyvek* membrane
- Printing/35# Kraft Paper/1.1 K-Resin from Phillips
- Nylon film 1 mil thick formed into a chub package.

It is very critical that the package and weight of salt solution be determined and designed properly for each food application. There are many variables, and we have found that if the detailed laboratory analysis and studies are completed at the start of an investigation, the actual results of the recommendations will be found to be acceptable and comparable to the laboratory testing.

The speed of water vapor entering and exiting the container holding the salt solution is a variable that can be matched with the desired function. If the moisture management system is to preserve the food product over an extended period of time, the permeability of this system can be at a slower speed. If it is to manage moisture levels while a package is being opened and reclosed numerous times, the moisture permeability rate needs to be faster so that the desired levels can be reached within 1–2 hours.

Once the product has been selected for a study and the A_w has been determined, the salt solution will be formulated that will keep the product being protected in its optimum state.

In order to determine and define the specifications for the package size and material, we need answers to a number of pertinent questions:

1. What is the water activity (A_w) of the product being managed?
2. What is the product weight?
3. How much water, in weight, is present in this quantity of product?
4. What tolerance does the product have to moisture gain and loss?
5. How long is the desired shelf-life before opening?
6. Are there conditions of reclosure that must be managed?
7. In what geographic area and environmental conditions is this product being distributed or held?

* Hytrel and Tyvek are trademarks of the Dupont Corporation.

8. What is the package profile and general construction?
9. What is the surface area of the exterior package holding the product?
10. What is the water vapor transmission rate of this outer package?
11. Are there any other product characteristics that might be unique?

With these answers, we can then select the pouch or container material to be used in this application, calculate the weight of salt solution required and conduct laboratory trials to observe the moisture management system's impact on shelf-life and product quality over time.

There is significant flexibility in designing the pouch or container to hold the salt solution. It must be designed to be cost-effective for the application while delivering on the performance required.

2.5.2 Other technologies

Moisture management has been limited to formulation, processing and packaging. Active systems to control moisture have been limited to desiccants such as silica gels, molecular sieves, certain clays and selected salts. Desiccants tend to reduce the A_w of the products to near zero and thus have limited application. Other active systems include a wide variety of devices to add moisture to the environment such as wicks, sponges and semi-permeable films. These hydration devices tend to increase the A_w to high levels, especially in humid environments.

Conditioned silica gels and some clays have the property of absorbing and desorbing moisture over a relatively narrow range of humidities. While effective, they suffer from a very limited capacity, often only 5–10% of their weight.

There are electronic devices to maintain a constant humidity in chambers, even buildings. However, these are rather expensive and not practical for consumer food packages. Also, they require significant, regular maintenance to be reliable.

2.6 Future trends

A number of food products can be greatly aided in their shelf-life if a moisture management system is designed to work in conjunction with temperature and in some instances headspace gas control. Some examples of situations where this technology will be commercialized within the next five years will potentially include:

1. International shipping of fresh fruits and vegetables. Moisture and humidity control has never been readily available to help prevent dehydration during ocean shipments and thus has forced some products into air freight. A moisture management system could offer cost savings while ensuring the delivery of a superior product.

2. The fresh produce sector has grown dramatically within the past five years with the many iterations of cut lettuce in various consumer pack varieties. Current systems do an excellent job of controlling the headspace gases but cannot impact the moisture and humidity within the package. As a result, packages moving from temperature controlled environments through the normal handling and consumer use systems will experience changes that can result in product deterioration and unacceptable products before the shelf-life limit has been reached. A moisture management system is a tool that could greatly aid this delivery system.
3. With improving economic status and all adults in a household in employment, the demand for ready-to-eat prepared vegetables and cut fruits or vegetables is increasing rapidly. The maintenance of constant humidity in the package is critical to preserving quality of appearance, flavor and texture.
4. Products in the home which are stored for future use could be better preserved if stored in a food protection system that included moisture management. Examples would be dried fruit, raisins, fresh fruits, etc.
5. Moisture management is an important component of controlling growth of microorganisms. With increasing public awareness of foodborne infections and microbial toxins, providing constant humidity in a package will contribute to the delivery of safe foods.

2.7 Bibliography

FENNEMA, O. 1996, Water and ice, in *Food Chemistry*, ed. Owen Fennema, Marcel Dekker, New York.

LEVINE, H., SLADE, L. and KAREL, M. 1988, *Moisture Management in Food Systems*, Center for Professional Advancement, PO Box H. East Brunswick, NJ08816, USA.

PARIZA, M. W. 1996, Toxic substances, in *Food Chemistry*, ed. Owen Fennema, Marcel Dekker, New York.

SAARI, A. and ESSE, R. 1999, US Patent Number 5,936,178, Humidity Control Device, issued Aug. 10, 1999, 35 claims.

SAARI, A. and ESSE, R. 2001, US Patent Number 6,244,432, Humidity Control Device for Gun Cases, issued June 12, 2001, 34 claims.

3

Temperature and food stability: analysis and control

P. S. Taoukis and M. C. Giannakourou, National Technical University of Athens, Greece

3.1 Introduction: temperature and the shelf-life of food

Modern consumers seek improved sensory quality, increased functional and nutritional properties combined with a traditional, wholesome image. They expect these qualities from food products of guaranteed safety but with less processing, fewer additives and 'technological' interventions. At the same time they expect extended shelf-life and high convenience in preparation and use.

The attainment of longer shelf-life with minimum processing requires not only optimization and control of all production and preservation parameters but often innovative techniques, to ensure safety and reduce food deterioration. The efforts of producers and regulators concentrate on the development and application of structured quality and safety assurance systems based on prevention through monitoring, recording and controlling of critical parameters through the product's entire life cycle. These systems include the post-processing phase and ideally extend to the consumer's table. The need for the inclusion of the post-processing phase is emphasized in the globally applied Hazard Analysis and Critical Control Point (HACCP) safety assurance system. Certain stages of the chill chain are recognized as important critical control points (CCPs) for minimally processed chilled products such as modified atmosphere packaged and other ready-to-eat chilled products.

Research and industrial studies show that chilled or frozen distribution and handling very often deviate from recommended temperature conditions. Temperature largely constitutes the determining post-processing parameter for shelf-life under Good Manufacturing and Hygiene Practices, and monitoring and controlling it is therefore of central importance. The complexity of the problem is highlighted when the variation in temperature exposure of single products within batches or transportation subunits is considered. A systematic study and kinetic modelling of the role of temperature in determining shelf-life is therefore indispensable in any quality management and optimization scheme. Based on reliable models of food product shelf-life, the effect of temperature can be monitored, recorded and translated from production to the consumer's table.

3.2 Quantifying the effect of temperature on food

Quality is a dynamic, complex attribute of food, that influences the degree of its acceptability by the consumer and is constantly moving towards lower levels post processing (Taoukis *et al.*, 1997). Through a careful study of the food components and the process, the reactions judged to have the most critical impact on the deterioration rate can be established. Based on this analysis and without underestimating the underlying complexity of food systems, the change of food quality can be represented by the loss of one or more quantifiable quality indices, symbolized by A (e.g. a nutrient or characteristic flavour), or by the formation of an undesirable product B (e.g. an off-flavour or discolouration). The rate of change of A and B can in general be represented by eq. (3.1):

$$r_{\mathrm{A}} = -\frac{\mathrm{d}[\mathrm{A}]}{\mathrm{d}t} = k[\mathrm{A}]^{m} \qquad [3.1]$$

$$r_{\mathrm{B}} = -\frac{\mathrm{d}[\mathrm{B}]}{\mathrm{d}t} = k'[\mathrm{B}]^{m'}$$

The quality factors [A] and [B] are usually quantifiable chemical, physical, microbiological or sensory parameters, selected so as to representatively describe the quality deterioration of the particular food system. The constants k and k' are the apparent reaction rate constants and m and m' are the apparent orders of the respective reactions. The use of the term 'apparent' indicates that eq. (3.1) does not necessarily describe the mechanism of the measured phenomenon. A general equation describing the loss of the quality factor A in a food system may be expressed as:

$$f_{\mathrm{q}}(\mathrm{A}) = k(C_i, E_j) \cdot t \qquad [3.2]$$

where f_{q} can be defined as the quality function of the food, and k, the apparent reaction rate constant, is a function of composition factors C_i, such as concentration of reactive compounds, inorganic catalysts, enzymes, reaction inhibitors, pH, water activity and microbial populations, and of environmental factors, E_j, such as temperature, relative humidity, total pressure and partial pressure of different gases, light and mechanical stresses.

The form of the quality function of the food for an apparent zero-, first-, second- and mth-order reaction is shown in Table 3.1. Methodology for determination of the apparent reaction order and reaction rate constant is described by Taoukis *et al.* (1997).

Table 3.1 Quality function based deterioration of quality index A

Apparent reaction order	Quality function $f_q(A_t)$
0	$A_0 - A_t$
1	$\ln(A_0 - A_t)$
2	$1/A_0 - 1/A_t$
$m(m \neq 1)$	$\frac{1}{m-1}(A_t^{1-m} - A_0^{1-m})$

The value of the quality index A_{t_s} that signals or corresponds to the limit of acceptability of the food can be translated to a value of the quality function, $f_q(A_{t_s})$. The time to reach this value at specified conditions, i.e. the shelf-life, t_s, is inversely proportional to the rate constant at these conditions (eq. 3.3):

$$t_s = \frac{f_q(A_{t_s})}{k} \qquad [3.3]$$

Most reactions responsible for shelf-life loss, based on a characteristic physicochemical, chemical or microbial index, have been classified as zero-order (e.g. frozen food overall quality, Maillard browning) or first-order (e.g. vitamin loss, oxidative colour loss, microbial growth). Kinetic equations for shelf-life estimation are specific to the food studied and the environmental conditions used. Among the environmental factors considered, the one being invariably emphasized and introduced in the shelf-life model is temperature. It strongly affects post-processing reaction rates and during subsequent handling, distribution and storage cannot be controlled *a priori* by means such as food packaging and depends on the imposed environmental (storage) conditions.

The prevailing effect of temperature on the rate of food-related reaction rates has long been the subject of research, and a significant number of kinetic studies of important indices of physical, chemical, microbiological or sensory deterioration of foods have been published. Of the mathematical equations that have been proposed to describe the temperature dependence of the quality loss rate, the Arrhenius relation, derived from thermodynamic laws and statistical mechanics principles, is the most widely used (Arrhenius, 1889). The Arrhenius relation, developed theoretically for reversible molecular chemical reactions, has been used to describe the effect of temperature on the rate of several reactions of quality loss, as follows (eq. 3.4):

$$k = k_A \exp\left(\frac{-E_A}{RT}\right) \qquad [3.4]$$

where k_A represents the Arrhenius equation constant and E_A, in joules or calories per mole, is defined as the activation energy, i.e. the excess energy barrier that quality parameter A needs to overcome to proceed to degradation products. R is the universal gas constant (1.9872 cal/mole K or 8.3144 J/mole K). To estimate the effect of temperature on the reaction rate of a specific quality deterioration mode, values of k are estimated at different temperatures in the range of interest, and $\ln k$ is plotted against the term of $1/T$ in a semilog graph. A straight line is obtained with a slope of $-E_A/R$ from which the activation energy is calculated.

In Fig. 3.1(a), the degradation of L-ascorbic acid in pasteurized orange juice is shown to be adequately described as an apparent first-order reaction, when stored in the range 0–15°C (Polydera *et al.*, 2003), allowing for the estimation of the apparent reaction rate k by linear regression. The temperature dependence of L-ascorbic acid oxidation is depicted in the respective Arrhenius plot (Fig. 3.1(b)) from which E_A was estimated at 44 kJ/mol.

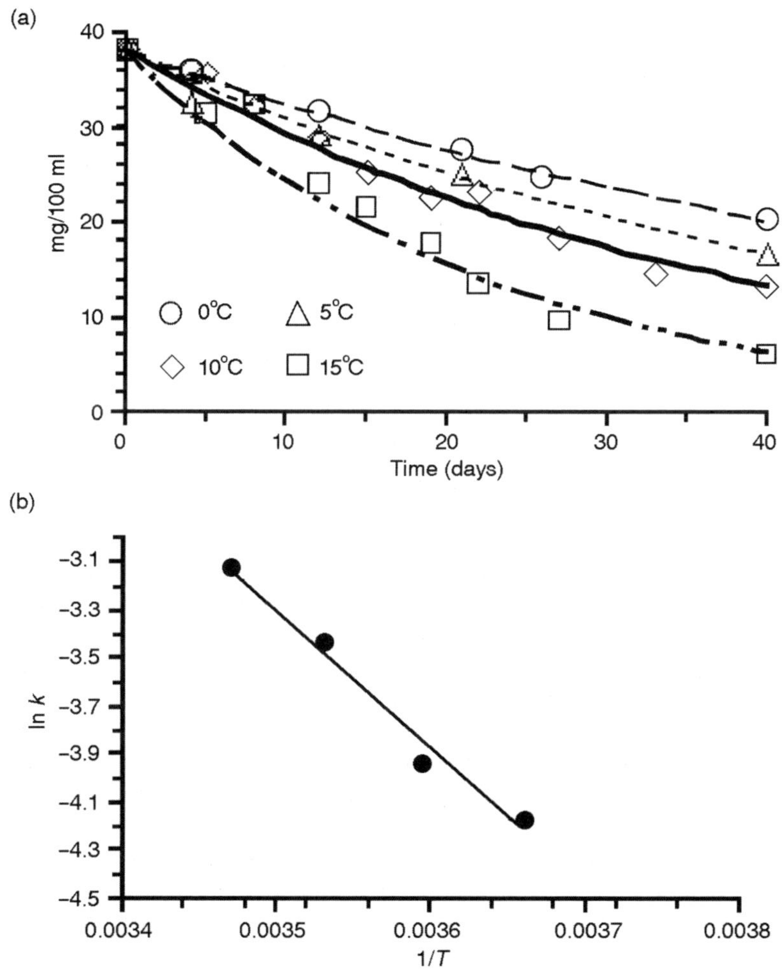

Fig. 3.1 (a) L-ascorbic acid loss during refrigerated storage of thermally pasteurized orange juice at four temperatures; (b) Arrhenius plot of the temperature dependence of L-ascorbic acid oxidation.

It should be noted that the Arrhenius equation implies that k_A is the value of the reaction rate at 0 K, which is of no practical interest. Alternatively, the use of a reference temperature, T_{ref}, is recommended, corresponding to a representative value in the temperature range of the process/storage of study. Equation (3.4) is then mathematically transformed as follows (eq. 3.5):

$$k = k_{ref} \exp\left[\frac{-E_A}{R}\left(\frac{1}{T} - \frac{1}{T_{ref}}\right)\right] \tag{3.5}$$

where k_{ref} is the rate constant at the reference temperature T_{ref}. In that case, the value of E_A is calculated from the linear regression of $\ln k$ vs. $(1/T - 1/T_{ref})$. Values of T_{ref} usually employed are 255 K for frozen, 273 K for chilled and 295 K for ambient temperature-stored food products. Besides giving the constant a practical physical meaning, the above transformation of the Arrhenius equation provides enhanced stability during numerical parameter estimation and integration.

When only three points in the Arrhenius plot are used for the determination of E_A, the 95% confidence limits of the Arrhenius parameters, as estimated with statistical tools, are usually wide. If narrower limits are required, experiments must include more temperatures in the range of interest. According to the optimum scheme of Lenz and Lund (1980), an optimum number of five or six isothermal experiments are adequate to obtain a satisfactory accuracy.

Alternatively to isothermal kinetic analysis, the study at a single non-isothermal temperature profile is proposed, where temperature varies with time following a predetermined function of $T(t)$. In that case, eq. (3.2) is modified as follows, assuming a first-order reaction (eq. 3.6):

$$A = A_0 \exp\left[-k_A \int_0^t \exp\left(-\frac{E_A}{R}\frac{1}{T(t)}\right) dt\right] \tag{3.6}$$

where the integral is calculated by numerical techniques. This approach requires very strict temperature control and is very sensitive to experimental error in concentration measurements. Additionally, the isothermal method at three different constant temperatures gave better results for the estimation of the Arrhenius parameters than the non-isothermal approach with a linearly increasing temperature at the same range and for an equal number of data points (Yoshioka *et al.*, 1987). The non-uniform temperature within the samples, as well as the difficulty in recognizing a deviation from an Arrhenius behaviour, should also be considered when applying the non-isothermal technique.

Numerous shelf-life models based on the Arrhenius equation have been published in order to describe either the temperature dependence of quality loss (Hertog *et al.*, 1997; Buedo *et al.*, 2001; Giannakourou and Taoukis, 2002) or the combined effect of temperature with other environmental factors (Seyhan *et al.*, 2002).

An alternative to the Arrhenius law to describe the temperature dependence of reaction rates is through the Q_{10} concept, a tool of practical importance to the food industry. Q_{10}, which was used in the early food science and biochemistry literature, is the ratio of the reaction rate constants at temperatures differing by 10ºC or, equivalently, it shows the reduction of shelf-life θ_s when the food is stored at a temperature 10ºC higher (eq. 3.7):

$$Q_{10} = \frac{k_{(T+10)}}{k_{(T)}} = \frac{\theta_s(T)}{\theta_s(T+10)} \qquad [3.7]$$

The Q_{10} approach in essence introduces a temperature dependence equation in the form of the following (eq. 3.8):

$$k(T) = k_0 e^{bT} \Rightarrow \ln k = \ln k_0 + bT \qquad [3.8]$$

which implies that if $\ln k$ is plotted against temperature (instead of $1/T$ of the Arrhenius equation), a straight line is obtained. Alternatively, shelf-life (t_s) can be plotted against temperature, as follows (eq. 3.9):

$$t_s(T) = t_{s0} e^{-bT} \Rightarrow \ln t_s = \ln t_{s0} - bT \qquad [3.9]$$

where the outcoming plots are often called shelf-life plots, where b is the slope of the shelf-life plot and t_{s0} is the intercept. In Fig. 3.2, the kinetic results of the shelf-life study of the conventionally thermally pasteurized orange juice of Figs 3.1(a) and (b) are alternatively demonstrated in a shelf-life plot based on 50% Vitamin C loss ($A_{t_s} = 0.5A_0$).

Shelf-life plots are practical and easier to understand as one can read directly the shelf-life of the food at any storage temperature. These plots are true straight lines only for narrow temperature ranges of 10–20ºC. Within this interval, Q_{10} and b are functions of temperature, correlated to the activation energy of the food quality deterioration reaction, following eq. (3.10):

$$\ln Q_{10} = 10b = \frac{E_A}{R} \frac{10}{T(T+10)} \qquad [3.10]$$

Shown in Table 3.2 are the dependence of Q_{10} on temperature and E_A and important types of food reactions that fall in the respective range of values of Q_{10} and E_A.

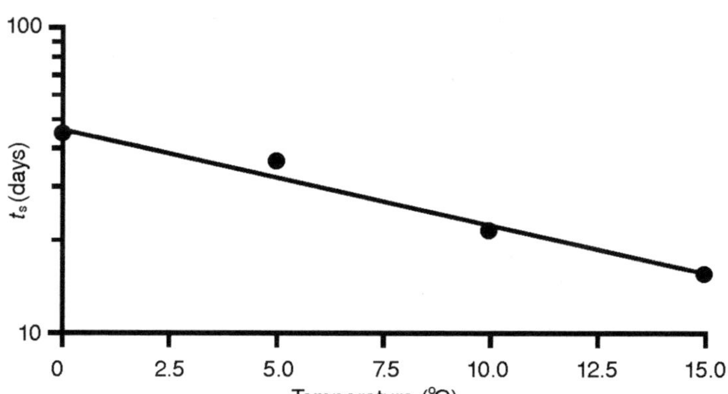

Fig. 3.2 Shelf-life plot of pasteurized orange juice based on 50% loss of Vitamin C.

Table 3.2 Q_{10} dependence on E_A and temperature

E_A (kJ/mol)	Q_{10} at 5°C	Q_{10} at 20°C	Q_{10} at 40°C	Typical food reactions
42	1.87	1.76	1.64	Diffusion controlled, enzymatic, hydrolytic
85	3.51	3.10	2.70	Lipid oxidation, nutrient loss
125	6.58	5.47	4.45	Non-enzymatic browning

The value of the quality function (eq. 3.2) at time *t*, when the food is exposed at a predetermined variable time–temperature condition *T*(*t*), can be estimated by calculating the integral of eq. (3.11):

$$f_q(A)_t = k_0 \int_0^t e^{bT} dt \qquad [3.11]$$

Alternatively to Q_{10}, sometimes Q_A is used instead, where 10°C is replaced by *A*°C (eq. 3.12):

$$Q_A = (Q_{10})^{\frac{A}{10}} \qquad [3.12]$$

When studying the kinetics of microbial inactivation, the *z*-value is frequently used to describe the temperature dependence of the rate of quality loss (Hayakawa, 1973). The value of *z* is the temperature range that causes a 10-fold change in the reaction rate constant. Similarly to the Q_{10} approach, *z*-value depends on the reference temperature and is related to *b* and E_A by eq. (3.13):

$$z = \frac{\ln 10}{b} = \frac{(\ln 10) \cdot RT^2}{E_A} \qquad [3.13]$$

In the field of predictive microbiology, several temperature dependence equations have been proposed for describing the *k*(*T*) function, where *k* is the exponential growth rate of the microorganism in question. Most prominently used in the suboptimal temperature range are the Arrhenius equation and the Belehradek equation (McMeekin *et al.*, 1993). The Belehradek equation (Belehradek, 1930) has the following generalized form (eq. 3.14):

$$k = a(T - T_0)^d \qquad [3.14]$$

where *a* and *d* are parameters to be fitted and T_0 is regarded as the 'biological zero', that is a temperature at and below which no growth is possible. To avoid this assumption, Ratkowsky *et al.* (1983) changed T_0 to T_{min} and proposed the square root model (eq. 3.15a):

$$\sqrt{k} = b(T - T_{min}) \qquad [3.15a]$$

$$\sqrt{k} = b(T - T_{min})\{1 - \exp[c(T - T_{max})]\} \qquad [3.15b]$$

where k is the growth rate, b is the slope of the regression line of $\sqrt{k}$ vs. T, and $T_{\min}$ is the hypothetical growth temperature where the regression line cuts the T-axis at $\sqrt{k} = 0$. $T_{\min}$ is usually 2–3°C lower than the temperature at which growth is actually observed and has been described as notional or conceptual temperature. Equation (3.15a) was later extended to the form of eq. (3.15b) to cover the entire biokinetic temperature range (Ratkowsky *et al.*, 1983). The terms $T_{\min}$ and $T_{\max}$ can be used to classify microbes in a more objective manner as psychrophiles, mesophiles or thermophiles (McDonald and Sun, 1999). However, it should be stressed that, since it is very difficult to obtain accurate data at extremely low growth rates, it should be pointed out that $T_{\min}$ and $T_{\max}$ may differ from the true temperature limits. Despite this problem, eq. (3.15a) and eq. (3.15b) have been successfully applied to many data sets in the predictive microbiology literature.

The Arrhenius equation, mostly applied for the description of the temperature dependence of chemical reactions, could also be used for microbial growth modelling, if all other environmental factors except for temperature conditions are kept constant. In that case, an apparent activation energy of bacterial growth is considered, assuming that one enzyme in a metabolic sequence is rate limiting both above and below the optimum temperature for growth and that this enzyme is reversibly denaturated by temperatures both higher and lower than the optimum (McMeekin *et al.*, 1993). Labuza *et al.* (1992) stated several cases of both successful and inadequate use of the Arrhenius equation and noticed that this relationship can also be applied as a model to describe mathematically the temperature dependence of the lag phase.

The relation between Q_{10} and $T_{\min}$ of the square root equation is (eq. 3.16):

$$Q_{10} = \left(\frac{T - T_{\min} + 10}{T - T_{\min}}\right)^2 \qquad [3.16]$$

In the literature, other forms of the $k(T)$ function have been introduced (Kwolek and Bookwalter, 1971), such as linear, power and hyperbolic equations, with the Arrhenius equation being predominantly applied.

Eyring's equation was applied in the pharmaceuticals sector (Kirkwood, 1977). It has the following mathematical expression (eq. 3.17):

$$\ln k = \ln\left(\frac{k_B}{h}\right) + \frac{S}{R} - \frac{H}{RT} + \ln T \qquad [3.17]$$

where H is the heat of activation, h is the Planck constant, k_B is the Boltzmann constant and S is the entropy. The significance of this equation lies in the fact that it relates the effect of temperature on the reaction rate constant to the fundamental terms of enthalpy and entropy changes and, consequently, one could estimate the enthalpy/entropy compensation in food reactions (Labuza, 1980). If, for instance, the enthalpy of activation has a high value, this would lead to a quite slow reaction rate at moderate temperatures; this could, however, be compensated by an increase in activation entropy such that the reaction can still proceed at a measurable rate (Van Boekel, 2000).

As has been often noted in the food literature, there are cases where the dependence on temperature of the rate of food quality loss can deviate from the Arrhenius equation (Labuza and Riboh, 1982). Phase change phenomena are often involved in such deviations. In the case of frozen foods, water is gradually being removed in the form of ice, causing a significant increase of the solute concentration. The freeze–concentration effect is predominant on the reaction rate in the immediate subfreezing temperature range, and the observed rate increase is especially notable for reactants of low initial concentration. The Arrhenius plot would therefore show an abrupt change in this range, and a single Arrhenius line should not be used in case the freezing point is crossed within the temperature range studied.

Depending on the conditions of the subsequent storage and handling, fats may change to the liquid state, leading to increased mobility of the organic reactants and vice versa (Templeman *et al.*, 1977) reflected in a non-Arrhenius temperature effect on the reaction rates. Other important phenomena include the crystallization of amorphous carbohydrates at lower temperatures, which alters significantly the food stability and quality (Kim *et al.*, 1981), and the denaturation of proteins that may modify their susceptibility to chemical reactions.

Glass transition is related to dramatic changes of food mechanical properties and molecular mobility and may occur in carbohydrate-containing foods when storage conditions are suddenly modified, such as during rapid cooling or solvent removal. Examples of the formation of metastable glasses that deteriorate following a kinetic pattern that deviates from the Arrhenius law include frozen carbohydrate-containing solutions or food products (Champion *et al.*, 2000; Carrington *et al.*, 1996; Biliaderis *et al.*, 1999; Blond and Simatos, 1991; Furuki, 2002), spray-dried milk (Bushill *et al.*, 1965), whey powder and dehydrated vegetables (Buera and Karel, 1993), osmotically dehydrofrozen fruits and vegetables (Chiralt *et al.*, 2001; Torreggianni *et al.*, 1999), etc.

In systems that are subject to glass transition, due to drastic acceleration of the diffusion-controlled reactions above T_g, the dependence of the rate of a food reaction on temperature cannot be described by a single Arrhenius equation. In the rubbery state above T_g, the activation energy is not constant, but is rather a function of temperature. This behaviour has been often described by an alternative equation, the Williams–Landel–Ferry (WLF) expression (eq. 3.18) that empirically models the temperature dependence of mechanical and dielectric relaxations in the range $T_g < T < T_g + 100$:

$$\log \frac{k_{ref}}{k} = \frac{C_1(T - T_{ref})}{C_2 + (T - T_{ref})} \qquad [3.18]$$

where k_{ref} is the rate constant at the reference temperature T_{ref} ($T_{ref} > T_g$) and C_1, C_2 are system-dependent coefficients. Williams *et al.* (1955), assuming $T_{ref} = T_g$ and applying the WLF equation for data available for various polymers, estimated mean values of the coefficients $C_1 = -17.44$ and $C_2 = 51.6$. However, the uniform application of these constants is often problematic (Peleg, 1992; Buera and Karel, 1993; Terefe and Hendrickx, 2002) and the calculation of system-specific values, whenever possible, should be preferred, using eq. (3.18) with $T_{\text{ref}} = T_g$ and rearranging the mathematical expression to the following form (eq. 3.19):

$$\left[\log \frac{k_g}{k}\right]^{-1} = \frac{-C_2}{C_1(T - T_g)} - \frac{1}{C_1} \qquad [3.19]$$

such that a plot of

$$\left[\log \frac{k_g}{k}\right]^{-1} \text{ vs } \frac{1}{T - T_g} \qquad [3.20]$$

gives a straight line with a slope equal to $-C_2/C_1$ and an intercept of $-1/C_1$, if the WLF model is successfully applied. Nelson (1993) suggested the determination of WLF coefficients when applying the WLF equation to describe ascorbic acid degradation in maltodextrin-based systems of various moisture levels. These coefficients were found to depend not only on the type of the matrix in question, but also on the water content. In the same study, it was observed in several cases that in the rubbery zone of the matrices, the Arrhenius equation described more adequately than the WLF model the temperature dependence of the loss of ascorbic acid.

It has been stated that the Arrhenius model is more adequate for describing the temperature dependence of reactions within the glassy state of food matrices, and also at 100°C above the glass transition temperature, but is not applicable within the rubbery state (Slade *et al.*, 1989). Terefe and Hendrickx (2002) studied the kinetics of the pectin methylesterase-catalyzed de-esterification of pectin in a frozen food model in a wide temperature range (−24 to 20°C) and observed that a single Arrhenius plot could not describe the temperature dependence in the whole range. The same observation was made by Terefe *et al.* (2002) for the alkaline phosphatase-catalyzed hydrolysis of disodium *p*-nitrophenyl phospate in the range from −28 to 20°C; the Arrhenius plot was curved as the temperature approached T'_g, in the rubbery state of the matrix, due to the extra temperature dependence which is not accounted for in the Arrhenius model. The shift in the slope of an Arrhenius plot observed in the aforementioned enzymatic reactions may indicate a change in the activation energy or a change in the controlling mechanism. Due to mechanical and mobility properties between glassy and rubbery systems, a break in the Arrhenius plot at or near the glass transition temperature of the system may be expected. However, from data of Lim and Reid (1991), the Arrhenius plot for an enzyme hydrolysis reaction within a partially frozen maltodextrin DE 25 above the estimated T'_g of the system was quite linear and the expected curvature was not observed. In order for this deviation from the Arrhenius linearity to be obvious, study over a broader temperature range is required; however, this is not possible for frozen systems since these matrices melt at slightly higher temperatures, and they are transformed to liquids above 0°C.

At this point, it should be stressed that in some cases, measuring the glass transition temperature may not be adequate for explaining the break in an Arrhenius plot (Terefe and Hendrickx, 2002; Manzocco *et al.*, 1999; Giannakourou and Taoukis, 2003c) and other changes should be taken into consideration.

Besides questions of theoretical validity of the Arrhenius equation in wide temperature ranges that include phase transition phenomena, most notably the frozen range, cautious application even on an empirical basis and within well-defined temperature limits of practical significance serves as a useful tool for shelf-life calculations and predictions. The Arrhenius parameters such as the E_A value give a well-comprehended measure of temperature dependence comparable to the respective ample information existing from kinetic modelling at most food systems in the frozen and non-frozen temperature ranges.

3.3 Shelf-life testing and indices

3.3.1 Shelf-life tests

When deciding on the important aspects of introducing a food product to market competition, knowledge of its shelf-life and, in particular, the dependence of quality loss on temperature conditions is essential (Fu and Labuza, 1993). The criterion for the end of shelf-life varies depending on the special characteristics of the product in question, and on consumer sensitivity. For most processed perishable or semiperishable foods, assuming that the essential requirement for product safety is met, their shelf-life is based mostly on sensory and microbiological quality (Fu and Labuza, 1993). For products with an extended shelf-life, as in the case of most frozen foods, slow chemical reactions are the main deterioration pathways that are frequently not readily recognizable by consumers. If the objective is to get a rapid estimation of the expected shelf-life of the product, then an Accelerated Shelf-Life Test (ASLT) can be used in the range of interest. Taking into account the limitations or possible deviations from the Arrhenius law, the ASLT involves the use of the Arrhenius equation at higher testing temperatures in a thorough shelf-life study and extrapolation of the kinetic results to normal, non-abusive storage conditions. This procedure is used to substantially reduce the experimental time through the acceleration of the quality deterioration reactions. The successive steps that outline shelf-life determination and ASLT methodology are presented in Taoukis *et al.* (1997) and Taoukis and Labuza (1996) and are detailed in Chapters 15 and 16.

Shelf-life studies and ASLT require a profound knowledge of the constituents of the food, the process, the microbiological safety factors, the main modes of quality deterioration and the intended storage conditions. With effective use of ASLT, an experiment that normally takes a year can be completed in about a month, if the testing temperature is raised by 20°C. The duration of the shelf-life determination by ASLT depends on the E_A of the quality deterioration phenomena as is shown in Table 3.3.

Table 3.3 Time to complete an ASLT test for a low moisture food product of two years targeted shelf-life at ambient storage depending on the temperature sensitivity (E_A) of the shelf-life determining reaction

E_A (kJ/mol)	Testing time at 40°C ASLT storage temperature	Testing time at 45°C ASLT storage temperature
45	224 days	171 days
85	78 days	47 days
125	28 days	13 days

3.3.2 Temperature dependence of main shelf-life indices

A prerequisite for the development of a reliable and accurate shelf-life model is the careful selection of the most appropriate index that reflects the quality deterioration of the food of interest. For the description of quality loss, chemical, microbiological and physical parameters are kinetically studied and the appropriate measuring techniques are applied. Based on relevant literature and the observations obtained by preliminary, accelerated tests, chemical and biological reactions and physical changes are carefully evaluated in order to distinguish the factors that are most significant to product safety, integrity and overall quality (Labuza, 1985). Kinetic data relevant to shelf-life have been compiled by Labuza (1982), covering most of the earlier literature. Ample information exists in the current literature. A lot of this data is not fully exploitable as it is not a result of a systematic kinetic approach that would allow modelling and prediction for other conditions and similar systems.

Microbiological spoilage, lipid oxidation, non-enzymatic browning and vitamin loss are among the most significant quality deteriorating factors that signal the end of food shelf-life. In Table 3.4, some recent representative examples of shelf-life studies are listed, based on the measurement of one of the aforementioned degradation modes.

A further step could focus on correlating the kinetics of the selected degradation reactions to sensory results for the same food and a preset limit that defines the lowest organoleptic quality marginally accepted by the consumer. At this point, it should be stressed that the correlation of specific chemical parameters to sensory data is frequently not straightforward, taking into account the complexity of overall quality, that is a composite of different parameters. The effect of each factor on the overall quality may be differentiated at different storage conditions and different quality levels.

3.4 Shelf-life prediction and management: time–temperature relationships

The end result of a thorough kinetic study of the behaviour of principal quality indices of a food, in a wide range of temperatures, is a validated, mathematical

Table 3.4 Representative shelf–life studies and kinetics of principal deterioration modes

Deterioration mode	Food studied	Quality index measured	Temperatures of study	Kinetic results	Reference
Microbiological spoilage	Fish (red mullet)	*B. thermosphacta*	0–15°C	Arrhenius eq., E_A =68.2 kJ/mol	Koutsoumanis *et al.* (2000)
		Pseudomonads, *S. putrefaciens*		E_A =65.4 kJ/mol	
	Fish (Mediterranean seabream)	Pseudomonads	0–15°C	Belehradek eq. (14) for μ_{max}, $b = 0.0193$, $T_{min} = -11.8$°C	Koutsoumanis (2001)
	Fish (sea bass)	Pseudomonads	0–15°C	E_A =74.0 kJ/mol	Koutsoumanis *et al.* (2002)
	Atlantic salmon fillets	Pseudomonads	2–10°C	Belehradek eq. (15b) $T_{min} = -6.1$°C, $T_{max} = 41$°C, $b = 0.1673$, $c = 0.192$	Rasmussen *et al.* (2002)
	Cod fillets (in modified atmosphere packaging)	*Photobacterium phosphoreum*	0–15°C	Belehradek eq. (15a) for μ_{max}, $b = 0.032$, $T_{min} = -8.8$°C	Dalgaard *et al.* (1997)

Non-enzymatic browning	Peach juice	Browning (spectrophotometric assay)	3–30°C	E_A = 88–135 kJ/mol	Buedo *et al.* (2001)
	Avocado purée	Chroma (CIELab scale)	5–25°C		López-Malo *et al.* (1998)
	Glucose syrups	Absorbance at 420 nm	25–55°C	E_A = 29.3–79.5 kJ/mol	Bostan and Boyacioglu (1997)
Lipid oxidation	Ground and roasted coffee	Sensory acceptance (Weibull Hazard Analysis)	4–35°C	E_A = 13 kJ/mol	Cardelli and Labuza (2001)
Vitamin loss	Citrus juice	Browning (spectrophotometric assay)	4, 20, 37, 76, 105°C	E_A = 20.5 kJ/mol	Roig *et al.* (1999)
	Orange juice	Ascorbic acid degradation (HPLC)	0–15°C	E_A = 43.8–61.1 kJ/mol	Polydera *et al.* (2003)

shelf-life model. The use of this model would then allow for a reliable estimation of the quality loss of the product in question at time–temperature conditions that differ from the experimental ones. One of the main factors that explain the necessity of full knowledge of the effect of temperature on the loss of quality parameters is the diversity of time–temperature history of a product during its actual distribution in the post-processing chain to the consumer end. Previous and recent surveys in the retail and the consumer stocking level confirm the detrimental temperature abuse (Giannakourou and Taoukis, 2003a) for chilled (Fig. 3.3) as well as for frozen foods (Fig. 3.4). Temperature data in domestic refrigerators (Fig. 3.3(b)) and domestic freezers (Fig. 3.4(b)) were obtained during surveys conducted on a representative sample of selected home equipment. Food products are exposed to a variable temperature environment

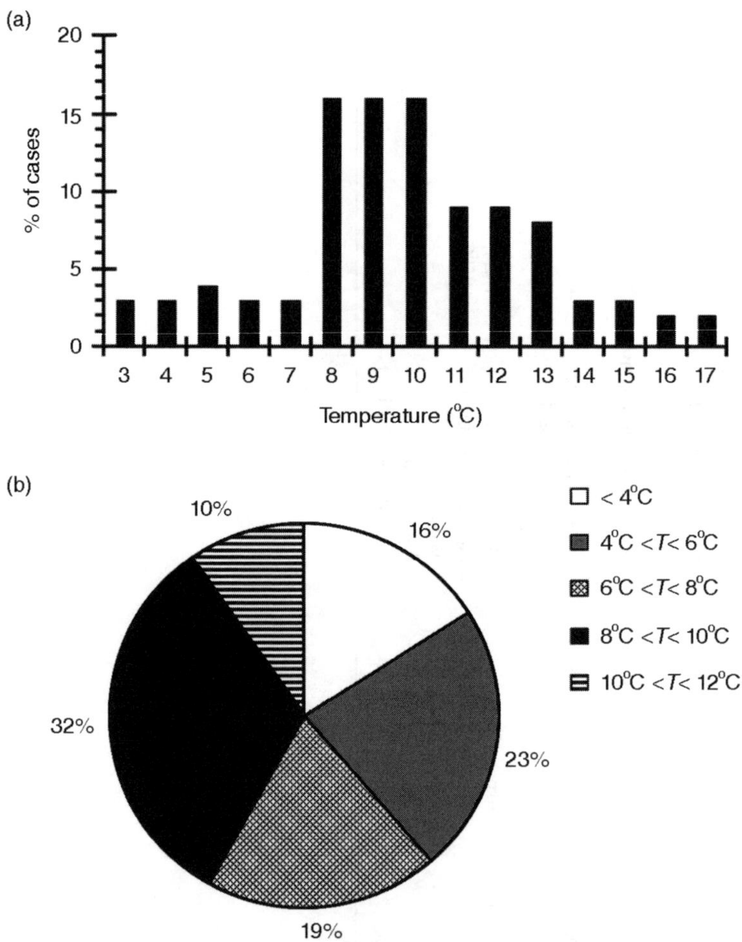

Fig. 3.3 Temperature distribution of (a) retail cooling cabinets and (b) domestic refrigerators.

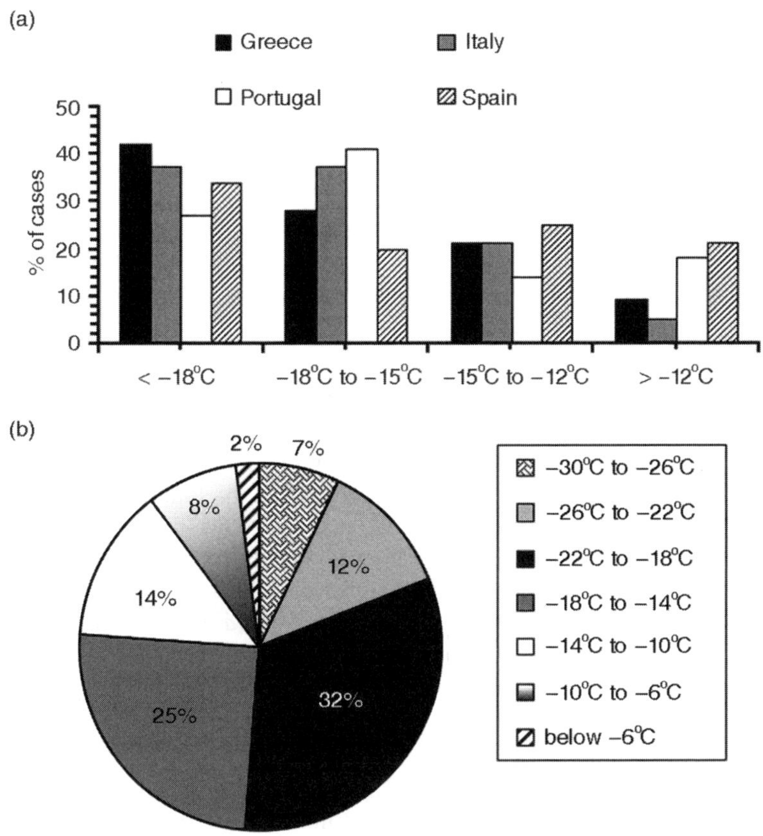

Fig. 3.4 Temperature distribution of (a) closed, vertical retail freezers in four Mediterranean countries and (b) domestic freezers.

that not infrequently includes stages of abusive storage or handling conditions. In a general form, the value of the quality function (eq. 3.2) at time t is calculated by the following integral (eq. 3.21), where $T(t)$ describes the change of temperature as a function of time:

$$f_q(\mathrm{A}) = \int k[T(t)]\mathrm{d}t \qquad [3.21]$$

To represent the integrated effect of the temperature variability on product quality degradation, the effective temperature, T_{eff}, can be introduced. T_{eff} is defined as the constant temperature that results in the same quality value as the variable temperature distribution over the same time period. This approach, representing the overall effect of non-isothermal handling with a single, constant value, simplifies eq. (3.21) to the following expression (eq. 3.22):

$$f_q(\mathrm{A}) = \int_0^{t_{\mathrm{tot}}} k[T(t)]\mathrm{d}t = k_{\mathrm{eff}} t_{\mathrm{tot}} \tag{3.22}$$

where k_{eff} is the value of the rate of the quality loss reaction at the effective temperature. If the $T(t)$ function can be described by a step sequence or, equivalently, can be discretized in small time increments t_i of constant temperature T_i (with $\Sigma t_i = t_{\mathrm{tot}}$), then eq. (3.22) can be expressed by eq. (3.23), assuming the applicability of the Arrhenius equation:

$$k_{\mathrm{ref}} \sum_i \left(\exp\left[-\frac{E_{\mathrm{A}}}{R} \left(\frac{1}{T_i} - \frac{1}{T_{\mathrm{ref}}} \right) \right] t_i \right) = k_{\mathrm{eff}} t_{\mathrm{tot}} \tag{3.23}$$

From eq. (3.23), the value of k_{eff} can be estimated, and subsequently, from the Arrhenius model, the effective temperature T_{eff} can be calculated.

Based on the aforementioned approach, in the real distribution chain that includes several stages of storage, transport and handling, one can estimate the extent of quality loss of a product, when its quality function and its time–temperature history are known.

To calculate the fraction of shelf-life consumed at the end of each stage, f_{con}, the time/temperature/tolerance (TTT) approach (Fu and Labuza, 1997; Van Arsdel *et al.*, 1969) can equivalently be used. According to this methodology, f_{con} is estimated as the sum of the times at each constant temperature segment t_i, divided by the shelf-life at that particular temperature θ_i (eq. 3.24):

$$f_{\mathrm{con}} = \sum_i \frac{t_i}{\theta_i} \tag{3.24}$$

where index i represents the different time–temperature steps within the particular stage of study. The remaining shelf-life of products can be calculated at a reference temperature, representative of their storage conditions, after each stage as $(1 - \Sigma f_{\mathrm{con}}) \cdot \theta$, where θ is the shelf-life at that reference temperature.

To demonstrate the usefulness of the TTT approach in assessing the effect of temperature conditions during transport and storage on food quality, an example in the real distribution chain is considered. Vitamin C loss, chosen as an important quality index, was measured for four popular frozen green vegetables, namely green peas, spinach, green beans and okra (Giannakourou and Taoukis, 2003b). A first-order apparent reaction order for Vitamin C oxidation was established, kinetic models for Vitamin C loss were developed and the applicability of the Arrhenius equation was validated under non-isothermal conditions. This study vividly stresses the care that should be exercised when making assumptions about the quality behaviour of seemingly similar systems, in this case the green frozen vegetables. Comparison between different green vegetables showed that the type of plant tissue affects significantly the rates of Vitamin C loss. Frozen spinach was found to be the most susceptible to Vitamin C degradation, peas and green beans demonstrated a moderate retention, whereas okra exhibited a substantially lower loss rate.

Table 3.5 Kinetic results for Vitamin C loss during frozen storage of four green vegetables in the temperature range −3 to −20°C

	Kinetic parameters			
	Green peas	Spinach	Green beans	Okra
E_A (kJ/mol)	98	112	102	106
k_{ref} (1/d)	0.00213	0.00454	0.00223	0.00105
Q_{10} (in the range −15 to −5°C)	5.5	7.0	5.8	6.3

The kinetic results of the use of the Arrhenius equation are summarized in Table 3.5 and subsequently used to predict the extent of Vitamin C loss at any point of the hypothetical distribution chain of the green vegetables, illustrated in Fig. 3.5.

This realistic distribution scenario includes an initial stage of 10 days' storage in the factory warehouses, intermediate transport, followed by 10 days' stocking in a distribution centre, where vegetables are distributed to different supermarkets (retail level) and exposed in closed vertical or open horizontal freezers for 15 days, before being purchased. The final stage of this 'time–temperature history simulation' is the 15 days of domestic storage, before final cooking and consumption. Temperature conditions for the initial stages were obtained from Jul (1984) and for the final stages of retail and domestic storage from the data shown in Fig. 3.4. When the time–temperature history of the products is

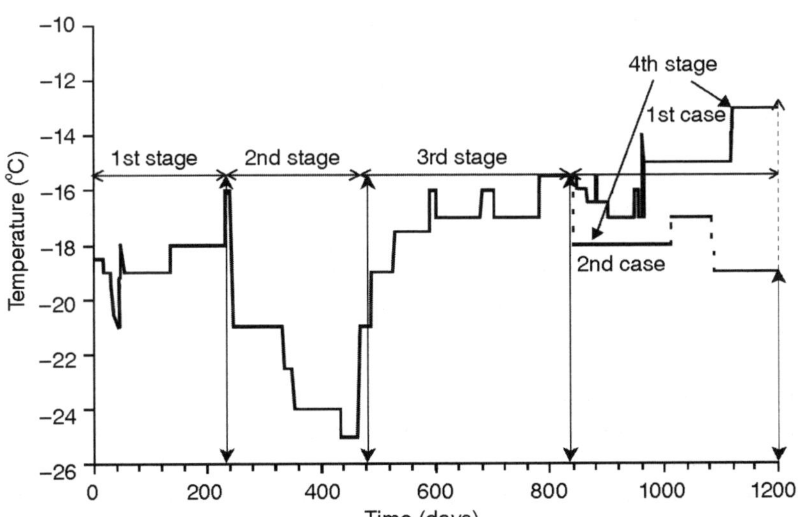

Fig. 3.5 Indicative distribution scenario (4th stage refers to domestic storage with rather abusive (1st case) or ideal conditions (2nd stage)).

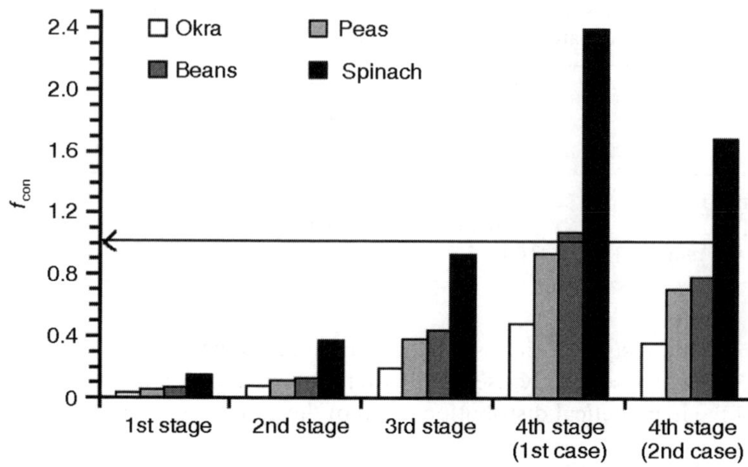

Fig. 3.6 Fraction of shelf-life (f_{con}) of frozen green vegetables consumed at −18°C, at the end of each stage of the distribution path presented in Fig. 3.5.

constantly monitored, one can assess the level of Vitamin C retention and the fraction of shelf-life consumed, f_{con}, after each distribution stage of the aforementioned marketing path (Fig. 3.4), for all frozen green vegetables, based on eq. (3.24).

When the reference temperature is −18°C, a common temperature of good practice, at the end of retail storage, the remaining shelf-lives for spinach, green beans, green peas and okra are 7, 119, 139 and 363 days, respectively, when 50% Vitamin C loss is considered as the acceptability limit. At the end of the entire distribution chain, under the specific time–temperature handling, spinach and green beans are already beyond acceptable Vitamin C level, whereas the remaining shelf-life is 13 and 234 days for green peas and okra. Results are illustrated in Fig. 3.6. The significance of handling at the consumer level is better observed if the previous scenario is slightly altered, with temperature conditions in the domestic freezer being ideal (near −18°C), as demonstrated by the dotted line in Fig. 3.5. It can be noted that when appropriate temperature conditions are maintained during domestic storage, green beans are acceptable at the time of consumption and the remaining shelf-life for okra and green peas is significantly increased.

Despite the usefulness of this method in the prediction of the shelf-life of a food, there are some particular cases where the effect of the different successive stages of food distribution is not additive but depends on the sequence of the time–temperature conditions encountered by the food, or even on the extent of temperature fluctuations. An example of such behaviour is the freeze concentration effect, which can cause significant chemical reaction acceleration just below the freezing point.

Assuming that the rule of additivity is valid for the majority of time–temperature scenarios of the distribution chain of processed foods and the history effect is negligible, Time Temperature Integrators (TTI) could be applied as reliable tools for continuous temperature monitoring and shelf-life prediction (Taoukis, 2001; Taoukis and Labuza, 2003). As will be discussed in a later chapter, TTI are devices with an easily measurable response, that reflect the accumulated time–temperature history of the product on which they are attached. Their operation is based on irreversible reactions that are initiated at the time of their activation and proceed with an increasing rate, as temperature is elevated, in a manner that resembles the temperature dependence of most quality loss reactions of foods. In essence, the usefulness of these tags in monitoring quality deterioration and estimating the remaining shelf-life at any point of the distribution chain of products depends on the successful simulation of the food quality loss kinetics. Thus, the TTI will reflect the quality status of the food only if the activation energy of the reaction that describes shelf-life loss is close to that of the TTI response, estimated by a thorough kinetic study of the behaviour of the particular tag (Taoukis and Labuza, 1989).

3.5 Future trends

One of the main aspects of current food research is oriented towards the extension of the shelf-life of foods and the optimized preservation of their initial high quality. This drive includes improvement of analytical tools and procedures to quantify quality loss and the effect of temperature on food deterioration, as well as the introduction of innovative preservation techniques.

In order to combine longer shelf-life with minimum processing, producers and regulators have concentrated on the development and application of structured quality and safety assurance systems, based mostly on prevention through monitoring, recording and controlling of crucial parameters through the entire life cycle of each product (Taoukis, 2001). Taking into consideration the significant temperature fluctuations during the distribution of the products and the frequently recorded deviations from the ideal handling conditions (Giannakourou and Taoukis, 2002), the use of TTI as quality monitoring and controlling tools is really intriguing. After validating the application scheme of TTI, that translates TTI response to the remaining quality status of the attached food package (Taoukis, 2001), these tags can be used to optimize distribution control and apply a more effective stock rotation system that takes into account the important effect of temperature on product quality. Such an inventory management and stock rotation tool was initially proposed by Labuza and Taoukis (1990), in order to replace the current practice which is coded as First In First Out (FIFO). According to FIFO, uniform handling is incorrectly assumed and quality is supposed to depend only on time. Alternatively, another method can be proposed based on the application of TTI in the real distribution chain, that takes into account the actual time–temperature history of each individual product. This approach is coded LSFO (Least Shelf-life First Out) and its structure is based on validated shelf-life modelling of the controlled food product, specification of the initial and the final, marginally accepted value of the selected quality parameters A_0 and A_s respectively, as well as careful temperature monitoring in the distribution chain with the appropriate TTI. LSFO aims at reducing the rejected products at the consumer end, by promoting, at selected decision-making points of the product life cycle, those product units with the shorter shelf-life, according to the response of the attached TTI (Taoukis *et al.*, 1998; Giannakourou and Taoukis, 2002).

A further improvement of the LSFO approach is a chill chain management system coded SLDS (Shelf-Life Decision System) that can be applied in the cold distribution chain of chilled food products (Giannakourou *et al.*, 2001). The main core of this method incorporates the building blocks of LSFO, which in this particular case are the predictive shelf-life model of the monitored food product, the kinetics of the response of the appropriate TTI and the determined limit of acceptability for the quality parameter A, taking additionally into account the realistic variability of the initial quality state A_0 of the product.

To demonstrate the effectiveness of SLDS on marine cultured chilled fish, a realistic distribution scenario of 134 hours maximum total time of such products is assumed, including the initial catch, all intermediate transportation and storage (Fig. 3.7). Numerous different distribution scenarios were assumed, based on data collected from temperature surveys and rapid microbiological techniques for the initial load of fish (Giannakourou *et al.*, 2001; Koutsoumanis *et al.*, 2002). To simulate the results of the developed SLDS and estimate the final microbiological quality of the product after the assumed distribution cycle, the Monte Carlo technique was used and the final quality status of fish for 5000 alternative scenarios was calculated.

In Fig. 3.8, the quality distribution, i.e. the percentage of products of a certain quality level, at consumption time is illustrated, showing simultaneously the probability for a unique product to reach a specific quality status at the final consumption time. The SLDS curve corresponds to a narrower distribution, reducing significantly the unacceptable products (left 'tail' of the curve), when compared to the FIFO curve. SLDS application, having the potential to be further expanded by the incorporation of other significant parameters that affect the product shelf-life, provides the advantage of using the appropriate TTI and the validated results of shelf-life tests as a chill chain management tool, which significantly reduces the probability of a product being rejected at the final stage of consumption, by rotating food based on its real remaining shelf-life, and not randomly.

The state of the TTI technology and of the scientific approach with regard to quantitative safety risk assessment in foods will also allow the undertaking of the next important step, i.e. the study and development of a TTI-based management system that will assure both safety and quality in the food chill chain. The development and application of such a system, coded with the acronym SMAS, is the target of a multipartner research project funded by the European Commission and titled 'Development and Modelling of a TTI based Safety Monitoring and Assurance System (SMAS) for Chilled Meat Products'

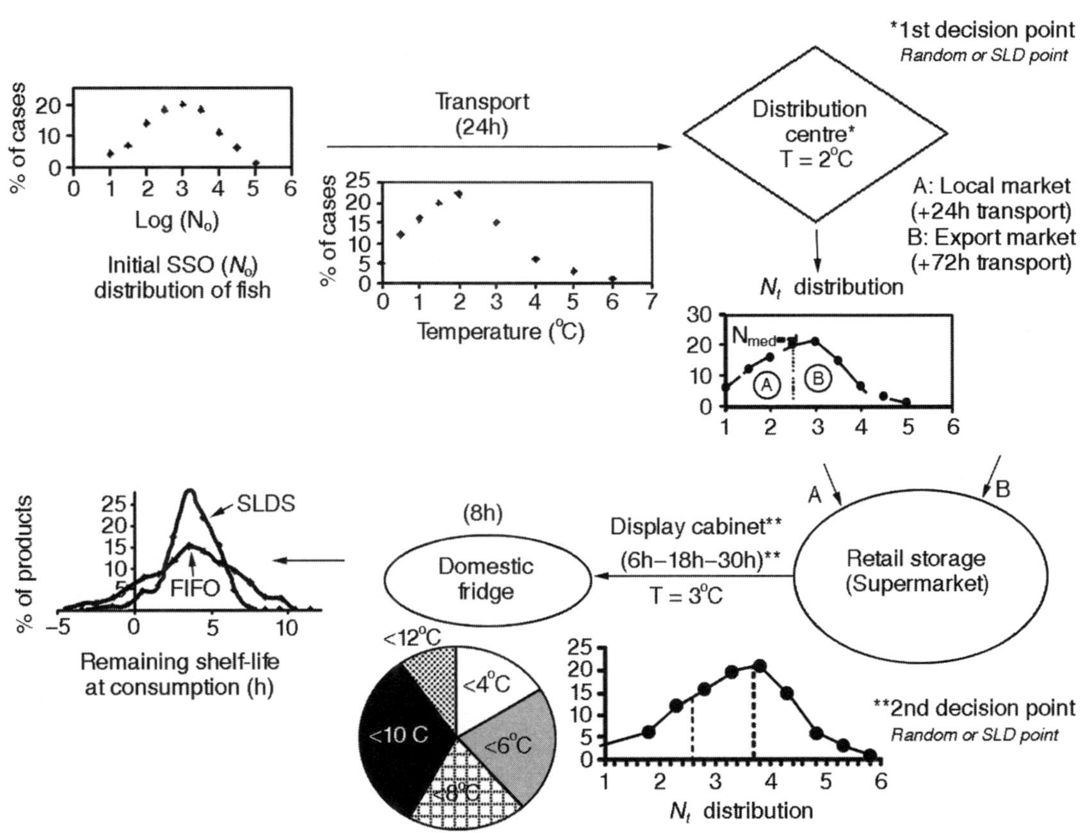

Fig. 3.7 Chill chain scenario indicating the consecutive stages, the conditions assumed, the two decision points selected and the final quality distribution during the distribution of fish products, using the SLDS and FIFO methods.

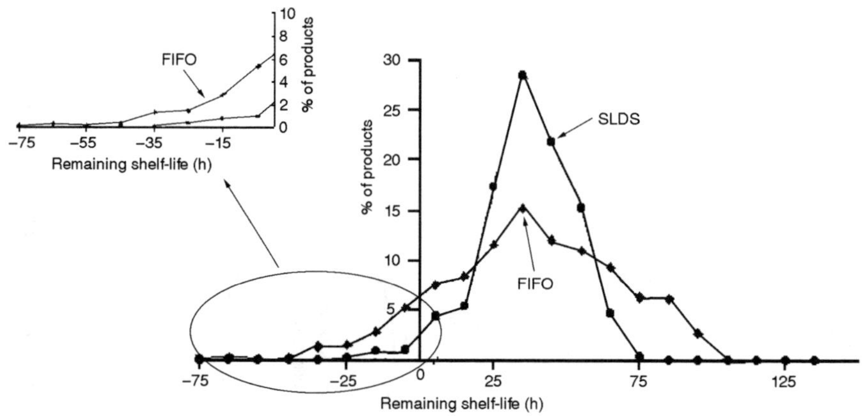

Fig. 3.8 Final quality distribution of fish product with the alternative chill chain management systems SLDS and FIFO, for route A of Fig. 3.7 (local market). Remaining shelf-life corresponds to the time the product would remain acceptable after the consumption time if stored isothermally at 0°C. Integration of the negative area (enlargement on the left) indicates the number of products being rejected before the time of consumption.

(project QLK1-CT2002-02545, 2003–2006). The main objectives of this project are as follows:

- Modelling the effect of food structure, microbial interactions and dynamic storage conditions on meat pathogens and spoilage bacteria.
- Combination of validated pathogen growth models with data on prevalence/concentration, dose–response and chill chain conditions for risk assessment with and without SMAS application.
- Development, modelling and optimization of TTI with accuracy to monitor microbiological safety of meat products.
- Development of SMAS into user-friendly computer software.
- Evaluation of the applicability and effectiveness of SMAS in real conditions of meat distribution.
- Assessment of the industry acceptance of the TTI and the concept of chill chain management and evaluation of the consumer attitude on use of TTI and correlation to quality.

Information on the outputs of this project is available on the project's website (http://smas.chemeng.ntua.gr).

Successful implementation of the developed innovative techniques for the production of minimally processed foods with extended shelf-life, and the assurance of continuous and reliable management of the problematic distribution chain of perishable products, will contribute to the improvement of consumer confidence concerning the safety and quality of foods.

3.6 References

ARRHENIUS S (1889), 'About the reaction rate of the inversion of non-refined sugar at souring', *Zeitschrift für Physikalische Chemie*, 4, 226–248.

BELEHRADEK J (1930), 'Temperature coefficients in biology', *Biol. Rev. Biot. Proc. Camb. Philos. Soc.*, 5, 30–60.

BILIADERIS C G, SWAN R S, ARVANITOYANNIS I (1999), 'Physicochemical properties of commercial starch hydrolyzates in the frozen state', *Food Chemistry*, 64, 537–546.

BLOND G, SIMATOS D (1991), 'Glass transition of the amorphous phase in frozen aqueous systems', *Thermochimica Acta*, 175, 239–247.

BOSTAN A, BOYACIOGLU D (1997), 'Kinetics of non-enzymatic colour development in glucose syrups during storage', *Food Chemistry*, 60(4), 581–585.

BUEDO A P, ELUSTONDO M P, URBICAIN M J (2001), 'Non-enzymatic browning of peach juice concentrate during storage', *Innovative Food Science and Emerging Technologies*, 1, 255–260.

BUERA P, KAREL M (1993), 'Application of the WLF equation to describe the combined effects of moisture, temperature and physical changes on non-enzymatic browning rates in food systems', *J. Food Proc. Preserv.*, 17, 31–47.

BUSHILL J H, WRIGHT W B, FULLER C H F, BELL A V (1965), 'The crystallization of lactose with particular reference to its occurrence in milk powders', *J. Sci. Food Agr.*, 16, 622.

CARDELLI C, LABUZA T P (2001), 'Application of Weibull hazard analysis to the determination of the shelf life of roasted and ground coffee', *Lebensmittel- Wissenschaft und- Technologie*, 34(5), 273–278.

CARRINGTON A K, GOFF H D, STANLEY D W (1996), 'Structure and stability of the glassy state in rapidly and slowly cooled carbohydrate solutions', *Food Research International*, 29(2), 207–213.

CHAMPION D, BLOND G, LE MESTE M, SIMATOS D (2000), 'Reaction rate modeling in cryoconcentrated solutions: alkaline phosphatase catalyzed DNPP hydrolysis', *J. Agric. Food Chem.*, 48, 4942–4947.

CHIRALT A, MARTINEZ-NAVARETTE N, MARTINEZ-MONZO J, TALENS P, MORAGA G, AYALA A, FITO P (2001), 'Changes in mechanical properties throughout osmotic processes. Cryoprotectant effect', *J. Food Engineering*, 49, 129–135.

DALGAARD P, MEJLHOLM O, HUSS H H (1997), 'Application of an iterative approach for development of a microbial model predicting the shelf-life of packed fish', *International J. of Food Microbiology*, 38, 169–179.

FU B, LABUZA T P (1993), 'Shelf-life prediction: theory and application', *Food Control*, 4, 125–133.

FU B, LABUZA TP (1997), 'Shelf-life testing: procedures and prediction methods'. In Erickson M C and Y H Hung, *Quality in Frozen Food*, New York: Chapman & Hall, 377–415.

FURUKI T (2002), 'Effect of molecular structure on thermodynamic properties of carbohydrates. A calorimetric study of aqueous di- and oligosaccharides at subzero temperatures', *Carbohydrate Research*, 337, 441–450.

GIANNAKOUROU M C, TAOUKIS P S (2002), 'Systematic application of Time Temperature Integrators as tools for control of frozen vegetable quality', *J. Food Sci.*, 67(6), 2221–2228.

GIANNAKOUROU M C, TAOUKIS P S (2003a), 'Application of a TTI-based distribution management system for quality optimisation of frozen vegetables at the consumer end', *J. Food Sci.*, 68(1), 201–209.

GIANNAKOUROU M C, TAOUKIS P S (2003b), 'Kinetic modelling of Vitamin C loss in frozen green vegetables under variable storage conditions', *Food Chemistry*, 83(1), 33–41.

GIANNAKOUROU M C, TAOUKIS P S (2003c), 'Stability of dehydrofrozen green peas pretreated with non-conventional osmotic agents'. *J Food Sci.*, 68(6), 2002–2010.

GIANNAKOUROU M C, KOUTSOUMANIS K, NYCHAS G J E, TAOUKIS P S (2001), 'Development and assessment of an intelligent shelf-life decision system for quality optimisation of the food chill chain', *J. Food Prot.*, 64(7), 1051–1057.

HAYAKAWA K (1973), 'New procedure for calculating parametric values for evaluating cooling treatments applied to fresh food', *Can. Inst. Food Sci. Technol.*, 69(3), 197.

HERTOG M L A T M, TIJSKENS L M M, HAK P S (1997). 'The effects of temperature and senescence on the accumulation of reducing sugars during storage of potato (*Solanum tuberosum* L.) tubers: a mathematical model', *Postharvest Biology and Technology*, 10, 67–79.

JUL M (1984), *The Quality of Frozen Foods*. Orlando: Academic Press.

KIM M N, SALTMARCH M, LABUZA T P (1981), 'Nonenzymatic browning of hygroscopic whey powders in open vs. sealed pouches', *J. Food Proc. Preserv.*, 5, 49.

KIRKWOOD T B L (1977), 'Predicting the stability of biological standards and products', *Biometrics*, 33, 736.

KOUTSOUMANIS K (2001), 'Predictive modeling of the shelf-life of fish under nonisothermal conditions', *Applied and Environmental Microbiology*, 67(4), 1821–1829.

KOUTSOUMANIS K P, TAOUKIS P S, DROSINOS E H, NYCHAS G J E (2000), 'Applicability of an Arrhenius model for the combined effect of temperature and CO_2 packaging on the spoilage microflora of fish', *Applied and Environmental Microbiology*, 66(8), 3528–3534.

KOUTSOUMANIS K, GIANNAKOUROU M C, TAOUKIS P S, NYCHAS G J E (2002), 'Application of shelf-life decision system (SLDS) to marine cultured fish quality', *International Journal of Food Microbiology*, 73, 375–382.

KWOLEK W F, BOOKWALTER G N (1971), 'Predicting storage stability from time–temperature data', *Food Technol.*, 25(10), 51.

LABUZA T P (1980), 'Temperature/enthalpy/entropy compensation in food reactions', *Food Technology*, 34(2), 67.

LABUZA T P (1982), *Shelf Life Dating*. Westport, CT: Food and Nutrition Press.

LABUZA T P (1985), 'An integrated approach to food chemistry: Illustrative cases, in Fennema, O R, *Food Chemistry*, second edition. New York: Marcel Dekker, 913–938.

LABUZA T P, RIBOH D (1982), 'Theory and applications of Arrhenius kinetics to the prediction of nutrient losses in food', *Food Technol.*, 36, 66–74.

LABUZA T P, TAOUKIS P S (1990), 'The relationship between processing and shelf life', in Birch G G, Campbell-Platt G, Lindley, M G, *Foods for the 90's*, London and New York Developments Series, Elsevier Applied Science Publishers, Ch. 6, 73–105.

LABUZA, T P, FU B, TAOUKIS P S (1992), 'Prediction for shelf life and safety of minimally processed CAP/MAP chilled foods: a review', *J. Food Prot.*, 55, 741–750.

LENZ M K, LUND D B (1980), 'Experimental procedures for determining destruction kinetics of food components', *Food Technol.*, 34(2), 51.

LIM M H, REID D S (1991), 'Studies of reaction kinetics in relation to the T_g' of polymers in frozen food model systems', in Levine H, Slade L, *Water Relationships in Foods*, New York: Plenum Press, 103–122.

LÓPEZ-MALO A, PALOU E, BARBOSA-CANOVAS G V, WELTI-CHANES J, SWANSON B G (1998),

'Polyphenolxidase activity and color changes during storage of high hydrostatic pressure treated avocado puree', *Food Research International*, 31(8), 549–556.

MANZOCCO L, NICOLI M C, ANESE M, PITOTTI A, MALTINI E (1999), 'Polyphenoloxidase and peroxidase activity in partially frozen systems with different physical properties', *Food Research International*, 31(5), 363–370.

McDONALD K, SUN D-W (1999), 'Review: Predictive food microbiology for the meat industry', *International Journal of Food Microbiology*, 52, 1–27.

McMEEKIN T A, OLLEY J N, ROSS T, RATKOWSKY D A (1993), *Predictive Microbiology: Theory and Application*, New York: John Wiley & Sons, 87–115.

NELSON K A (1993), 'Reaction kinetics of food stability: Comparison of glass transition and classical models for temperature and moisture dependence', PhD thesis, University of Minnesota.

PELEG M (1992), 'On the use of WLF model in polymers and foods', *Crit. Rev. Food Sci.*, 32, 59–66.

POLYDERA A C, STOFOROS N G, TAOUKIS P S (2003), 'Comparative shelf-life study and Vitamin C loss kinetics in pasteurized and high pressure reconstituted orange juice', *Journal of Food Engineering*, 60, 21–29.

RASMUSSEN S K J, ROSS T, OLLEY J, McMEEKIN T (2002), 'A process risk model for the shelf-life of Atlantic salmon fillets', *Int. J. Food Microbiology*, 73, 47–60.

RATKOWSKY D A, LOWRY R K, McMEEKIN T A, STOKES A N, CHANDLER R E (1983), 'Model for bacterial culture growth rate throughout the entire biokinetic temperature range', *J. Bacteriol.*, 154, 1222–1226.

ROIG M G, BELLO J F, RIVERA Z S, KENNEDY J F (1999), 'Studies on the occurrence of non-enzymatic browning during storage of citrus juice', *Food Research International*, 32, 609–619.

SEYHAN F, TIJSKENS L M M, EVRANUZ O (2002), 'Modelling temperature and pH dependence of lipase and peroxidase activity in Turkish hazelnuts', *J. Food Engineering*, 52, 387–395.

SLADE L, LEVINE H, FINEY J (1989), 'Protein–water interactions: water as a plasticizer of gluten and other protein polymers', in Phillips R D, Finlay J W, *Protein Quality and the Effects of Processing*, New York: Marcel Dekker, 9–123.

TAOUKIS P S (2001), 'Modelling the use of time-temperature indicators', in Tijskens L M M, Hertog M L A T M, Nicolaï B M, *Food Process Modeling*, New York: CRC Press, 402–431.

TAOUKIS P S, LABUZA T P (1989), 'Applicability of time-temperature indicators as shelf life monitors of food products', *J Food Sci.*, 54(4), 783–789.

TAOUKIS P S, LABUZA T P (1996) 'Summary: Integrative concepts (shelf life testing and modeling)', in Fennema O, *Food Chemistry*, third edition, New York: Marcel Dekker, Ch. 17, 1013–1042.

TAOUKIS P S, LABUZA T P (2003), 'Time-Temperature Indicators (TTIs)', in Ahvenainen R, *Novel Food Packaging Techniques*, Cambridge, UK: Woodhead Publishing, Ch. 6, 103–126.

TAOUKIS P S, LABUZA T P, SAGUY I S (1997), 'Kinetics of food deterioration and shelf-life prediction', in Valentas K J, Rotstein E, Singh R P, *Handbook of Food Engineering Practice*, New York: CRC Press, 361–403.

TAOUKIS P S, BILI M, GIANNAKOUROU M (1998), 'Application of shelf life modelling of chilled salad products to a TTI based distribution and stock rotation system', *Acta Hort.*, 466, 131–140.

TEMPLEMAN G, SHOLL J J, LABUZA T P (1977), 'Evaluation of several NMR techniques for solids in fat determination in commercial fats', *J Food Sci.*, 42, 432.

TEREFE N S, HENDRICKX M E (2002), 'Kinetics of the pectin methylesterase catalyzed de-esterification of pectin in frozen food model systems', *Biotechnol. Prog.*, 18, 221–228.

TEREFE N S, MOKWENA K K, VAN LOEY A, HENDRICKX M E (2002), 'Kinetics of the alkaline phosphatase catalyzed hydrolysis of disodium *p*-nitrophenyl phosphate in frozen model systems', *Biotechnol. Prog.*, 18, 1249–1256.

TORREGGIANI D, FORNI E, GUERCILENA I, MAESTRELLI A, BERTOLO G, ARCHER G P, KENNEDY C J, BONE S, BLOND G, CONTRERAS-LOPEZ E, CHAMPION D (1999), 'Modification of glass transition temperature through carbohydrate additions: effect upon colour and anthocyanin pigment stability in frozen strawberry juices', *Food Research International*, 32, 441–446.

VAN ARSDEL W B, COPLY M J, OLSON R L (1969), *Quality and Stability of Frozen Foods*, New York: Wiley-Interscience.

VAN BOEKEL M A J S (2000), 'Kinetic modeling', in Tijskens L M M, Hertog M L A T M and Nicolaï B M, *Food Process Modeling*, New York: CRC Press, 35–59.

WILLIAMS M L, LANDEL R F, FERRY J D (1955), 'The temperature dependence of relaxation mechanisms in amorphous polymers and other glass-forming liquids', *J. Chem. Eng.*, 77, 3701–3707.

YOSHIOKA S, ASO Y, UCHIYAMA M (1987), 'Statistical evaluation of nonisothermal prediction of drug stability', *J. Pharmaceutical Sci.*, 76(10), 794–798.

4

Genetic and physiological factors affecting colour and firmness

R.E. Schouten and O. van Kooten, Wageningen University and H. Jalink, I.F. Kappers, J.F.H. Snel and W. Jordi, Plant Research International, The Netherlands

4.1 Introduction

Quality is an important property for agricultural produce. Defining quality is not easy; a suitable definition might be 'fitness for use' (Juran, 1974) or 'fitness for purpose' (Simmonds, 1979). Quality is also a difficult property, as everyone uses a slightly different set of criteria to interpret the quality of a product. In order to have some practical grip on quality-related issues, the concept of acceptability was introduced (Tijskens, 2000). When somebody decides on the acceptability of a product, the quality is compared to a criterion, the quality limit. If the quality exceeds that limit, the product is accepted, otherwise it is rejected (Wilkinson and Tijskens, 2002). So, acceptability of a product depends on product quality and on the level of the acceptance limit. The acceptance limit is primarily defined by economic and psychological factors; the quality of a product is largely defined by its intrinsic properties. Acceptability is directly related to the keeping quality of a product. For fruits and vegetables, product properties such as colour, firmness and taste change over time. Keeping quality is the time before the product attribute drops below the acceptance limit. So, keeping quality combines two aspects of product acceptance, the acceptance limit and product quality, into a generally applicable index of quality (Tijskens and Polderdijk, 1996). Keeping quality and shelf-life are terms often used interchangeably and are indeed closely connected. Shelf-life is the keeping quality under standardised storage conditions (Tijskens, 2000).

The concept of keeping quality enables quality research to be formally separated into two fields. Consumer attitudes towards regional food preferences

(Verlegh and Steenkamp, 1999), sensory evaluation (Munoz, 2002) and customer value processes (Payne and Holt, 2001) are recent examples of research efforts aimed at the acceptance limit. Here the focus will be on the other field, the factors that influence product behaviour. Product behaviour shows generally a decay in quality attributes after harvest. Within the time the product remains acceptable, it travels through the horticultural chain. Knowledge of the expected keeping quality at the start of the horticultural chain will greatly benefit participants, whether consumers are getting a guaranteed high quality product or the producer is able to export his product instead of supplying it to the domestic market.

It is increasingly possible to manipulate keeping quality. This can be done either by a number of postharvest methods, by breeding, or by genetic modification. Here, the focus will be on arguably the most important quality attributes, firmness and colour. The aim is to report on recent developments in firmness and colour research from both a physiological and a genetic viewpoint and to investigate whether bottlenecks exist that hamper further development.

4.2 Physiology of firmness: fruits and vegetables

Firmness is the textural property sensed in the mouth. According to Van Dijk and Tijskens (2000) firmness can be generated from different sources:

- Turgor pressure inside living cells and the associated tissue tension
- Special compounds inside cells possibly generating strength such as starch
- Cohesive forces within a cell, e.g. chemical properties of the cell wall
- Adhesive forces between cells, e.g. chemical properties of pectin
- Overall structure and shape of separate cells
- Overall structure and shape of tissue, e.g. strength and distribution of vascular tissue.

With only turgor as major firmness contributor, products are soft and juicy (e.g. raspberry); when pectin forces are primarily providing firmness, products are crispy and juicy (e.g. apples); when cell wall forces dominate, products are essentially mealy and dry (e.g. overripe tomatoes); and when vascular tissue is important, products are essentially tough and fibrous (e.g. asparagus). Firmness in a product results from a combination of firmness generating sources and this is probably why there is still a poor understanding of what causes firmness variation, and consequent variation in keeping quality. Sams (1999) reviews a number of factors influencing firmness. They range from the effect of water stress, (intense) solar radiation and temperature during growth (environmental factors) to nitrogen, phosphorus, potassium and especially calcium nutrition (cultural factors) and fruit size, maturity at harvest and fibre content (physiological factors). In general decrease in firmness seems to be characterised by three distinct phases for a number of fruits, including apple, melon and tomatoes. Fruits soften slowly during the first phase, more rapidly during the second phase and again slowly during the last phase. Apples soften by 25–50%

to a final firmness of 25–50 N whereas tomatoes and kiwifruit soften by 75–100% during ripening to a final firmness of 0–10 N. Once the second phase is induced the subsequent softening is difficult to slow. Thus, the first phase of softening needs to be prolonged for achieving longer keeping quality. However, little is known about the cellular mechanism that regulates the onset and rate of softening of either phase (Johnston *et al.*, 2002).

Firmness is often assessed by a puncture test. The relation between this test and sensory perception is not always unambiguous. On the one hand a rather large firmness difference of 6 N in apples was required before the average trained panellist could detect a difference in fruit, indicating that a puncture test is a suitable firmness-measuring device. On the other hand, some textural differences between apples were not always adequately predicted by instrumental tests (Harker *et al.*, 2002). Minimal firmness limits as determined by puncture tests could be applied to achieve acceptable eating quality for different apple cultivars, 44 N for 'Golden Delicious' and 56 N for 'Gala' (Hoehn *et al.*, 2003).

Tijskens (1979) stored 'Golden Delicious' apples under low temperature air storage and CA storage. Interestingly, the firmness decrease stopped at around 5.5 kg for the air-stored apples and at 7 kg for the CA-stored apples. Firmness (F) development during storage was described by a simple first-order reaction with a final firmness F_{fix}. When applying this model to the stored apples it seems that F_{fix} is different for each storage method. However, it might be that the CA storage prevents another firmness-decreasing process to occur, and that the two processes combined describe the firmness decrease for air-stored apples. So, during air storage the final firmness, after decay of F_1 and F_2, consists of only F_{fix} and during CA storage it consists of F_{fix} and the initial amount of firmness to be broken down by the second process (Fig. 4.1). Although the second process has a reaction rate (k_2) 10 times smaller than the reaction rate

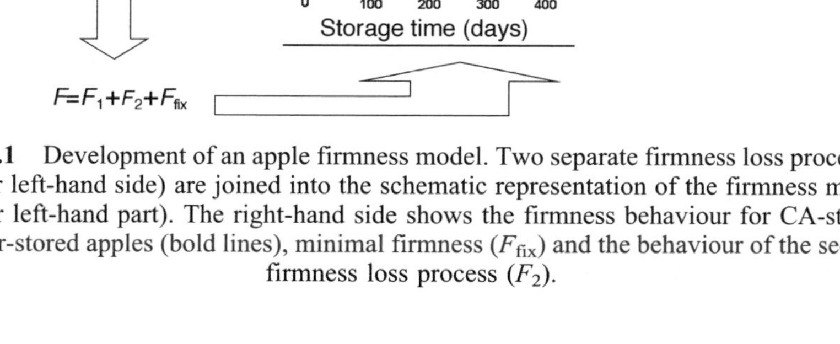

Fig. 4.1 Development of an apple firmness model. Two separate firmness loss processes (upper left-hand side) are joined into the schematic representation of the firmness model (lower left-hand part). The right-hand side shows the firmness behaviour for CA-stored and air-stored apples (bold lines), minimal firmness (F_{fix}) and the behaviour of the second firmness loss process (F_2).

(k_1) of the first process, it is still possible to detect this process by looking at the different apparent F_{fix} values. This behaviour is especially visible when products are stored at different temperatures (Tijskens *et al.*, 1999) and might be a method to identify and describe the different firmness-generating processes.

4.3 Methods of improving and maintaining firmness

4.3.1 Physiology and firmness

Apples treated with 1-methylcyclopropene (MCP), an inhibitor of ethylene action, softened more slowly and had reduced internal ethylene concentrations compared to untreated apples. Re-treatment with MCP during storage resulted in firmer apples compared to apples treated only at harvest (Mir *et al.*, 2001). It was shown that ethylene is needed for the expression of PG, even in a late ripening stage (Hiwasa *et al.*, 2003). Treatment of tomato fruit with MCP delayed colour development, softening and ethylene production in mature green, breaker and orange stages. Even at advanced stages of maturity, MCP inhibited ripening, lasting 5–7 days, and the effect could be prolonged by renewed exposure (Hoeberichts *et al.*, 2002). The fact that MCP can be applied to inhibit ripening at different maturity stages and is not considered toxic makes it very attractive to be developed and used in practice. MCP is marketed as SmartFresh, and just passed EPA. Another method aimed at increasing firmness and colour retention of tomatoes has been presented by Maharaj *et al.* (1999). A delay of 7 days was found in the climacteric response in addition to a lower rate of respiration and ethylene production after treatment with a UV radiation dose. The retardation of colour and softness development was attributed to the maintenance of a high level of putrescine which has a physiological function opposite to that of ethylene.

Modified atmosphere packaging (MAP), applying elevated concentrations of carbon dioxide and reduced levels of oxygen and ethylene, can inhibit the loss of quality during the postharvest life of a wide range of products (reviewed by Kader *et al.*, 1989). Atmosphere modification depends on film permeability, respiration rate, gas diffusion characteristics and atmospheric composition within the package. Hertog *et al.* (2001) found a close relationship between rate of softening and rate of gas exchange of 'Braeburn' apples when applying a range of MA conditions. This indicates that fruit softening is directly metabolically linked to gas exchange. A combination of MAP with a natural clay adsorbent that takes up water and volatiles resulted in better firmness retention of raspberries compared to application of only MAP (Toivonen *et al.*, 2002).

4.3.2 Genetics and firmness

Fruits are generally classified either by increased respiration and ethylene biosynthesis (climacteric fruits) or without ethylene biosynthesis (non-climacteric fruits). Although non-climacteric fruits may respond to ethylene,

ethylene is not required for ripening. In contrast ethylene is necessary for the coordination and completion of ripening in climacteric fruits. This was demonstrated by examining transgenic (climacteric) tomato plants blocked in ethylene synthesis. Ethylene is synthesised from S-adenosyl-L-methionine through the activity of two enzymes: ACC synthase and ACC oxidase. Reduced ethylene evolution resulted in ripening inhibition of ACC synthase and ACC oxidase antisense lines (Oeller *et al.*, 1991). A clear link between firmness and ethylene production was found with the appearance of the Nr (never-ripe) mutation in tomato. The Nr gene has been cloned and encodes a protein with homology to the Arabidopsis ethylene receptor ETR1, but is lacking the response regulator domain found in ETR1 and related signal transducers. A single amino acid change confers ethylene insensitivity when expressed in transgenic tomato plants (Wilkinson *et al.*, 1995).

The Nr tomato mutant is part of a class of single gene mutations. The rin (ripening-inhibitor), nor (non-ripening), Nr and Cnr (colourless non-ripening) are the best characterised. Fruit of the rin and nor tomato mutants fail to produce ethylene, do not ripen, and also do not ripen in response to exogenous applied ethylene. However, exogenous ethylene displays signs of ethylene sensitivity and signalling, including the induction of ethylene regulated genes (Yen *et al.*, 1995). So it seems that some aspects of ripening operate independently of ethylene. Both rin and nor mutations represent genes upstream of ethylene control. The gene corresponding to the rin mutation has been cloned and it appears that ripening is, in part, regulated by a MADS-box transcription factor (Moore *et al.*, 2002). Interestingly, a homologue of the RIN gene is isolated in strawberry, a non-climacteric fruit, suggesting that common ripening regulation is present in all fruits (White, 2002). The mutations rin and nor have been exploited extensively in order to produce tomato hybrids with improved firmness characteristics (Seroczynska and Niermirowicz, 1998). The Cnr mutant results in firm tomato fruits with much reduced cell-to-cell adhesion, resulting in a mealy texture, and complete absence of carotenoid biosynthesis (Fraser *et al.*, 2001). The mealy phenotype of Cnr must reflect significant changes in the cell wall structure of the pericarp and specific changes to the pectin structure (Orfila *et al.*, 2002). It appears that the changes during ripening in Cnr fruit result in a shortage of calcium-cross-linkable pectins, influencing the capacity to maintain cell-to-cell adhesion. Ripe Cnr fruit cells separate at the middle lamella rather than rupture, resulting in the mealy texture (Seymour *et al.*, 2002).

Transgenic research may provide valuable insight into whether certain enzymes are important in contributing to firmness generation. Pectin polymers are the main constituents in the middle lamella, a region considered important for maintaining cell-to-cell adhesion and cell packing in fruit tissues (Wakabayashi, 2000). Softening is usually associated with increased content of water-soluble pectins. The enzyme initially considered for pectin solubilisation and therefore softening was polygalacturonase (PG). Electron microscopy revealed that treatment of unripe discs with PG caused similar disruption of the middle lamella to that observed in ripe apples (Ben-Arie *et al.*,

1979). However, gene isolation and functional characterisation of tomato PG in transgenic plants (rin mutant) indicated that PG alone is insufficient to initiate softening and is likely to function with additional factors (Giovannoni *et al.*, 1989). Another enzyme involved in cell wall metabolism is pectin-methyl-esterase (PME) which may either increase accessibility of PG to its pectin substrate (Wakabayashi, 2000) or de-esterificate highly methoxylated pectin regions which could result in swelling and solubilisation of pectins (Yoshioka *et al.*, 1992). Antisense repression of a tomato fruit PME resulted in decreased pectin degradation but consistently with PG repression did not alter additional ripening characteristics, including softening (Tieman *et al.*, 1992). The role of additional enzymes in apple is reviewed in Johnston *et al.* (2002), but it seems that no one enzyme is primarily responsible for softening. Expansins are proteins that cause cell wall loosening, and are involved in many aspects of cell wall modification during development (Cosgrove, 2000). Tomato and strawberry expansin genes upregulated during fruit ripening have been isolated (Rose *et al.*, 1997; Civello *et al.*, 1999). Repression of a fruit ripening-specific expansin, Exp 1, in tomato resulted in reduced softening, and overexpression resulted in enhanced softening, including softening of mature green fruit (Brummell *et al.*, 1999). Transgenic silencing of the expansin gene LeExp1 resulted in firmer tomato fruit. No effects on fruit size and fruit number per plant were noticed and shelf-life was extended by 5–10 days for tomatoes suppressed in expression of LeExp1 (Brummell *et al.*, 2002). Small GTPases may be involved in the control of cell wall-modifying enzymes. Antisense tomato fruit changed colour as expected but failed to soften normally. This was accompanied by reduced levels of two cell wall hydrolases, pectinesterase and polygalacturonase (Lu *et al.*, 2001).

In soft fruit like strawberry, PG is hardly detectable (Nogata *et al.*, 1993) and enzymes like pectate lyase (PL) are more important with regard to softening. Jimenez-Bermudez *et al.* (2002) used transgenic plants that incorporated an antisense sequence of a strawberry PL gene under the control of the 35S promoter and noticed reduced softening compared to control fruits. Analysis showed a lower degree of *in vitro* swelling and a lower amount of ionically bound pectins for the transgenic fruits. Recently, Benitez-Burraco *et al.* (2003) isolated two strawberry PL clones. CO_2 enriched atmospheres promote strawberry fruit firmness (Siriphanich, 1998). Interestingly, expression of both PL genes was strongly reduced in harvested fruit kept in high CO_2 storage.

QTL (Quantitative Trait Loci) analysis can indicate regions in the genome that contribute to trait variation. QTL analysis of tomatoes has shown that a few chromosome regions control the variation in organoleptic quality as determined by both physical and sensory measures. A significant effect for juiciness occurs close to the Cnr locus on chromosome 2 (Causse *et al.*, 2002). King *et al.* (2000) accounted for genetic variation in firmness in apple fruits as determined by penetrometer, stiffness determined by acoustic resonance and sensory attributes. Fruit firmness as measured by the penetrometer measures a combination of compression stiffness and shear strength. This complex measurement was reflected in the QTL analyses, where significant QTL were observed for three linkage groups from a previously prepared linkage map (Maliepaard *et al.*, 1998). A highly significant QTL was detected for crispness, juiciness, sponginess and overall liking. Further investigation confirmed that this significant single QTL correlated well with juiciness and crispness (King *et al.*, 2001).

4.4 Physiology of colour: fruits and vegetables

Colour is not directly a quality attribute, but it is strongly related to physiological maturity and can be perceived nondestructively. The colour of many products changes, from green to red (tomatoes, cherries, strawberries), or to yellow (bananas, cucumbers, broccoli). Colour measurements are usually performed using devices that are based on the CIE chromaticity colour space (e.g. Minolta Chromameter), RGB colour space (e.g. video camera) or reflectance spectra (spectrophotometer with an integrating sphere). Using the RGB system, colour is expressed as a combination of R (red), G (green) and B (blue) values. Not all possible colours can be expressed this way, a disadvantage not present using the CIE Lab system (Williamson and Cummins, 1983). A practical disadvantage using the CIE system is that often only very small surfaces can be measured, while a video camera can easily measure large surfaces consisting of several hundred thousands of RGB measurements. Additionally, RGB images can be edited routinely using colour recognition software (Schouten *et al.*, 2002a). Reflectance spectra are increasingly used to assess the plant's physiological status (Penuelas and Filella, 1998) with the appearance of fast and cheap spectrophotometers. Polder *et al.* (2002) showed that spectral images offer more discriminating power than standard RGB images for measuring ripeness stages of tomatoes, and Merzlyak *et al.* (2003) linked spectral reflectance spectra of several apple cultivars to chlorophylls, carotenoids and anthocyanins.

Recently, a colour model was presented that describes the colour development of cucumbers during dark storage. The model is based on the combined literature regarding the processes of synthesis and degradation of chlorophyll in terms of colour compounds. Applying knowledge regarding chlorophyll degradation might be obvious, but the application of knowledge regarding chlorophyll synthesis is, at first sight, not. However, close inspection of nondestructive, repeated colour measurements on individual cucumbers revealed that cucumbers sometimes increase in (green) colour during the first days of dark storage. It was therefore assumed that protochlorophyllide (Pchl), the colourless precursor of chlorophyll and chlorophyllide, can still be transformed into colour compounds during the postharvest period. During chlorophyll synthesis, Pchl is transformed into chlorophyllide (chl) and later into chlorophyll (CHL) (Porra, 1997). The most likely pathway for chlorophyll degradation in fruits and vegetables is cleavage by chlorophyllase resulting in the formation of chlorophyllide. Chlorophyllide will be converted into brown compounds and finally into colourless compounds (reviewed in Heaton and Marangoni, 1996). Interestingly, chl plays a role in both chlorophyll degradation and chlorophyll synthesis (Fig. 4.2). This enabled the formation of a colour model based on physiological processes and nondestructive colour measurements that describes the development of the colour of cucumbers as function of time and temperature very well (Fig. 4.2). The limiting factor for the keeping quality, here the time before the colour limit (Fig. 4.2) is reached, could be assessed as being the amount of Pchl present at harvest (Schouten *et al.*, 2002b). It may be expected that identical colour processes in cucumber, save for the numerical values of the kinetic parameters, also occur in other non-climacteric products that experience the same colour development (e.g. green beans, Brussels sprouts).

For climacteric fruits like banana and climacteric-like products like broccoli the green to yellow colour development is different. Broccoli is harvested at an immature stage and this causes considerable stress due to disruption in energy, nutrient and hormone supplies, resulting in a very short shelf-life. In asparagus, harvesting in a stage of rapid growth causes the tips to lose large amounts of sucrose and undergo major changes in gene expression. This leads to a markedly altered metabolism including protein and lipid loss and amino acid accumulation which may be described as a starvation response (Page *et al.*, 2001). Ethylene

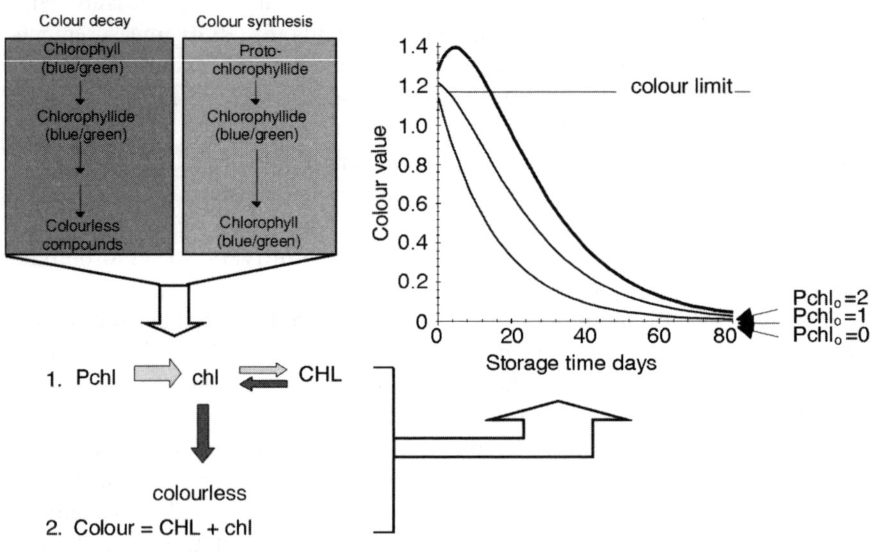

Fig. 4.2 Development of the cucumber colour model. Separate literature sources describing chlorophyll degradation and chlorophyll synthesis (upper left-hand side) are joined into the schematic representation of the colour model (lower left-hand side). The right-hand side shows the colour behaviour for three cucumbers differing in initial amount of Pchl ($Pchl_0$) stored at 20°C. The colour limit, indicative of the end of keeping quality, is shown.

has an important role in regulating the yellowing of broccoli since chlorophyll loss is associated with an increase in floret ethylene synthesis. Interestingly, treatment with the plant hormone cytokinin resulted in longer postharvest life as less chlorophyll loss and a delay in asparagine accumulation was observed, although sucrose loss was unaffected (Downs *et al.*, 1997). For banana, a climacteric fruit, the green to yellow colour development is different as described for cucumber. For instance, neither chlorophyllide nor pheophorbide accumulated during ripening and, most interesting, chlorophyll retention in the banana peel was larger at 35°C than at 20°C. This may be explained by a temperature sensitive dissociation of chlorophyll from the thylakoid membrane (Drury *et al.*, 1999) or perhaps by the existence of another chlorophyll degradation pathway. It was shown by Janave (1997) that chlorophyll may be degraded by oxidative enzymes *in vitro*. In contrast, colour measurements of bananas during ripening indicated that colour decay is higher with increasing temperature (Chen and Ramaswamy, 2002).

Red colour development may also be divided into climacteric and non-climacteric cases. Postharvest colour development of e.g. non-climacteric strawberries and cherries ranges from light red to deep red, indicative of anthocyanin production. A mechanistic physiological colour model was built that linked the description of colour development to rot incidence during dark storage of strawberries. This was possible as the precursor of the anthocyanins and the proanthocyanins (unspecific enzyme inhibitors that keep the fungi responsible for rot development in a quiescence state) have a common precursor. The limiting factor for the keeping quality, the time before a strawberry shows signs of rot, depended on the amount of precursor present at harvest (Schouten *et al.*, 2002a).

The green to red colour development in climacteric tomato, red pepper and apple is largely due to the transition of chloroplasts to chromoplasts. During photosynthetic membrane degradation chlorophyll is metabolised and detoxified (reviewed in Matile and Hörtensteiner, 1999) and large amounts of carotenoids, mainly β-carotene and lycopene, are synthesised. Geranyl-geranyldiphosphate (GGPP) is the precursor of the carotenoids and the conversion of GGPP to phytoene is the first step in the carotenoid biosynthesis. The next steps desaturate phytoene and ξ-carotene and produce lycopene, responsible for the red colour of ripe tomato fruit. Lycopene may undergo cyclisation to either β- or α-carotenes. For apple, it was found that five anthocyanin enzymes are coordinately expressed during red colouration in the skin (Honda *et al.*, 2002). Lycopene accumulation for tomato arises during ripening as a consequence of reduced lycopene cyclisation and the presence of ripening enhanced phytoene synthase (Fraser *et al.*, 2002).

4.5 Methods of improving and maintaining colour

4.5.1 Physiology and colour

Improvement of the keeping quality for non-climacteric products with a green to yellow colour development may be achieved by combining nondestructive measurements and the mechanistic physiological model described in Fig. 4.2. The concentration of Pchl at harvest, $Pchl_0$, determines the keeping quality for individual cucumbers. When the $Pchl_0$ values are expressed per batch (cucumbers having the same growth history) it seems that the shape of the distributions varies between two limits, in which vicinity they are skewed (Fig. 4.3). One of the borders is nil, indicative of no colour precursor left, and consequent quick colour loss for the cucumbers belonging to that batch. The other border is identical for batches of the same cultivar. This cultivar-dependent border can be viewed as a cultivar-specific identifier of colour keeping quality (Schouten *et al.*, 2003). It is clearly different for two commercial cultivars (Fig. 4.3). This technique might be used to select for cultivars with high keeping quality.

MCP inhibits ripening for a host of climacteric fruits and climacteric-like products, specifically firmness but also colour retention. For instance, application of MCP increased the colour retention of broccoli over 20% (Able *et al.*, 2002). Lower ethylene production and respiration, slower loss of firmness and acidity, and less of a change in peel colour from green to yellow was observed for a rapidly ripening summer apple during shelf-life and storage (Pre-Aymard *et al.*, 2003). Other colour development retardants with a similar function as MCP can also be used. Gibberellic acid, a plant hormone, can retain green colour, whether applied as preharvest spray or postharvest dip treatment in citrus fruit (Porat *et al.*, 2001). Ethylene oxide (EO) and sulphur dioxide (SO_2) may be used to prevent ripening in bananas. Treatment with EO and SO_2 was efficient in extending the shelf-life of bananas, showing a fresh appearance, good colour and minimal mould development (Williams *et al.*, 2003).

For products that show a non-climacteric red colour development caused by anthocyanin production like strawberry, postharvest light or heat treatment might be used to increase the keeping quality. Light treatment was able to overcome poor red colour and 'white shoulders' in two strawberry cultivars while also diminishing fruit rot at the same time, thereby increasing the keeping quality (Saks *et al.*, 1996). Interestingly, Vicente *et al.* (2003) applied heat treatments (45°C for 3 h) to strawberry (cv Selva) in combination with MAP and found a reduced fungal decay, softening and red colour development after a market simulation period, especially when the CO_2 produced during heating was allowed to retain in the package. So, red colour development and reduced fungal decay seem to be equally enhanced by light treatment, while heat treatment specifically increases the resistance against fungal decay and not so much the red colour development.

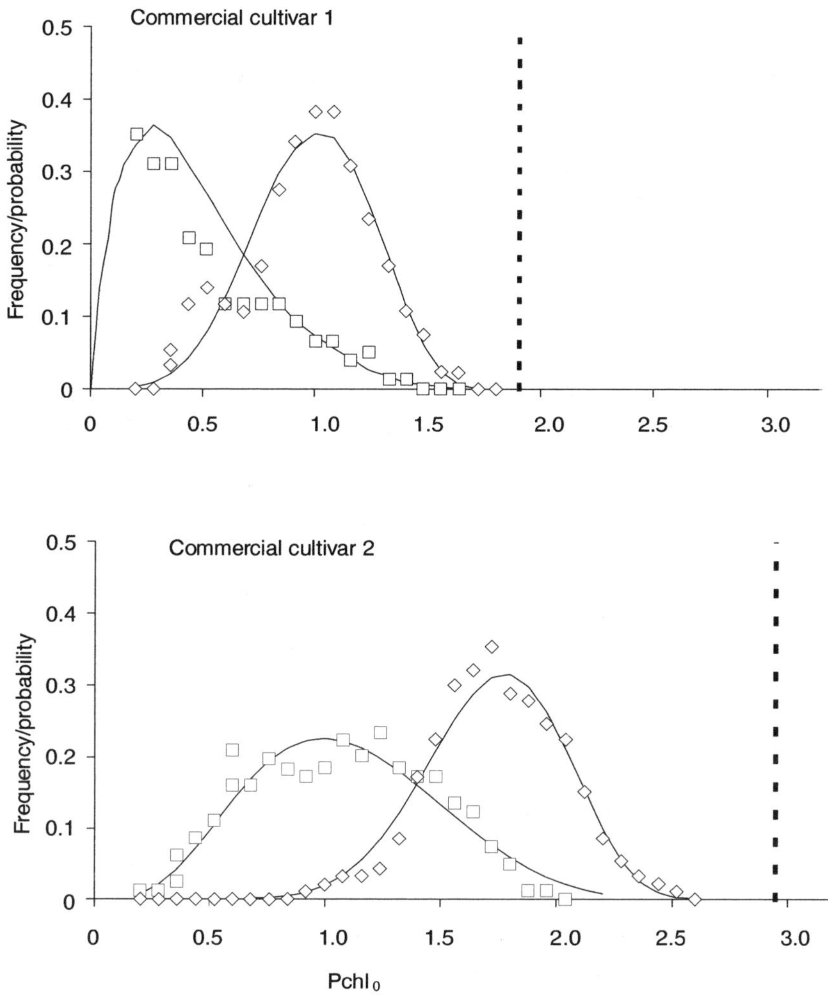

Fig. 4.3 $Pchl_0$ distributions for an autumn batch (□) and a spring batch (◇) of cucumbers per cultivar. The cultivar-dependent keeping quality limit is indicated by a dashed line.

4.5.2 Genetics and colour

Cytokinins are believed to delay senescence by maintaining cellular integrity, preventing proteases from the vacuole to leak into the cytoplasm hydrolysing proteins of, for example, the chloroplast and thereby retaining chlorophyll. Initially, transgenic tobacco was engineered overexpressing the ipt gene. The ipt gene encodes isopentenyl phosphotransferase, the enzyme that catalyses the rate-limiting step for cytokinin synthesis. Overexpressing the ipt gene resulted in delayed leaf senescence but unfortunately also in delayed growth and fertility (Garratt *et al.*, 2002). Gan and Amasino (1995) devised a strategy of auto-regulated cytokinin production using the highly senescence-specific SAG12 promoter fused to the ipt gene in transgenic tobacco. The chimaeric P_{SAG12}-IPT gene was only activated at the onset of senescence, thereby preventing over-expression. This strategy has, next to tobacco, also been carried out for lettuce cv Evola (McCabe *et al.*, 2001). Mature plants exhibited normal morphology with no differences in fresh weight and head diameter but with retention of chlorophyll in the leaves. In the transgenic lines no senescent leaves were present at the seedling stage nor during later development. Interestingly, with the induction of senescence by nitrogen starvation, total nitrogen, nitrate and growth of transgenic and control plants decreased rapidly, but chlorophyll was retained in the lower (outer) leaves of the transgenic plants. Next to the SAG12 promoter, also the pSG529 and pSG766A promoters have been used in conjunction with the ipt gene. Chen *et al.* (2001) applied these chimaeric genes to transform broccoli, resulting in transgenic broccoli with 50% chlorophyll retention after four days' storage at 25°C. About 31% of the transformants exhibited the effect of retarding yellowing in detached leaves with 16% having the effects on florets and 7% on both leaves and florets. Application of the ipt gene together with other promoters was generally not successful due to the overproduction of endogenous cytokinins which resulted in developmental abnormalities (Garratt *et al.*, 2002). However, also other genes may be employed aimed to generate higher internal cytokinin concentrations. PetE-KNAT1, a kn1–like homologue under the control of the pea plastocyanin promoter PetE, is a chimaeric gene successfully used to delay leaf senescence in lettuce (Frugis *et al.*, 2001). kn-1 genes are normally expressed in shoot meristems and overexpressing them resulted in traits characteristic of altered cytokinin physiology.

Both cytokinins and phytochrome activation affect quality of plants and plant organs and cause similar morphogenic and biochemical responses (Thomas *et al.*, 1997). Cytokinins consist chemically of a purine structure with different side chains. The naturally occurring cytokinins with a hydroxylated aromatic side chain, topolins, strongly delay senescence processes in older and harvested plants (Strnad, 1997). For this group of cytokinins it has recently been demonstrated, using LCMS/MS identification, that phytochrome activation by a short red light pulse causes within 70 minutes the transient accumulation of the meta-hydroxylated cytokinin, metatopolin (riboside), and not of other isoprenoid cytokinis (Fig. 4.4) (Kappers, 1998). Recent experiments using various phytochrome mutants of tomato demonstrated that phytochrome B and not phytochrome A activation causes the accumulation of metatopolin in tomato leaf (unpublished observation, Davelaar and Jordi).

For tomato, numerous colour mutants exist (reviewed by Gray and Picton, 1994). Cnr mutant tomato fruits have low levels of carotenoids and of phytoene and lycopene. Extracts from ripe fruit showed a reduced ability to synthesise the carotenoid precursor GGPP, but also a lack of phytoene synthase (Fraser *et al.*, 2001). Ronen *et al.* (2000) investigated two pigmentation mutants in tomato, β and og. β is a single dominant gene that increases carotene and og is a recessive mutation that lacks β-carotene and increased lycopene. Cloning of both genes revealed that β encodes lycopene-ϵ-cyclase, a key enzyme that converts

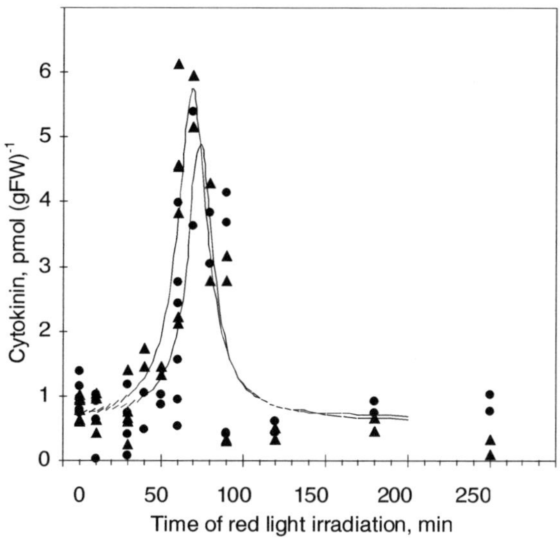

Fig. 4.4 Cytokinin concentrations (pmol (g FW)$^{-1}$) in alstroemeria leaves placed in darkness for 24 hours and subsequently irradiated with 5 μmol m^{-2}s^{-1} red light: metatopolin (triangles) and metatopolin riboside (circles).

lycopene to β-carotene. During fruit development the mRNA levels of lycopene-producing enzymes phytoene and phytoene desaturase increase, while the mRNA levels of the genes for the lycopene cyclases decline or completely disappear (Ronen *et al.*, 1999). Other interesting mutations towards carotenoid accumulation are the high pigment mutations, hp-1 and hp-2. These mutants show exaggerated photoresponses during de-etiolation and higher lycopene and β-carotene levels in combination with higher chlorophyll levels in immature fruit (Kerckhoffs *et al.*, 1997). Tomato seedling de-etioliation is a phytochrome (red light) response, which can be enhanced by blue light, suggesting that hp-1 may influence phytochrome. Probably, the hp mutants function as negative regulators of light signalling (Giovannoni, 2001).

Combinations of mutants could have interesting consequences with regard to tomato keeping quality. Both hp and og are colour intensifier mutants. The double mutant hp/hp og^{c}/og^{c} tomato fruits showed improved red colour and increased shelf-life but also several undesirable effects (de Araújo *et al.*, 2002). The effect on yield, firmness and colour development by combining the mutants, hp, og^{c} and alc to create all possible homozygous and heterozygous combinations was recently investigated for tomato fruits. The inclusion of alc should improve firmness as mutant alc fruit does not fully ripen off the vine unless picked beyond breaker stage and does not show a climacteric pattern resulting in increased storability up to 300% for homozygous alc/alc fruit. Analysis of intra-allelic additive and dominant interactions within these three loci and their interallelic interactions resulted in a number of genotypic combinations that represented a good compromise between yield, shelf-life and colour development (de Araújo *et al.*, 2002).

Transgenic research on ripening often affects firmness and colour development alike, especially when related to ethylene production. For some type of products, however, firmness retention is less of an issue. For example, in ACO antisense Cantaloupe melons, the major pigments, chlorophylls and carotenoids, remained undegraded during ripening compared to wild-type melons (Flores *et al*., 2001). Henzi *et al*. (2000) evaluated 12 transgenic broccoli lines containing a tomato antisense ACC oxidase gene. Three lines performed within the limits of acceptability and showed less ethylene production and improvement in head colour compared to controls after 98 and 48 hours of postharvest storage, respectively. Transgenic research aimed specifically at colour retention has also been carried out. Bacterial phytoene desaturase expressed in tomato did not elevate total carotenoid levels, but did increase the fraction of β-carotene threefold (Romer *et al*., 2000). Another approach used bacterial phytoene synthase to be overexpressed in tomato applying the PG promoter. It resulted in about a two-fold increase of phytoene, lycopene and β-carotene (Fraser *et al*., 2002).

4.6 Future trends

Transgenic research has brought promising means of enhancing colour and firmness retention. Applications to market genetically modified (GM) foods, however, have turned out to be troublesome, as exemplified by the FlavrSavr tomato. The FlavrSavr tomato was modified by reverse orientation of the PG gene, thereby retaining firmness during ripening. Although evaluated as safe for public consumption, public acceptance was too low. Acceptance may change, however, when more consumer-oriented traits in GM foods are introduced (Kleter *et al*., 2001). For now, the role of transgenic research is more orientated as a tool, elucidating important factors in colour and firmness development. A number of genetic factors have been identified and progress is swift due to molecular approaches such as positional cloning, QTL mapping and genetic engineering (White, 2002). From QTL analysis it is clear that, for example, firmness-generating processes constitute a number of probably interacting factors. To understand which factors are important during ripening and how they interact, colour and firmness measurements seem essential. For instance, the lack of mechanical studies of tomato texture regarding the rin and nor mutations makes it difficult to compare information on cell wall strength and to link physical properties to enzyme activities and transcriptional regulators (Seymour *et al*., 2002). However, it is questionable whether firmness and colour measurements combined with transgenic research alone will give sufficient insight into the processes occurring. Current black-box models, or models where physiological factors are linked on an *ad hoc* basis (semi-black-box models), do not generate an understanding of the underlying processes. Mathematical descriptions that describe the actual colour and firmness synthesis and decay processes and their interactions, or in other words mechanistic physiological models, are needed.

One example of a mechanistic physiological model has been mentioned earlier, that of the cucumber colour model. The cucumber colour model describes only a small number of processes. However, for products that show firmness or climacteric colour decay, the number of interacting processes is much larger and extensive mechanistic models do not exist yet. That is not to say that, for example, colour models for climacteric products do not exist at all. For instance, the green to yellow development of banana has been described as a combination of a logistic and a linear function (Chen and Ramaswamy, 2002). The green to red development of tomato fruits has been modelled using a logistic function (Tijskens and Everlo, 1994). However, colour is described, not the processes underlying colour development. Knowledge of the development processes in climacteric fruits and climacteric-like vegetables is currently insufficiently developed to generate mechanistic physiological models.

The possible bottleneck in the creation of mechanistic physiological models is most likely the application of suitable colour and firmness measurements. Colour and firmness development of fruits and vegetables would, ideally, be measured nondestructively. Nondestructive firmness measurements enable repeated measurements on the same fruit or vegetable, instead of always taking new samples. Using, for example, the commonly used puncture test, the effects of a firmness factor, for instance genetically up and downregulated, can be hard to observe as every sample will be in a slightly different stage of maturity. So, such measurements will show each time a slightly different combination of firmness-generating factors, complicating the elucidation of the genetic firmness factor on physiology. Colour measurements are mostly nondestructive. For practical reasons, RGB measurements are probably a better method for colour measurement than the widely used Lab measurements. Assessment of reflectance spectra has enabled the measurement of the separate colour generating sources (Merzlyak *et al.*, 2003). The effect of, for example, transgenic up or downregulation of colour genes on the concentration of colour compounds can be monitored via reflectance spectra and the overall effect on the colour can then be measured using RGB measurement. Development of nondestructive firmness measurements is progressing and is based on forced vibrations through a large range of frequencies (Peleg, 1993; Liljedahl and Abbott, 1994), striking with a mechanical impulse and measuring the audible resonant frequency with a microphone (Chen and De Baerdemaeker, 1995) or applying a small plunger that penetrates slightly into fruits (Lesage and Desdain, 1996).

A new development is the use of multiple imaging sensors for nondestructive measurement of physiological processes (Chaerle and Van Der Straeten, 2000). The Multiple Imaging Plant Stress (MIPS) facility at Plant Research International in Wageningen, The Netherlands, combines chlorophyll fluorescence, heat and colour imaging sensors for nondestructive image analysis of vitality and quality of plants. These combined sensors provide detailed information on the level and type of plant response to abiotic and biotic stress as experienced during transport and storage. The sensors are mounted on an industrial robot for automated high-throughput screening of physiological traits. Figure 4.5A–D show a typical

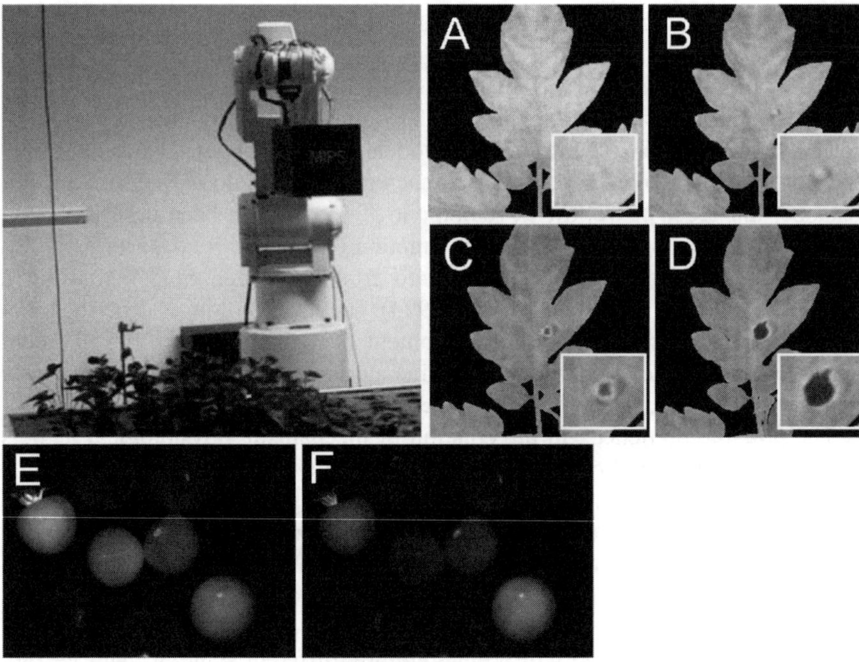

Fig. 4.5 Photograph of the MIPS robot in action (left image). Images A to D show an example of the use of the MIPS in measuring the development of a Botrytis infection on a tomato leaf by imaging of photosystem II efficiency by means of chlorophyll fluorescence. Time after inoculation 3 h (A), 5 h (B), 29 h (C) and 53 h (D). Images E and F show the chlorophyll fluorescence images of tomato fruit directly after harvesting (E) and after 70 h storage in darkness (F).

application of the MIPS system (leftmost image) in the detection of quality loss related to pathogen infection. Young tomato plants were infected by inoculation with *Botrytis infestans* and photochemical efficiency of photosystem II was monitored by means of chlorophyll fluorescence (Van Kooten and Snel, 1990). The development of the infestation could be monitored in this way. One important aspect is the time at which the first symptoms are recognised. In this experiment the effect of *Botrytis infestans* on the plant could be detected at a very early stage (<5 h after inoculation, Fig. 4.5B), well before visual symptoms appeared. A second, equally important parameter is the rate of growth of the infestation. From the image the area where the plant responds to the presence of the pathogen can be calculated. Under standardised conditions using a given pathogen strain, this parameter reflects the resistance of the plant to the growth of the pathogen once it has succeeded in entering the leaf and is an indication of the effectiveness of the plant's defence system to the pathogen. The MIPS technique is also applicable in the measurement of pigment changes associated with ripening. Figure 4.5E and F show the effect of dark storage on the chlorophyll fluorescence intensity of tomato fruit. The lower intensity of the signal in Fig. 4.5F reflects the loss of chlorophyll, especially when combined with non-invasive detection of fluorescent reporters associated with critical genes involved in colour and firmness of the produce. Reporters like GFP, RFP and DsRed are examples of reporters that could be exploited. This novel MIPS approach has great potential for breeding or for GM programmes on new colour and firmness traits. MIPS allows rapid screening of plants for loss of chlorophyll and important additional traits like resistance to abiotic and biotic factors during production, transport and storage.

4.7 Acknowledgements

The authors wish to thank Ad Schapendonk and Rob van der Schoor for their contribution to the MIPS facility and Jan Kodde for his help in the botrytis experiment. Furthermore, Mirek Strnad and Alena Sulcova are acknowledged for collaboration on the topoline experiment. We also thank Dick Kendrick for supplying tomato seeds for the phytochrome mutants.

4.8 References

ABLE A J, WONG L S, PRASAD A, O'HARE T J (2002), '1–MCP is more effective on a floral brassica (*Brassica oleracea* var. *italica* L.) than a leafy brassica (*Brassica rapa* var. *chenensis*)', *Postharv Biol Technol*, 26, 147–155.

BEN-ARIE R, KISLEV N, FRENKEL C (1979), 'Ultrastructural changes in the cell walls of ripening apple and pear fruit', *Plant Physiology*, 64, 197–202.

BENITEZ-BURRACO A, BLANCO-PORTALES R, REDONDO-NEVADO J, BELLIDO M L, MOYANO E, CABALLERO J L, MUNOZ-BLANCO J (2003), 'Cloning and characterization of two ripening-related strawberry (*Fragaria* × *ananassa* cv. Chandler) pectate lyase genes', *J Exp Bot*, 54, 633–645.

BRUMMELL D A, HARPSTER M H, CIVELLO P M, PALYSD J M, BENNET A B (1999), 'Modification of expansin protein abundance in tomato fruit alters softening and cell wall polymer metabolism during ripening', *Plant Cell*, 11, 2203–2216.

BRUMMELL D A, HOWIE W J, MA C, DUNSMUIR, P (2002), 'Postharvest fruit quality of transgenic tomatoes suppressed in expression of a ripening-related expansin', *Postharv Biol Technol*, 25, 209–220.

CAUSSE M, SALIBA-COLOMBANI V, LECOMTE L, DUFFÉ P, ROUSELLE P, BURET M (2002), 'QTL analysis of fruit quality in fresh market tomato: a few chromosome regions control the varaition sensory and instrumental traits', *J Exp Bot*, 53, 2089–2098.

CHAERLE L, VAN DER STRAETEN D (2000), 'Imaging techniques and the early detection of plant stress', *Trends in Plant Sciences*, 5, 495–501.

CHEN C R, RAMASWAMY H S (2002), 'Color and texture change kinetics in ripening bananas', *Lebensm-Wiss u-Technol*, 35, 415–419.

CHEN H, DE BAERDEMAEKER J (1995), 'Total least squares technique for estimating the vibration parameters of the apple from the time domain impulse response signal', *J Agric Eng Res*, 61, 283–290.

CHEN L F O, HWANG J Y, CHARNG Y Y, SUN C W, YANG S F (2001), 'Transformation of broccoli (*Brassica oleracea* var. *italica*) with isopentenyltransferase gene via *Agrobacterium tumefaciens* for post-harvest yellowing retardation', *Molecular Breeding*, 7, 243–257.

CIVELLO P M, POWELL A L T, SABEHAT A, BENNET A B (1999), 'An expansion gene expressed in ripening strawberry fruit', *Plant Physiol Rockville*, 121, 1273–1279.

COSGROVE D J (2000), 'New genes and new biological roles for expansins', *Curr Opin Plant Biol*, 3, 73–78.

DE ARAÚJO M L, MALUF W R, GOMES L A A, OLIVEIRA A C B (2002), 'Intra and interlocus interactions between alcobaca (alc), crimson (ogc), and high pigment (hp) loci in tomato *Lycopersicon esculentum* Mill.', *Euphytica*, 125, 215–226.

DOWNS C G, SOMERFIELD S D, DAVEY M C (1997), 'Cytokinin treatment delays senescence but not sucrose loss in harvested broccoli', *Postharv Biol Technol*, 11, 93–100.

DRURY R, HÖRTENSTEINER S, DONNISON I, BIRD C R, SEYMOUR G B (1999), 'Chlorophyll catabolism and gene expression in the peel of ripening banana fruits', *Physiologia Plantarum*, 107, 32–38.

FLORES F B, MARTINEZ-MADRID M C, SANCHEZ-HIDALGO F J, ROMOJARO F (2001), 'Differential rind and pulp ripening of transgenic antisense ACC oxidase melon', *Plant Physiol Biochem*, 39, 37–43.

FRASER P D, BRAMLEY P, SEYMOUR G B (2001), 'Effect of the Cnr mutation on carotenoid formation during tomato fruit ripening', *Phytochemistry*, 58, 75–79.

FRASER P D, ROMER S, SHIPTON C A, MILLS P B, KIANO J W, MISAWA N, DRAKE R G, SCHUCH W, BRAMLEY P M (2002), 'Evaluation of transgenic tomato plants expressing an additional phytoene synthase in a fruit-specific manner', *Proc Natl Acad Sci USA*, 99, 1092–1097.

FRUGIS G, GIANNINO D, MELE G, NICOLODI C, CHIAPPETTA A, BITONTI M B, INNOCENTI A M, DEWITTE W, VAN ONCKELEN H, MARIOTTI D (2001), 'Overexpression of KNAT1 in lettuce shifts leaf determinate growth to a shoot-like indeterminate growth associated with an accumulation of isopentenyl-type cytokinins', *Plant Physiol Rockville*, 126, 1370–1380.

GAN S, AMASINO R M (1995), 'Inhibition of leaf senescence by autoregulated production of cytokinin', *Science*, 270, 1986–1988.

GARRATT L C, POWER, J B, DAVEY, M R (2002), 'Improving the shelf-life of vegetables by genetic modification', in Jongen W, *Fruit and Vegetable Processing: Improving Quality*, Cambridge, UK: Woodhead Publishing, 267–287.

GIOVANNONI J (2001), 'Molecular biology of fruit maturation and ripening', *Ann Rev Plant Physiol Plant Mol Biol*, 52, 725–749.

GIOVANNONI J J, DELLAPENNA D, BENNET A B, FISHER R L (1989), 'Expression of a chimeric polygalacturonase gene in transgenic rin (ripening inhibitor) tomato fruit results in polyuronide degradation', *Plant Cell*, 1, 53–64.

GRAY J E, PICTON S (1994), 'The use of transgenic and naturally occurring mutants to understand and manipulate tomato fruit ripening', *Plant Cell Environ*, 17, 557–571.

HARKER F R, MAINDONALD J, MURRAY S H, GUNSON F A, HALLETT I C, WALKER S B (2002), 'Sensory interpretation of instrumental measurements 1: Texture of apple fruit', *Postharv Biol Technol*, 24, 225–239.

HEATON J W, MARANGONI, A G (1996), 'Chlorophyll degradation in processed foods and senescent plant tissues', *Trends in Food Science and Technology*, 7, 8–15.

HENZI M X, CHRISTEY M C, MCNEIL D L (2000), 'Morphological characterisation and agronomic evaluation of transgenic broccoli (*Brassica oleracea* L. var. *italica*) containing an antisense ACC oxidase gene', *Euphytica*, 113, 9–18.

HERTOG M L A T M, NICHOLSON S E, BANKS N H (2001), 'The effect of modified atmospheres on the rate of firmness change in Braeburn apples', *Postharv Biol Technol*, 23, 174–184.

HIWASA K, KINUGASA Y, AMANO S, HASHIMOTO A, NAKANO R, INABA A, KUBO Y (2003), 'Ethylene is required for both the initiation and progression of softening in pear (*Pyrus communis* L.) fruit', *J Exp Bot*, 383, 771–779.

HOEBERICHTS F A, VAN DER PLAS L H W, WOLTERING E J (2002), 'Ethylene perception is required for the expression of tomato ripening-related genes and associated physiological changes even at advanced stages of ripening', *Postharv Biol Technol*, 26, 125–133.

HOEHN E, GASSER F, GUGGENBUHL B, KUNSCH U (2003), 'Efficacy of instrumental measurements for determination of minimum requirements of firmness, soluble solids, and acidity of several apple varieties in comparison to consumer expectations', *Postharv Biol Technol*, 27, 27–37.

HONDA C, KOTODA N, WADA M, KONDO S, KOBAYASHI S, SOEJIMA J, ZHANG Z L, TSUDA T, MORIGUCHI T (2002), 'Anthocyanin biosynthetic genes are coordinately expressed during red coloration in apple skin', *Plant Physiol Biochem*, 40, 955–962.

JANAVE M T (1997), 'Enzymic degradation of chlorophyll in cavendish bananas: *in vitro* evidence for two independent degradative pathways', *Plant Physiol Biochem*, 35, 837–846.

JIMENEZ-BERMUDEZ S, REDONDO-NEVADO J, MUNOZ-BLANCO J, CABALLERO J L (2002), 'Manipulation of strawberry fruit softening by antisense expression of a pectate lyase gene', *Plant-Physiol*, 128, 751–759.

JOHNSTON J W, HEWETT E W, HERTOG M L A T M (2002), 'Postharvest softening of apple fruit: a review', *New Zealand Journal of Crop and Horticultural Science*, 30, 145–160.

JURAN, J M (1974), 'Basic concepts', in Juran J M, Goyna F M Jr, Bingham R S Jr, *Quality Control Handbook*, New York: McGraw-Hill.

KADER A A, ZAGORY D, KERBEL E L (1989), 'Modified atmosphere packaging of fruits and vegetables', *Crit Rev Food Sci Nutr*, 28, 1–30.

KAPPERS I F (1998) 'Leaf senescence in alstroemeria. Regulation by phytochrome, gibberellins and cytokinins', PhD thesis, ISBN 90–5485–919–9.

KERCKHOFFS L H J, DEGROOT N A M A. VAN TUINEN A, SCHREUDER M E L, NAGATANI A, KOORNNEEF M, KENDRICK, R E (1997) 'Physiological chracterization of exaggerated-photoresponse mutants of tomato', *J Plant Physiol*, 150, 578–587.

KING G J, MALIEPAARD C, LYNN J R, ALSTON F H, DUREL C E, EVANS K M, GRIFFON B, LAURENS F, MANGANARIS A G, SCHREVENS E, TARTARINI S, VERHAEGH J (2000), 'Quantitative genetic analysis and comparison of physical and sensory descriptors relating to fruit flesh firmness in apple (*Malus pumila* Mill.)', *Theor Appl Genet*, 100, 1074–1084.

KING G J, LYNN J R, DOVER C J, EVANS K M, SEYMOUR G B (2001), 'Resolution of quantitative trait loci for mechanical measures accouting for genetic variation in fruit texture of apple (*Malus pumila* Mill.)', *Theorl Appl Genet* 100, 1227–1235.

KLETER G A, VAN DER KRIEKEN W M, KOK E J, BOSCH D, JORDI W, GILISSEN L J W J (2001), 'Regulation and exploitation of genetically modified criops', *Nature Biotechnol*, 19, 1105–1110.

LESAGE P, DESDAIN M-F (1996), 'Measurement of tomato firmness by using a non-destructive mechanical sensor', *Postharv Biol Technol*, 8, 45–55.

LILJEDAHL L A, ABBOTT J A (1994), 'Changes in sonic resonance of "Delicious" and "Golden Delicious" apples undergoing accelerated ripening', *Trans ASEA*, 37, 907–912.

LU C G, ZAINAL Z, TUCKER G A, LYCETT G W (2001), 'Developmental abnormalities and reduced fruit softening in tomato plants expressing an antisense Rab11 GTPase gene', *Plant Cell*, 13, 1819–1833.

MAHARAJ R, ARUL J, NADEAU P (1999), 'Effect of photochemical treatment in the preservation of fresh tomato (*Lycopersicon esculentum* cv. Capello) by delaying senescence', *Postharv Biol Technol*, 15, 13–23.

MALIEPAARD C, ALSTON F H, VAN ARKEL G, BROWN L M *et al.* (1998), 'Aligning male and female linkage maps of apple (*Malus pumila* Mill.) using multi-allelic markers'. *Theor Appl Genet*, 97, 60–73.

MATILE P, HÖRTENSTEINER S (1999), 'Chlorophyll degradation', *Ann Rev Plant Physiol Plant Mol Biol*, 50, 6795.

MCCABE M S, GARRATT L C, SCHEPERS F, JORDI W J R M, STOOPEN G M, DAVELAAR E, VAN RHIJN J H A, POWER J B, DAVEY M R (2001), 'Effects of PSAG12–IPT gene expression on development and senescence in transgenic lettuce', *Plant Physiol Rockville*, 127, 505–516.

MERZLYAK M N, SOLVCHENKO A E, GITELSON A A (2003), 'Reflectance spectral features and non-destructive estimation of chlorophyll, carotenoid and anthocyanin content in apple fruit', *Postharv Biol Technol*, 27, 197–211.

MIR N A, CURELL E, KHAN N, WHITAKER M, BEAUDRY R M (2001), 'Harvest maturity, storage temperature, and 1–MCP application frequency alter firmness retention and chlorophyll fluorescence of "Redchief Delicious" apples', *J Amer Soc Hort Sci*, 126, 618–624.

MOORE S, VREBALOV J, PAYTON P, GIOVANNONI J (2002), 'Use of genomics tools to isolate key ripening genes and analyse fruit maturation in tomato', *J Exp Bot*, 53, 2023–2030.

MUNOZ A M (2002), 'Sensory evaluation in quality control: an overview, new developments and future opportunities', *Food Quality and Preference*, 13, 329–339.

NOGATA Y, OHTA H, VORAGEN A (1993), 'Polygalacturonase in strawberry fruit', *Phytochemistry*, 34, 617–620.

OELLER P W, WONG L M, TAYLER L P, PIKE D A, THEOLOGUS A. (1991), 'Reversible inhibition of tomato fruit senescence by antisense 1-aminocyclopropane-1-carboxylate synthase', *Science*, 254, 427–439.

ORFILA C, HUISMAN M M H, WILLATS W G T, VAN ALEBEEK G J W M, SCHOLS H A, SEYMOUR G B, KNOX J P (2002), 'Altered cell wall disassembly during ripening of Cnr tomato fruit: implications for cell adhesion and fruit softening', *Planta*, 215, 440–447.

PAGE T, GRIFFITHS G, BUCHANON-WOLLESTON V (2001), 'Molecular and biochemical characterisation of postharvest senescence in broccoli', *Plant Physiol*, 125, 718–727.

PAYNE A, HOLT S (2001), 'Diagnosing customer value: integrating the value process and relationship marketing', *British Journal of Management*, 12, 159–182.

PELEG K (1993), 'Comparison of non-destructive and destructive measurements of apple firmness', *J Agric Eng Res*, 55, 227–238.

PENUELAS J, FILELLA I (1998), 'Visible and near-infrared reflectance techniques for diagnosing plant physiological status', *Trends Plant Sci*, 3, 151–156.

POLDER G, VAN DER HEIJDEN G W A M, YOUNG I T (2002), 'Spectral images for measuring ripeness of tomatoes', *Trans ASEA*, 45, 1155–1161.

PORAT R, FENG X Q, HUBERMAN M, GALILI D, GOREN R, GOLDSCHMIDT E E (2001), 'Gibberellic acid slows postharvest degreening of "Oroblanco" citrus fruits', *Hortscience*, 36, 937–940.

PORRA R J (1997), 'Recent advances in porphyrin and chlorophyll synthesis', *Photochem Photobiol*, 65, 492–516.

PRE-AYMARD C, WEKSLER A, LURIE S (2003), 'Responses of "Anna", a rapidly ripening summer apple, to 1–methylcyclopropene', *Postharv Biol Technol*, 27, 163–170.

ROMER S, FRASER P D, KIANO J W, SHIPTON C A, MISAWA N, SCHUCH W, BRAMLEY P M (2000), 'Elevation of the provitamin A content of transgenic tomato plants', *Nature-Biotechnology*, 18, 666–669.

RONEN G, COHEN M, ZAMIR D, HIRSCHBERG J (1999), 'Regulation of carotenoid biosynthesis during tomato fruit development: expression of the gene for lycopene epsilon-cyclase is down-regulated during ripening and is elevated in the mutant Delta', *Plant J*, 17, 341–351.

RONEN G, CARMEL M, ZAMIR D, HIRSCHBERG J (2000), 'An alternative pathway to beta carotene formation in plant chromoplasts by map-based cloning og Beta and old-gold color mutations in tomato. *Proc Natl Acad Sci USA*, 97, 11102–11107.

ROSE J K C, LEE H H, BENNET A B (1997), 'Expression of a divergent expansin gene is fruit specific and riening regulated', *Proc Natl Acad Sci USA*, 94, 5955–5960.

SAKS Y, COPEL A, BARKAI-GOLA R (1996), 'Improvement of harvested strawberry quality by illumination: colour and Botrytis infection', *Postharv Biol Technol*, 8, 19–27.

SAMS C E (1999), 'Preharvest factors affecting postharvest texture', *Postharv Biol Technol*, 15, 249–254.

SCHOUTEN R E, KESSLER D, ORCARAY L, VAN KOOTEN O (2002a), 'Predictability of keeping quality for strawberry batches', *Postharv Biol Technol*, 26, 35–47.

SCHOUTEN R E , TIJSKENS L M M , VAN KOOTEN O (2002b), 'Predicting keeping quality of batches of cucumber fruit based on a physiological mechanism. *Postharv Biol Technol*, 26, 209–220.

SCHOUTEN R E, JONGBLOED G, VAN KOOTEN O (2003), 'Relation between cultivar and keeping quality for batches of cucumbers', *Acta Hort*, 59, 451–459.

SEROCZYNSKA A, NIERMIROWICZ S K (1998), 'Genetic analysis of selected tomato (*Lycopersicon esculentum* Mill.) traits in crosses between cultivated lines and the nor mutant', *J Appl Gen*, 39, 259–273.

SEYMOUR G B, MANNING K, ERIKSSON, EM, POPOVICH, A H, KING G J (2002), 'Genetic identification and genomic organisation of factors affecting fruit texture', *J Exp Bot*, 53, 2065–2071.

SIMMONDS (1979), *Principles of Crop Improvement*, London: Longman.

SIRIPHANICH J (1998), 'High CO_2 atmosphere enhances fruit firmness during storage', *J Jap Soc Hort Sci*, 6, 1167–1170.

STRNAD M (1997), 'The aromatic cytokinins', *Physiol. Plant*, 101, 674–688.

THOMAS T H, HARE P D, VAN STADEN J. (1997). 'Phytochrome and cytokinin responses', *Plant Growth Regulation* 23, 105–122.

TIEMAN D M, HARRIMAN R W, RAMAMOHAN G, HANDA A K (1992), 'An antisense pectin methylesterase gene alters pectin chemistry and soluble solids in tomato fruit', *Plant Cell*, 4, 667–679.

TIJSKENS L M M (1979), 'Texture of golden delicious apples during storage', *Lebensm-Wiss u-Technol*, 12, 138–142.

TIJSKENS L M M (2000), 'Acceptability', in Shewfelt R L, Brückner B, *Fruits and Vegetables Quality: An integrated view*, Lancaster: Technomic Press, 125–143.

TIJSKENS L M M, EVERLO R G (1994), 'Modelling colour of tomatoes during postharvest storage', *Postharv Biol Technol*, 4, 85–98.

TIJSKENS L M M, POLDERDIJK J J (1996), 'A generic model for keeping quality of vegetable produce during storage and distribution', *Agric Syst*, 51, 431–452.

TIJSKENS L M M, HERTOG M L A T M, VAN SCHAIK A C R, DE JAGER A (1999), 'Modelling the firmness of Elstar apples during storage and transport', *Acta Hort*, 485, 363–372.

TOIVONEN P M A, KEMPLER C, STAN S J (2002), 'The use of a natural clay adsorbent improves quality retention in three cultivars of raspberries stored in modified atmosphere packages', *Food Quality*, 25, 385–393.

VAN DIJK C, TIJSKENS L M M (2000), 'Mathematical modeling of enzymatic reactions as related to the texture of vegetables after storage and mild heat treatments', in Alzamora S M, Tapia A, López-Malo A, *Design of Minimal Processing Technologies for Fruit and Vegetables*, Gaithersburg, MD: Aspen Publishers.

VAN KOOTEN O, SNEL J H F (1990), 'Modulated Chlorophyll Fluorescence nomenclature', *Photosynth Res*, 25, 147–150.

VERLEGH P W J, STEENKAMP J B E M (1999), 'A review and meta-analysis of country-of-origin research', *Journal of Economic Psychology*, 20, 521–546.

VICENTE A R, MARTINEZ G A, CHAVES A R, CIVELLO P M (2003), 'Influence of self-produced CO_2 on postharvest life of heat-treated strawberries', *Postharv Biol Technol*, 27, 255–264.

WAKABAYASHI K (2000), 'Changes in cell wall polysaccharides during fruit ripening', *Journal of Plant Research*, 113, 231–237.

WHITE J W (2002), 'Recent advances in fruit development and ripening: an overview', *J Exp Bot*, 53, 1995–2000.

WILKINSON, E C, TIJSKENS, L M M (2002), 'Modeling food quality', in Tijskens, L M M, Hertog, M L A T M, Nicolaï B M, *Food Process Modelling*, Cambridge: Woodhead Publishing, 367–382.

WILKINSON J Q, LANAHAN M B, YEN H C *et al.* (1995), 'An ethylene inducible component of signal-transduction encoded by Never-ripe', *Science*, 270, 1807–1809.

WILLIAMS O J, RAGHAVAN G S V, GOLDEN K D, GARIEPY Y (2003), 'Postharvest storage of Giant Cavendish bananas using ethylene oxide and sulphur dioxide', *J Sci Food Agric*, 83, 180–186.

WILLIAMSON S J, CUMMINS H Z (1983), *Light and Color in Nature and Art*, New York, John Wiley & Sons.

YEN H, LEE S, TANKSLEY S, LANAHAN M, KLEE H J, GIOVANNONI J J (1995), 'The tomato never-ripe locus regulates ethylene-inducible gene expression and is linked to a homologue of the Arabidopsis ETR1 gene, *Plant Physiology*, 107, 1807–1809.

YOSHIOKA H, AOBA K, KASHIMURA Y (1992), 'Molecular weight and degree of methyloxylation in cell wall polyuronide during softening in pear and apple fruit', *J Amer Soc Hort Sci*, 117, 600–606.

5

Spoilage yeasts

T. Deak, Budapest University of Economic Science, Hungary

5.1 Introduction

Yeasts represent a small group of fungi, only about 1% of all fungal species known. Their significance to mankind, however, far surpasses their proportion. Yeasts that bring about the leavening of bread and the fermentation of wine and beer are essential parts of our everyday life. While their outstanding role in these processes resulting in great economic gains is widely appreciated, much less is known about the losses they cause by the spoilage of foods and the health risk posed by a few yeast species.

Apart from the few yeasts that are intimately associated with human activities and which are, in fact, the most domesticated of all microorganisms, hundreds of yeast species thrive in nature. Various yeast species can be found in a range of habitats, in soil and water, on plants and animals, and in food production systems ranging from vineyards to processed foods. For yeasts, foods are natural enough habitats as they are rich in nutrients and generally provide a suitable environment for growth. Depending on composition, processing, preservation, packaging and storage, each type of food represents a specific niche in which the best-fitting group of yeasts develops in association with other microorganisms. If factors controlling growth are not strong enough, yeasts start propagating and cause spoilage. Spoilage by yeasts manifests itself in gaseous emissions, turbidity, textural change, film formation, discolouration and change of taste and flavour that render foods unpalatable. Fortunately, no food-borne yeast is known to cause food infection or poisoning. It is the task of food technologists to control the microbial ecology of foods to prevent contamination, inhibit growth or inactivate yeast cells in order to protect them from spoilage.

5.2 Characteristics and classification of yeasts

Yeasts are a phylogenetically diverse group of fungi that may belong to either Ascomycetes (e.g. *Saccharomyces*, *Candida*) or Basidiomycetes (e.g. *Rhodotorula*, *Cryptococcus*). The small group of *Schizosaccharomyces* is regarded as a member of Archiascomycetes. Notwithstanding their taxonomic position, most yeasts are unicellular fungi characterized by their typical way of vegetative propagation: budding. In contrast to the predominantly hyphal (mycelial) moulds and other fungi, only some yeasts produce filamentous growth (develop true hyphae or pseudohyphae), and when they propagate sexually, their spores are not enclosed in a fruiting body.

Yeasts are classified primarily by their mode of sexual reproduction, i.e. sporulation. In ascomycetous yeasts spores are formed through conjugation of opposite mating types, and spores develop after meiotic division in the diploid cell that serves as sporangium (Fig. 5.1). This typical way of sporulation is frequently modified, and conjugation may occur between mother cell and its bud, or diploid cells can transform directly into sporangium (Fig. 5.2). In basidiomycetous yeasts, a complex sexual cycle results in the formation of basidiospores which take on a budding yeast form. Sporulation occurs rarely in basidiomycetous yeasts. Moreover, with about half of described yeast species, no sexual process is known. These occur in anamorphic (imperfect) state as opposed to the sexually propagating teleomorphic (perfect) states. Unfortunately, anamorphic and teleomorphic states often bear different names, e.g.

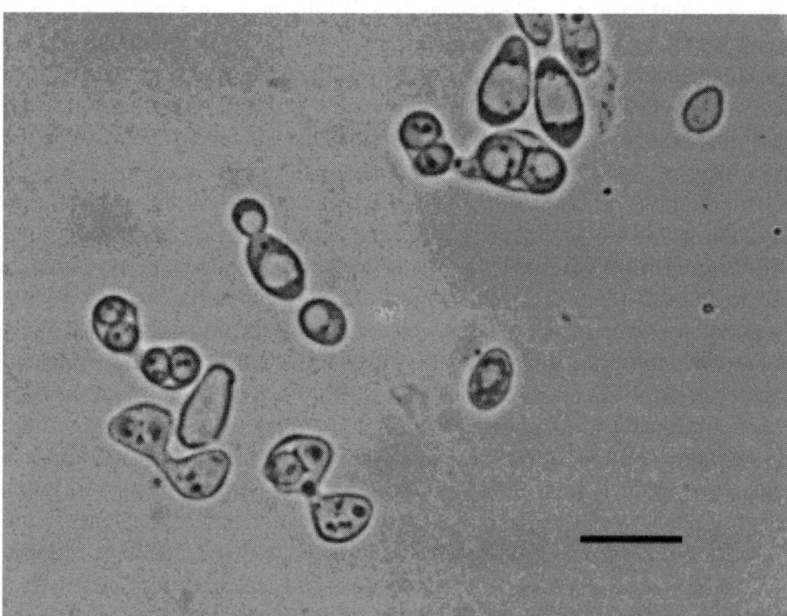

Fig. 5.1 Conjugating cells and spores of *Zygosaccharomyces bailii*. Bar = 10 μm.

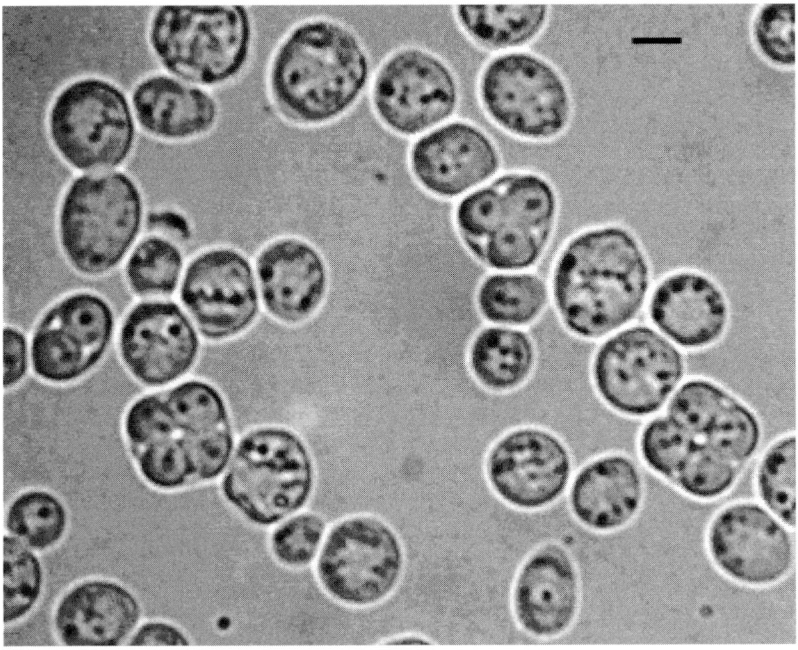

Fig. 5.2 Sporulating cells of *Saccharomyces cerevisiae;* asci (sporangia) with two, three and four spores. Bar = 5 μm.

Hanseniaspora uvarum has its better known anamorph named *Kloeckera apiculata.* In addition, numerous synonyms have been created by nomenclatural changes as a consequence of taxonomic rearrangements and the discovery of perfect–imperfect relationships. For example, the widely known old name *Torula utilis* later became *Torulopsis utilis* and both are synonyms with *Candida utilis* which in turn is the anamorph of *Hansenula jadinii,* now a synonym of *Pichia jadinii*. This range of synonyms presents a problem to lay people and casts some doubts on the reliability of taxonomy. For the orientation of the reader, Table 5.1 lists some corresponding names of frequent food-borne yeasts cited in this chapter. Taxonomy of yeasts is indeed in constant flux. Current trends, based on molecular studies, lend a solid phylogenetic framework for their classification. For the discrimination and identification of species, traditional morphological and physiological criteria are still useful. In the next section, a short overview is provided on the main groups of important food spoilage yeasts.

5.1.2 Main groups of food-borne yeasts

The majority of yeasts most frequently occurring in foods typically comprise oval to round-shaped single, budding cells. Most of them are related to ascomycete fungi and form spores endogenously in a cell (class Hemiascomycetes, order Saccharomycetales). Six groups are distinguished here.

Table 5.1 Corresponding names of some food-borne yeast species

Teleomorph	Anamorph	Synonym[1]
Debaryomyces hansenii	*Candida famata*	*Torulopsis candida*
Dekkera anomala	*Brettanomyces anomalus*	*Brettanomyces claussenii*
Galactomyces geotrichum	*Geotrichum candidum*	*Oospora lactis*
Hanseniaspora uvarum	*Kloeckera apiculata*	
Issatchenkia orientalis	*Candida krusei*	
Kluyveromyces marxianus	*Candida kefyr*	*Saccharomyces fragilis*
Pichia anomala	*Candida pelliculosa*	*Hansenula anomala*
Pichia mambranifaciens	*Candida valida*	*Candida mycoderma*
Rhodosporidium toruloides	*Rhodotorula glutinis*	
Saccharomyces cerevisiae	*Candida robusta*	*Saccharomyces ellipsoideus*
Saccharomyces exiguous	*Candida holmii*	*Torulopsis holmii*
Torulaspora delbrueckii	*Candida colliculosa*	*Saccharomyces fermentati*
Yarrowia lipolytica	*Candida lipolytica*	*Saccharomycopsis lipolytica*
Zygosaccharomyces rouxii	*Candida mogii*	*Saccharomyces rouxii*

[1] Of several existing synonyms only the most common is given.

Strongly fermentative yeasts

Yeasts belonging to the genera *Saccharomyces*, *Zygosaccharomyces*, *Torulaspora* and *Kluyveromyces* are closely related, and both the genera and the species have been reclassified frequently. At one time some 40 species of *Saccharomyces* were listed; a number of them became amalgamated into a single species, *Saccharomyces cerevisiae,* and recently the number of recognized species has again increased. Most industrial beer, wine, spirit, and bakers' yeast strains belong to this species. At the same time, it is one of the most frequent spoilage yeasts in foods. *Zygosaccharomyces bailii* is characterized by its extreme resistance to acetic acid and preservatives, whereas *Zygosaccharomyces rouxii* tolerates high concentrations of sugar and salt. Both are widely occurring spoilage yeasts, together with *Torulaspora delbrueckii* of diverse physiological properties. *Kluyveromyces marxianus* and *K. lactis* ferment lactose, exhibit rather high thermotolerance, and are frequent spoilage species in dairy products.

A fermentative species, described among the first yeasts, is *Schizosaccharomyces pombe.* Its distinguishing character is the division of cells instead of budding. In fact, phylogenetically it seems to be distantly related to the budding 'true' yeasts. It may cause spoilage in sugar-containing products less commonly than the species mentioned above.

Weakly fermenting yeasts

A distinguishing property of a number of typical budding and sporulating yeasts is that they ferment poorly or not at all. The most important representative of this group is *Debaryomyces hansenii*, which occurs in a variety of food products and is salt tolerant. Due to their mostly aerobic metabolism, many of these species, such as *Pichia membranifaciens* and *Issatchenkia orientalis*, develop a film on liquids. Other frequent spoilage species, e.g. *Pichia anomala,* may ferment weakly.

Apiculate yeasts
Budding localized at the two poles of cells lends these yeasts a special cell shape, often called apiculate. A typical representative is *Hanseniaspora uvarum.* This and related species commonly occur on grapes and other fruits, and initiate their fermentation; however, their alcohol tolerance is lower than that of *Saccharomyces* yeasts.

Hyphal yeasts
A number of yeasts develop filamentous cells also bearing buds, sometimes, however, breaking into fragments (arthroconidia), such as *Galactomyces geotrichum* (Fig. 5.3). Some hyphal yeast is quite frequent in foods, e.g. *Yarrowia lipolytica,* a cause of spoilage in meat and dairy products due to its strong lipolytic and proteolytic activity. With several other yeasts, cells elongate after budding and remain together forming a pseudohypha (Fig. 5.4); the lack of cell walls dividing true hyphae into compartments is a distinguishing trait but hard to observe.

Imperfect yeasts
As mentioned before, many yeast species are unable or have lost the capability to form sexual spores; these propagate themselves with budding, and are called

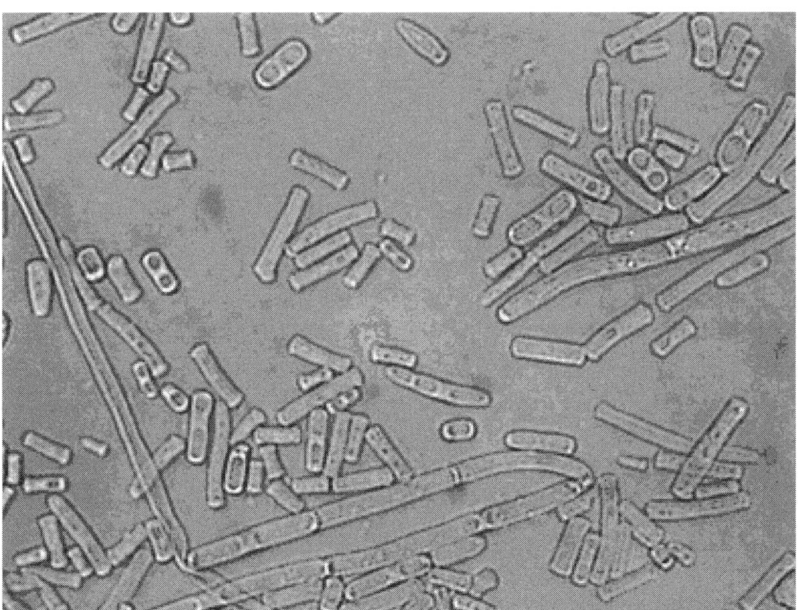

Fig. 5.3 True hyphae and arthroconidia of *Galactomyces geotrichum.* Bar = 20 μm.

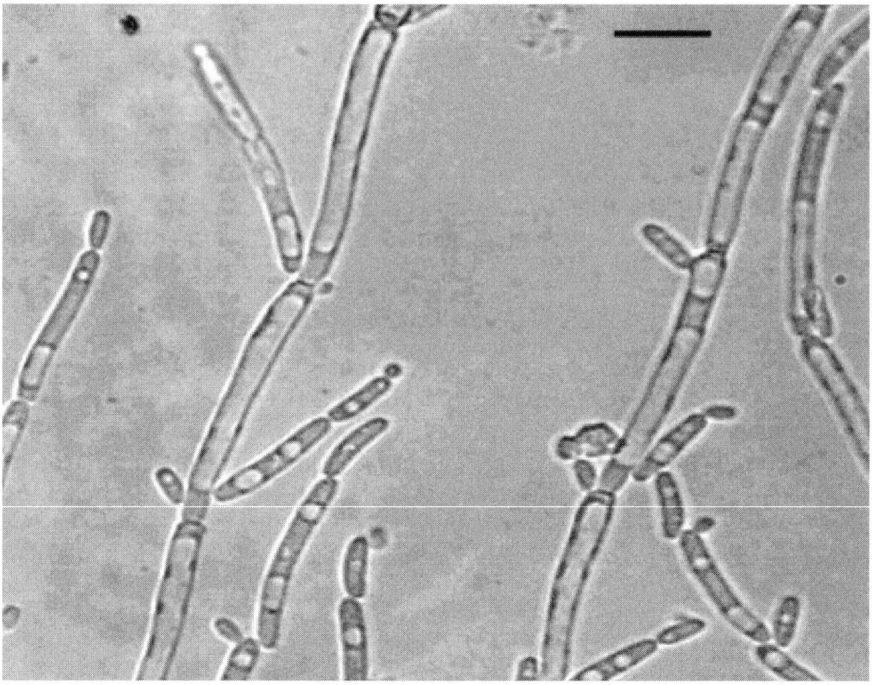

Fig. 5.4 Pseudohyphae of *Issatchenkia orientalis* (anamorph *Candida krusei*). Bar = 10 μm.

anamorphs (imperfect forms). For many of them the related sporulating (teleomorph) state has been found. One of the largest genera of yeasts, comprising over 100 species, is *Candida.* A number of common and frequent spoilage species belong to this genus, e.g. *C. tropicalis*, *C. stellata* and *C. zeylanoides*, and teleomorphs are often better known in their *Candida* anamorphs (e.g. *C. krusei*; see also Table 5.1).

Red yeasts

The most common representatives of basidiomycetous yeasts in foods form typical coloured colonies of various shades of orange to red, such as species of *Rhodotorula* and *Sporobolomyces.* These are names of anamorphic forms; some of them are also known as teleomorphs (*Rhodosporidium*, *Sporidiobolus*). Other basidiomycetous yeasts are not red but rather white, such as *Cryptococcus albidus* and *C. laurentii,* the two most frequent species on natural substrates and also in foods. *Trichosporon* is the basidiomycetous counterpart of the hyphal, arthroconidia-forming yeasts, such as the ascomycete *Geotrichum.* The basidiomycetous yeasts, with few exceptions, do not ferment but are equipped with hydrolytic activities and capable of growth at low temperatures.

Table 5.2 Some examples of frequent yeast species found in natural habitats

Habitat	Yeast species
Soil	*Lipomyces lipofer*, *Debaryomyces occidentalis*, *Cryptococcus terreus*
Fresh water	*Rhodotorula glutinis*, *Pichia anomala*, *Issatchenkia orientalis*
Plants	*Metchnikowia pulcherrima*, *Sporobolomyces roseus*, *Cryptococcus laurentii*
Dust in air	*Rhodotorula mucilaginosa*, *Cryptococcus albidus*, *Debaryomyces hansenii*

5.2.2 Natural habitats

Yeasts are widely distributed in nature. They thrive on plant leaves, flowers, and especially fruits. They are commonly associated with insects, and many occur on the skin, hide, and feathers and also in the alimentary tracts of vertebrates. Soil is an important reservoir of yeasts from where they can get into surface waters. Some yeasts are regular inhabitants of seawater. Some yeast species characteristic of specific habitats are included in Table 5.2. These natural habitats are important vehicles for carrying yeasts into food materials and processing facilities.

5.3 Factors affecting the growth and survival of spoilage yeasts

The main factors affecting the growth and survival of yeasts in foods include the inherent properties of the products (intrinsic factors), the effects of the surrounding environment (extrinsic factors), and the physiological characteristics and biological interactions of yeasts themselves as well as with other microorganisms (implicit factors).

5.3.1 Metabolic activities

In common with fungi, yeasts are aerobic organisms, and metabolize their nutrients oxidatively. Contrary to common belief, only about half of all yeast species can ferment sugars. These are facultative anaerobic. Alcoholic fermentation is the most noticeable feature of a number of yeasts. Species of *Saccharomyces* ferment most vigorously and tolerate ethanol up to 15% (v/v). Ethanol is a highly valuable product of various food and chemical industries, whereas the other main end product, carbon dioxide, may contribute to food spoilage.

Under aerobic conditions yeast can utilize most simple constituents of foodstuffs such as mono- and disaccharide sugars, organic acids, alcohols and amino acids, but only a limited number of sugars (mostly hexoses and some disaccharides) can be used for substrates of fermentation. Ability to hydrolyse macromolecules is also limited among yeasts. In particular, few yeast species possess amylolytic enzymes for hydrolysis of starch. A special variety of *Saccharomyces cerevisiae,* often considered to be a separate species called *Saccharomyces diastaticus*, is capable of partial hydrolysis of starch. The most active amylolytic yeasts, *Debaryomyces occidentalis* and *Saccharomycopsis fibuligera*, may be responsible for spoilage of bread and other bakery products ('chalky bread'). Lipolytic and proteolytic enzymes occur widely in yeast cells but are rarely active enough to cause extensive hydrolysis. *Yarrowia lipolytica,* some *Candida* and *Rhodotorula* species may be responsible for lipolytic and proteolytic spoilage in meat and dairy products.

5.3.2 Growth requirements

Populations of yeasts that propagate by budding or splitting show a typical growth curve, similar to that of bacteria, that can be divided into phases of differing growth rate such as lag, exponential, stationary and declining or death phase. The length of these phases depends largely on internal and external factors, but the growth rate is highest in the exponential phase. In general, the growth rate of yeasts is slower than that of bacteria, but faster than the filamentous growth of moulds. This provides a competitive advantage for bacteria over yeasts in most natural habitats and foods. Sometimes, however, when conditions are favourable for yeasts, they can easily outgrow bacteria. Under optimal conditions, yeasts may have a generation time of about 1 to 2 h (corresponding to a growth rate of 0.70 to 0.35 h^{-1}). Certain yeasts, e.g. *Zygosaccharomyces* and *Galactomyces* species, grow at a slower rate; their generation time may be 4 to 8 h or longer (Table 5.3). Under unfavourable conditions the growth rate decreases and the length of the lag phase increases.

Yeasts are able to grow within certain limits of environmental conditions, such as temperature, moisture, pH and others. Outside these limits they stop growing but may survive. Yeasts differ widely in their tolerance to unfavourable conditions; studies on these are important as they provide the basis of the methods employed for protection of foods from spoilage by yeasts.

The temperature range for yeast growth in general extends from several degrees below 0°C to a few degrees over 40°C. Most of them are mesophilic and grow best at 25–30°C, whereas psychrotrophic yeasts have their optimal growth

Table 5.3 Growth characteristics of some food spoilage yeasts[1]

Species	Lag time (hours)	Generation time (hours)
Debaryomyces hansenii	3	0.8
Hanseniaspora uvarum	3	0.9
Zygosaccharomyces bailii	4	1.4
Schizosaccharomyces pombe	4	2.4
Galactomyces geotrichum	10	3.6

[1] Aerated cultures in broth at pH 4 and 30°C.

Table 5.4 Maximum growth temperatures of some food-borne yeasts[1]

Species	Temperature (°C)
Kluyveromyces marxianus	46
Pichia guilliermondii	41
Metschnikowia pulcherrima	38
Candida zeylanoides	33
Leucosporidium sottii	23

[1] Average values of 7–16 strains measured in synthetic medium.

Table 5.5 Effect of temperature on the growth rates of wine yeast

Yeast species	Doubling time (hours) at temperatures		
	10°C	15°C	25°C
Saccharomyces cerevisiae	17.3	8.7	4.1
Hanseniaspora uvarum	11.6	5.8	3.3
Candida stellata	17.3	5.8	5.3
Torulaspora delbrueckii	23.1	9.9	4.3
Issatchenkia orientalis	69.3	23.1	8.7

Source: adapted from Charoenchai *et al.* (1998).

temperature below 20°C. A few yeasts can grow only at temperatures up to 45–47°C, and not even these could be considered true thermophiles (Table 5.4). The growth rate decreases with temperatures departing from optimum (Table 5.5).

Most yeasts are more tolerant to reduced water activity (a_w) than bacteria, and the majority of food spoilage yeasts possess a minimum a_w value of 0.90 to 0.95 for growth. However, some species, e.g. *Zygosaccharomyces rouxii*, can grow at a_w below 0.70, in the presence of a high concentration of either sugar or salt. These are usually referred to as osmophilic yeasts; a more accurate term is xerotolerant (Table 5.6).

Table 5.6 Minimum water activity values for growth of some food-borne yeasts[1]

Species	Limiting water activity (a_w) in solution of	
	Glucose	NaCl
Candida versatilis	0.79	0.84
Debaryomyces hansenii	0.84	0.84
Hanseniaspora uvarum	0.90	0.95
Torulaspora delbrueckii	0.86	0.90
Zygosaccharomyces rouxii	0.79	0.86

[1] Measured in broth cultures incubated at room temperatures for up to 120 days.

Yeasts tolerate a wide range of pH and grow readily at values between 3 and 10. In general, yeasts prefer a slightly acidic medium and have an optimum pH between 4.5 and 5.5. They show a remarkable tolerance of pH, and several species can grow at acidic pH values as low as 1.5.

The impact on growth of each ecological factor is influenced by others, e.g. at low a_w the minimum temperature for growth increases, or at low temperatures both the minimum pH and the minimum a_w permitting growth are higher. This is the basis for various combined preservation methods. In this respect, the interactions of yeasts with other organisms in microbial communities also play an important role, although the practical significance of biological interaction is not yet fully appreciated. Yeasts and lactic acid bacteria occur simultaneously in many natural habitats and food systems because they have many common ecological determinants.

5.4 Diversity and frequency of food spoilage yeasts

Yeasts are a frequent cause of spoilage of foods rich in fermentable carbohydrates, such as fruits, fruit juices, and soft drinks. Foods with properties controlling bacterial activity may be spoiled by yeasts, e.g. alcoholic beverages, and low pH or high sugar (low a_w). The type of foods and their intrinsic and extrinsic characteristics determine the species of yeasts of appropriate physiological attributes that may occur on or in them.

By and large, some 50 to 100 species of yeast occur most frequently in foods. However, their diversity and frequency are different in various foods. The largest variety of yeasts, some 100 species, is found on fruits, fruit juices, soft drinks and wine; 60 to 80 different species have been isolated from vegetables, dairy and meat products, whereas a restricted species diversity, 30 to 40 species, can be found in fermented, acidified, brined, salted and high sugar products (Table 5.7). These are, however, overall figures on species diversity, and each given food actually contains a much smaller number of yeast species, 10 to 20, and of these just a single or a few species are dominant and can be responsible for spoilage.

Based on extensive surveys, the two most frequent yeast species in foods are *Saccharomyces cerevisiae* and *Debaryomyces hansenii* (or its anamorph *Candida famata*), making up about 7% each of yeasts isolated from various products. Further, very frequent species are *Pichia anomala*, *P. membranifaciens*, *Torulaspora delbrueckii* and *Issatchenkia orientalis*, each found on average in 3–5% of cases. In foods, these species mostly occur in their anamorphic forms, respectively *Candida pelliculosa*, *C. valida*, *C. colliculosa* and *C. krusei.* Of the basidiomycetous yeasts, the most frequently observed are the red yeasts, *Rhodotorula glutinis* and *R. mucilaginosa.* Some yeasts are ubiquitous and cosmopolitan, such as *Saccharomyces cerevisiae* and *Pichia anomala,* whereas others are typical of specific types of food, e.g. *Debaryomyces hansenii* in meat and meat products, and *Zygosaccharomyces*

Table 5.7 Most frequent spoilage yeasts in foods[1]

Species	Calculated frequencies (%) of occurrence in				
	All foods	Fruits, beverages	Meat, dairy products	Low a_w products	Low pH products
Saccharomyces cerevisiae	7.0	6.4	6.7	7.5	7.4
Debaryomyces hansenii	6.7	4.6	8.7	5.6	4.7
Pichia anomala	4.6	4.3	4.2	3.5	5.5
Pichia membranifaciens	4.3	4.4	3.3	3.5	4.0
Rhodotorula glutinis	4.2	2.6	5.9	1.6	1.8
Torulaspora delbrueckii	3.6	4.7	2.0	7.5	2.1
Kluyveromyces marxianus	3.4	2.2	3.9	4.1	0.8
Issatchenkia orientalis	3.2	3.2	2.6	1.8	4.0
Zygosaccharomyces bailii	3.0	4.8	1.1	4.9	5.5
Zygosaccharomyces rouxii	2.7	3.2	1.7	9.4	4.2

[1] Frequencies calculated from records on 62–100 species.
Source: adapted from Deák and Beuchat (1995).

rouxii in foods of low a_w. *Hanseniaspora uvarum* (anamorph *Kloeckera apiculata*) can be found almost exclusively on fruits.

Yeast spoilage manifests itself by visible growth, slime or film on the surface of products, by production of gas, developing off-odours, off-flavours, seldom softening in texture, and causing turbidity and sediment in liquid products.

5.4.1 Fruit juices and soft drinks

For their low pH (2.5 to 4.0), these products are subjected to spoilage most frequently by yeasts. Besides yeasts, certain moulds and lactic acid bacteria may cause spoilage in juices and soft drinks. The composition and production of these beverages are, however, very different. They may be preserved by pasteurization, addition of preservatives, or aseptic bottling. Fruit-based juices are rich in nutrients and are more susceptible to yeast spoilage; carbonated beverages, on the other hand, are deficient in growth factors, and carbon dioxide also exerts a protective effect. The most frequent spoilage species are *Saccharomyces cerevisiae*, the carbonate-resistant *Dekkera anomala* and *Brettanomyces naardenensis*, and the preservative-resistant *Zygosaccharomyces bailii*.

5.4.2 Alcoholic beverages

From a spoilage point of view, the most decisive property of alcoholic beverages is the concentration of ethanol. That is 8–14% (v/v) in wines and 3–5% (v/v) in beers. Another important property is their low pH (3.0–4.0). In addition, wines may contain some sulphur dioxide and/or sorbic acid as a preservative, whereas in beers the constituents of hops exert a preservative effect. Beers are usually pasteurized after bottling or canning. Nevertheless, some yeast may cause turbidity or settlement in bottles. *Saccharomyces cerevisiae* and *Zygosaccharomyces bailii* are most frequent in wine, whereas *S. cerevisiae* as well as *Dekkera* and *Brettanomyces* yeasts are most frequent in beer. On the surface of wine in barrels when not filled completely, a film may form of mainly aerobic yeasts such as *Candida vini*, *Issatchenkia orientalis* and *Pichia membranifaciens*, that oxidatively degrade ethanol and produce by-products of unpleasant flavour. In the surface film, acetic acid bacteria join frequently with yeasts.

5.4.3 Dried foods and high-sugar foods

Though these groups represent highly varied products, their common attribute is their low water content (10–35%) resulting in low a_w (0.70–0.85). At these water activities, growth of most bacteria is inhibited. Hence, in products such as jams, syrups, condiments, honey, dehydrated or dried fruits and vegetables, as well as spices, seasonings, dried soups, flours, etc., shelf-life is threatened only by certain xerotolerant moulds and yeasts. Of the latter, the leading species is *Zygosaccharomyces rouxii*; in addition, *Candida etchellsii*, *C. lactis-condensi*, *C. versatilis*, *Torulaspora delbrueckii* and *Debaryomyces hansenii* may be able to grow in foods of low a_w.

5.4.4 Acidic foods

In addition to fermented pickles and acetic acid-preserved vegetables, mayonnaise, ketchup and salad dressings as well as certain delicatessen products and *hors d'ouevres* may belong to this group of foods whose main preserving factor is their low pH adjusted by the addition of acetic, lactic or citric acid. Low pH is, however, less inhibitory to lactic acid bacteria and yeasts. In these products, the most frequent spoilage yeast species are *Candida etchellsii*, *C. parapsilosis*, *Pichia anomala*, *Issatchenkia orientalis*, *Debaryomyces hansenii*, and the most notorious of all, *Zygosaccharomyces bailii.*

5.4.5 Dairy products

Varied as they are, dairy products represent specific ecological niches for yeasts. Factors determining growth are the relatively low pH, high salt concentration and low storage temperature. Under these conditions, only yeasts can grow that are able to use lactose and possess proteolytic and/or lipolytic activity. Species with these properties are *Debaryomyces hansenii*, *Kluyveromyces marxianus*, *Yarrowia lipolytica*, *Trichosporon moniliforme* and *Galactomyces geotrichum.* These species are frequent contaminants of fermented dairy products and cheeses, and contribute to spoilage although they are not its primary cause.

5.4.6 Meat, meat products, fish and seafoods

These foods are rich in nutrients, have high water content and neutral pH, and consequently are spoiled usually by various bacteria. Yeasts may prevail under special conditions, in particular in marinated, fermented or ripened products as well as in vacuum-packed and chilled processed products. Frequent species that may sometimes initiate spoilage are *Candida zeylanoides*, *C. intermedia*, *C. parapsilosis*, *Debaryomyces hansenii* and *Trichosporon moniliforme*.

5.5 Factors affecting the inactivation of spoilage yeasts

A wide variety of physical and chemical agents can be used to bring about inhibition of growth or inactivation and death of yeasts. The effectiveness of individual agents is markedly dependent on the dose (concentration) and time of exposure. The agents also interact with each other and influence their efficacy. All these factors are of great practical importance.

5.5.1 Heating

High temperatures bring about inactivation of yeast cells. Yeasts are usually killed within a few minutes at temperatures over 55°C. Unlike bacterial endospores, the ascospores or basidiospores of yeasts are only slightly more resistant to heat than vegetative cells. The decimal reduction time (*D* value) at 55°C is about 5 to 10 min, at 65°C less than 1 min. The death rate increases tenfold when the temperature rises by 4 to 5°C, i.e. the *z*-value is 4 to 5°C. These are average values only, and the composition of the food has a substantial effect on the rate of inactivation. For example, the *D* values of *Pichia anomala* in fruit juices were about 6 min at pH 3.95, 3 min at pH 3.0, and 2 min at pH 2.62. The heat resistance of cells is smaller at acidic pH values, and increases with lowering a_w. Hence, fruit juices (pH about 3.5) can be preserved by pasteurization temperatures, whereas the same heat treatment may be insufficient for the preservation of jams of the same pH but containing 55% or more sugar.

5.5.2 Freezing

Freezing does not cause immediate death of yeast cells, though the number of survivors in a frozen state decreases with time. The degree and rate of death caused by freezing depend on a number of factors, such as the temperature of freezing, the rate of temperature decrease, the time spent in the frozen state, and the conditions of thawing. In general, the faster the rate of freezing and thawing, and the lower the temperature of the frozen state, the higher the ratio of survivors. This is due to the formation of ice microcrystals that lessen the destruction of cells. Under these conditions the cell membranes are exposed for a shorter time to the destructive effect of increased osmotic pressure. Hence, in practice, strains in culture collections can be stored at −80°C for a long time, and an even better way to preserve strains is by very fast deep freezing at the temperature of vapour of liquid nitrogen (−196°C).

5.5.3 Dehydration

Decreasing water activity beyond limits permitting growth can be brought about in many ways. Drying is one of the oldest methods of food preservation, and it is still used successfully to inhibit growth of yeasts on fruits and vegetables (e.g. raisins, nuts and pulses). High concentrations of sugar (50–60%) or salt (5–10%) also bring about binding of free water and prevent yeasts from multiplying (e.g. in jams, syrups, ham and soy sauce). Some xerotolerant (osmophilic) yeasts, such as species of *Zygosaccharomyces*, may grow slowly, particularly if moisture is absorbed at the surface of products.

The water activity falls below 0.70 in these dried and concentrated foods. A great variety of foods, e.g. cheeses, sausages and bakery products, have a_w values in the range 0.85 to 0.95. These are called intermediate moisture foods, and their keeping quality depends on the application of other preservation methods such as chilling, vacuum packaging, modified atmosphere, addition of preservatives and combinations thereof.

5.5.4 Irradiation

Various kinds of ionizing radiation such as electron beam, X-rays and gamma rays from ^{60}Co have high energy and cause extensive death of cells. Irradiation of foods has long been recognized as a potent preservation technique. Extensive research and technical experiences have clearly established its benefits and limitations. Its use has been approved in an increasing number of countries, although there remains a certain degree of consumer resistance towards irradiated foods.

No other preservation technology has been evaluated so thoroughly as irradiation, and a large body of literature exists concerning its biological effects and applications. Only a few data concerning yeasts will be mentioned here. The radiation resistance of some yeasts is given in Table 5.8. Yeasts are more resistant to ionizing radiation than the most vegetative bacteria. Their decimal reduction value falls in the range 0.1–0.5 kGy, and a radiation dose of about

Table 5.8 Radiation resistance of some yeasts

Yeast species[1]	Irradiation medium	Irradiation dose (kGy)[2]
Sporidiobolus pararoseus	Nutrient broth	5
Issatchenkia orientalis	Phosphate buffer	5.5
Debaryomyces hansenii	Grape juice	7.5
Cryptococcus albidus	Grape juice	10
Torulaspora delbrueckii	Grape juice	15
Saccharomyces cerevisiae	Grape juice	18
Trichosporon pullulans	Phosphate buffer	20

[1] Initial cell number 10^6–10^7 per ml.
[2] Dose required to prevent growth for at least 15 days.
Source: data adapted from Farkas (1999).

5 kGy would reduce their number by 10 log cycles. However, survival curves often show 'shoulders' that render the calculation of the inactivation dose less straightforward. Some yeasts, e.g. *Trichosporon* species, and some strains of a given species appear to bear higher radiation resistance.

In commercial application gamma irradiation at relatively mild doses (1–3 kGy) decreases significantly the yeast contamination of perishable fruits and increases their storage time. Irradiation is especially suited for extending the shelf-life of soft fruits (e.g. strawberry, raspberry) that are sensitive to other preservation methods.

5.5.5 Preservatives

Weak organic acid preservatives (sorbic, benzoic and propionic acids as their salts), in concentrations permitted in foods, inhibit growth rather than kill cells. Effectiveness of these preservatives is greatest at low pH values. Potassium sorbate is more effective against yeasts than is sodium benzoate. Some yeast species, notably *Zygosaccharomyces bailii*, possess resistance to preservatives, being able not only to adapt to high concentrations of preservatives but even to abolish inhibition by metabolizing and decomposing them. Sulphur dioxide can be used for temporary preservation of fruit pulps. In wine fermentation sulphite is added for the inactivation of wild yeasts (certain species of *Pichia* and *Candida*) in must, because wine yeast, *S. cerevisiae*, is less sensitive to SO_2.

5.5.6 Combined preservation systems

Preservative factors can be combined using milder treatments to retain the nutritional value and organoleptic quality while assuring the required safety of products. The combination of preservative factors is often called hurdle technology and is illustrated with several 'hurdles' that the microorganisms are not able to overcome. This analogy is inadequate in that 'hurdles' are combined simultaneously and interact synergistically. The combination of treatments has resulted in innovative techniques and products including minimal processing, modified atmosphere packaging and intermediate moisture foods.

Minimally processed foods are mildly heated, contain less acid, salt and sugar, and in particular less preservatives such as sulphite, nitrite, benzoic or sorbic acid, but are usually refrigerated (Ohlsson, 1994). The combination of factors should protect the product from growth of pathogenic bacteria but may not inhibit spoilage microorganisms including yeasts. Typical examples are ready-to-use fruits and vegetables processed by trimming, peeling or cutting, washing, disinfecting and packaging. Blanching may be used to control enzymatic browning and to reduce counts of yeasts, moulds and bacteria (Nguyen-the and Carlin, 1994).

Intermediate moisture foods (IMF) generally have an a_w below 0.90 and are shelf-stable when combined with low pH, pasteurization or refrigeration. Besides various meat products and cheeses, fruit preserves are good examples of IMF products. Whole fruits or halves, slices or purées can be preserved by blanching and adjusting the a_w in the range 0.91–0.98 with sugar. The pH of most fruits is between 3.1 and 3.5; it can be reduced if necessary with the addition of citric acid. Sulphite and sorbate can be also added. Under these conditions, the potential spoilage organisms are yeasts, in particular osmophilic species.

Modified atmosphere packaging (MAP) is frequently used for the preservation of fresh and minimally processed foods. For meat products, reducing the oxygen content within the package can be achieved by applying vacuum, whereas the atmosphere is modified by the ongoing respiration of fresh fruits and vegetables after sealing in semipermeable film or shrink wrapping. Levels of oxygen may fall to 3–5% while CO_2 concentration increases to 5–10%. In such an atmosphere growth of aerobic bacteria and moulds is greatly reduced but activity of yeasts and lactic acid bacteria can continue at a slower rate even if MAP is combined with refrigeration.

5.5.7 Injury and repair

An important consideration for the application of milder treatments in combination is the possibility of survival of microorganisms. Sublethal treatments may not kill yeast cells but may only cause injury to cell structure and/or metabolism. Cell membranes are subjected to damage causing leakage of cell constituents and loss of transport processes. Enzyme activities, pathways of enzyme synthesis and regulation of metabolism, as well as nucleic acid replication and transcription, may be the target of injuries. Cells suffering injury become more susceptible to environmental factors and can be inactivated by further treatment. However, when conditions permit and with time, cells may be able to repair injury and start to propagate.

The practical side of these phenomena is manifold. Firstly, they provide the basis for combination technologies, i.e. the simultaneous and joint application of mild preservative treatments in doses that alone would not be effective. Secondly, for preparing bulk amounts of industrially used microorganisms (e.g. dried bakers' yeast, frozen dough) methods should be used to avoid cell injuries. Thirdly, at the assessment and control of microbial quality of food, false results may be obtained when cultivation conditions (e.g. direct plating on selective medium) would not permit the growth of injured cells (see Chapter 13).

5.6 Future trends: alternative technologies

In recent years, the growing demands for safer, fresher, more nutritious and novel food products have stimulated research into alternative preservation technologies. Most of these apply physical principles other than heat treatment, such as high pressure, pulsed electric field, oscillating magnetic field and ultrasound, alone or in combination with pasteurization, refrigeration or freezing. Certain chemical treatments such as use of chitosans, bacteriocins and other antimicrobials have been tested as well. So far, experiments have been conducted on a laboratory or pilot scale with little commercial application as yet. The microbiological studies have focused mostly on pathogenic bacteria but yeast cells often serve as eukaryotic model organisms.

5.6.1 Novel non-thermal physical treatments

High hydrostatic pressure

High pressure over 100 MPa (megapascals) damages cell membranes and may inactivate vegetative cells. Eukaryotic yeast cells are more sensitive than prokaryotic bacteria, bacterial spores being particularly resistant. Pressures of 400–600 MPa bring about inactivation of microorganisms, but this is incomplete and some cells may survive injury. Water activity is a critical factor in the inactivation of yeasts while pH does not influence survival (Table 5.9). Although equipment cost and throughput limit commercial application, use of high-pressure processing is increasing (Yuste *et al.*, 2001). Application of high-pressure technology is most advanced for preservation of fruit juices, jams, jellies and sauces, products that are prone to spoilage by yeasts.

Pulsed electric field

Electric fields with intensities between 5 and 25 kV/cm, applied for several tens of pulses each with a duration of microseconds, can inactivate microorganisms. Yeasts are killed more easily than bacteria by electric fields and inactivation increases greatly with field strength and number of pulses. Electric treatment causes clear damage in cell structure as evidenced in electronmicroscopic photographs. The technique has potential in preservation of liquid foods such as fruit juices and milk, without causing adverse organoleptic changes.

Oscillating magnetic fields

Magnetic fields with intensities high enough to inactivate microorganisms can be generated by electric coils under current. For satisfactory preservation by

Table 5.9 Decimal reduction time (*D*) values for high-pressure inactivation of *Saccharomyces cerevisiae* ascospores

Pressure (MPa)	*D*-values (min.)			
	Orange juice pH 3.8	Apple juice pH 3.9	Model juice pH 3.5	Model juice pH 5.0
500	0.18	0.15	0.14	0.19
450	0.50	0.48	0.38	0.49
400	0.97	0.88	0.72	0.98
350	2.80	2.51	2.15	2.84
300	10.81	9.97	7.21	9.42

Source: adapted from Zook *et al.* (1999).

magnetic fields, foods of high electrical resistivity are best suited. Successful experiments have been conducted with orange juice and yoghurt inoculated with *Saccharomyces cerevisiae.* However, more detailed studies are needed before this technology can be developed commercially.

Light pulses

Ultraviolet light has an antimicrobial effect due to damage of the DNA bringing about lethal mutation. UV irradiation has been used in particular for the sterilization of air and for the treatment of drinking water. An extension of ultraviolet light to the near-infrared region of wavelengths applied in intense short pulses constitutes the basis of a technology that can be used for sterilization of food surfaces and packaging materials as well as transparent liquids in thin layers. Yeast cell counts can be reduced by five to six log cycles using full-spectrum light with intensities of 0.1–0.4 J/cm^2 for only two to four flashes. Moulds and bacteria are apparently more resistant. The technology is an attractive alternative in aseptic processing.

Ultrasound

Ultrasound technology (ultrasonics) applies high energy sound waves of 20 000 or more vibrations per second (kHz). High-power ultrasound has a microbicid effect caused by intracellular cavitation that disrupts cellular structure and function. Food materials may interfere with sound wave penetration, and ultrasound must be combined with other preservation methods for safe application. Ultrasonics appears to have the greatest potential for product decontamination and cleaning of processing equipment. The cleaning action of cavitation appears to remove cells from the surface of raw and minimally processed fruits and vegetables. No commercial products have as yet been released.

5.6.2 Antimicrobial chemicals

In addition to novel physical methods to inactivate microorganisms in technological use, the search for new chemical and biochemical compounds is proceeding with promising results. Natural antimicrobials such as lysozyme, nisin and other bacteriocins demonstrate antagonistic effects on bacteria but generally have no action against fungi. Reuterin, a metabolite of *Lactobacillus reuteri,* has a broad spectrum of activity and inhibits both Gram-positive and Gram-negative bacteria as well as fungi. With the exception of nisin that has been used as a biopreservative in a range of foods, no other bacteriocin has yet found application.

Chitosan, the deacylated derivative of chitin, exhibits antimicrobial activity in particular against filamentous moulds and yeasts. Its use in concentrations of 0.3–1.0 g/L has been investigated as a possible means of extending the postharvest shelf-life of fresh fruits and vegetables. In fruit juices and mayonnaise-based salads, however, yeast growth was not completely suppressed (Rhoades and Roller, 2000; Roller and Covill, 2000).

Several reports refer to the antimicrobial effects of plant extracts containing various natural substances such as essential oils and oleoresins (Nychas, 1995). Extracts from garlic, onion and horseradish, as well as spices (cinnamon, clove, mustard, oregano and others) may show antibacterial or antifungal effect or both. However, they impart strong flavour at effective concentrations which is not compatible with most food products. The effectiveness of plant extracts can be enhanced when they are used in combination with other preservative factors. For example, Cerutti and Alzamora (1996) demonstrated that vanillin in 0.2% concentration inhibited the growth of important spoilage yeasts (*Saccharomyces cerevisiae*, *Zygosaccharomyces bailii*, *Z. rouxii* and *Debaryomyces hansenii*) in apple purée at pH 3.5 and a_w 0.95 for 40 days' storage at 27ºC.

5.6.3 Novel combinations

Many of the emerging alternative methods are being developed for use as part of a hurdle technology that involves combining more than one method with synergistic effect. As in the case of irradiation, alternative physical treatments in combination with other preservation methods can be used in lesser doses than are required alone to obtain optimal product quality and microbiological safety. High pressure, pulsed electric fields, ultrasound and oscillating magnetic fields have been suggested in combination with pasteurization or freezing without decreasing the efficacy of treatment. Combined technologies such as mano-thermo-sonication (low pressure + mild heat + ultrasonics) or osmo-dehydro-freezing (sugar addition + air drying + freezing) have been proposed for protection of fruit juices and fruit slices from yeast spoilage while retaining flavour, colour and functional properties. The utilization of novel interactions to prevent the growth of spoilage as well as of pathogenic microorganisms and to assure the quality and safety of foods is a complex and challenging task in food technology.

5.7 Sources of further information

BARBOSA-CÁNOVAS G V, POTHAKAMURY U R, PALOU E and SWANSON BG (1998) *Nonthermal Preservation of Foods.* Marcel Dekker, New York.

DEÁK T and BEUCHAT L R (1995) *Handbook of Food Spoilage Yeasts.* CRC Press, Boca Raton, FL.

GOULD G W, ed. (1995) *New Methods of Food Preservation.* Blackie Academic and Professional, London.

5.8 References

CERUTTI P and ALZAMORA S M (1996) 'Inhibitory effects of vanillin on some food spoilage yeasts in laboratory media and fruit purées'. *Int. J. Food Microbiol.* 29, 379–386.

CHAROENCHAI C, FLEET G H AND HENSCHKE P A (1998) 'Effects of temperature, pH and sugar concentration on the growth rates and cell biomass of wine yeasts'. *Am. J. Enol. Vitic.* 49, 283–288.

DEÁK T and BEUCHAT L R (1995) *Handbook of Food Spoilage Yeasts.* CRC Press, Boca Raton, FL.

FARKAS J (1999) 'Microbiological consideration'. In: *High-dose Irradiation: Wholesomeness of Food Irradiated with Doses above 10 kGy* (Report of a joint FAO/IAEA/WHO study group), World Health Organization, Geneva, 49–77.

NGUYEN-THE C and CARLIN F (1994) 'The microbiology of minimally processed fresh fruits and vegetables'. *CRC Crit. Rev. Food Sci. Nutr.* 34, 371–401.

NYCHAS G J E (1995) 'Natural antimicrobials from plants'. In: Gould G W (ed.) *New methods of food preservation.* Blackie Academic and Professional, London, 58–89.

OHLSSON T (1994) 'Minimal processing – preservation methods of the future: a review'. *Trends Food Sci. Technol.* 5 (12), 341–344.

RHOADES J and ROLLER S (2000) 'Antimicrobial actions of degraded and native chitosan against spoilage organisms in laboratory media and foods'. *Appl. Environ, Microbiol.* 66, 80–86.

ROLLER S and COVILL N (2000) 'The antifungal properties of chitosan in laboratory media and apple juice'. *Int. J. Fd. Microbiol.* 47, 67–77.

YUSTE J, CAPELLAS M, PLA R, FUNG D Y C AND MOR-MUR M (2001) 'High pressure processing for food safety and preservation: a review'. *J. Rapid Methods Automation Microbiology,* 9, 1–10.

ZOOK C D, PARISH M E, BRADDOCK R J AND BALABAN M O (1999) 'High pressure inactivation kinetics of *Saccharomyces cerevisiae* ascospores in orange and apple juices'. *J. Food Science* 64, 533–535.

6

Factors affecting the Maillard reaction

A. Arnoldi, University of Milan, Italy

6.1 Introduction: the Maillard reaction (MR)

Contrary to most animals, which consume only fresh foods, human societies are mainly based on foods that are produced a long time before they are consumed. Therefore, it is not a surprise that the preservation of food from spoilage is a major problem. The use of fire represented a tremendous innovation in human history, because it permitted an increase in the shelf-life of foods by reducing drastically contamination by microorganisms. In addition it enlarged dramatically the variety of starting materials available as foodstuffs, by reducing the presence of many antinutritional factors and by increasing digestibility, and has provided in addition pleasant new flavours and tastes.

Many chemicals induced in foods by heating are related to the Maillard reaction (MR), a term which indicates all the modifications involving free or protein-linked amino acids and sugars that affect to a great extent the sensory and nutritional characteristics of foods (Arnoldi, 2001). This chapter will consider the MR, underlining mainly the aspects that are related to food shelf-life. After two sections dedicated to the general mechanism and the factors that mainly affect the kinetics of MR, the negative effects of the MR on food spoilage will be extensively discussed. The following two sections will be dedicated to some favourable effects of the MR, in particular the formation of antioxidative compounds, which greatly contribute to reducing lipid autoxidation, and antimicrobial compounds, which may reduce the microbial spoilage of foods.

6.1.1 Mechanisms underlying the Maillard reaction

The term Maillard reaction (MR) is used to indicate a cascade of complex competitive reactions involving on one side amino acids, peptides and proteins and, on the other, reducing sugars. Another interchangeable term used for indicating the same phenomenon is non-enzymatic browning, since colour formation is a major consequence of these processes.

The first observations on the MR were made about 90 years ago by Maillard (1912). After about 20 years, Amadori was able to identify a rearranged stable product deriving from the Schiff base **1** coming from the first sugar/amino acid interaction, which was named Amadori rearrangement product **2** (Fig. 6.1) (ARP) (Amadori, 1931), whereas the corresponding compound from fructose **3** was described only 30 years after by Heyns and Noack (1962). A detailed description of the synthetic procedures, physico-chemical characterisation, properties and reactivity of the ARPs may be found in an excellent review by Yaylayan and Huyggues-Despointes (1994). The first general review on the MR was published by Hodge (1953), while one of the most detailed descriptions of the pathways leading to the main MR products (MRPs) can be found in the excellent review by Ledl and Schleicher (1990).

Reducing sugars are indispensable for the MR: pentoses, such as ribose, arabinose or xylose, although generally not very abundant in foods, are very effective in non-enzymatic browning, whereas hexoses, such as glucose or fructose, are less reactive, and reducing disaccharides, such as maltose or lactose, react rather slowly. Sucrose as well as bound sugars (for example glycoproteins, glycolipids, and flavonoids) are involved only after hydrolysis, induced by heating or, quite often, by fermentation, as in dough leavening or

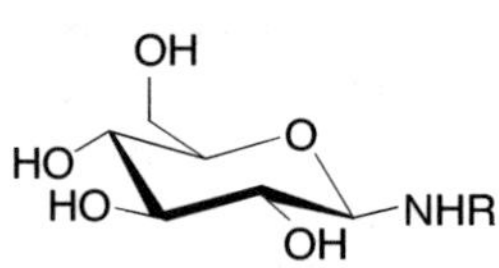

1-amino-1-deoxyaldose

1

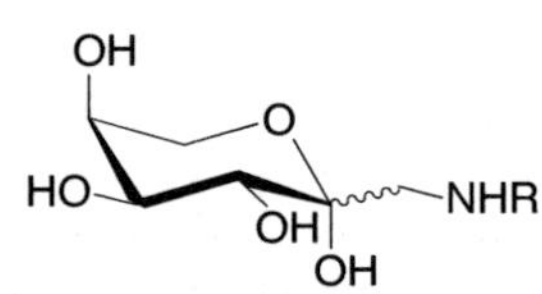

1-amino-1-deoxyketose
Amadori rearranged product

2

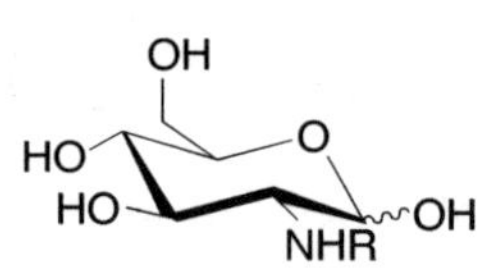

2-amino-2-deoxy-D-glucose
Heyns rearranged product

3

Fig. 6.1 First sugar/amino acid interactions: 1 Schiff base, 2 Amadori rearranged product, 3 Heyns rearranged product.

cocoa bean preparation before roasting (Ledl and Schleicher, 1990). The counterparts are proteins or free amino acids, already present in the raw material or produced in their turn by fermentation. In some cases (e.g. cheese) biogenic amines can react as amino compounds, whereas small amounts of ammonia may be produced from amino acids during the MR.

In proteins the most relevant effect of the MR is non-enzymatic glycosylation, which involves mostly lysine and takes place easily even at physiological temperature and has important health consequences (Arai, 2002; Baynes, 2001; Faist and Erbersdobler, 2002; Heidland *et al.*, 2001; Kirkland, 2002). The first glycosylation products are then converted to the Amadori product fructosyllysine that can cross-link intra- or intermolecularly. The resulting polymeric aggregates are called Advanced Glycation End products (AGEs).

With low water content and pH values in the range 3 to 6, ARPs are considered the main precursors of reactive intermediates in model systems, whereas below pH 3 and above pH 8 or at temperatures above 130ºC (caramelisation), sugars degrade also in the absence of amines (Ledl and Schleicher, 1990). Ring opening followed by 1,2 or 2,3–enolisation are crucial steps in the degradation of ARPs and are followed by dehydrations and fragmentations with the formation of many very reactive dicarbonyl fragments. This is considered the intermediate stage of the MR.

One of the first observations of Maillard was the production of CO_2, which derives from the Strecker degradation (Fig. 6.2). An amino acid reacts with an α-dicarbonyl compound to produce an azovinylogous β-ketoacid **4** that undergoes decarboxylation. By this process amino acids are converted to aldehydes containing one fewer carbon atom, which are very reactive and often have very peculiar sensory properties, not always particularly pleasant. The most important consequence of the Strecker reaction is the incorporation of nitrogen in very reactive low-molecular-weight compounds deriving from sugars, such as **5**, which are intermediates for the formation of many heterocyclic compounds, with distinctive aromas and low odour thresholds, such as pyrazines (Ledl and Schleicher, 1990).

Depending on food composition and the process applied, the MR produces thousands of different end products, which may be classified taking into account their role in foods. Thus very volatile compounds, such as pyrazines, pyridines, furans, thiophenes, thiazoles, thiazolines, and dithiazines, are relevant for aroma, some low-molecular-weight compounds are relevant for taste (Frank and Hofmann, 2002; Ottiger *et al.*, 2001), others behave as antioxidants, a few are mutagenic (Jägerstad *et al.*, 1998), whereas the brown polymers named melanoidins, that in some foods such as coffee, roasted cocoa beans, malt, or soy sauce, are the major MRPs, are responsible for food colour (Arnoldi *et al.*, 2002). The nutritional consequences of the MR have been recently reviewed in another book of this series (Arnoldi, 2002). This chapter will consider only compounds related to the shelf-life of foods.

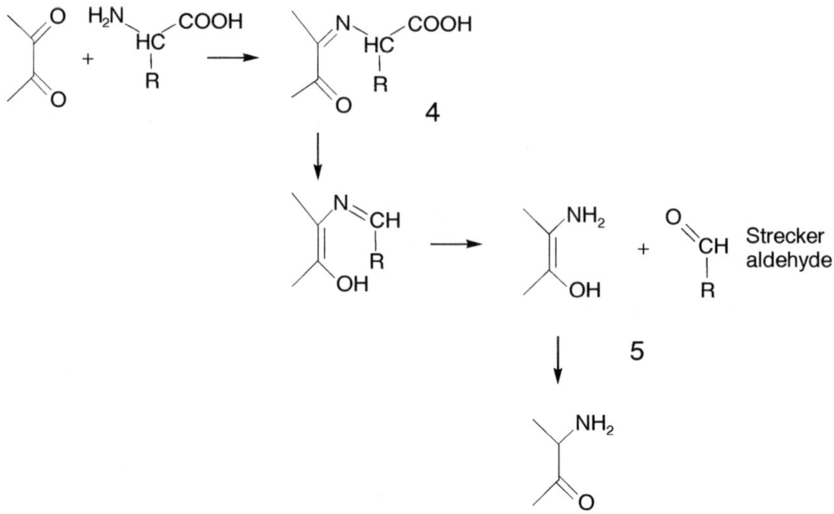

Fig. 6.2 Strecker degradation.

6.2 Factors affecting the Maillard reaction

The most important parameters that affect the MR are the structure of amino acids and sugars involved, the temperature, the pH value, and water activity. In a very complex process, such as the MR, each food feature, for example colour, flavour or the nutritional value, is affected in a different way by these factors. An important parameter is the structure of the reactants, sugars and amino acids. The relative reactivity of reducing sugars is pentoses > hexoses > disaccharides, and aldoses > ketoses. Non-reducing sugars, such as sucrose, dextrines and bound sugars, can be involved only after hydrolysis.

6.2.1 Amino acids

Although in most foods free amino acids are not very abundant, they react easily with sugars. Aspartic and glutamic acid react relatively slowly, whereas arginine and lysine are very fast. The structure of the side chain of the amino acids determines the formation of particular MRPs through the Strecker degradation. In particular cysteine and methionine give rise to sulfur compounds (Fig. 6.3), characterised by very distinctive aromas, not always pleasant, with very low odour thresholds. In some processes, such as dough leavening, the presence of free amino acids may be enhanced by fermentation. Proteins participate in the MR owing to the glycosylation of the reactive side-chains of lysine and arginine, lysine in particular.

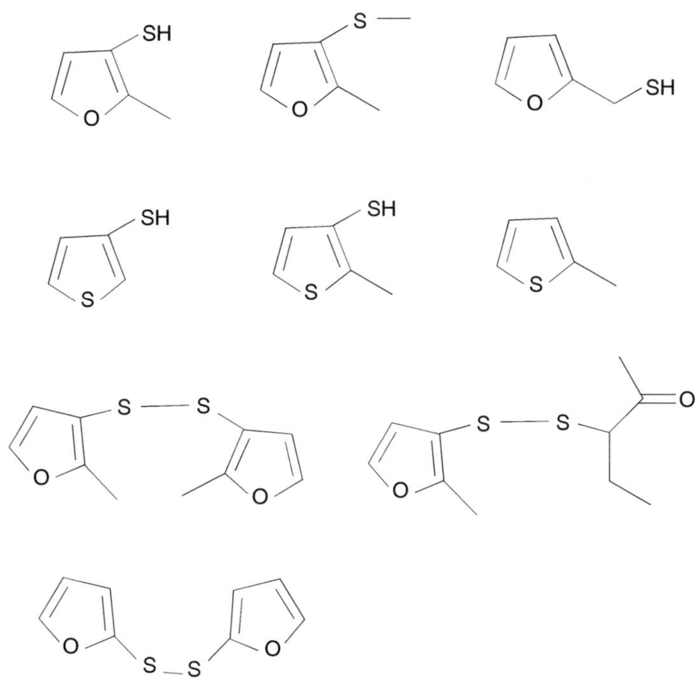

Fig. 6.3 Some sulfur compounds from the Maillard reaction relevant for food aroma.

6.2.2 Temperature

The effect of temperature is particularly strong and involves every aspect of the MR. Numerous experiments have shown that an increase in temperature and/or time of heating leads to an increase of browning and aroma profile. Not only the amount of the MRPs is increased, but also their nature is modified. For example it has been demonstrated that the carbon-to-nitrogen ratio, the degree of unsaturation and the chemical aromaticity of the melanoidins formed in model systems increase with temperature and time (Ledl and Schleicher, 1990). The formation of highly undesirable compounds, such as mutagenic heterocyclic amines (HAs) (Jägerstad *et al.*, 1998), requires high temperatures, which means that a careful choice of the cooking procedures minimises their formation. However, contrary to popular opinion, the MR does not requires high temperature; thus sugars and amino acids even stored at refrigerated temperature can show signs of non-enzymatic browning during storage (Whitfield, 1992), which indicates how relevant the MR may be for food shelf-life.

6.2.3 Water activity

As far as the water content is concerned, the MR proceeds much more readily at low moisture level and it is generally accepted that moisture values corresponding to a water activity around 0.65–0.75 are the most favourable. The differences in colour and aroma between the outer and inner parts of baked or roasted foods correspond to their different dehydration rates. However, not all the MRPs are sensitive in the same way and studies on flavour have demonstrated that different classes of volatile components are more or less sensitive depending on whether or not water is required for their formation. The complex network of reactions that produce browning is far from disclosure, but certainly many condensations and dehydrations are involved and water activity is a critical parameter for their kinetics (Ledl and Schleicher, 1990).

6.2.4 pH

Several studies on MR model systems have demonstrated that the pH of the reaction medium results in qualitative and quantitative changes in volatiles and coloured products (Shibamoto and Bernhard, 1977; Shu and Ho, 1988; Bemis-Young *et al.*, 1993; Meynier and Mottram, 1995). Browning is faster in neutral foods and decreasing pH can reduce the rate of colour formation. The qualitative composition of flavour is influenced, too. For example, pyrazines, reductones and fission products are preferred at high pH values, while furans, especially 2-furancarboxaldehydes, prevail at lower pH.

6.3 The Maillard reaction and spoilage: flavour deterioration

Food flavour is not stable and its deterioration is a major matter of concern for food manufacturers: food staling can be described as a change in the aroma profile due to the loss of low boiling compounds and degradation processes. Storage temperature, penetration of oxygen in the package and loss of volatiles through diffusion are crucial for rancidity, and moisture can accelerate the staling process. All these processes are fundamental for food preservation. Lipid autoxidation is generally considered the main source of off-flavours in food and many MRPs are considered to have a role in its prevention for their antioxidant properties (Alaiz *et al.*, 1997; Antony *et al.*, 2000; Lignert and Eriksson, 1981). However, in specific cases also the MR is a source of off-flavours. These negative aspects of the MR during either food processing or storage have been investigated much less extensively than lipid autoxidation.

The actual impact of these compounds on the sensory features of foods depends not only on their concentration, but also on their specific sensory thresholds, which are distributed over a wide range of values (Fors, 1983). As far as flavour spoilage is concerned, it is useful to divide food items in two groups:

1. Those in which a high level of quality requires that flavour is identical to unprocessed food, although thermal treatments are required for microbiological stabilisation.

2. Those in which the MR is intrinsically necessary for obtaining the typical flavour and texture.

6.3.1 Fruit juices

Fruit juices are a typical example of the former class, in which during storage a slow MR at room temperature is very detrimental for flavour quality. General information about the flavour changes during processing and storage of fruit juices may be found in a review by Askar (1999). For example, the deterioration of flavour accompanied by browning has been a major problem throughout the history of citrus processing (Handwerk and Coleman, 1988). Maintaining the product at low temperature is still the major means of avoiding flavour and colour deterioration. After some time the sensory response becomes typical of aged or heat-abused juice. Although the low pH of these beverages is not particularly favourable for the MR, at least 14 out of 21 compounds detected in the flavour of old orange juice certainly derive from the MR, for example 5-methylfurfural, furfural, 5-hydroxymethylfurfural, 2-(hydroxyacetyl)furan, 2-acetylfuran, 2-acetylpyrrole and 5-methylpyrrole-2-carboxaldehyde (Handwerk and Coleman, 1988). The presence of free amino acids in the juice has a relevant role in this phenomenon and their removal by ion-exchange resins has been proposed to increase stability against deterioration changes.

6.3.2 Dairy products

Another relevant example of the first class of food items is dairy products. Milk processing is necessary to assure microbiological safety and an acceptable shelf-life, but it is also detrimental to flavour quality. The well-known differences between the flavour and taste of pasteurised and UHT-treated milk are related to the MR, and research into milder and milder technologies in milk processing has been promoted to improve both the nutritional and sensory characteristics of drinkable milk. However, the MR is responsible also for the deterioration of milk flavour during storage. Comparison of the flavour of different commercial UHT milk samples, either whole or skimmed, during four months, showed that the sensory characteristics of the latter were slightly worse and that many of the new components are related to both lipid oxidation and the MR, which gives especially furanic compounds (Valero *et al.*, 2000). Other dairy products investigated for the formation of off-flavours from the MR are whey protein concentrates (Morr and Ha, 1991) and milk powder (Stapelfeldt *et al.*, 1997; Renner, 1988).

6.3.3 Beer and lager

The deterioration of the flavour of naturally aged or forced-aged lagers has been studied by chromatographic olfactometry techniques: some compounds responsible for beer ageing are related to the MR, for example methional and phenylacetaldehyde (Evans *et al.*, 1999). Other work (Rangel-Aldao *et al.*, 2002) has shown that beer ageing is related to the accumulation of α-dicarbonylic compounds, which are intermediates of the MR. These intermediates can be reduced in wort, or in finished beer, by a NADPH-dependent oxidoreductase enzyme from brewer's yeast. The pure enzyme displayed activity for α-dicarbonyls, such as 2,3-hexanedione, methylglyoxal and diacetyl, and for oxidative-stress inducing compounds. Thus careful selection and management of yeasts may become a tool for directing the MR during storage of fermented foods.

6.3.4 Roasted peanuts

Considering now food items that intrinsically require thermal processing, an example may be roasted nuts and seeds. An interesting paper by Warner *et al.* (1996) considered the problem of flavour fade and off-flavour formation in ground roasted peanuts, which is a major problem in the confectionery industry. The authors decided to monitor in the headspace for three months some selected pyrazines, such as methylpyrazine, 2,6-dimethylpyrazine, and 2,3,5-trimethylpyrazine, which are key impact compounds of roasted peanut flavour, and some aldehydes produced by lipid autoxidation, such as pentanal, hexanal, heptanal, octanal, and nonanal (peanut oil is rich in polyunsaturated fatty acids, linoleic acid in particular, and the formation of fatty aldehydes, mainly hexanal, was expected). This investigation demonstrated that the concentration of pyrazines remains practically constant, whereas the concentration of aldehydes increases very quickly (for example, the concentration of hexanal in 68 days became 10 times higher than at time zero). Sensory evaluation was completely in agreement with a prevailing contribution of lipid autoxidation to the formation of peanut off-flavour. This investigation may also represent an example for off-flavour formation during storage of other roasted nuts.

6.3.5 Coffee

Another typical example is coffee. Cappuccio *et al.* (2001) investigated the rate of staling in roasted and ground coffee at different temperatures after package opening and tried to correlate the most significant chemical data, selected by discriminant analysis, with a sensory evaluation. Among the most significant compounds, there are some MRPs, such as H_2S, methanethiol, 2-methylpropanal, diacetyl, 2-butanone, 2-methylfurane, 3-methyl- and 2-methylbutanal, 3-methylfuranthiol, and 2-furfurylthiol. There is a decrease especially of sulfur compounds that starts immediately after package opening, while compounds deriving from the degradation of lipids appear only after some days, depending on the temperature. The same results have also been observed in the aroma of coffee brew, which changes shortly after preparation. Similar observations have been made in the manufacture of instant coffee, in the heat sterilisation of coffee beverages, and in keeping freshly prepared coffee brews warm in a thermos flask.

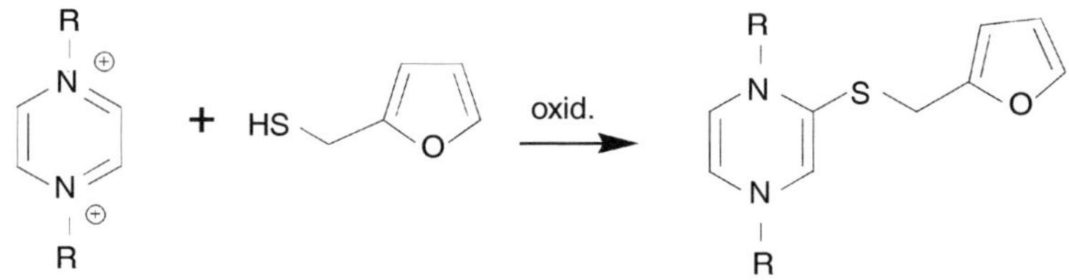

Fig. 6.4 Possible mechanism for the trapping of thiol compounds in melanoidins (Hofmann *et al.*, 2001).

A recent investigation, in which strategies combining instrumental analysis with olfactometry perception have been applied, has revealed a rapid decrease in the concentration of odorous thiols when coffee brews are stored or processed (Hofmann *et al.*, 2001). The results have shown that, in particular, the key coffee odorants 2-furfurylthiol and 2-methyl-3-furanthiol (Fig. 6.4) are significantly reduced (Hofmann and Schieberle, 2002), causing a strong decrease of the sulfury-roasty odour quality in the overall aroma of the coffee beverages. With a very elegant experiment Hofmann *et al.* (2001) have shown that the addition of melanoidins isolated from coffee powder to an aqueous aroma recombinate prepared using 25 coffee aroma compounds in the same concentration as determined in the original coffee brew, reduced the intensity of sulfury-roasty odour quality in the head space. Compounds particularly affected were 2-furfurylthiol, 3-mercapto-3-methylbutyl formate and 3-methyl-2-butene-1-thiol, known as the key thiols in coffee aroma. It was possible to demonstrate that these thiols were covalently bound to melanoidins via Maillard-derived pyrazinium compounds formed as oxidation products of 1,4-*bis*-(5-amino-5-carboxy-1-pentyl)pyrazinium radical cations. A scheme of this reaction is shown in Fig. 6.4 (Hofmann and Schieberle, 2002). This is the first demonstration that flavour staling may be due not only to the loss of key odorants and the formation of rancidity or undesired MRPs at room temperature, but also to the binding of specific compounds to food melanoidins or other polymers.

6.4 The Maillard reaction and spoilage: nutritional losses and browning

Flavour deterioration is certainly the main cause of concern for food manufacturers, though other aspects, such as the decrease of the nutritional value and browning, are relevant too. Fig. 6.5 shows the structures of the main molecular markers that have been used in the literature for measuring the MR.

6.4.1 Milk

Information about the chemical changes occurring during the storage of UHT milk (the MR included) and their importance for sensory properties and

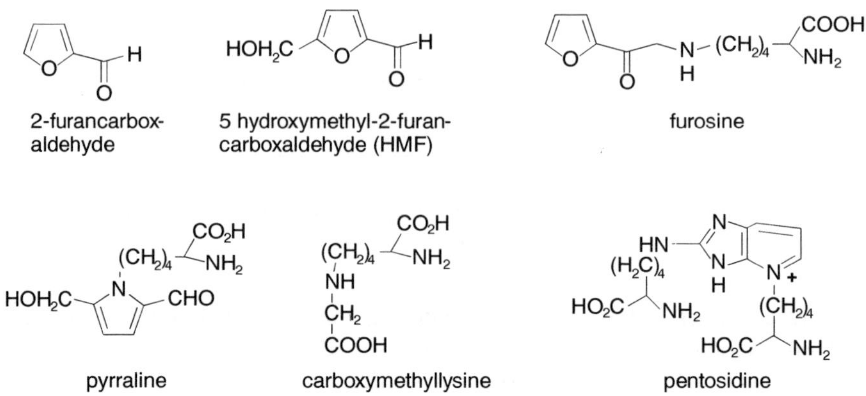

Fig. 6.5 Main molecular markers of the Maillard reaction in food.

nutritional value, in connection with the establishment of uniform legislation within the European Community regarding UHT-milk shelf-life, may be found in a review by Glaeser (1989). The consequences of the MR during storage of processed cheese and ready-made fondue have also been recently reviewed (Schär and Bosset, 2002).

Two recent papers have investigated the progress of the MR in stored infant formulas. In the first (Albalà-Hurtado *et al.*, 1998) a liquid and a powdered infant milk were monitored for nine months by using 2-furancarboxaldehyde, 5-hydroxymethylfurancarboxaldehyde (HMF) and available lysine, as markers. Available lysine is a very important parameter in this kind of foodstuff because these products are the only source of lysine for infants (Erbersdobler and Hupe, 1991). The trends of these parameters, even at 20°C, indicated a continuous deterioration of the quality of the proteins. From these results it was possible to derive an equation for the prediction of the shelf-life of these products.

In the second paper (Guerra-Hernandez *et al.*, 2002), besides furfural and HMF, also furosine and browning, measured by the absorbance at 284 and 420 nm, were used as parameters for measuring the MR during the storage of two differently processed samples of liquid infant milk formulae. Furosine, which is an indirect index of protein-bound lactulosyl-lysine (Erbersdobler and Hupe, 1991), is an useful indicator of heat damage in milk, because it is increased by prolonged heating or inadequate storage. Formula A was prepared by blending, pasteurisation, spray drying, reconstitution, packaging and sterilisation by rotary autoclaving; sample B by blending, UHT, packaging and sterilisation by rotary autoclaving. The target of the work was to correlate the deterioration during storage with the processing and storage conditions, in particular the exclusion of oxygen. Sample B was less damaged during both processing and storage.

6.4.2 Eggs

Furosine was proposed as an index for egg freshness (Hidalgo *et al.*, 1995). Fresh eggs from different hen breeds were used to assess HPLC method repeatability, to investigate natural variability of furosine content in fresh shell eggs, and to study furosine formation during storage at 5, 20, 30, and 38°C. Furosine, present in fresh egg albumen at a level of about 10 mg per 100 g of protein, increased during storage with dependence on temperature, reaching after 40 days at 20°C a level of about 100 mg per 100 g of protein. The modification of furosine content in yolk during storage at 20°C was minimal, whereas furosine, when measured in albumen, resulted in a very reliable index for egg freshness.

6.4.3 Royal jelly

Royal jelly is a viscous substance secreted by the hypopharyngeal and mandibular glands of nurse worker bees (*Apis mellifera* L.) that constitutes the essential food for the larvae of the queen bee. In modern diet the use of royal jelly has significantly increased among demanding consumers who require more 'natural' and healthier foods and nutraceuticals. Furosine has been applied as a suitable marker for assessing the quality and freshness of royal jelly (Marconi *et al.*, 2002). Its value, starting from about 50–80 mg per 100 g protein, increases in 10 months to about 500 mg per 100 g protein, indicating that also in this food, characterised by a high content of reducing sugars and free amino acids and low water activity, the MR is of main concern during storage.

6.4.4 Browning

Browning is another important consequence of the MR. The most important contribution to colour during food processing comes from melanoidins, a polymeric material whose structure is still rather elusive. Browning during storage represents a problem especially in those foods in which an 'unprocessed appearance' is required, again especially fruit juices, preserved vegetables or fruits, and dairy products. However, it is important to say that, as far as food shelf-life is concerned, although browning during storage is very detrimental for the quality of these food items, generally flavour becomes unacceptable before a perceptible colour change takes place (Handwerk and Coleman, 1988). The relevance to dairy products, such as infant milk formulae (Guerra-Hernandez *et al.*, 2002), and in fruit juices (Handwerk and Coleman, 1988) has already been commented on above.

The kinetics of non-enzymic color development in glucose syrups have been studied during 13-week storage periods at 23, 35, 45 and 55°C at three different pH values (4.0, 4.5 and 5.0). Storage at 25° and 35°C produced only minor changes, whereas higher temperatures damaged the appearance of glucose syrup (Bostan and Boyacioglu, 1997). The kinetic model permitted the calculation of the longest acceptable shelf-life at any temperature.

A very recent paper (Vercet, 2003) has considered the browning of white chocolate during storage, because browning is one of the main problems that limit the shelf-life of confectionery products manufactured with white chocolate or white chocolate substitutes. In both products browning occurs mainly on the surface of the product. Comparison of lipid oxidation indices (peroxide value, acidity index, and saturated/unsaturated fatty acid ratio) and MR indices (browning, fluorescence, and HMF) indicated that the MR is mainly responsible for this deterioration in both white chocolate and white chocolate substitutes and that the environmental conditions during storage are important.

6.5 Improving shelf-life: antioxidative Maillard reactions

Another very important reaction during food storage involves lipids that can degrade by autoxidation (Belitz and Grosch, 1999), giving in turn reactive intermediates, mainly saturated or unsaturated aldehydes or ketones, but also glyoxal and methylglyoxal (in common with the MR) and malondialdehyde. The main aldehydes from oleic acid are octanal and nonanal, from linoleic acid hexanal, (E)-2-heptenal, (Z) and (E)-2-octenal, (E,Z) and (E,E)-2,4-decadienal, whereas linoleic acid gives a complex mixture very rich in (E,Z)-2,4-heptadienal (Belitz and Grosch, 1999).

Lipid autoxidation may be delayed or slackened by antioxidants. The overall antioxidant capacity of foods derives from the concerted contributions of all the antioxidants present in foods. In raw materials, the main contribution derives from ascorbic acid and polyphenols, such as tocoferols, flavanones or functionalised cinnamic acid derivatives, whereas in processed foods the MR must be taken into consideration, while some polyphenols may have been destroyed by heating.

In fact it has been demonstrated that some compounds deriving from the MR have antioxidative properties. Their formation has been investigated in several different model systems, for example sugar/amino acids model systems (Lignert and Eriksson, 1981) and honey/lysine model systems (Antony *et al.*, 2000). These experiments have been recently reviewed by Manzocco *et al.* (2001) with the aim of finding a relationship between the development of colour and antioxidant capacity, because high antioxidant capacity is generally associated with the formation of brown melanoidins.

However, the most interesting work has been done on heated foods, such as tomato juice (Anese *et al.*, 1999a), malt (Woffenden *et al.*, 2002), or pasta (Anese *et al.*, 1999b) processed in different time–temperature conditions. A positive correlation between colour and antioxidant properties can be found. It must be pointed out that the correlation was found in foods where this phenomenon is the sole or prevailing event. This generally occurs in food products with no or low content of naturally occurring antioxidants, such as pasta, or in those foods in which naturally occurring antioxidants are very stable, such as tomato. In these cases in fact the changes in the antioxidant capacity upon processing are due only to the formation of heat-induced antioxidants.

In other foods more complex phenomena occur, especially in foods with a high content of polyphenols and in foods submitted to strong roasting treatment, such as coffee. During the first minutes of roasting the antioxidant capacity (measured as chain-breaking activity) increases up to the medium–dark roasted stage, then it decreases with further roasting (Nicoli *et al.*, 1997). Therefore in this case the relationship between colour and antioxidant activity does not exist. This experimental observation has been explained by partial pyrolysis of the polyphenols and perhaps also of MRPs.

A recent investigation has characterised one of the substructures of bread melanoidins that is responsible for the antioxidant activity (Lindenmeier *et al.*, 2002). In this study, the application of an *in vitro* antioxidant assay to solvent fractions isolated from bread crust, bread crumb, and flour, respectively, revealed the highest antioxidative potential for the dark brown, ethanol extracts of the crust, whereas corresponding crumb and flour fractions showed only minor activities. In a second stage of the work, antioxidant screening of Maillard-type model mixtures, followed by structure determination, revealed that the pyrrolinone reductones **6** and **7** (Fig. 6.6) are the key antioxidants formed from the hexose-derived acetylformoin and $N(\alpha)$-acetyl-L-lysine methyl ester or glycine methyl ester, chosen as model substances to mimic non-enzymatic browning reactions with the lysine side chain or the N terminus of proteins, respectively. Finally, quantitation of protein-bound pyrrolinone reductonyl-lysine, abbreviated to pronyl-lysine **8**, revealed high amounts in the bread crust (62.2 mg/kg), low amounts in the crumb (8.0 mg/kg), and the absence of this compound in untreated flour. These contents correlate well with the overall antioxidant capacity of the different parts of bread, though other substructures in melanoidins may be important too.

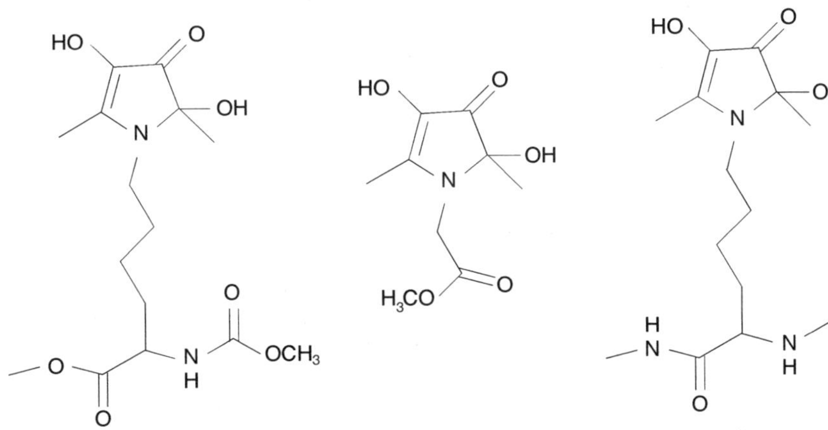

6 7 8

Fig. 6.6 Antioxidant residues in melanoidins (adapted from Lindenmeier *et al.*, 2002).

6.6 Improving shelf-life: the Maillard reaction and microbial spoilage

There are also a few data on the possible role of the MR in inhibiting the growth of microorganisms in food. The direct antibiotic activity of MRPs against both pathogenic and spoilage organisms, including *Lactobacillus*, *Proteus*, *Salmonella* and *Streptococcus faecalis* and others, has been investigated (Einarsson *et al.*, 1983; Einarsson *et al.*, 1988). MRPs were obtained by refluxing solutions containing either arginine and xylose (AX) or histidine and glucose (HG). The solutions tested were either unfractionated or partly purified by dialysis through a membrane with a cut-off at 1000 daltons. The MIC values obtained showed that the inhibitory effect of MRP is dependent on the type of MRP and type of bacteria used. The lag phase of growth was prolonged with increased concentrations of MRP. The high-molecular-weight fraction (>1000 daltons) was more inhibitory than the low-molecular-weight fraction (<1000 daltons), when tested with *Bacillus subtilis*, *Escherichia coli* and *Staphylococcus aureus*.

Other authors (Fadel *et al.*, 1991) have produced MRPs by refluxing solutions containing asparagine and xylose and have tested their activity on both pathogenic and spoilage bacteria frequently found in food. Some inhibitory effects on *S. aureus* were observed, whereas an opposite trend was observed for *B. subtilis* and *Escherichia coli*. *Candida tropicalis* was almost totally insensitive. In addition, as melanoidins possess the ability to bind metals, such as copper and zinc (O'Brien and Morrissey, 1989; Andrieux *et al.*, 1980; Furniss *et al.*, 1986), they may indirectly impair the multiplication of microorganisms.

6.7 Conclusion

In conclusion, as far as the shelf-life of foods is concerned, the MR during processing produces antioxidative products that may be useful for the stabilisation of easily oxidisable lipids during storage, whereas during storage it impairs the quality of flavour and colour and slowly reduces the nutritional value of several food items.

6.8 References

ALAIZ M, HIDALGO F J, ZAMORA R (1997) 'Antioxidative activity of non-enzymatically browned proteins produced in oxidised lipid/protein reaction', *J Agric Food Chem* 45, 3250–3254.

ALBALÀ-HURTADO S, VECIANA-NOGUES M T, MARINE-FONT A, VIDAL-CAROU M C (1998) 'Changes in Furfural Compounds during Storage of Infant Milks. *J Agric Food Chem* 46, 2998–3003.

AMADORI M (1931) 'Condensation products of glucose with *p*-toluidine', *Atti R Accad Naz Lincei Mem Cl Sci Fis Mat Nat* 13, 72.

ANDRIEUX C, SACQUET E, GUEGUEN L (1980) 'Interactions between Maillard's reaction products, the microflora of the digestive tract and mineral metabolism', *Reprod Nutr Develop* 20, 1061–1069.

ANESE M, MANZOCCO L, NICOLI M C, LERICI C R (1999a) 'Antioxidant properties of tomato juice as affected by heating' *J Sci Food Agric 79*, 750–754.

ANESE M, NICOLI M C, MASSINI R, LERICI C R (1999b) 'Effects of drying processing on the Maillard reaction in pasta'. *Food Res Int* 32, 193–199.

ANTONY S M, HAN I Y, RIECK J R, DAWSON P L (2000) 'Antioxidative effect of Maillard reaction products formed from honey at different reaction times', *J Agric Food Chem,* 48, 3985–3989.

ARAI M (2002) 'Advanced glycation end products and their receptor: do they play a role in diabetic cardiomyopathy?', *J Molec Cell Cardiol* 34, 1305–1308.

ARNOLDI A (2001) 'Thermal processing and foods quality: analysis and control', in Richardson P, ed., *Thermal Technologies in Food Processing*, Woodhead Publishing, Cambridge UK, 138–159.

ARNOLDI A (2002) 'Thermal processing and nutritional quality'. *The Nutrition Handbook for Food Processors*, Richardson P. ed., Woodhead Publishing, Cambridge UK, 265–292.

ARNOLDI A, BOSCHIN G, D'AGOSTINA A (2002) 'Melanoidins in foods', *Res Advanc Food Sci* 3, 1–10.

ASKAR A (1999) 'Flavor changes during processing and storage of fruit juices. Part 1. Markers for processed and stored fruit juices', *Fruit Process* 9, 236–244.

BAYNES J W (2001) 'The role of AGEs in aging: causation or correlation', *Exper Gerontol* 36, 1527–1537.

BELITZ H D, GROSCH W (1999) 'Lipids', in Beliyz H D, Grosch W, eds, *Food Chemistry. Second Edition*, Springer, Berlin, 152–236.

BEMIS-YOUNG G L, HUANG J, BERNHARD R A (1993) 'Effect of pH on pyrazine formation in glucose–glycine model systems', *Food Chem* 46, 383–387.

BOSTAN A, BOYACIOGLU D (1997) 'Kinetics of non-enzymic color development in glucose syrups during storage', *Food Chem* 60, 581–585.

CAPPUCCIO R., FULL G, LONZARICH V, SAVONITTI O (2001) 'Staling of roasted and ground coffee at different temperatures: combining sensory and GC analysis', *Colloq Sci Int Cafe 19th*, on CD, 11 pages.

EINARSSON H, EKLUND T, NES I F (1988) 'Inhibitory mechanisms of Maillard reaction products', *Microbios* 53, 27–36.

EINARSSON H, GORAN S, SNYGG B G, ERIKSSON C (1983) 'Inhibition of bacterial growth by Maillard reaction products', *J Agric Food Chem* 31, 1043–1047.

ERBERSDOBLER H F, HUPE A (1991) 'Determination of lysine damage and calculation of lysine bio-availability in several processed foods', *Z Ernalhrungswiss*, 30, 46–49.

EVANS, D J, SCHMEDDING D J M, BRUIJNJE A, HEIDEMAN T, KING B M, GROESBEEK N M (1999) 'Flavour impact of aged beers', *J Inst Brew* 105, 301–307.

FADEL H M, HEBASH K A H, SOLIMAN M M A (1991) 'Effect of varying molar ratio of heated asparagine-xylose products on the growth of some microorganisms', *Indian J Technol* 29, 511–12.

FAIST V, ERBERSDOBLER H F (2002) 'Health impact of food-derived Maillard reaction products', *Kieler Milchwirtschaft Forsch* 54, 137–147.

FORS S (1983) 'Sensory properties of volatile Maillard reaction products and related compounds. A literature review', in Waller G R, Feather M S, eds, *The Maillard Reaction in Food and Nutrition*. ACS Symp. Ser. 215, American Chemical

Society, Washington DC, 185–286.

FRANK O, HOFMANN T (2002) 'Reinvestigation of the chemical structure of bitter-tasting quinizolate and homoquinizolate and studies on their Maillard-type formation pathways using suitable (13)C-labeling experiments', *J Agric Food Chem* 50, 6027–6036.

FURNISS D E, HURRELL R F, FINOT P A (1986) 'Modification of urinary zinc excretion in the rat associated with the feeding of Maillard reaction products', *Acta Pharmacol Toxicol* 59, 188–190.

GLAESER H (1989) 'Criteria for the establishment of a ''best-before'' date in ultra-high-temperature milk', *Deuts Molkerei-Zeit* 110, 580–585.

GUERRA-HERNANDEZ E, GOMEZ C, GARCIA-VILLANOVA B, SANCHEZ N C, GOMEZ J M R (2002) 'Effect of storage on non-enzymatic browning of liquid infant milk formulae', *J Sci Food Agric* 82, 587–592.

HANDWERK R L, COLEMAN R L (1988) 'Approaches to the citrus browning problem. A review', *J Agric Food Chem* 36, 231–236.

HEIDLAND A, SEBEKOVA K, SCHINZEL R (2001) 'Advanced glycation end products and the progressive course of renal disease', *Am J Kidney Dis* 38 (Suppl. 1) S100–S106.

HEYNS K, NOACK H (1962) 'Die Umzetzung von d-fructose mit L-Lysine and L-Arginin und deren Beiziehung zu nichtenenzymatischen Bräunungsreaktionen', *Chem Ber* 720–727.

HIDALGO A, ROSSI M, POMPEI C (1995) 'Furosine as a Freshness Parameter of Shell Eggs', *J Agric Food Chem* 43, 1673–7.

HODGE J E (1953) 'Chemistry of browning reactions in model systems', *J Agric Food Chem* 1, 928–943.

HOFMANN T, SCHIEBERLE P (2002) 'Chemical interactions between odor-active thiols and melanoidins involved in the aroma staling of coffee beverages'. *J Agric Food Chem* 50, 319–326.

HOFMANN T, CZERNY M, CALLIGARIS S, SCHIEBERLE P (2001) 'Model studies on the influence of coffee melanoidins on flavor volatiles of coffee beverages', *J Agric Food Chem* 49, 2382–2386.

JÄGERSTAD M, SKOG K, ARVIDSSON P, SOLYAKOV A (1998) 'Chemistry, formation, and occurrence of genotoxic heterocyclic amines identified in model systems and cooked foods', *Z Lebensm Unters Forsch A* 207, 419–427.

KIRKLAND J L (2002) 'The biology of senescence: potential for prevention of disease', *Clin Geriatr Med* 18, 383–405.

LEDL F, SCHLEICHER E (1990) 'New aspects of the Maillard reaction in foods and in the human body', *Angew Chem Int Ed* 29, 565–706.

LIGNERT H, ERIKSSON C E (1981) 'Antioxidative effect of Maillard reaction products', *Prog Food Nutr Sci* 5, 453–466.

LINDENMEIER M, FAIST V, HOFMANN T (2002) 'Structural and functional characterization of pronyl-lysine, a novel protein modification in bread crust melanoidins showing *in vitro* antioxidative and phase I/II enzyme modulating activity', *J Agric Food Chem 50*, 6997–7006.

MAILLARD A C (1912) 'Action des acides amines sur les sucres. Formation des melanoidines par voie methodologique', *C R Acad Sci* 154, 66–68.

MANZOCCO L, CALLIGARIS S, MASTROCOLA D, NICOLI M C, LERICI C R (2001) 'Review of non-enzymatic browning and antioxidant capacity in processed foods', *Trends Food Sci Technol* 11, 340–346.

MARCONI E, CABONI M F, MESSI M C, PANFILI G J (2002) 'Furosine. A suitable marker for assessing the freshness of royal jelly', *J Agric Food Chem* 50, 2825–2829.

MEYNIER A, MOTTRAM D S (1995) 'The effect of pH on the formation of volatile compounds in meat related model systems', *Food Chem* 52, 361–366.

MORR C V, HA E Y W (1991) 'Off-flavors of whey protein concentrates: a literature review', *Int Dairy J* 1, 1–11.

NICOLI M C, MARZOCCO L, LERICI C R, ANESE M (1997) 'Antioxidant properties of coffee-brews in relation to roasting', *Lebensm Wiss u-Technol* 30, 292–297.

O'BRIEN J, MORRISSEY P A (1989) 'Nutritional and toxicological aspects of the Maillard browning reaction in foods', *Crit Rev Food Nutr* 28, 211–248.

OTTIGER H, BARETH A, HOFMANN T (2001) 'Characterization of natural "cooling" compounds formed from glucose and L-proline in dark malt by application of taste dilution analysis', *J Agric Food Chem* 49, 1336–1344.

RANGEL-ALDAO R, BRAVO A, GALINDO-CASTRO I, SANCHEZ B, REVEROL L, SCHERER E, MADRID J, RAMIREZ J L, HERRERA J, PENTTILA M, VEHKOMAKI M-L, VIDGREN V, VIRTANEN H, HOME S (2002) *Beer flavor stabilization through the control of Maillard reaction intermediates*. Monograph, European Brewery Convention, 31 (Flavour and Flavour Stability), 111–124.

RENNER E (1988) 'Storage stability and some nutritional aspects of milk powders and ultra-high temperature products at high ambient temperatures', *J Dairy Res* 55, 125–142.

SCHÄR W, BOSSET J O (2002) 'Chemical and physico-chemical changes in processed cheese and ready-made fondue during storage', *Lebens-Wiss u-Technol* 35, 15–20.

SHIBAMOTO T, BERNHARD R A (1977) 'Investigation of pyrazine formation pathways in sugar-ammonia model systems', *J Agric Food Chem* 25, 609–614.

SHU C-K, HO C-T (1988) 'Effect of pH on the volatile formation from the reaction between cysteine and 2,5–dimethyl-4-hydroxy-3(2H)-furanone', *J Agric Food Chem* 36, 801–803.

STAPELFELDT H, NIELSEN B R, SKIBSTED L H (1997) 'Effect of heat treatment, water activity and storage temperature on the oxidative stability of whole milk powder', *Int Dairy J* 7, 331–339.

VALERO E, VILLAMIEL M, MIRALLES B, SANZ J, MARTINEZ-CASTRO I (2000) 'Changes in flavor and volatile components during storage of whole and skimmed UHT milk', *Food Chem* 72, 51–58.

VERCET A (2003) 'Browning of white chocolate during storage' *Food Chem* 81, 371–377.

WARNER K J H, DIMICK P S, ZIEGLER G R, MUMMA R O, HOLLENDER R (1996) ' "Flavor-fade" and off-flavors in ground roasted peanuts as related to selected pyrazines and aldehydes', *J Food Sci* 61, 469–472.

WHITFIELD F B (1992) 'Volatiles from the interactions of the Maillard reaction and lipids', *Crit Rev Food Sci Nutr* 31, 1–58.

WOFFENDEN H M, AMES J M, CHANDRA S, ANESE M, NICOLI M C (2002) 'Effect of kilning on the antioxidant and pro-oxidant activities of pale malts', *J Agric Food Chem* 50, 4925–4933.

YAYLAYAN V A, HUYGGUES-DESPOINTES A (1994) 'Chemistry of Amadori rearrangement products' *Crit Rev Food Sci Nutr* 34, 321–369.

7

Factors affecting lipid oxidation

M. H. Gordon, University of Reading, UK

7.1 Introduction: mechanisms of autoxidation

The oxidation of lipids constitutes an important reaction that limits the shelf-life of many foods. Lipids occur in nearly all food raw materials with the major classes being triglycerides (also known as triacylglycerols), which occur in fat storage cells of plants and animals, and phospholipids, which occur in biological membranes. In the processing of a wide range of foods, fats may be added as part of the food formulation. The added fats are a major component of many foods including mayonnaise, margarine, and frying oils. These fats are almost completely triglycerides, and it is these components that are of most significance as potential sources of oxidative off-flavours in these foods. In plant or animal tissues used as foods, the phospholipids present in all biological membranes may also be an important substrate for oxidative deterioration. This chapter discusses environmental and compositional variables that affect the susceptibility of lipids to oxidation, and discusses the methods that are available for measuring the extent of lipid oxidation and predicting the shelf-life of foods subject to lipid oxidation.

7.1.1 Mechanism of autoxidation

Deterioration of foods by lipid oxidation generally displays an induction period during which very little oxidation occurs (Fig 7.1). This is followed by a stage when deterioration proceeds rapidly. The length of the induction period is shortened dramatically by low concentrations of metals such as iron or copper, and these are described as pro-oxidants, but it can be extended considerably by low concentrations of antioxidants such as α-tocopherol. The rate of

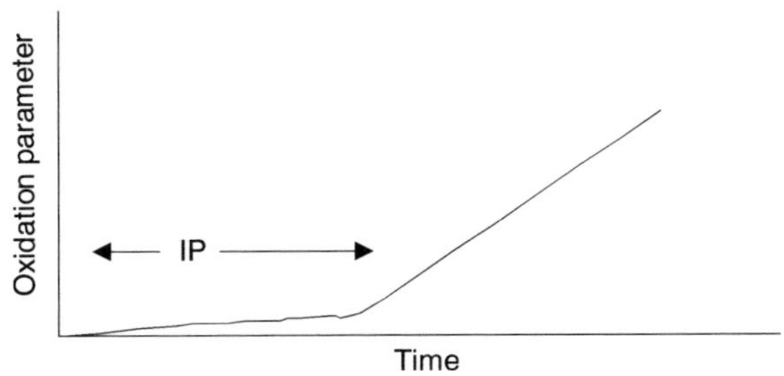

Fig. 7.1 Typical change in a fat during oxidation showing the presence of an induction period (IP).

deterioration increases markedly with an increase in temperature. These characteristics have led researchers to conclude that the reaction is a free-radical chain reaction.

As a free-radical reaction, autoxidation proceeds in three distinct steps (Fig. 7.2). The first step is initiation in which lipid radicals are formed from lipid molecules. Abstraction of a hydrogen atom by a reactive species such as a hydroxyl radical may lead to initiation of lipid oxidation. However, in oils there is often a trace of hydroperoxides, which may have been formed by lipoxygenase action in the plant prior to, and during, extraction of the oil. Secondary initiation by homolytic cleavage of hydroperoxides is a relatively low energy reaction, and is normally the main initiation reaction in edible oils. This reaction is commonly catalysed by metal ions. After initiation, propagation reactions occur in which one lipid radical is converted into a different lipid radical. These reactions commonly involve abstraction of a hydrogen atom from a lipid molecule or addition of oxygen to an alkyl radical. The enthalpy of reaction is relatively low compared with that of the initiation reactions, so propagation reactions occur rapidly compared with initiation reactions. At normal atmospheric pressure of oxygen, the reaction of alkyl radicals with oxygen is very rapid, and the peroxyl radicals are present at much higher concentrations than the alkyl radicals.

Alkoxy radicals formed by hydroperoxide decomposition can decompose to release volatile hydrocarbons, alcohols or aldehydes, that are no longer bound to

Initiation	$X^{\cdot} + RH$	→	$R^{\cdot} + XH$
Propagation	$R^{\cdot} + O_2$	→	$ROO^{\cdot}$
	$ROO + R^{\cdot}H$	→	$ROOH + R^{\cdot}$
Termination	$ROO^{\cdot} + ROO^{\cdot}$	→	$ROOR + O_2$
	$ROO^{\cdot} + R^{\cdot}$	→	$ROOR$
	$R^{\cdot} + R^{\cdot}$	→	RR
Secondary initiation	$ROOH$	→	$RO^{\cdot} + {}^{\cdot}OH$
	$2\ ROOH$	→	$RO^{\cdot} + ROO^{\cdot} + H_2O$

Fig. 7.2 Mechanism of autoxidation.

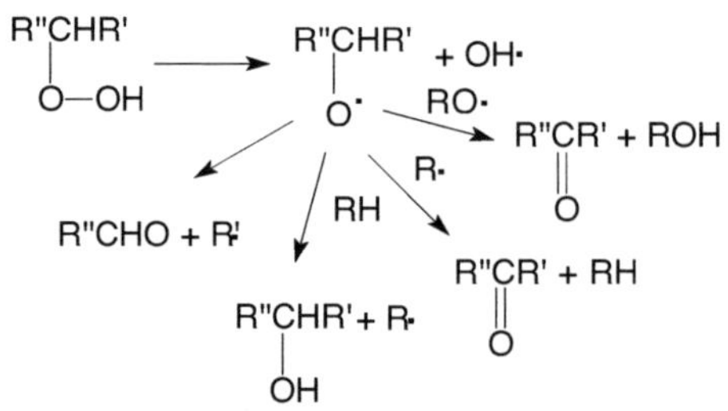

Fig 7.3 Formation of secondary oxidation products by hydroperoxide decomposition.

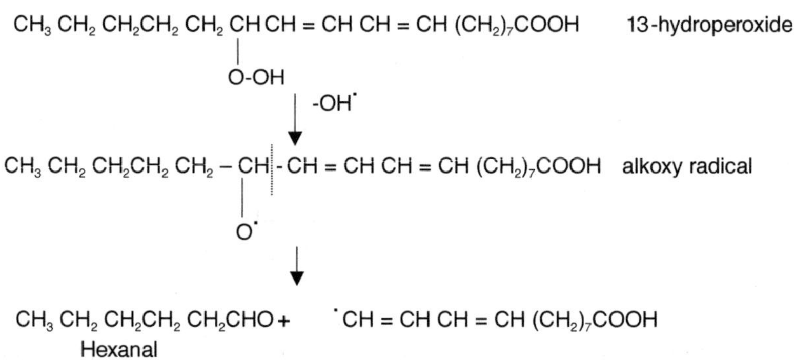

Fig 7.4 Decomposition of 13-hydroperoxide from linoleic acid to form hexanal.

the glycerol backbone when the fatty acid is present as a glyceride. Non-volatile alcohols and ketones may also be formed as shown in Fig. 7.3. Volatile aldehydes are particularly important as contributors to the aroma of oxidised oils, and hexanal is commonly monitored in assessing the formation of secondary oxidation products during lipid oxidation. Hexanal is normally formed in relatively large amounts during the oxidation of lipids via linoleic acid 13-hydroperoxide (Fig. 7.4), although it is not one of the aldehydes to which the palate is most sensitive. Consequently, other volatile products may contribute

Table 7.1 Flavour thresholds of possible linoleic acid oxidation products in paraffin oil (Forss, 1973)

Compound	Threshold (mg/kg)
Hexanal	0.08–0.6
Heptanal	0.04–0.055
Octanal	0.04–0.6
Trans-2-nonenal	0.04–0.4
Cis-2-decenal	0.1
Trans, *trans*-2,4-nonadienal	0.46
Trans, *cis*-2,4-decadienal	0.02

more than hexanal to the off-flavour perceived in the sensory assessment of oxidised oils. The flavour thresholds of some aldehydes formed in the autoxidation of linoleic acid are shown in Table 7.1.

7.2 Factors influencing the rate of lipid oxidation

7.2.1 Effect of temperature

An increase in temperature causes a very strong reduction in the length of the induction period. In principle the rate of oxidation increases exponentially with an increase in temperature but the effect of temperature is complicated by a reduction in oxygen solubility in liquids at increased temperature and by changes in partitioning of antioxidants between phases if more than one phase is present. Commonly, the rate-limiting step in the autoxidation pathway may change with an increase in temperature. For a sunflower oil-in-water emulsion from which tocopherols had been removed, the time to a peroxide value of 50 meq/kg decreased from 8 days at 30ºC to 3 days at 50ºC in one of our experiments.

7.2.2 Effect of fatty acid composition

Abstraction of hydrogen in the propagation phase of autoxidation takes place preferentially at carbon atoms where the bond dissociation energy is low. Saturated fatty acids are very stable and do not oxidise at a significant rate. Since the bond dissociation energy of the C–H bond is reduced by neighbouring alkene functionality, abstraction of hydrogen takes place most rapidly at the methylene group between two alkene groups in a polyunsaturated fatty acid (PUFA). Consequently, the rate of oxidation is much faster when polyunsaturated fatty acids are present in the food. The relative rate of oxidation of oleic acid (18:1) and linoleic acid (18:2) has been reported to be between 1:12 and 1:40 from different studies in the literature, but the increase in rate with additional double bonds in the fatty acid is normally roughly in proportion to the number of methylene groups between pairs of double bonds. Thus, the relative rate of oxidation of 18:2, 18:3 and 20:4 is roughly 1:2:3.

As well as increasing the rate of oxidation, the polyunsaturated fatty acids present in a food will produce different volatiles. Commonly, it is found that fatty acids with an *n*-3 structure such as linolenic acid produce volatiles on oxidation that are perceived as off-flavours at significantly lower levels than volatiles from an *n*-6 fatty acid such as linoleic acid.

7.2.3 Effect of antioxidants

Antioxidants may act by various mechanisms. The scavenging of lipid free radicals to produce less reactive species, and hence to interrupt the propagation stage of lipid autoxidation, is the main antioxidant mechanism by which phenolic antioxidants such as α-tocopherol act. However, metal chelation by antioxidants such as citric acid is also an effective mechanism of antioxidant action. Reducing compounds such as vitamin C also contribute to the total antioxidant potential of a food.

7.2.4 Effect of metals

Metals such as iron or copper are very effective pro-oxidants even if present at part per million levels or less.

Metals are particularly effective at catalysing the decomposition of hydroperoxides by mechanisms involving one electron transfer.

$$M^{n+} + ROOH \rightarrow M^{(n+1)+} + RO^{\bullet} + OH^{-}$$

$$M^{(n+1)+} + ROOH \rightarrow M^{n+} + ROO^{\bullet} + H^{+}$$

7.2.5 Enzyme-catalysed reactions

Lipoxygenase is present in plant tissues including those of soybean, pea and tomato, as well as fish and animal. The enzyme catalyses the reaction between polyunsaturated fatty acids and oxygen to produce hydroperoxides, and other enzymes within the plant or animal tissue may contribute to the formation of volatiles from the hydroperoxides during storage. Lipoxygenase can be denatured by heating to extend the shelf-life of fish or plant foods.

7.2.6 Effect of water

The concentration required for a volatile component to be detected as a contributor to the flavour depends on the medium. Normally non-polar components have a higher flavour threshold in non-polar media such as edible oils than in water. The phases present in the food will also affect the rate of oxidation by affecting the activity of the antioxidants present, and by partitioning of pro- and antioxidants between oil and aqueous phases. The term polar paradox has been applied to the phenomenon whereby polar antioxidants are most effective in oils, whereas non-polar antioxidants are more effective in emulsions. Normally, metal chelation is less effective as an antioxidant mechanism in water-containing foods than in oils.

7.3 Methods of measuring oxidation in an oil or food

Several methods can be applied to assess the current state of an oil or food sample. The principles of these methods are described below.

7.3.1 Sensory analysis

For the food industry, the detection of oxidative off-flavours by taste or smell is the main method of deciding when a lipid-containing food is no longer fit for consumption. The ability of individuals to describe the nature of the aroma is useful, and the sensitivity of a trained panel to oxidative off-flavours may allow detection of oxidative deterioration at a stage when common chemical methods, such as peroxide value measurements, are unable to detect any deterioration. The main problems with sensory evaluation are that different individuals vary in their sensitivity to these off-flavours, and their performance may vary depending on their state of health and other variables. Trained panellists are much more reliable than untrained panellists, but the reproducibility of sensory analysis is normally worse than that of chemical or instrumental methods.

7.3.2 Headspace analysis

Although they represent only a small proportion of the oxidation products, volatile lipid decomposition products are the products that are perceived by the consumer as off-flavours. Consequently, it is tempting to monitor these volatile oxidation products in order to have an instrumental method that correlates well with consumer perception of the extent of deterioration of an oil. The application is normally a correlation between an individual aroma component, or the total volatile concentration, and sensory assessment of oil deterioration. However, it should be remembered that the aroma of a sample includes contributions from many different compounds. Individual aroma compounds vary in their contributions to the aroma with different flavour thresholds and concentration dependence of the aroma (Table 7.1). Nevertheless, analysis of volatiles has been widely used as a method of monitoring oxidative deterioration of oil samples. Several procedures have been developed.

Static headspace analysis

The mass of each component in the headspace of a sample in a sealed vial depends on the vapour pressure of the pure component, the sample temperature and the concentration of the component in the sample. Although it is possible to sample the headspace with a gas syringe and inject it onto a GC column, it is difficult to avoid problems due to adsorption of trace components, condensation of volatiles in the syringe and leakage from the syringe during transfer to the GC. Consequently, automated headspace injectors are normally used. In a typical commercial headspace analyser, the sample is sealed in a vial closed with a septum and crimped aluminium cap. Vials are held in a temperature-controlled autosampler for equilibration. A sample needle penetrates the septum when the vial is in the position for analysis, and the vial is pressurised for a time before injection in order to raise the pressure within the vial to that of the column head. The carrier gas supply to the column is then switched off by a solenoid valve, the pressure at the head of the column falls and sample vapour flows via the sample needle on to the column. After the injection period, which is typically 5 seconds, the carrier gas flow to the column is restored and injection of vapour ceases, after which time the injection needle is withdrawn.

The temperature used for analysis of headspace volatiles in an edible oil may vary between 40ºC and 180ºC. Above 150ºC, hydroperoxides are unstable and the measured headspace volatiles are due to the volatiles in the sample when inserted into the vial and those formed by hydroperoxide decomposition. Even at temperatures as low as 90ºC, partial decomposition of hydroperoxides will occur during the equilibration time. The time required for equilibration is typically 10–20 minutes. Static headspace analysis is a relatively quick and simple procedure, and no solvents are used. It is less sensitive than dynamic headspace analysis and the procedure detects mainly the very volatile components, while dynamic headspace analysis detects components with a wider range of volatility. When monitoring the oxidative deterioration of edible oils, either the hexanal, pentane or total volatile concentration is normally monitored for oil containing linoleic acid or other polyunsaturated fatty acids with an *n*-6 structure. Propanal, 2-hexenal, 3-hexenal and 2,4-heptadienal are formed from α-linolenic acid or other polyunsaturated fatty acids with an *n*-3 structure.

Solid Phase Microextraction (SPME) is an alternative technique for static headspace analysis. SPME uses a special syringe with a short length of fused silica optical fibre externally coated with a polymeric GC stationary phase (eg polydimethylsiloxane). The coated fibre is shielded by an adjustable needle guide as it is pushed through a septum into the headspace above a sample in a sealed sample vial. After penetrating the septum, the needle guide is withdrawn to allow the coated fibre to be exposed to the headspace. After allowing equilibrium to be achieved, or after a defined time, the fibre is again shielded by the needle guide. The syringe is withdrawn, and transferred to a GC injection port. The fibre is exposed after puncturing the septum, and the analytes are thermally desorbed to introduce them onto the GC column.

SPME has the advantage that the technique is easy to use and quick, requiring about 30 minutes for trapping the volatiles. No special injector is required, so there are no capital costs involved, and no artifacts are introduced. However, the technique is less sensitive than other headspace techniques, and the fibres are fragile and require periodic replacement.

Dynamic headspace analysis

An alternative method for the analysis of volatile components is dynamic headspace analysis, which involves purging the sample with nitrogen or helium for a period whilst continuously trapping the volatiles. The volatiles are trapped on a porous polymer trap (often Tenax™) held at room temperature. The most common method of transferring the volatiles onto a GC column involves placing the trap in the inlet of a gas chromatograph, then heating the trap to desorb the volatiles. Solvent extraction of volatiles from the trap and injection of a solution may be used as an alternative transfer procedure, but this is much less sensitive than thermal desorption. Dynamic headspace analysis allows components with a wider range of volatility to be detected than in the case of static headspace analysis, but poorly adsorbed components may be lost by passing through the trap (breakthrough) before the trapping period is complete.

Direct injection method
The direct injection method involves applying a sample at the inlet of a GC column, and then passing carrier gas through the sample to sweep the volatiles onto the column, which is often cooled to allow efficient trapping as a narrow band. The method allows the collection of volatiles covering a wide range of volatility. However, a much smaller sample size is used than in dynamic headspace analysis. The method can only be applied directly to oils whereas static and dynamic headspace analysis can be applied to more complex foods.

7.3.3 Peroxide value (PV)

The PV is still the most common chemical method of measuring oxidative deterioration of oils. Although hydroperoxides decompose to a mixture of volatile and non-volatile products and they also react further to endoperoxides and other products, the PV measurement is a useful method of monitoring oxidative deterioration of oils, although it should normally be combined with a method of monitoring secondary oxidation products to provide a fuller picture of the progress of oxidation. A high PV value may reflect either increased formation of hydroperoxides or reduced decomposition. Consequently, antioxidants may improve the flavour stability of an oil without it being evident from PV measurements.

The traditional method of determining PV involves a titration of the oil containing potassium iodide in a chloroform–acetic acid mixture. The hydroperoxides oxidise the iodide to iodine, which is determined by titration with sodium thiosulfate. In order to avoid the use of chloroform, the AOCS has developed an alternative method which uses isooctane as solvent, although the method is limited to PV < 70 meq/kg (AOCS, 1998).

The PV at which oxidation of oils can be detected as an off-flavour varies widely depending on the nature of the oil. Samples of olive oil may not be perceived as rancid till the PV reaches 20 meq/kg whereas fish oil may develop off-flavours at PV < 1 meq/kg.

7.3.4 Conjugated dienes

Formation of hydroperoxides from PUFA leads to conjugation of the pentadiene structure. This causes absorption of UV radiation at 233–234 nm. This represents a simple and rapid method of monitoring oxidative deterioration of an oil. Although the absorbance is mainly a measure of hydroperoxide content, some products formed following hydroperoxide decomposition such as 9-hydroxyoctadeca-10,12-dienoic acid and 13-hydroxyoctadeca-9,11-dienoic acid retain this conjugated structure and will contribute to the absorbance. The method is therefore less specific than PV measurement.

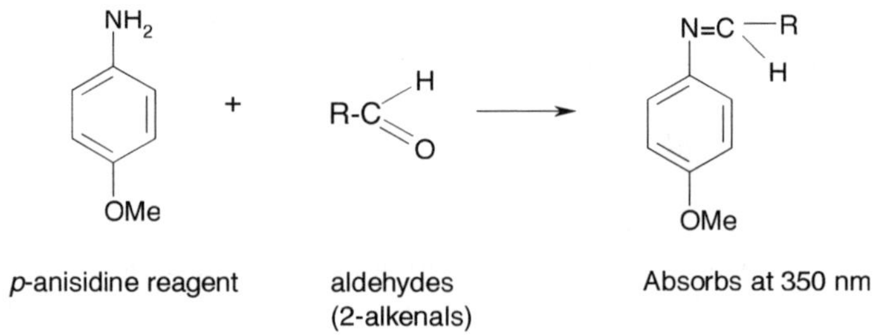

Fig. 7.5 Reaction of *p*-anisidine with an aldehyde to form a coloured product.

7.3.5 *Para*-anisidine value

Para-anisidine is a reagent that reacts with aldehydes to give products that absorb at 350 nm (Fig. 7.5). The *p*-anisidine value is defined as the absorbance of a solution resulting from the reaction of 1g fat in isooctane solution (100 ml) with *p*-anisidine (0.25% in glacial acetic acid). The products formed by reaction with unsaturated aldehydes (2–alkenals) absorb more strongly at this wavelength, and consequently the test is particularly sensitive to these oxidation products. Although the test does not distinguish between volatile and non-volatile products, the palate is generally more sensitive to unsaturated volatile aldehydes than to saturated volatile aldehydes, so the test is a reasonable way to assess secondary oxidation products. Measurements of *p*-anisidine value are commonly used together with peroxide value measurements in describing the total extent of oxidation by the Totox value, which equals the sum of the *p*-anisidine value plus twice the peroxide value. However, the Totox value is an empirical parameter since it corresponds to the addition of two parameters with different units.

7.3.6 Thiobarbituric acid value (TBA)

Malonaldehyde may be formed from polyunsaturated fatty acids with at least three double bonds. The concentration of this product may be assessed by reaction with thiobarbituric acid which reacts with malonaldehyde to form red condensation products that absorb at 532–535 nm with molar absorptivity of 27.5 absorbance units/μmol. However, the reaction is not specific, and reaction with a wide variety of other products may contribute to the absorbance. 2,4-Alkadienals such as 2,4-decadienal also react with TBA to show strong absorption at 532 nm. Saturated aldehydes normally absorb at lower wavelengths after reaction with TBA. Several food components including proteins, Maillard browning products and sugar degradation products affect the determination. In order to emphasise the lack of specificity, the values obtained in the test are commonly described as TBARS (TBA reactive substances). The TBA test has recently been reviewed (Guillen-Sans and Guzman-Chozas, 1998).

7.3.7 Octanoate value

The octanoate value is a measure of the bound octanoate present in an oil. Octanoate is formed from the decomposition of linoleic acid 9-hydroperoxide (Peers and Swoboda, 1979). The method involves *trans*-methylation of an oil with a base such as sodium methoxide, and GC analysis of the methyl octanoate formed.

7.3.8 Conjugable oxidation products

Analysis of conjugable oxidation products is based on the fact that hydroperoxides from polyunsaturated fatty acids and some decomposition products may be reduced with sodium borohydride and dehydrated to give conjugated trienes and tetraenes (Parr and Swoboda, 1976). Triene and tetraene concentrations are determined from the absorbance values at 268 nm and 301 nm respectively.

7.3.9 Infrared spectroscopy

Fourier transform infrared spectroscopy (FTIR) has considerable potential for the analysis of hydroperoxides in oils. Some progress was made in analysing the content of hydroperoxides and other oxidation products by direct determination in oils (van de Voort *et al.*, 1994). The method involved calibration with known standards, and is an attractive approach due to its speed, once calibrated, and avoidance of chemicals for the analysis. However, improved specificity and precision can be achieved by addition of triphenylphosphine (TPP) to an oil containing hydroperoxides (Ma *et al.*, 1997). This causes the formation of triphenylphosphine oxide (TPPO) which has an intense absorption band at 542 cm^{-1}. In the paper describing the method, PV was determined in the range 0–15 meq/kg by calibrating with TPPO as the standard. The resulting calibration was linear over the analytical range. The analytical procedure involved addition of a 33% (w/w) stock solution of TPP in hexanol (0.2 g) to melted fat or oil (30 g). The mixture was shaken and transferred to a 100 μm IR transmission cell maintained at 80°C. The reaction and FTIR analysis required about 2 min per sample. The method was validated by comparing the analytical results of the AOCS PV method to those of the automated FTIR procedure by using both oxidised oils and oils spiked with *tert*-butyl hydroperoxide. The reproducibility of the FTIR method is superior to that of the standard chemical method.

7.4 Monitoring changes in oxidation and the use of predictive methods

All of the above methods may be applied to assess the state of oxidation of oils, but other methods can only be used to monitor changes in oils. These include the following.

7.4.1 Loss of polyunsaturated fatty acids

Analysis of changes in fatty acid composition is always an insensitive way of assessing oxidative deterioration. Monitoring the formation of end-products of oxidation by methods such as PV determination is a much more sensitive method of detecting oxidative changes in food. This is in line with the general scientific principle that it is much more difficult to measure a small change in a large number than the same change in a very small number.

7.4.2 Weight gain

Edible oils increase in weight during the early stages of lipid oxidation as fatty acids combine with oxygen during the formation of hydroperoxides. The increase in weight of a heated sample during storage can be used to determine the induction time of the fat. Rapid weight gain occurs after the induction period, or the time for a certain weight increase can be determined. However, decomposition of hydroperoxides leads to a weight reduction, and the end of the induction time may not be clearly detected by this method until the oil is severely oxidised.

7.4.3 Predictive methods

Predictive methods are methods in which samples are continuously monitored during accelerated oxidation conditions. They can be useful to monitor the effect of changes in raw material quality on the shelf-life of a food. However, the rate-limiting step in the mechanism of oxidation commonly changes with temperature, and some volatile components such as butylated hydroxyanisole are rapidly lost at high temperatures, so the predictions from accelerated test methods should only be considered as a guide to the shelf-life of a food at lower temperatures.

Differential scanning calorimetry (DSC)

DSC is an instrumental method that monitors exothermic or endothermic changes due to phase changes or chemical reactions in samples. The end of the induction time is marked by an increased heat of reaction due to more rapid reaction of unsaturated lipids reacting with oxygen (Cross, 1970). Very small sample sizes are used so the method is limited to oils rather than complex foods, and the use of very high temperatures to get a reasonably short induction period limits the value of the method.

Oil stability index (OSI)

The OSI is an automated development of the AOM (active oxygen method). In the AOM, the time for an oil to reach a PV of 100 meq/kg during oxidation at 97.8°C, with an air flow of 2.33 mL per tube per second, is determined. Instruments for determining the OSI are the Rancimat™, manufactured by Metrohm, Basel, or the Oxidative Stability Instrument™, manufactured by

Omnion, Rockland, USA. These instruments depend on the increase in electrical conductivity, when effluent from oxidising oils in passed through water. Volatile carboxylic acids are generated in the oxidising oil and these cause the increase in electrical conductivity. The samples, assessed by the OSI methods, are held at 100ºC, 110ºC, 120ºC, 130ºC, or 140ºC. The temperature may be adjusted to allow the oxidation time to fall within the range 4–15h. The sample size is 2.5 g or 5 g depending on the instrument used. Although these instruments are useful for quality control of oils, they are not recommended for the assessment of antioxidant effectiveness for several reasons. The high temperatures used do not allow reliable predictions of antioxidant effectiveness at lower temperatures. Volatile antioxidants may be swept out of the oil by the air flow under the test conditions, and also the oils are severely deteriorated when the end-point is reached.

Oxipres

The Oxipres™, manufactured by Mikrolab Aarhus, is a method for examining the oxidative stability of heterogeneous products such as potato crisps, margarine or mayonnaise. Oxidation is accelerated by heating and by the use of oxygen under pressure. The pressure drop in a glass pressure vessel, containing the sample (up to 100 mL), and filled with oxygen at pressures up to 10 bar (1 MPa), is monitored. The instrument consists of a control unit, a block heater, which can heat two samples at temperatures up to 150ºC, and a bomb into which the sample is introduced in a glass bottle. The pressure in the bomb is measured electronically and recorded on a multichannel recorder or transferred to a PC.

Oxidograph

The Oxidograph™, manufactured by Mikrolab Aarhus, is an instrument based on the FIRA–Astell apparatus which employs the principle of the Sylvester test. The sample of oil or fat is exposed to oxygen or air at an elevated temperature, with stirring, to accelerate the test. As the sample absorbs oxygen, the pressure drop is measured electronically by means of pressure transducers. The aluminium heating block has spaces for six sample tubes. An analogue signal is recorded for each sample on a six-channel recorder. The sample tubes are glass, and designed to be leakproof, when connected, with no grease required.

Applications to particular foods

Most of the methods mentioned above can be applied to a wide range of foods. Additional steps, such as isolation of fat, may be required in some cases. The standard PV determination and some other analyses cannot be applied to samples with significant water contents. One approach is to separate the fat phase and perform the standard tests. For butter this can be achieved by melting the sample, and for model emulsions freezing overnight at –70ºC may allow the oil to separate on thawing due to its density being lower than that of water. However, for commercial emulsions, such as margarine and mayonnaise, addition of solvent, normally hexane, is required to extract the oil and the standard tests can then be performed after evaporation of the solvent. Confectionery products and biscuits also commonly require solvent extraction of the fat before analysis.

Sensory evaluation is a good method for monitoring deterioration of meat and fish products. For fish oil in particular, the human olfactory system is a very sensitive detector of off-flavours. Off-flavours are normally detected at very low PV values. Headspace analysis represents a useful instrumental method, and the TBA method is also commonly applied as an objective method of detecting deterioration in these foods.

7.5 Future trends

Instrumental methods of assessing the state of oxidation of oils are likely to become more important in the future. Methods that do not require the use of chemical reagents or solvents reduce waste disposal problems. Headspace analysis and FTIR spectroscopy in particular are likely to become more widespread. The trend towards miniaturisation of chromatographic equipment is likely to allow the development of small-scale GC equipment that can be transported for headspace analysis of samples to be carried out with reduced space requirements. The use of autosamplers allows many samples to be analysed by headspace analysis with minimal operator attention. Developments in robotics are likely to allow some of the chemical assays to be automated too.

7.6 Sources of further information and advice

There are many commercial sources of food antioxidants. Details of the chemical properties of food antioxidants and their suppliers are given in *The Index of Antioxidants and Antiozonants* (Ash and Ash, 1997). The following books are recommended for consultation:

ALLEN J C and HAMILTON R J (eds) (1994), *Rancidity in Foods*, 3rd edition, London: Chapman & Hall.

FRANKEL E N (1998), *Lipid Oxidation*, Dundee: The Oily Press.

POKORNY J, YANISHLIEVA N and GORDON M H (eds) (2001), *Antioxidants in Food Practical Applications*, Cambridge: Woodhead Publishing Limited.

SCOTT G (ed.) (1993), *Atmospheric Oxidation and Antioxidants*, Amsterdam: Elsevier Science Publishers.

7.7 References

AOCS (1998), Method Cd 8b-90, *AOCS Official Methods and Recommended Practices of the AOCS*, 5th edition, Champaign, IL: AOCS.

ASH M and ASH I (1997), *The Index of Antioxidants and Antiozonants*, Aldershot: Gower.

CROSS C K (1970), Oil stability. A DSC alternative for the Active Oxygen Method, *J Amer Oil Chem Soc,* **64**, 993–996.

FORSS D A (1973), Odor and flavor compounds from lipids, *Prog Chem Fats and Other Lipids*, **13**, 177–258.

GUILLEN-SANS R and GUZMAN-CHOZAS M (1998), The thiobarbituric acid (TBA) reaction in foods: a review, *Crit Rev Food Sci Nutrition*, **38**, 315–330.

MA K, VAN DEVOORT F R, SEDMAN J and ISMAIL A A (1997), Stoichiometric determination of hydroperoxides in fats and oils by Fourier transform infrared spectroscopy, *J Amer Oil Chem Soc*, **74**, 897–906.

PARR L J and SWOBODA P A T (1976), The assay of conjugable oxidation products applied to lipid deterioration in stored foods, *J Food Technol*, **11**, 1–12.

PEERS K E and SWOBODA P A T (1979), Octanoate: an assay for oxidative deterioration in oils and fats, *J Sci Food Agric*, **30**, 876–880.

VAN DE VOORT F R, ISMAIL A A, SEDMAN J and EMO G (1994), Monitoring the oxidation of edible oils by Fourier transform infrared spectroscopy, *J Amer Oil Chem Soc,* **71**, 243–253.

8

Lipolysis in lipid oxidation

C. Davies, Unilever R & D, Sharnbrook, UK

8.1 Introduction

Lipid oxidation is often a determining factor in the shelf-life of foods, leading to adverse changes in the sensory properties (flavour, colour, texture), nutritive value and possibly production of toxic compounds (Jensen *et al.* 1998a, van Ruth *et al.* 1999). This can be influenced and even rate-limited by the extent of lipolysis as this is the first committed step in the deteriorative pathway and enzymes, such as lipoxygenases (LOX), that catalyse the oxidation of fats, prefer or exclusively act on free fatty acids (FFA).

The stability of the food material towards lipolytic breakdown will be a reflection of the (bio)chemistry of the enzymes, cofactors and lipid substrates involved. Lipids are not water soluble and tend to aggregate, forming an interface with the aqueous environment. Lipases and phospholipases have the characteristic feature in that they work at the lipid–water interface. For this reason, susceptibility to lipolysis, and consequent lipid oxidation, will also be governed by the physicochemical properties of this unique two-dimensional environment. In the case of natural products, derived from plant and animal sources, their nutritional and physiological status prior to harvest/slaughter might influence subsequent product quality and shelf-life parameters.

This chapter describes the role of lipolysis in food deterioration. It describes the enzymes involved in the context of their interfacial nature and the importance of the molecular configuration, organisation and dynamics of the lipid environment in which they work. It describes the role of lipolysis in the shelf-life of various food materials – milk and dairy products, meat and fish, and plant-based foods, such as grains, legumes and leafy vegetables – and looks at methods to mitigate the effects of lipolytic deterioration and extend the shelf-life of the food material. It does not cover the effect of lipolytic microorganisms in food spoilage. Fabricated foods, other than those of dairy origin, are usually free of (phospho)lipases, as are oils and fats used as food or food ingredients. For this reason, these also are not included.

8.2 Lipolytic enzymes, lipids and food spoilage

Lipolytic enzymes will not be reviewed here in detail, as they are described elsewhere (Huang 1987, Derewenda 1994, Ransac *et al.* 1996, Nigam *et al.* 2000).

Lipases (acylglycerol acyl hydrolases, EC 3.1.1.3) hydrolyse the ester bonds in tri-, di-, and monoacylglycerols, releasing free fatty acids. Some have a broader specificity and will hydrolyse other ester linkages. Lipases are important in the mobilisation of storage triacylglycerides (TAG), e.g. in oil seeds and adipose tissue. A lipoprotein lipase is present in milk, requiring an apoprotein for activation.

Phospholipases can be split broadly into two groups: the acyl hydrolases (phospholipases A and B) and the phosphodiesterases (phospholipases C and D).

- PLA1 (EC 3.1.1.32) hydrolyses the acyl group attached to the sn-1 position of the phospholipid.
- PLA2 (EC 3.1.1.4) selectively cleaves at the sn-2 position of phosphoglycerides.
- PLB (EC 3.1.1.5) acts on lysophospholipids (monoacylphospholipids).

PLA1 and PLA2 are present in both meat and fish. Some PLAs are Ca^{2+} dependent.

- PLC (EC 3.1.4.3) hydrolyses the glycerophosphorus bond, leaving a diacylglyceride.
- PLD (EC 3.1.4.4) acts at the phosphorus-nitrogen bond, resulting in phosphatidic acid and the terminal moiety of the phospholipid, e.g. choline, ethanolamine, etc.

PLD is important in the senescing plant and in developing seeds. PLC has also been described in plants used as foods. Other acyl hydrolases that we are concerned with are more general in nature, e.g. acting on glycolipids, sulpholipids, and di- and monoacyl glycerides. 'Galactolipases' in plants are examples of these. The most common phospholipids (PL) are phosphatidylcholine (PC) and phosphatidylethanolamine (PE). Galactolipids predominate in chloroplast membranes of plants.

Lipases have an unusual feature, in that they display enhanced activity when their lipid substrates are aggregated and form an interface with the aqueous medium (Verger 1997). This phenomenon of 'interfacial activation', which usually involves an allosteric change in the enzyme, has also been observed in phospholipases (Derewenda 1994).

Some (phospho)lipases require a second component (a colipase or activator) in order to bind effectively at the lipid–water interface, e.g. lipoprotein lipases in milk can only work when the triacylglyceride substrate forms a complex with a lipoprotein component from serum (Korn 1962). A further consequence of this substrate aggregation is that the two-dimensional surface of the lipid represents the available substrate concentration rather than its three-dimensional volume (Blow 1991). This means that storage lipid, e.g. in oil bodies in seeds and adipose tissue in animals, has a relatively small surface area compared with phospholipids arranged as a membrane bilayer. This is important at high reaction rates, when the low substrate availability will rate-limit the reaction. This constraint might not hold true at lower temperatures when the reaction rate falls, which can lead to underestimates of the role played in rancidity by hydrolysis of storage lipids. This is an important consideration for many foods that are stored at low temperatures for extended periods of time, including in the frozen state.

In terms of rancidity, the degree of unsaturation of the fatty acid chains in the lipid is crucial (Labuza 1971); as a general rule, the more double bonds, the more susceptible it is to oxidation. In nature, polyunsaturated fatty acids (PUFA, two or more double bonds) tend to be located at the sn-2 position in both triacylglycerides and phospholipids. It has been proposed that this arrangement of fatty acids stabilises triacylycerides towards autoxidation (Raghuveer and Hammond 1967). Enzymes that act at the sn-2 position may contribute more to oxidative rancidity by providing FFA that are unsaturated. Generally PUFAs are more susceptible to oxidation when in the free form rather than esterified and it is due to this that lipid hydrolysis or lipolysis can play a key deleterious role.

In the case of the phospholipid, the degree of unsaturation of FA chains will influence a number of properties that will be important to membrane integrity and potentially to stability in the natural raw materials used as foods.

The lipids of membrane phospholipids with highly unsaturated acyl chains, such as in fish, cannot pack closely together due to steric hindrance caused by *cis* double bonds. This has an impact on the fluidity and phase transition temperature of the lipid, as well as on penetration of catalyst/enzyme into the lipid interface (van den Berg *et al.* 1993, Ashton 2002, attributed to Abousalham and Verger (2004)).

Lipids undergo a phase change in response to temperature change. Above the phase transition temperature they are in the liquid crystalline state and are fluid, and below they are in the gel phase and are rigid. *In vivo*, membrane lipids are normally in the liquid crystalline phase. As the temperature drops and they undergo a phase transition, the membrane becomes leaky to solutes (Simon 1974). As in tissue wounding, where this can also occur, this can lead to activation of enzymes, including calcium-dependent phospholipases, which can lead to autolysis of the membrane. As integrity is lost, further reactions occur due to the mixing of enzyme and their co-factors with substrates (Schwimmer 1981, Haard 1995).

Phospholipid membranes with a high degree of unsaturation will stay fluid at lower temperatures. This could be an important factor in the storage/freezing conditions of intact raw materials (i.e. when enzymes present have not been heat inactivated).

Oxidation has the reverse effect to unsaturation in terms of membrane dynamics in that it causes a decrease in fluidity (Borst *et al.* 2000). Abousalham *et al.* (2000), using lipid monolayers, have shown that FFAs 'expand' on oxidation. On application of pressure, the oxidised lipid is preferentially squeezed out. It is well established that oxidised fatty acids are more soluble than their non-oxidised counterpart. These observations demonstrate that this phenomenon is due purely to the physicochemical properties of the monolayer. It is interesting to speculate on whether this at least partly explains examples in both animal and plant systems, where fatty acids oxidised *in situ* in membranes are selectively removed (Yasuda and Fujita 1977, van Kuijk *et al.* 1987, Banaś *et al.* 1992, van den Berg *et al.* 1993, Feussner *et al.* 1995, Sevanian and Kim 1985).

As reactions catalysed by (phospho)lipases involve interfacial adsorption and subsequent catalysis, these physicochemical properties must have some bearing on lipolysis in food materials. It is noteworthy that oxidation is enhanced at lower pressures, when the monolayer is in the expanded state (Ashton 2002, attributed to Abousalham and Verger (2004)). This occurs after lipolysis and suggests how lipolysis and lipid oxidation could work together (Abousalham and Verger, 2004). It also points to the importance of the physicochemical environment in addition to properties of the enzymes involved *per se*.

In the case of natural products, such as fish, meat or vegetables, it may also be appropriate to consider lipolytic enzymes in their normal physiological environment, as some are under hormonal or signal-mediated control, or are involved in defence mechanisms which can result in rapid mobilisation of lipids and their subsequent breakdown.

8.3 Lipolysis in particular foods: dairy, meat and fish products

8.3.1 Milk and dairy products

Raw milk contains lipoprotein lipase (LPL) and a complex mix of fatty acids, esterified predominantly as triacylglyceride. This 'butterfat' constitutes 2.6–6% milk and $\geq$80% butter (Padley *et al.* 1994). Butyric (C4:0) and caproic (C6:0) acids esterified at the *sn*-3 position, if released by hydrolysis, are volatile and could lead to taint (rancid butter and goat-like notes, respectively). However, despite the presence of LPL, there is little lipolysis-induced rancidity. This is because the milk fat is protected by a milk fat globule membrane (Fox and Morrissey 1981). This membrane is composed of glycerolipids, phospholipids, cholesterol and its esters, free fatty acids, squalene, carotenoids and some proteins (Dimick *et al.* 1970). In order for the lipase to hydrolyse the fat, it has to bind to the fat–aqueous interface. This requires the apoprotein apoC11, which anchors the enzyme to the substrate particle in the correct orientation. This apoprotein, present in serum, is absent from milk. However, any event that modifies or disrupts the fat–water interface, e.g. mechanical agitation during milking, churning, creaming, etc., where the globule structure can be compromised, can lead to lipolysis and spoilage by the native enzyme in raw milk. These factors need to be taken into consideration when introducing changes to the milk production system.

Lipoprotein lipase, isolated by Egelrud and Olivecrona (1972) as a 62–66 kDa protein, is denatured at 55–60°C and so heat treatment, such as pasteurisation at 72°C, is successful at inactivating it. (Lipolysis can also occur as a result of microbial contamination after heat treatment.) Skimming can also remove fat globules, making oxidative rancidity less likely.

In goats' milk and milk products, the lipolytic system plays an important part in the development of the characteristic goat flavour. Despite the levels of the lipoprotein lipase being lower, the rate of lipolysis is higher. This is because more of the enzyme is bound to the fat globule than in bovine milk (Chilliard *et al.* 2003).

Lipolysis can play a role in flavour development in cheese. If pasteurised milk is used, lipase can be added during ripening. When raw milk is used, the native lipase present in the milk or curd loses its potential for lipolytic activity during the ripening process due to curd scalding, whey removal and the drop in pH, though fresh or soft ripening cheeses can be affected by taint (Guerts 2003). Skim milk can also be recombined with unsaturated fats, e.g. from vegetable origin, in the production of soft ripening cheeses. Again, both lipolysis and the oxidative stability of the added lipid fraction need to be considered in respect of shelf-life (During *et al.* 2000, Gonzalez *et al.* 2003).

Milk fat can also be modified, in the manufacture of ice cream, butter and other products with altered properties (spreading, melting, etc.). The C18 polyunsaturated fatty acid content in bovine milk is low (C18:2, C18:3 is typically 1–3%, showing seasonal variation). This is because microorganisms in the rumen partially hydrogenate dietary C18 fatty acids before adsorption. However, biohydrogenation can be overcome, for example, by feeding encapsulated polyunsaturated fatty acids, which can increase C18:2 content to 35%, with subsequent products having oxidative stability problems (Edmondson *et al.* 1974). The oxidised taint can be prevented by addition of α-tocopherol to the fresh milk.

Dietary supplementation with α-tocopherol can also improve the oxidative stability of milk and can overcome the pro-oxidant effects of added copper (Chawla *et al.* 2003).

8.3.2 Meat and meat products

Lipid oxidation is a principal factor in quality deterioration in meat and meat products (Jensen *et al.* 1998a). Free fatty acids continue to increase with meat ageing (Valin *et al.* 1975) and it also appears to be a key quality limiting parameter in the dry-curing process of hams where phospholipases are thought to be responsible for thc loss of PC and PE (Toldrá and Flores 1998, Hernández *et al.* 1999).

The nature and extent of lipolysis appear to depend on the metabolic muscle type. This is due to levels of both substrates and enzymes, which differ with the oxidative activity of the muscle, and (though phospholipid breakdown might be more important to oxidative rancidity, especially in the short term) to the efficacy of mobilisation of lipids from depot fats post-mortem.

Hernández *et al.* (1998), working on different muscles in pork, showed that red (oxidative) muscle had higher levels of phospholipid (and consequently PUFA content) as well as higher levels of phospholipase and lipase activity, compared with white (glycolytic) muscle. The level of PUFA content increased with the oxidative activity of the muscle.

In later work on a white muscle type (porcine *Longissimus dorsi* muscle, vacuum packed and stored at −18°C for six months) it appears that the source of FFA could be wholly attributed to phospholipase activity on phospholipid, principally from the loss of linoleic acid (C18:2) from PE, with no change in the non-polar fraction. Consistent with this trend, Alasnier *et al.* (2000) showed that phospholipid was a more important source of FFA than TAG in the white muscle of rabbit, whereas, in red muscle, TAG contributed similar or higher levels of FFA than phospholipid. This reflects the metabolic enzyme profile in the different muscle types, as red (oxidative) muscle, which is rich in iron-containing enzymes, is better equipped at drawing energy from depot fats, whereas white (glycolytic) muscle is more suited for drawing energy from glucose. A similar observation has been made in stored turkey meat (Sklan *et al.* 1983) where oxidation was more pronounced in the leg muscle than in the breast muscle. In this case, the authors attributed this to the higher proportions of phospholipid and PUFA content.

The efficacy of iron-containing proteins present in red muscle, in promoting oxidation, might also be an important factor in oxidative stability. Nevertheless, red muscle meat and products made from these parts of the animal are more likely to be susceptible to oxidative rancidity compared with white muscle and its products (see also Shahidi 1992).

Poultry meats, such as chicken, turkey, duck, etc., contain a higher proportion of PUFA compared with red meats, but more of the fat is found under the skin rather than distributed in the meat (Scott 1956) and can be trimmed. Cooked chicken breast contains 1.3% fat, compared with 13–30% for beef (Mountney 1976). The unsaturated fat content of adipose tissue in ruminants at different depths below the outer surface has been looked at by Body (1988).

Lipolytic enzymes present in adipose tissue include lipoprotein lipase, hormone-sensitive lipase (hydrolysing TAG to DAG) and monoacylglyceride lipase (Toldrá and Flores 1998). Since at least some of the events are controlled by hormones, the physiological and nutritional status of the animal pre-slaughter might affect the final meat quality and its stability on storage.

As is the case with milk, fat from various sources on the animal can be influenced by the degree of fatty acid saturation in their diet. The breed of animal can also influence the fat content of meat (Bass *et al.* 1990). Supplementation with vitamin E can protect against oxidative reactions in stored muscle tissue and provide a nutritional advantage (Jensen *et al.* 1997, 1998b, Granit *et al.* 2001). See also Russel *et al.* (2003).

8.3.3 Fish

The lipids in fish are different from that of most other animals in that they are longer chained and have a higher degree of unsaturation. These polyenoic acids originate in phytoplankton and algae at the base of the marine food chain. The two major fatty acids, eicosapentaenoic acid (EPA) (C20:5 *n*-3) and docosahexaenoic acid (DHA) (C22:6 *n*-3), are thought to be beneficial to human health and are associated with a reduced risk of coronary vascular disease and atherosclerosis (Dyerberg 1978, Barlow 1980, Dyerberg and Jorgensen 1982). The unsaturated nature of the lipids, however, has a penalty post-mortal with regards oxidative stability.

In frozen cod, hept-*cis*-4-enal is associated with a 'cardboardy' off-flavour (Hardy *et al.* 1979) and 2-*trans*, 4-*cis*, 7-*cis* decatrienal, which gives a strong fishy flavour, has been shown to originate from EPA (Meijboom and Stroink 1972).

FFA have reported to be responsible for taint themselves, with DHA, palmitoleic acid, linoleic acid and EPA released from TAG being responsible for train oil taste, bitterness and metal taste in frozen salmon (Refsgaard *et al.* 2000). FFA are also suspected of being involved in cross-linking of proteins (Dyer and Fraser 1959, Hanson and Olley 1964) and spoilage of texture. FFA are generally accepted to be the better substrates for oxidation in fish, though there is some debate (Labuza 1971, Shewfelt and Hultin 1983).

Body and Vlieg (1989) looked at the distribution of lipid at different sites in the blue mackerel, an oily fish. The dark meat (the first 0.5 cm depth under the skin), representing 20% of the fillet, contained 20% lipid; the white meat, representing 80% of the fillet, contained 4% lipid. The white meat contained twice as much PL as the dark (19% compared with 10%) and had more of the *n*-3 fatty acids EPA and DHA (approximately double).

Bosund and Ganrot (1969) found that, in frozen Baltic herring, most FFA came from TAG in dark muscle (55%) and most from the PL fraction in white muscle (75%). Tsukada (1976) reported the same trend in frozen stored tuna. This is similar to the situation in meat of mammalian origin (described in Section 8.3.2) in that PL was the major source of FFA in light muscle, whilst red meat was competent at breaking down storage TAG and PL. In the case of Bosund and Ganrot, the dark muscle (and therefore TAG) was the most important with respect to lipid hydrolysis as the initial levels of lipid and the rate of hydrolysis were higher than in the light muscle. The FFA content was 1000 mg per 100 g and 280 mg per 100 g in dark muscle and light muscle respectively after 12 weeks at −15°C. The situation is not simple, however. In frozen milkfish, another member of the herring family, hydrolysis of neutral lipid only occurred after hydrolysis of PL. In frozen cod fillets, a lean fish which is not particularly regarded as having a rancidity problem (Hardy *et al.* 1979), FFA came from PL only.

It is the oily fish, like herring, salmon, pilchard, menhaden, etc., with a lipid content of $\geq$5% total weight of flesh, that are regarded as suffering from rancidity problems. The biggest variation in lipid content is accounted for by triacylglyceride and can be up to 25–30%. It will vary according to a number of factors, including species, feeding habit, season, gender, etc. Despite this, most work has been done on phospholipids, though here triacylglycerides and lipases will be dealt with first.

In fish triacylglycerides, polyunsaturated fatty acids occur at the sn-2 position, with a saturated or monounsaturated fatty acid at the sn-1 position and a monounsaturated fatty acid at *sn*-3 (Brockerhoff *et al.* 1968, Malins and Wekell 1970, Litchfield 1972).

Few TAG lipases have been reported in fish muscle, but these include a short-chain lipase from mackerel (George 1962) and a long-chain lipase from rainbow trout (Bilinski *et al.* 1971) which was later shown to be of lysosomal origin which was released on slow freezing and temperature fluctuations. Fast to intermediate freezing rates and low storage temperatures ameliorated this affect (Geromel and Montgomery 1980). An enzyme has also been partially purified from the adipose tissue of steelhead trout (Sheridan and Allen 1984).

There is some evidence that fish muscle lipases involved in the mobilisation of storage TAG are hormone sensitive and so nutritional status and stress on capture could have important shelf-life considerations (Ashton 2002). Lipases from the digestive system could also cause problems, particularly in small fish species, e.g. sardines and anchovies, or when degutting has not been successful or when rupture has occurred during post-catch storage and transport of the ungutted fish. The latter can be controlled to some extent by storage at low temperature.

Phospholipids contain proportionately more EPA/DHA compared with triacylglycerides, mostly at the sn-2 position of phosphatidylcholine and phosphatidylethanolamine. Both phospholipases A1 and A2 have been found in fish. The phosphodiesterases PLC and PLD have not been identified. PLA2 have been reported from a number of sources, in both lean and oily fish (reviewed by Ashton 2002). Some, as in pollack muscle, have a requirement for Ca^{2+} (Audley *et al.* 1978); others are calcium independent, e.g. in trout (Bilinski and Jonas 1966). In trout and salmon, there is evidence that PLA2 is involved in eicosanoid production in the inflammatory response (Pettit *et al.* 1989, Bell *et al.* 1992). If parallels are drawn with the mammalian system, tissue damage and stress caused on capture could cause elevated levels of this enzyme which could have shelf-life implications (Ashton 2002). A calcium-independent PLA1 has been found in the white muscle of tuna and in Atlantic mackerel (Hirano *et al.* 1997, Ashton 2002). Phospholipase activity in lean fish has been associated with toughened texture caused by the interaction of FFA and protein muscle (Dyer and Fraser 1959, Hanson and Olley 1964, Sikorski *et al.* 1976).

In addition to enzyme-catalysed events, the molecular configuration of these polyenoic fatty acyl chains when present esterified as phospholipids in biomembranes could also contribute to instability, as *cis* double bonds cause kinks which will impede tight packing of the acyl chains and influence membrane organisation and integrity (Yasuda and Fujita 1977, van Kuijk *et al.* 1987, van den Berg *et al.* 1993, Sevanian and Kim 1985).

On cold adaptation, the saturated fatty acid at the sn-1 position can be replaced by an unsaturated fatty acid and this will affect the liquid crystalline/gel phase transition temperature, e.g. the transition temperature for cold-adapted carp can shift from −8°C to −13°C (Fodor *et al.* 1995; see also Hazel *et al.* 1998). These phase changes cause the membrane to become leaky to solutes including Ca^{2+}, important in the activation of some phospholipases. In some systems the lipid phase could also affect the efficacy of binding of the lipolytic enzyme to the interface which could be more efficient in the gel phase, i.e. at lower temperature (Menashe *et al.* 1986).

De Koning and Mol (1990) made an interesting comparison between loss of PL and TAG in frozen Cape hake at different temperatures. They showed that the rate of hydrolysis of PL declined more sharply with decreasing temperature than that of neutral lipids. They calculated that, above −12°C, PL was hydrolysed faster than neutral lipid, and below −12°C TAG was hydrolysed faster. This might be a reflection of the available substrate which for PL is given by the surface provided by the bilayer of the membrane, whereas in the case of storage TAG, only the lipid at the surface of the oil globule will be available for hydrolysis. This could rate-limit the hydrolysis of the TAG and provide an explanation for this phenomenon. However, it is interesting that this calculated cross-over point for the rates of hydrolysis of PL and TAG (at −12°C) approximates to the phase transition temperature of fish PL (see above) where a number of other factors could come into play. These are important considerations for fish which is often stored on ice for days or frozen for extended periods in the raw state. De Koning and Mol deduced that above −12°C, lean fish would develop a higher level of FFA than oily fish and the reverse would be the case below −12°C.

The effects of dietary levels of fat, α-tocopherol and astaxanthin, a naturally occurring derivative of vitamin A, have been looked at with respect to storage stability of frozen rainbow trout and smoked fillets stored chilled (Jensen *et al.* 1998c, Ingemansson *et al.* 1993). Increasing levels of dietary astaxanthin appeared to protect against lipid oxidation in frozen fish (−28°C), whereas in smoked fillets, stored chilled at 3°C, increasing supplementation with α-tocopherol counteracted lipid oxidation. Ruff *et al.* (2002) looked at the shelf-life of Atlantic halibut and turbot and found that halibut, which had a higher level of α-tocopherol, was more resistant to oxidation and colour change than turbot.

8.4 Lipolysis in particular foods: cereals and vegetables

Lipolytic enzymes play a role in the mobilisation of oil reserves in seeds on germination and in post-germination growth. Consequently this has implications for the shelf-life of seeds used as human foods, such as grains and legumes, which make up a significant part of the human diet. Lipolysis is also involved in the yellowing of leafy vegetables and other parts of the plant used as foods.

Cereal grains contain about 2–10% lipid, most in the form of TAG of which 80–90% of the FA are C18:2 and C18:1 (Huang 1987). When the wholemeal or bran fraction is ground to a flour, the baked product can suffer from the development of undesirable flavours, as well as reduced baking quality because of the presence of lipid metabolising enzymes in the bran and germ component (Galliard 1986a). This is a storage-dependent phenomenon. The TAG lipase in the bran component is active at low water activity and leads to the slow accumulation of free PUFA in the dry wholemeal over several weeks. This is rapidly oxidised (within minutes) by lipoxygenases (LOX), present in the germ on addition of water.

The lipolytic step, whilst slow, is likely to be rate limiting, as the germ lipoxygenase has a preference for free fatty acids and will be the main factor in the fluctuation of the flavour (rancidity) as well as in loaf volume and crumb texture (Galliard 1986b, Tait and Galliard 1988). LOX related co-oxidation reactions will also be affected (Galliard 1986b, Tait and Galliard 1988, O'Connor *et al.* 1992).

The lipase content can be used as a predictive tool for the deterioration rate for wholemeal during storage whilst oxygen uptake measurements of FFA levels can inform on the storage history and extent of the deterioration of the flour. The effect on baking quality is linked to the ability of fatty acid to 'bind' to non-lipid components during dough mixing and can be restored by addition of fat (TAG) and emulsifiers.

Lipase present in the bran of wheat grain is stable for up to 80°C for at least 7 days (Galliard 1987). Temperature inactivation is important in some grain processes, for example, in the production of oat products for human consumption where post-harvest treatment and storage and the processing of the oat kernels cause tissue damage. Both high temperature and good heat transport into the sample are needed for enzyme inactivation (Ekstrand *et al.* 1992). In the case of rice bran, where rancidity has been demonstrated during storage (Hirayama and Matsuda 1975), microwave heating has been shown to be effective (Ramezanzadeh *et al.* 1999).

Some grains, e.g. maize, rice bran and wheat germ, are used in the production of oil commercially. The cereal grain is often first processed by heating and drying to inactivate the enzyme and prevent the oil being hydrolysed. Oils of good quality are characterised by the presence of tocopherols and in some cases naturally occurring antioxidants such as esters of ferulic acid (rice bran) (Padley *et al.* 1994).

In 1992 it was estimated that commercial losses due to lipase activity in wheat amounted to about 5% of the wholemeal flour operation. Off-flavours in the production of malt and beers have also been attributed to lipases present in barley (O'Connor *et al.* 1992). In the case of germinating barley, such as is used in brewing, lysophospholipases are present in the endosperm that appear to have a role in releasing lysophospholipid from starch during germination (Huang 1987).

Two other lipolytic enzymes important to the shelf-life of plant foods are phospholipase D (PLD) and so-called 'galactolipase'. PLD is involved in membrane lipid hydrolysis in a number of physiological processes in plants, e.g. senescence and response to stress injuries, as well as lipid mobilisation during seed germination (Dyer *et al.* 1996). PLD is also involved in phospholipid degradation when plant cells suffer freezing injury (Yoshida 1979). The enzyme has been found in a number of plant sources relevant as foods including peas, beans, carrots, spinach leaves, sprouts, cabbage, etc. (Lee 1989, Wang *et al.* 1993, Abousalham *et al.* 1995). Moreau (1987) has compared the levels of this enzyme in various plant tissues. PLD is highest in rapidly growing plant tissues such as storage tissues and seeds (Moreau 1987). Peas and beans deteriorate rapidly as a result of tissue damage caused by mechanical harvesting. Plant PLD is strictly dependent on Ca^{2+} (Novotná *et al.* 1999) and so perturbations (or phase changes) in the membrane, causing solute leakage and loss of cellular integrity, could lead to rapid activation of PLD, not only at the site of the wounding, but also in undamaged parts of the tissue (Ryu and Wang 1996).

Phosphatidic acid released on PLD-catalysed hydrolysis can be further broken down to diacylglycerides (DAG) by phosphatidic acid phosphatase. DAG can then be hydrolysed by a non-specific acyl hydrolase, releasing FFA which is dioxygenated by lipoxygenases to produce hydroperoxides. This ultimately leads to the formation of volatile short-chain aldehydes and alcohols, resulting in both desirable aromas and undesirable taints which can be a quality-limiting factor in many fruits and vegetables (Pendlington 1965, Holdsworth 1969, Luning *et al.* 1995, Robinson *et al.* 1995, Fan *et al.* 1997). Vegetables, such as peas and beans, which have high levels of LOX and suffer post-harvest damage due to the mechanical harvesting methods used, are usually blanched to inactivate enzymes prior to freezing, so that they are stable on prolonged storage.

Soyabeans which are often stored for long periods of time before processing or for production of soyabean oil, if stored under poor conditions or are damaged due to disease or insect attack, or cracked, have elevated levels of phosphatidic acid (PA) and lysophosphatidic acid (LPA) which lead to quality problems (Simpson 1991). Microwave heat and live steam have been demonstrated to be effective in the inactivation of PLD in soyabean and soyflake (List *et al.* 1990).

Lipoxygenases are often thought of as the main culprit in the oxidative degradation of vegetables, but as they tend to have a preference for FFA (Todd *et al.* 1990), the lipolytic steps in the pathway are likely to be rate limiting. There is some evidence, however, that microsomal plant PLD prefers PC with an oxygenated acyl group, though this varies between different plant species (Banás *et al.* 1992) and presumably with developmental stage. LOX could therefore play a role in the early stage of this process in some circumstances.

Yellowing of the foliage of leafy vegetables, which can occur soon after harvesting and is a key factor in quality loss, has also been attributed to lipolytic enzymes. In this case it is principally due to deacylation of glycolipids

(Yamauchi *et al.* 1986, Zhuang *et al.* 1994). The enzyme responsible is often referred to as 'galactolipase', though it tends to be non-specific in nature, capable of hydrolysing PL and other acyl molecules, but usually not TAG. More than one enzyme with differing substrate specificities can be present, as is the case in the beans *Phaseolus multifloris* and *P. vulgaris* (Burns *et al.* 1979, 1980, Mukerjee 1994). See also Galliard (1971).

In green leaves, galactolipids and sulpholipids are the main lipid components of chloroplasts. At 25°C, galactolipids, mainly in the form of monogalactosyldiglyceride (MGDG) and digalactosyldiglyceride (DGDG), together with phosphatidylglycerol (PG), show a considerable loss. These chloroplast lipids have a high proportion of PUFA (Gounaris *et al.* 1986). Deterioration of the chloroplast thylakoids leads to the degradation of chlorophyll and yellowing in leaf tissues and also in other green parts of the plant, e.g. the florets of broccoli (Zhuang *et al.* 1994).

MGDG is also susceptible to hydrolysis in potato tubers where there can be a loss of membrane integrity on storage which is a common phenomenon (Galliard 1970, Kumar and Knowles 1993). The lipid acylhydrolase in potato tubers (patatin) constitutes 20–40% of the total soluble protein or ~2% of the dry weight of the tubers (Racusen and Foote 1980, Racusen 1983). Patatin also hydrolyses PL and monoacylglyceride (MAG), but is ineffective on TAG (Galliard 1970, Andrews *et al.* 1988). PLC has also been reported in plant foods, e.g. celery and oat roots (McMurray and Irvine 1988, Huang *et al.* 1995), though little work has been done.

8.5 Controlling lipolysis to improve shelf-life

Nature has a number of strategies for protecting against unwanted lipolytic reactions, including deposition of lipid in globules, oil bodies and adipose tissue, etc., thus restricting the surface area available for lipolysis; separation of lipolytic enzymes from their cofactors and activators to prevent interfacial activation and catalysis; cold adaptive strategies to modulate lipids, ensuring membrane fluidity and preservation of (sub)cellular integrity; and use of α-tocopherol/vitamin E effective in the membrane environment which has both biochemical and biophysical antioxidant properties (Fryer 1995). Broadly speaking, these can be classified as (i) conservation of lipids and membranes, (ii) ensuring lipolytic enzymes are not active, and (iii) exploitation of the special role that α-tocopherol/vitamin E fulfils. Strategies to improve the shelf-life of foods containing lipids generally fall in line with these criteria.

The first steps to improving shelf-life can be made prior to harvest/slaughter. In some animals and fish, lipolytic enzymes are under hormonal control and so their physiological and nutritional status is important. Dietary lipid, supplementation with antioxidants, e.g. vitamin E, and breeding for lean animals can also play a role. In plants, drought, in-field freezing injury and events triggering senescence could play a role.

Stress on wounding, caused by harvesting/slaughter, can cause a cascade of deleterious reactions and should be minimised. As lipolysis is the first committed step and is likely to rate-limit lipid oxidation, it is desirable in most cases to act quickly to inactivate or restrict the activity of enzymes. Peas and beans develop rancidity quickly after wounding, presumably due to Ca^{2+} mediated lipolytic events and LOX catalysed oxidation, and need to be blanched within a short period of time. Action should certainly be taken before there are signs that lipid oxidation has occurred. A good case in point is seen in the storage of grains, where under the dry conditions of storage, lipases are active, but oxidative rancidity is not apparent until water is added when LOX can rapidly oxygenate the accumulated FFA.

Heat inactivation is a commonly used strategy, e.g. in the pasteurisation/sterilisation/UHT treatment of milk, blanching of vegetables and steam/microwave inactivation of enzymes in grains and legumes such as soyabeans. Prior to heat treatment of milk, it is important to maintain the integrity of the milk fat globule by minimising agitation in the milk/dairy process. When heat treatment to inactivate enzymes is not possible, reduction in temperature/freezing to slow down the reaction rates of enzymes is another strategy. Temperature fluctuations should be minimised and appropriate rates of freezing should be employed in order to prevent additional problems due to loss of membrane integrity, lysosomal disruption with subsequent release of enzymes and solute leakage due to phase transition or ice crystal damage.

As well as dietary routes, α-tocopherol can be added to foods such as milk during processing, e.g. in the production of dairy products. Chelation of cofactors and reduction of pH (as seen in the cheese production process) could also be successful in some circumstances. Consideration can also be given to the tissue localisation of lipolytic enzymes, e.g. in some grains, lipase is found predominantly in the bran component and not in the germ. Removal of fat from meat or deep skinning in the case of fish are also options.

8.6 Future trends

The *n*-3 PUFAs, EPA and DHA, present in the lipids of fish and other marine sources, are essential fatty acids in the human diet, as they cannot be synthesised in the body. The presence of these fatty acids in the diet is thought to be beneficial in reducing cardiovascular disease and potentially in reducing the potency of some inflammatory conditions, such as rheumatoid arthritis and asthma. Despite this, the current intake of these lipids, in the UK at least, is considerably short of government recommendations. There is some evidence that the balance between *n*-6 and *n*-3 fatty acids is the important criterion. As dietary *n*-6 lipids have increased since the 1970s with use of vegetable-sourced oils as alternatives to animal/dairy fats, this continues to be an issue. Reintroduction of the *n*-3 PUFA into the diet is likely to be a continuing trend. Current examples include *n*-3 enriched eggs, brought about by modification of the diet of laying hens, and the (proposed) introduction of DHA into infant formula milks.

Alpha-linolenic acid (C18:3), the other member of the *n*-3 family, present in nuts (especially walnuts) and some oilseeds/cereals and leafy vegetables, such as spinach, can be converted to DHA in the body. This *n*-3 fatty acid is also present in grass-fed animals, such as beef cattle and lamb. Feeding strategies could lead to increased levels in these meats, though consumption of red meats has declined in recent years.

As *n*-3 fatty acids are susceptible to oxidative rancidity, any measure that reduces lipolysis and subsequent oxidation is important. An obstacle to using marine-sourced *n*-3 oils as an ingredient in foods or as a heart health supplement is the fishy odour brought about by oxidative rancidity. It is anticipated that deodorisation, encapsulation and methods of protection, maybe with naturally occurring antioxidants, could be developed in due course, with research input. Many naturally occurring antioxidants occur in herb extracts and plant-derived oils, which are currently not fully exploited, and it is possible that lipolytic inhibitors are present. The need to obtain raw materials from sustainable fish sources might also give rise to the use of alternative fish species as foods and to the farming of oily fish species. Use of hoki as an alternative to cod (both lean species) is an example of the former.

A further crucial health trend in Europe and the US is the drive towards reducing levels of obesity in both adults and children. This, if successful, could result in a shift in the levels of different food groups consumed and could lead food manufacturers to develop foods with less fat, or altered PUFA or saturated fatty acids (SFA) ratios, with consequences for lipolytic and oxidative stability. Consumption of foods with an increased wholemeal or bran component (containing lipase) could also increase.

There is also a growing market for convenience foods, which can be eaten immediately (snacking) or with minimal preparation/cooking time. Many of these are ambient stable/chilled and have a shorter shelf-life than frozen, canned or dried foods and so spend a shorter time in the food chain. Desire to extend the shelf-life of these products might lead to rethinking of microbial and enzyme-related deterioration, particularly in foods that cannot be heat-treated or are processed subsequent to enzyme denaturation. More attention is likely to be directed towards preventing the oxidative steps rather than lipolytic steps as this is better understood, is easier to measure and does not need to take into account difficulties due to the interfacial character of lipolysis and the lack of solubility of the lipid substrate in the aqueous phase of food materials.

8.7 Sources of further information and advice

Most literature in the area of lipolysis and shelf-life focuses on lipid oxidation or analysis of lipids/lipid fragments where lipolysis is implicated. Much of this is covered in the references in Section 8.8. Whilst the Internet is an excellent resource for searching online peer-reviewed papers and searchable indexes, such as the Institute of Scientific Information's ISI Web of Science, a comprehensive site serving lipolysis and foods has not been identified.

The single most important piece of advice with respect to lipolysis is always to consider the implications of the substrate (and sometimes the product) not being water soluble. This is particularly the case when performing measurements of lipolytic activity. A large number of assays are available and the following reference is comprehensive, providing methodology and recommendations:

Beisson F *et al.* (2000), 'Methods for lipase detection and assays: a critical review', *Eur J Lipid Sci Technol*, 2, 133–153.

8.8 References

ABOUSALHAM A and VERGER R (2004), 'Continuous measurement of the lipoxygenase-catalyzed oxidation of unsaturated lipids using the monomolecular film technique', (in press).

ABOUSALHAM A *et al.* (1995), 'Phospholipase D from soybean (*Glycine max* L.) suspension-cultured cells: purification, structural and enzymatic properties', *Plant Cell Physiol*, 36 (6), 989–996.

ABOUSALHAM A *et al.* (2000), 'Surface properties of unsaturated non-oxidized and oxidized free fatty acids spread as monomolecular films at an argon/water interface', *Chem Phys Lipids*, 104, 93–99.

ALASNIER C *et al.* (2000), 'Lipolysis in muscles during refrigerated storage as related to the metabolic type of the fibres in the rabbit', *Meat Sci,* 54 (2), 127–134.

ANDREWS D L *et al.* (1988), 'Characterization of the lipid acyl hydrolase activity of the major potato (*Solanum tuberosum*) tuber protein, patatin, by cloning and abundant expression in a baculovirus vector', *Biochem J*, 252, 199–206.

ASHTON I P (2002), 'Understanding lipid oxidation in fish', in Bremner H (ed.), *Safety and Quality Issues in Fish Processing*, Cambridge, Woodhead, 254–285.

AUDLEY A *et al.* (1978), 'Isolation and properties of phospholipase A from pollock', *J Food Sci*, 43, 1771–1775.

BANAŚ A *et al.* (1992), 'Plant microsomal phospholipases exhibit preference for phosphatidylcholine with oxygenated acyl groups', *Plant Science*, 84, 137–144.

BARLOW S M (1980), *Nutritional Evaluation of Long-chain Fatty Acids in Fish Oil*, London, Academic Press.

BASS J J *et al.* (1990), 'Practical methods of controlling fatness in farm animals', in Wood J D and Fisher A V, *Reducing Fat in Meat Animals*, London and New York, Elsevier Applied Science, 148–154.

BELL J G *et al.* (1992), 'Effects of increasing dietary linoleic acid on the phospholipid fatty acid composition and eicosanoid production in leucocytes and gill cells of Atlantic Salmon', *Prostaglandins, Leukotriens and Essential Fatty Acids*, 45, 197–206.

BILINSKI E R *et al.* (1971), 'Lysosomal triglyceride lipase from the lateral line of rainbow trout', *J Fish Res Bd Canada*, 26, 1857–1866.

BILINSKI E R and JONAS R E E (1966), 'Distribution of lecithinase in the subcellular fractions of rainbow trout lateral line', *J Fish Res Bd Canada,* 23, 1811–1813.

BLOW D (1991), 'Lipases reach the surface', *Nature*, 351, 444–445.

BODY D R (1988), 'The lipid composition of adipose tissue', *Prog Lipid Res*, 27, 39–60.

BODY D R and VLIEG P (1989), 'Distribution of the lipid classes and eicosapentaenoic (20:5) and docosahexaenoic (22:6) acids in different sites in blue mackerel (*Scomber australasicus*) fillets', *J Food Sci*, 54 (3), 569–572.

BORST J W *et al.* (2000), 'Oxidation of unsaturated phospholipids in membrane bilayer mixtures is accompanied by membrane fluidity changes', *Biochem Biophys Acta*, 1487, 61–73.

BOSUND I and GANROT B (1969), 'Lipid hydrolysis in frozen baltic herring', *J Food Sci* 34, 13–18.

BROCKERHOFF H *et al.* (1968), 'Positional distribution of fatty acids in depot triglycerides of aquatic animals', *Lipids*, 3, 24–29.

BURNS D D *et al.* (1979), 'Purification of acyl hydrolase enzymes from the leaves of *Phaseolus multiflorus*', *Phytochem*, 18 (11), 1793–1797.

BURNS D D *et al.* (1980), 'Properties of acyl hydrolase enzymes from *Phaseolus multiflorus* leaves', *Phytochem*, 19 (11), 2281–2285.

CHAWLA R *et al.* (2003), 'Influence of vitamin E and beta-carotene supplementation on spontaneous oxidised flavour of milk in dairy cows', *Indian J Anim Sci*, 73 (7), 807–811.

CHILLIARD Y *et al.* (2003), 'A review of nutritional and physiological factors affecting goat milk lipid synthesis and lipolysis', *J Dairy Sci*, 86 (5), 1751–1770.

DE KONING A J AND MOL T H (1990), 'Rates of free fatty acid formation from phospholipids and neutral lipids in frozen cape hake (*Merluccius* spp) mince at various temperatures', *J Sci Food Agric*, 50, 391–398.

DEREWENDA Z S (1994), 'Structure and function of lipases', *Adv Protein Chemistry*, 45, 1–52.

DIMICK P S *et al.* (1970), in Phillipson A T (ed.), *Physiology of Digestion and Metabolism of the Ruminant*, Newcastle upon Tyne, Oriel Press, 529.

DURING A *et al.* (2000), 'Lipolysis and oxidative stability of soft ripening cheeses containing vegetable oils', *J Dairy Research*, 67 (3), 461–466.

DYER J H *et al.* (1996), 'Structural heterogeneity of phospholipase D in 10 dicots', *Biochem Biophys Res Commun*, 221, 31–36.

DYER W J AND FRASER D I (1959), 'Proteins in fish muscle. 13. Lipid hydrolysis', *J Fish Res Bd (Canada)*, 23, 1025.

DYERBERG J (1978), 'Eicosapentaenoic acid and prevention of thrombosis and atherosclerosis', *Lancet*, 2, 117–119.

DYERBERG J and JORGENSEN K A (1982), 'Marine oils and thrombogenesis Prog', *Lipid Res*, 21, 255–267.

EDMONDSON L F *et al.* (1974), 'Feeding encapsulated oils to increase polyunsaturation in milk and meat fat', *J Amer Oil Chem Soc*, 51 (3), 72–76.

EGELRUD T and OLIVECRONA T (1972), 'The purification of a lipoprotein lipase from bovine skim milk', *J Biol Chem*, 19 (10), 6212–6217.

EKSTRAND B *et al.* (1992), 'Lipase activity in oats – distribution, pH dependence, and heat inactivation', *Cereal Chem*, 69 (4), 379–381.

FAN *et al.* (1997), 'Antisense suppression of phospholipase Dα retards abscisic acid- and ethylene-promoted senescence of postharvest *Arabisopsis* leaves', *Plant Cell*, 9, 2183–2196.

FEUSSNER I *et al.* (1995), 'Lipoxygenase-catalyzed oxygenation of storage lipids is implicated in lipid mobilization during germination', *Proc Natl Acad Sci USA*, 92, 11849–11853.

FODOR E *et al.* (1995), 'Molecular architecture and biophysical properties of phospholipids during thermal adaptation in fish – an experimental and model study', *Lipids*, 30 (12), 1119–1126.

FOX P F and MORRISSEY P A (1981), 'Indigenous enzymes in bovine milk', in Birch G G, Blakebrough N and Parker K J (eds), *Enzymes and Food Processing*, London, Applied Science Publishers, ISBN 0–85334–935–5.

FRYER M J (1995), 'The mechanism of apotopsis, cell-membrane lipid-peroxidation and a novel *in vivo* function for antioxidant vitamin E (alpha tocopherol)', *Redox Report*, 1 (2), 159–161.

GALLIARD T (1970), 'The enzymic breakdown of lipids in potato tuber by phospholipid- and galactolipid-acyl hydrolase activities and by lipoxygenase', *Phytochemistry*, 9, 1725–1734.

GALLAIRD T (1971), 'The enzymic deacylation of phospholipids and galactolipids in plants – purification and properties of a lipolytic acyl-hydrolase from potato tubers', *Biochem J*, 121, 379–390.

GALLIARD T (1986a), 'Hydrolytic and oxidative degradation of lipids during storage of wholemeal flour: effects of bran and germ components', *J Cereal Sci*, 4, 179–192.

GALLIARD T (1986b), 'Oxygen consumption of aqueous suspensions of wheat wholemeal, bran and germ: involvement of lipase and lipoxygenase', *J Cereal Sci*, 4, 33–50.

GALLIARD T (1987), in Morton I D (ed.), *Cereals in a European Context*, Chichester, Ellis Horwood, 462–479.

GEORGE J C (1962), 'A histophysiological study of the red and white muscle of mackerel', *Am Midl Nat,* 68, 487–494.

GEROMEL E J and MONTGOMERY M W (1980), 'Lipase release from lysosomes of rainbow trout (*Salmo gairdneri*) muscle subjected to low temperatures', *J Food Sci*, 45, 412–419.

GONZALEZ S *et al.* (2003), 'Oxidation and textural characteristics of butter and ice cream with modified fatty acid profiles', *J Dairy Sci*, 86 (1), 70–77.

GOUNARIS K *et al.* (1986), 'The thylakoid membranes of higher plant chloroplasts', *Biochem J*, 237, 313–326.

GRANIT R *et al.* (2001), 'Effects of vitamin E supplementation on lipid peroxidation and color retention of salted calf muscle from a diet rich in polyunsaturated fatty acids', *J Agric Food Chem*, 49 (12), 5951–5956.

GUERTS T J (2003), 'Lipase in milk, curd and cheese', *Milchwissenschaft*, 58 (1–2), 61–65.

HAARD N F (1995), 'Food as cellular systems: impact on quality and preservation. A review', *J Food Biochem*, 19, 191–238.

HANSON S W F and OLLEY J (1964), 'Observations on the relationship between lipids and protein deterioration', *The Technology of Fish Utilization*, Int symp Husum, 111.

HARDY R *et al.* (1979), 'Lipid and autoxidative changes in cold stored cod (*Gadus morhua*)', *J Sci Food Agric*, 30, 999–1006.

HAZEL J R *et al.* (1998), 'Thermal acclimation of phase behavior in plasma membrane lipids of rainbow trout hepatocytes', *Am J Physiol Regul Integr Comp Physiol*, 275 (3), R861–R869.

HERNÁNDEZ P *et al.* (1998), 'Lipid composition and lipolytic enzyme activities in porcine skeletal muscles with different oxidative patterns', *Meat Sci*, 49 (1), 1–10.

HERNÁNDEZ P *et al.* (1999), 'Lipolytic and oxidative changes in two Spanish pork loin products: dry-cured loin and pickled-cured loin', *Meat Sci*, 51, 123–128.

HIRANO K *et al.* (1997), 'Properties of phospholipase A1 transacylase in the white muscle of bonito *Euthynnus pelamis* (Linnaeus)', *J Biochem*, 122, 1160–1166.

HIRAYAMA O and MATSUDA H (1975), 'Purification and characterization of lipolytic acyl-hydrolases from rice bran', *J Agr Chem Soc Jpn*, 49 (11), 569–576.

HOLDSWORTH S D (1969), 'Handling vined peas and beans', *Process Biochemistry*, 4, 26–34.

HUANG A H C (1987), 'Lipases', in Stumpf P K (ed.), *The Biochemistry of Plants*, New York, Academic Press, Vol 9 (Lipids: structure and function), 91–119, ISBN 0–12–675409–8.

HUANG C H *et al.* (1995), 'Multiple phosphoinositide-specific phospholipases-C in oat roots – characterization and partial-purification', *Plant J*, 8 (2), 257–267.

INGEMANSSON T *et al.* (1993), 'Lipid hydrolysis and oxidation related to astaxanthin content in light and dark muscle of frozen stored rainbow-trout (*Oncorhynchus mykiss*), *J Food Sci*, 58 (3), 513–518.

JENSEN C *et al.* (1997), 'Influence of the oxidative quality of dietary oil on broiler meat storage stabilty', *Meat Sci*, 47 (3–4), 211–222.

JENSEN C *et al.* (1998a), 'Dietary vitamin E: Quality and storage stability of pork and poultry', *Trends in Food Sci and Technol*, 9 (2), 62–72.

JENSEN C *et al.* (1998b), 'Oxidative stability of frozen-stored raw pork chops, chill-stored pre-frozen raw pork chops, and frozen pre-cooked sausage in relation to dietary $CuSO_4$, rapeseed oil and vitamin E', *Z-Lebensm Unters Forsch*, 207 (5), 363–368.

JENSEN C *et al.* (1998c), 'Effect of dietary levels of fat, alpha-tocopherol and astaxanthin on colour and lipid oxidation during storage of frozen rainbow trout (*Oncorhynchus mykiss*) and during chill storage of smoked trout', *Z Lebensm Unters Forsch A-Foo* 207 (3), 189–196.

KORN E D (1962), 'The lipoprotein lipase of cow's milk', *J Lipid Res*, 3, 246–250.

KUMAR G N M and KNOWLES N R (1993), 'Changes in lipid peroxidation and lipolytic and free-radical scavenging enzyme activities during aging and sprouting of potato (*Solanum tuberosum*) seed-tubers', *Plant Physiol*, 102, 115–124.

LABUZA T P (1971), 'Kinetics of lipid oxidation in foods', *CRC Critical Reviews in Food Technology*, October 1971, 355–405.

LEE M H (1989), 'Phospholipase D of rice bran. II. The effects of the enzyme inhibitors and activators on the germination and growth of root and seedling in rice', *Plant Sci*, 59, 35–43.

LIST G R *et al.* (1990), 'Effect of moisture, microwave heating, and live steam treatment on phospholipase D activity in soybeans and soy flakes', *J Am Oil Chem Soc*, 67 (11), 867–871.

LITCHFIELD C (1972), *Analysis of Triglycerides*, New York, Academic Press, 104–138.

LUNING P A *et al.* (1995), 'Characterization and occurrence of lipoxygenase in bell peppers at different ripening stages in relation to the formation of volatile flavor compounds', *J Agric Food Chem*, 43, 1493–1500.

MALINS D and WEKELL J C (1970), in Holman R T (ed.), *Progress in the Chemistry of Fats and Other Lipids*, Oxford, Pergamon, 10, 337–363.

McMURRAY W C and IRVINE R F (1988), 'Phosphatidylinositol 4,5-bisphosphate phosphodiesterase in higher-plants', *Biochem J*, 249 (3), 877–881.

MEIJBOOM P W and STROINK J B A (1972), '2-*trans*, 4-*cis*, 7-*cis*-decatrienal, the fishy off-flavour occurring in strongly autoxidised oils containing linolenic acid or ω 3,6,9 etc. fatty acids', *J Am Oil Chem Soc*, 49, 555–558.

MENASHE M *et al.* (1986), 'Hydrolysis of dipalmitoylphosphatidylcholine small unilamellar vesicles by porcine pancreatic phospholipase A2', *J Biol Chem*, 261 (12), 5328–5333.

MOREAU R A (1987), 'Autolysis of phospholipids in homogenates of various plant tissues',

Phytochem, 26 (7), 1899–1902.

MOUNTNEY G J (1976), 'Chemical and nutritive characteristics', in *Poultry Products Technology*, 2nd edition, Westport, CT, AVI Publishing, 54–55.

MUKERJEE K D (1994), 'Plant lipases and their applications in lipid biotransformations', *Prog Lipid Res*, 33 (1/2), 165–174.

NIGAM S, WILTON D C and DENNIS E A (eds) (2000), 'PLA2', *BBA - Molecular and Cell Biology of Lipids,* 1488 (1–2).

NOVOTNÁ Z *et al.* (1999), 'Purification and characterisation of rape seed phospholipase D', *Plant Physiol Biochem*, 37 (7/8), 531–537.

O'CONNOR J *et al.* (1992), 'A comparison of lipase activity in various cereal grains', *J Cereal Sci*, 16, 153–163.

PADLEY F B *et al.* (1994), 'Occurrence and characteristics of oils and fats', in Gunstone F D, Harwood J L and Padley F B, (eds), *The Lipid Handbook*, 2nd edition, London, Chapman & Hall, 47–223.

PENDLINGTON S (1965), *Proc 1st Int Congress on Food Sci and Tech 1961*, 4, 107 (published 1965).

PETTIT T R *et al.* (1989), 'Synthesis of leukotriene-B and other conjugated triene lipoxygenase products by blood-cells of the rainbow trout (*Salmo gairdneri*)', *Biochem Biophys Acta*, 1003, 397–402.

RACUSEN D (1983), 'Occurrence of patatin during growth and storage of potato-tubers', *Can J Bot*, 61 (1), 370–373.

RACUSEN D and FOOTE M (1980), 'A major soluble glycoprotein of potato-tubers', *J Food Biochem*, 4 (1), 43–52.

RAGHUVEER K G and HAMMOND E G (1967), 'The influence of glyceride structure on the rate of autoxidation', *J Am Oil Chem Soc*, 44, 239–243.

RAMEZANZADEH F M *et al.* (1999), 'Prevention of oxidative rancidity in rice bran during storage', *J Agric Food Chem*, 47 (8), 2997–3000.

RANSAC S *et al.* (1996), 'The kinetics, specificities and structural features of lipases', in Op den Kamp J A F (ed.), *NATO ASI Series H96 Molecular Dynamics of Biomembranes*, Berlin, Heidelberg, Springer-Verlag, 265–304.

REFSGAARD H H F *et al.* (2000), 'Free polyunsaturated fatty acids cause taste deterioration of salmon during frozen storage', *J Agric Food Chem*, 48, 3280–3285.

ROBINSON D S *et al.* (1995), 'Lipoxygenase and the quality of foods', *Food Chem*, 54, 33–43.

RUFF N *et al.* (2002), 'Comparative composition and shelf-life of fillets of wild and cultured turbot (*Scophthalmus maximus*) and Atlantic halibut (*Hippoglossus hippoglossus*)', *Aquaculture International*, 10 (3), 241.

RUSSEL E A *et al.* (2003), 'Quality and shelf life of duck liver paté as influenced by dietary supplementation with alpha-tocopheryl acetate and various fat sources', *J Food Sci*, 68 (3), 799–802.

RYU S B and WANG X (1996), 'Activation of phospholipase D and the possible mechanism of activation in wound induced lipid hydrolysis in castor bean leaves', *Plant Physiol*, 111 (2) Suppl, 153.

SCHWIMMER S (1981), 'Cell disruption and its consequences in food processing', *J Food Sci*, 37, 530–534.

SCOTT M L (1956), 'Composition of turkey meat', *J Am Dietet Assoc*, 32, 941–944.

SEVANIAN A and KIM E (1985), 'Phospholipase A2 dependent release of fatty acids from peroxidised membranes', *J Free Radicals Biol Med*, 1, 263–271.

SHAHIDI F (1992), 'Prevention of lipid oxidation in muscle foods by nitrite and nitrite-free composition', in St Angelo A J (ed.), *Lipid Oxidation in Food*, Washington, DC, American Chemical Society.

SHERIDAN M A and ALLEN M V (1984), 'Partial purification of a triacylglycerol lipase isolated from steelhead trout (*Salmo gairdneri*) adipose tissue', *Lipids*, 19, 347–352.

SHEWFELT R L and HULTIN H O (1983), 'Inhibition of enzymic and non-enzymic lipid peroxidation of flounder muscle sarcoplasmic reticulum by pretreatment with phospholipase A2', *Biochem Biophys Acta*, 751, 432–438.

SIKORSKI Z *et al.* (1976), 'Protein changes in frozen fish', *Food Sci and Nutr,* 8, 97–104.

SIMON E W (1974), 'Phospholipids and plant membrane permeability', *New Phytol*, 73, 377–420.

SIMPSON T D (1991), 'Phospholipase D activity in hexane', *J Am Oil Chem Soc*, 68 (3), 176–178.

SKLAN D *et al.* (1983), 'The effect of dietary fat and tocopherol on lipolysis and oxidation in turkey meat stored at different temperatures', *Poultry Sci*, 62, 2017–2021.

TAIT S P C and GALLIARD T (1988), 'Effect on baking quality of changes in lipid composition during wholemeal storage', *J Cereal Sci*, 8, 125–137.

TODD J F *et al.* (1990), 'Characteristics of a membrane-associated lipoxygenase in tomato fruits', *Plant Physiol*, 94, 1225–1232.

TOLDRÁ F and FLORES M (1998), 'The role of muscle proteases and lipases in flavor development during processing of dry-cured ham', *Critical Reviews in Food Sci*, 38 (4), 331–352.

TSUKADA N (1976), 'Changes in the lipids of skipjack during frozen storage', *Bull Toakai Reg Fish Res Lab,* 84, 31–41.

VALIN C *et al.* (1975), 'Etude de la qualité de la viande bovine. I Etude biochimique de la maturation des viandes de taurillons', *Annales de Technologie Agricole,* 24, 47–64.

VAN DEN BERG J J M *et al.* (1993), 'Conformational changes in oxidised phospholipids and their preferential hydrolysis by phospholipase A_2: a monolayer study', *Biochemistry*, 32, 4962–4967.

VAN KUIJK P J G M *et al.* (1987), 'A new role for phospholipase A2: protection of membranes from lipid peroxidation damage', *Trends Biochem Sci*, 12, 31–34.

VAN RUTH S M *et al.* (1999), 'Volatile composition of sunflower oil-in-water emulsions during initial lipid oxidation: influence of pH', *J Agric Food Chem*, 47, 4365–4369.

VERGER R (1997), 'Interfacial activation of lipases: facts and artifacts', *TIBTECH*, 15, 33–38.

WANG X *et al.* (1993), 'Purification and immunological analysis of phospholipase D from castor bean endosperm', *Arch Biochem Biophys*, 306 (2), 486–494.

YAMAUCHI N *et al.* (1986), 'Polar lipids content and their fatty acid composition with reference to yellowing of stored spinach leaves', *J Japan Soc Hort Sci*, 55 (3), 355–362.

YASUDA M and FUJITA T (1977), 'Effect of lipid peroxidation on phospholipase A2 activity of rat-liver mitochondria', *Jpn J Pharmacol*, 27 (3), 429–435.

YOSHIDA S (1979), 'Freezing injury and phospholipid degradation *in vivo* in woody plant cells. 1. Subcellular localization of phospholipase D in living bark tissues of the black locust tree (*Robinia pseudocacia* L.)', *Plant Physiol*, 64, 241–246.

ZHUANG H *et al.* (1994), 'Packaging influenced total chlorophyll, soluble protein, fatty acid composition and lipoxygenase activity in broccoli florets', *J Food Sci*, 59 (6), 1171–1174.

Part II

Measuring shelf-life and spoilage

9

Ways of measuring shelf-life and spoilage

T. K. Singh and K. R. Cadwallader, University of Illinois, USA

9.1 Introduction: understanding and estimating the shelf-life of food

Before attempting to understand shelf-life of foods, it is important to realize that foods are diverse, complex and active systems in which microbiological, enzymatic and physicochemical reactions are simultaneously taking place. These reactions have major consequences in relation to flavor, texture and shelf-life. Food preservation is dependent on the understanding of mechanisms of these reactions and the successful limitation of those most responsible for loss or spoilage of desirable characteristics and sometimes the channeling of other reactions towards beneficial changes. Essentially, the shelf-life of a food can be defined as the period for which it will retain an acceptable level of eating quality, from a safety and sensory point of view. There are four critical factors in this endeavor:

- Formulation
- Processing
- Packaging
- Storage conditions.

All four factors are critical but their relative importance depends on the food. An understanding of the interplay between these factors is key to shelf-life estimation and testing. For example, a change in a single processing parameter may lead to undesirable chemical or physical changes in a product, or it may require reformulation or a change in packaging in order to attain the required shelf-life. Similarly, the very act of processing may subject the formulated materials and ingredients to conditions that are unfavorable or inhibitory to undesirable deteriorative reactions and promote desirable physical and chemical changes, thus giving the food product its final form and characteristics. And once the food leaves the processing stage, its keeping properties and the extent to which it will retain its intended attributes is a function of its micro-environment (packaging and storage conditions), i.e. gas composition (oxygen, carbon dioxide, inert gases, ethylene, and so on), relative humidity (RH), pressure or mechanical stresses, light and temperature.[1]

Some of the contributing factors in the search for improved shelf-life include increased consumer demand for fresh, convenient, safe and superior quality foods available year-round, and the continued globalization of food distribution systems. This has placed pressure on the food industry to ensure shelf stability and storage times as products travel further and further from their point of origin. For the food industry, meeting these ostensibly contrary objectives of consumer demand for longer shelf-life and at the same time minimizing processing require the implementation of enhanced preservation parameters, improvement in testing and analytical procedures, a better understanding of food quality factors as related to their sensory characteristics, and continued education of scientists in food quality modeling and accelerated shelf-life testing procedures. One such trend is worth noting: the continued introduction of legal drivers for shelf-life testing. Although there is no federally mandated, uniform open dating system, many US government organizations have ruled that certain foods must have some type of open date.[2,3] The European Union also has such legislation in place for all food products.

This chapter will briefly review the current state of approaches to shelf-life determination testing methods, tools and technologies employed to ensure that consumers receive high quality food products with the added convenience of extended shelf-life.

9.1.1 Approaches to shelf-life estimation of food products

To attain knowledge of a food's expected shelf-life, one must (1) understand the concerted series of biochemical/physicochemical reactions taking place in any given food, and (2) identify the mechanisms responsible for spoilage or loss of desirable characteristics such as texture, flavor, odor and/or nutrients. With better understanding of the underlying mechanism of reactions, food scientists will be able to quantify and model the data on the relevant biochemical/ physicochemical changes taking place in foods for estimating the period for which they will retain an acceptable level of eating quality from a safety and sensory perspective.

Selecting an appropriate, reliable approach to modeling quality loss of a food product is an important first step in estimating shelf-life, and allows for the efficient design of appropriate shelf-life tests. Shelf-life predictions are based on fundamental principles of food quality loss modeling, primarily kinetic modeling of different deterioration mechanisms that occur in food systems, which have been detailed extensively in the literature.[4–12] Several established approaches can be utilized for gathering of shelf-life data of food products:

- Estimating shelf-life based on published data
- Utilizing known distribution times for similar products on the market
- Using consumer complaints as the basis for determining whether a problem is occurring
- Accelerated shelf-life testing (ASLT).

These methods have their downsides, however, including the fact that most shelf-life data on specific engineered foods is proprietary, similar products to benchmark against do not exist, or there is no information on actual consumer home storage times. If one is confident in a product's shelf-life or it is already in the marketplace, one can use a distribution test method. Product is collected at supermarket sites and stored in the lab under home-use conditions. Only one such study has been reported in the literature, although this method has been used by others, especially in cases when states or countries instituted new open dating legislation.[13] This method results in the product shelf-life based on both distribution and home storage conditions.

In order to provide a good estimate of product shelf-life, food scientists involved in product development and quality control must have detailed information on the specific processing method to be employed and the types of raw materials and functional ingredients used, must have prior experience with similar formulations, packaging, and so on, and must perform confined experiments on the finished product/package combination under some abuse condition (ASLT methods), examine the product periodically up to the end of its shelf-life, and then use this data to project shelf-life under true distribution conditions.[14] Designing shelf-life and ASLT tests are described in Chapters 16 and 15 respectively.

9.2 Key factors influencing the shelf-life of food

Food quality loss can be described in terms of a number of *compositional factors*, such as concentration of reactive species, microorganism levels, catalysts, reaction inhibitors, pH and water activity, as well as *environmental factors* which include temperature, relative humidity, light, mechanical stress and total pressure.[1] According to Troller and Christian,[15] water activity, temperature and pH are the most important factors that control rates of deteriorative changes and microbial growth in foods. These parameters were also referred to as 'hurdles' by Leistner.[16]

Environmental factors can significantly affect the rates of the reactions and need to be defined and closely monitored during kinetic experiments. A kinetic model for quality loss is particular not only to the studied food system, but also to the set of environmental conditions of the experiment, including the permeability of the packaging material. It would be desirable to generalize the models so that they include, as parameters, the environmental factors that more strongly affect the quality loss rates and that are susceptible to variation during the life of the food. Some of these important factors in food preservation and quality are detailed below.

9.2.1 Temperature

The important effect of temperature on reaction rates has long been recognized. Raising the storage temperature will accelerate many ageing processes and this is the basis of many of the accelerated methods. The most prevalent and widely used model is the Arrhenius relation, derived from thermodynamic laws as well as statistical mechanics principles.[9] The Arrhenius relation, developed theoretically for reversible molecular chemical reactions, has been experimentally shown to hold empirically for a number of more complex chemical and physical phenomena (e.g., viscosity, diffusion, and sorption). Food quality loss reactions described by the kinetic models have also been shown to follow an Arrhenius behavior with temperature.

An alternative way of expressing temperature dependence which has been extensively used by the food industry is the Q_{10} approach. Q_{10} is defined as the ratio of the reaction rate constants at temperatures differing by 10ºC. This model can be used to describe how much faster a reaction will go if the product is held at some other temperature, including high abuse temperatures. If the temperature-accelerating factor is given, then extrapolation to lower temperatures, such as those found during distribution, may be used to predict expected product shelf-life. This is the principle behind ASLT. As described previously, ASLT involves the use of higher testing temperatures in food quality loss and shelf-life experiments and extrapolation of the results to regular storage conditions through the use of the Arrhenius equation, which cuts down testing time substantially. A reaction of average activation energy (E_a) of 20 kcal/mol may be accelerated by 9 to 13 times with a 20ºC increase in the testing temperature, depending on the temperature zone.[1] This principle and the methodology in conducting effective ASLTs are described in the literature and in Chapter 15.[10,12,17] However, caution should be exercised when interpreting results and extrapolating data to other conditions. For example, when the product/package system is tested, the package also controls shelf-life so that the true shelf-life of the food itself is unknown; thus, if a new package with different permeabilities to oxygen, water, or carbon dioxide is chosen, the prior results may not be applicable. Package geometry and shaking rate must also be closely controlled as these parameters significantly affect product movement and shearing.

If the ASLT conditions are chosen properly and the appropriate algorithms for extrapolation are used, then shelf-life under any 'known' distribution conditions should be predictable. A few other practical problems that may arise in the use of ASLT conditions include, but are not limited to, the following.

- Error may occur in analytical or sensory evaluation. Generally, any analytical measurement should have a variability of less than ±10% to minimize prediction errors.
- Generally, as temperature rises, phase changes (such as fat changing from solid to liquid) may occur which can accelerate certain reactions. Thus, the actual shelf-life at the lower temperature may be longer than predicted. Similarly, it has been known since 1990 that for dry foods with a given moisture content put at a higher temperature (above the glass transition temperature), the projection of shelf-life using a shelf-life to room temperature storage plot could be wrong, resulting in a prediction of either longer or shorter time than the actual shelf-life. This has opened the door for a whole set of new laboratory testing procedures such as measuring the glass transition using differential scanning calorimetry (DSC) or by some thermal rheological method like dynamic mechanical analysis (DMA) or dynamic mechanical thermal analysis (DMTA).
- Upon freezing, such as used for storage of control samples, reactants are concentrated in the unfrozen liquid, creating a higher rate of quality loss at certain temperatures, which is unaccounted for by the Q_{10} value and will cause prediction errors.
- Storage defects in frozen products can be accelerated by storage at higher than normal temperatures. For example, more rapid changes will occur at −18°C compared with normal long-term storage temperatures of below −25°C, still faster changes will occur at −10°C. Certain forms of deterioration, such as ice crystal growth and freezer burn (i.e., sublimation of ice as water vapor from the surface of the frozen food), will also be accelerated if the temperature is made to fluctuate while the food remains frozen.[18]
- Cycling the product between 0°C and room temperature will accelerate watery separation in starch-thickened foods. Absence of any separation after 30 cycles over two months normally suggests that product will be stable for two years at ambient temperature.[18]
- If high enough temperatures are used, proteins may become denatured. This can result in both increases and decreases in the reaction rate of certain amino acid side-chains and thus cause either under- or over-prediction of true shelf-life.
- The solubility of gases, especially oxygen, in fat or water decreases by almost 25% for each 10°C rise in temperature. An oxidative reaction (loss of Vitamin E, A, C or linoleic acid) can decrease in rate if oxygen availability is the limiting factor. Thus, at the higher temperature, the rate will be lower than the theoretical rate, resulting in under-prediction of true shelf-life at normal storage temperature.

The potential problems and possible errors that can arise in the use of accelerated techniques have been described in greater detail by Robertson.[19] Further details on the effects of temperature on spoilage reaction can be found in Chapter 3.

9.2.2 Water activity

Water is present in all foods, ranging from trace amounts in dried food products to very high amounts in beverages. The stability and shelf-life of foods are highly dependent on the water content since it directly affects the rate of food deterioration reactions.[1] Water activity describes a thermodynamic energy property of water in the food, and in part, water acts as a solvent and participates in chemical reactions.[1] The relative humidity of the immediate environment which directly affects the moisture content and water activity (a_w) of a food is the second most important environmental factor. Rates of deteriorative changes and microbial growth under normal food storage conditions often depend on water content and a_w. Food deterioration due to microbial growth is not likely to occur at $a_w < 0.6$.[20] However, chemical reactions and enzymatic changes may occur at considerably lower a_w values. Although a higher a_w does not necessarily mean a faster reaction rate, critical levels of a_w can be established above which undesirable factors that lead to the deterioration of food occur, such as microbial growth or textural changes. Controlling the a_w is the basis for preservation of dry and intermediate moisture foods.[21] Besides the specific critical a_w limits, water activity has a pronounced effect on chemical reactions in these foods. Generally, the ability of water to act as a solvent, a reaction medium and a reactant itself increases with increasing a_w up to a point, where other factors decrease reaction rates. As a result, the rate of many deteriorative reactions increases exponentially with increasing a_w above the value corresponding to the monolayer moisture, the value at which most reactions have a minimum rate. For example, if one has a wet food and tries to dehydrate it to different water activities from the wet state, the reaction rate will increase, reach a maximum, and then decrease. This is a key concept from the standpoint of intermediate moisture foods, for example, especially with regard to semi-soft food bars, which are generally in a water activity range where rates of deterioration are very high. The rate of lipid oxidation increases again as the a_w decreases below the monolayer, and for most aqueous phase reactions, one rate decreases again above a certain a_w in the 0.6 to 0.8 range.

Mathematical models that incorporate the effect of a_w as an additional parameter can be used for shelf-life predictions of moisture-sensitive foods. Also, ASLT methods have been used to predict shelf-life under normal conditions based on data collected at high temperature and in high humidity conditions.[22]

Further details on the effects of water activity and food stability can be found in Chapter 2.

9.2.3 Other environmental factors

Understanding *gas composition*, which is also a factor that may play a significant role in some quality loss reactions, is important but not clearly understood or researched. Oxygen availability is very important for oxidative reactions and can affect both the rate of reaction and apparent reaction order, depending upon whether it is limiting or in excess.[5] It also affects the respiration rates and senescence of plant materials and microbial growth, depending on the redox potential. Vacuum packaging and nitrogen flushing are based on slowing down undesirable reactions by limiting the availability of O_2. Controlled and modified atmosphere packaging (MAP) is based on these principles. Further, the presence and relative amount of other gases, especially carbon dioxide, strongly affect biological and microbial reactions in fresh meats, fruits and vegetables. The mode of action of CO_2 has not been completely elucidated, but is partly connected with surface acidification.[1] Different commodities have different optimum O_2–CO_2–N_2 gas composition requirements for maximum shelf-life. Excess CO_2 in many cases is detrimental since the aqueous solution is a strong oxidant, causing bleaching of color and generation of off-flavor. Gamma-irradiation in combination with MAP has been used to control color bleaching in cooked pork sausages.[23] Spencer and Humphreys describe a novel method of packaging of different food products using the inert gas argon, which is more efficient in excluding O_2 than N_2 due to its physicochemical properties.[24] Excellent results have been achieved with high concentrations of argon and the results were even better for krypton, xenon and sometimes neon. Argon is a safe inert gas and has GRAS (generally recognized as safe) status in the USA, as does nitrogen. This system of packaging with argon was shown to inhibit oxidation and microbial growth, leading to significant extension of shelf-life and improvement in quality parameters such as flavor, aroma, color and overall consumer acceptability. Successful application of inert gas MAP technology was demonstrated using a diverse range of foodstuffs, such as respiring products (salads and produce), non-respiring products (containing oil; potato chips and nuts), and chilled products (processed meats, fresh pasta and fresh meats). Other important gases are ethylene and propylene oxides (epoxides) and ozone.[19]

Proper selection of packaging material with the desirable permeation properties, the concentration of gases and the RH inside the package can be kept within predictable limits determined by the conditions set at processing. One key problem in gas composition analysis is that analysts will flush with a certain gas, but because films are permeable to gases and some gases may react with a food (e.g., CO_2 may dissolve in the food or oxygen may be used up in oxidation reactions), it is very possible for the gas composition to change over time. Therefore, it is important to know the gas composition profiles throughout the shelf-life of the product.

9.3 Quality indices for testing the shelf-life of food

Obtaining a reliable approach to modeling quality loss of a food product is based on defining an appropriate index that measures, or directly corresponds to, food quality. Again, shelf-life can be defined as the time until a product becomes unacceptable to consumers under a given storage condition. These indices include sensory evaluation, as well as chemical, microbiological and physical testing through instrumental or classical methods. The quality indices used most commonly today in shelf-life and storage studies are detailed below.

9.3.1 Physical changes

Physical changes are caused by mishandling of foods during harvesting, processing and distribution; these changes lead to reduced shelf-life of foods. Some of the common physical changes are as follows:

- Bruising of fruits and vegetables during harvesting and post-harvest handling which leads to development of rot, browning, or even off-flavor generation (e.g. production of methanethiol and dimethyl disulfide in broccoli)[25]
- Wilting or loss of moisture from fruits and vegetables during storage in low-humidity environments
- Absorption of moisture by a dry food leading to sogginess, e.g. bakery, dried milk powders, infant formulas, dry pet foods, etc.
- Growth of ice crystals due to temperature fluctuation in frozen stored food products, e.g. ice cream, frozen fruits and vegetables
- Freeze damage to fruits (causes emission of volatiles such as ethanol, ethyl butanoate/octanoate, methyl hexanoate)[26]
- Freezer burn
- Textural and flavor changes due to thawing/refreezing
- Melting and solidifying of fat, e.g. confectionery/bakery products
- Viscosity changes, e.g. mayonnaise viscosity changes in salad
- Phase separation, e.g. separation of whey or syneresis in yoghurt.

Many of the above-mentioned physical changes can be prevented by careful handling, proper packaging and stricter control of storage temperature.

9.3.2 Chemical changes

During the processing and storage of foods, several chemical changes occur that involve the internal food constituents and the external environmental factors. These changes may cause food deterioration and reduce the shelf-life. The most important chemical changes are associated with the enzymatic reactions, oxidative reactions, particularly lipid oxidation, and non-enzymatic browning.

The least stable macro-constituents in foods are the lipids. Depending on the degree of unsaturation, lipids are highly susceptible to oxidation, resulting in the development of oxidative rancidity. When this occurs, the food becomes unacceptable and is rejected by consumers. While the development of off-flavor is markedly noticeable in rancid foods, the generation of free radicals during the autocatalytic process leads to other undesirable reactions, e.g. loss of vitamins, alteration in color, and degradation of proteins.[9] In addition to the development of off-flavors, many of the oxidized products of rancidity are now considered to be unhealthy.[27]

The rates of lipid oxidation are influenced by several factors. Presence of oxygen in the vicinity of food and temperature play critical roles in influencing the rate of reaction. Similarly, water plays an important role; lipid oxidation occurs at high rates at very low water activities.[28] In determining the shelf-life of foods that contain lipid, especially higher concentrations of unsaturated fatty acids, the important reaction mechanisms and their rates of reaction must be known.[9] A number of techniques have been used to study oxidation of lipids, namely peroxide value, the 2-thiobarbituric acid method and gas chromatographic methods to monitor volatiles.[29] In recent times, static or dynamic headspace or solid-phase microextraction sampling followed by GC separation of volatiles generated during lipid oxidation has been the preferred method due to relatively simple sample preparation, sensitivity and fast analysis time. In this method, the primary lipid oxidation marker compounds can be separated and quantified, e.g. aldehydes (pentanal, hexanal, heptanal, octanal, nonanal, and decadienals), ketones (2,3-octanedione and 2-heptanone) and hydrocarbons (propane, pentane).[29] Stored boiled potatoes develop an off-flavor that can be described as cardboard-like (also called potato off-flavor, POF). This is a serious problem in food service systems where this is a common item. Analyses of this off-flavor in precooked vacuum-packed potatoes resulted in the identification of (E,E)-2,4-nonadienal and (E,E)-2,4-decadienal as the most potent of the POF compounds.[30] Kim and Morr reported major volatile compounds recovered and identified from 12-hour light-exposed milk (200 foot-candle fluorescent light, 0–5°C, simulating commercial milk storage conditions), resulting from photosensitization and decomposition of riboflavin and oxidation of lipids and sulfur compounds, including hexanal, pentanal, dimethyl disulfide, 2-butanone and 2-propanol.[31] Under fluorescent light of the type used for market display, exposure of goat cheese was found to increase off-flavor which was described as rotten, soapy, rancid and cheesy. This light-induced off-flavor in cheese was the result of production of compounds such as 1-heptanol (chemical), heptanal (fatty), nonanal (soapy), and 2-decenal (tallowy) from singlet oxygen oxidation of oleic acid.[32] Similarly, lipid oxidation generated aldehydes and ketones in butter oil[33] and non-fat dry milk[34, 35] that were found to be the responsible off-flavors. According to Ullrich and Grosch,[36] who followed the formation of intense aroma compounds during the autoxidation of linoleic acid, hexanal, (Z)-2-octenal and (E)-2-nonenal exhibited the highest flavor dilution factor (highest dilution at which a substance can still be smelt) in GC-olfactometry analysis. Hexanal has been used as a marker oxidation product in various studies.[29,37–41]

In addition to lipid oxidation, there are other chemical reactions that are induced by light such as loss of vitamins and browning of meat.[9] Light sensitivity of riboflavin in milk has been previously reported.[31]

Specific knowledge on the loss/formation of flavor or off-flavor ingredients during storage is slowly developing. Thiols contribute to the flavor of most foods, but tend to be somewhat unstable especially in aqueous solutions, and in the presence of oxygen. Instability can be due to oxidation,[42–44] but the instability of 2-methyl-3-furanthiol (MFT), an important flavor impact component for meat, appears to be due to electrophilic mechanisms.[45] Cysteine was found to increase the stability of thiols, but to decrease the stability of some other important flavor components such as 4-hydroxy-2,5-dimethyl-3-furanone and sotolone. Thiamin (vitamin B1), the second most abundant water-soluble vitamin for which orange juice is the best source, can thermally degrade to produce the potent sulfur aroma compound, e.g. MFT and *bis*(2-methyl-3-furyl)disulfide (dimmer of MFT). The formation of MFT and 3-(methylthio) propanal (Strecker aldehydes from methionine) in orange juice was found to contribute off-flavor to the stored juice.[46] The oxidative degradation of white wines rapidly leads to loss of sensory qualities, which was mainly attributed to 3-(methylthio) propanal,[47] a compound that has also been found as a precursor of dimethyl trisulfide which is known to cause an onion-like off-flavor in aged beer.[48] 2-Aminoacetophenone, a degradation product of tryptophan, is known to cause off-flavors, described variously as stale/grape-like/foxy note/animal/wet-dog-like, in milk products, corn products, wine and beer.[34,35]

Non-enzymatic browning is a major cause of quality change and degradation of nutritional content in many foods. This type of browning occurs due to the interaction between reducing sugars and amino acids. These reactions result in the loss of protein solubility, darkening of light-colored dried products and the development of bitter flavors. Environmental factors such as temperature, water activity and pH have an influence on non-enzymatic browning.

At favorable temperatures such as room temperature, many enzymatic reactions proceed at rapid rates, altering the quality attributes of the foods. For example, fruits upon cutting tend to brown rapidly at room temperature due to the reaction of phenolase with the cell constituents released due to the cutting of the tissue in the presence of oxygen. Enzymes such as lipoxygenase, if not denatured during the blanching process, can influence the food quality even at sub-freezing temperatures.[49] In addition to temperature, other factors such as oxygen, water and pH induce deleterious changes in foods that are catalyzed by enzymes.

9.3.3 Microbiological changes

Spoilage of food and beverages is the result of microbial activity of a variety of microorganisms. The microflora that colonizes a particular food or beverage depends on the characteristics of the product (e.g. composition, pH, etc.) and the way it is processed and stored. Some of the consequences of microbial growth in food products are changes in pH, formation of toxic compounds, gas production, slime formation and off-flavor production.

Classical microbial evaluation of perishable foods is of limited value for predictive prognoses since these foods are sold or eaten before the microbiological results become available. As a result of these problems, many workers in recent years have increasingly investigated techniques that measure chemical changes produced by bacteria rather than measuring the total numbers of bacteria. A popular technique is electrical impedance, which has been applied to shelf-life evaluation of milk and milk products.[50] A widely used rapid technique for determining the number of bacteria present in milk is the measurement of adenosine triphosphate (ATP), using luciferase and cofactor to produce light.[51] However, this test correlates better with the total bacterial counts than do the counts with actual product shelf-life, because it does not necessarily measure the direct cause of off-flavor formation (e.g. malodorous bacterial metabolites) and the end of shelf-life. During extended refrigerated storage various psychrotrophic bacteria can grow and produce heat-stable proteinases (*Pseudomonas*, *Aeromonas*, *Serratia* and *Bacillus* species) and lipases (*Pseudomonas*, *Flavobacteria* and *Alcaligenes* species).[52,53] Once present these heat-stable enzymes could continue to cause defects even in UHT-treated milks.[52,53]

New, highly sensitive and specific microbiological methods based upon immunological and molecular techniques have already been developed for the detection of pathogenic microorganisms.[54] These techniques could also be applied for the early detection of specific spoilage organisms. However, before such techniques could be used for the detection of specific organisms, those microorganisms must be identified for each type of product and their effect on spoilage characteristics must be determined.[55,56]

An alternative way is to pinpoint the specific microbial metabolite formed during the metabolism/growth of the specific microorganism causing spoilage or off-flavour in the food product.[57] Formation of such metabolite could be followed upon isolation by sensitive instrumentation, such as high performance liquid chromatography/mass spectrometry (HPLC/MS) or gas chromatography–olfactometry (GCO) and gas chromatography/mass spectrometry (GC/MS). One promising non-microbiological technique that has been shown to be a better predictor of shelf-life than microbial plate counts is the determination of volatiles in milk by dynamic headspace capillary gas chromatography (DH-GC) followed by multivariate interpretation of GC peaks.[58,59] Solid-phase microextraction-GC/MS with multivariate analysis (MVA) has also been successfully used to classify abused milk samples as to the cause of off-flavors, i.e. oxidation (light- or copper-induced) and microbial spoilage.[37,38] DH-GC/MS was applied as a non-destructive method for the analysis of volatile compounds produced by *Escherichia coli*, both pathogenic (O157:H7) and non-pathogenic strains.[60] All the strains of *E. coli*, in the model system, were found to produce indole and lesser amounts of other components including methyl ketones (2-heptanone, 2-nonanone, 2-undecanone and 2-tridecanone). Strawberries were found to be a suitable host for strain O157:H7, but the headspace analysis from inoculated fruits showed no detectable quantity of indole. Hamilton-Kemp *et al.*[61] and Yu *et al.*[60] further demonstrated that strawberries readily absorbed volatile compounds produced by the bacteria and in some cases metabolized the compounds into new volatile compounds. This technique has potential application for the detection of these bacteria in foods, especially fresh produce, which do not absorb and metabolize marker compounds as readily as strawberries, for example hamburger, salmon and grapes.

But the major problem is to find the relation between microbial composition and presence of microbial metabolites, with the presence and possible prediction of microbial spoilage. Table 9.1 summarizes a number of such metabolites, which could be used as quality indices.[62–69] This kind of unifying description of the interaction between the microflora developing in the product and the chemical changes in the same product yields important information, which can be used for developing newer methodologies for shelf-life evaluation of foods. Such an integrated understanding of each type or group of food product would indeed be beneficial for the overall understanding and help in increasing interest in developing newer ingredients and processing techniques to make foods safer and with longer shelf-lives.

9.3.4 Sensory quality changes

When spoilage is due to changes in texture or the development of off-flavor, caused by physicochemical and biochemical or microbial reactions, the underlying mechanism might be difficult to identify. Therefore, the evaluation of spoilage will always, directly or indirectly, be related to a sensory assessment. Sensory evaluation by a trained panel usually gives a good estimate of the overall quality of a food. One approach in sensory testing is to try to determine, at a certain level of probability, whether a product has changed (difference tests). Hence, this approach gives 'endpoint' information and does not allow for modeling quality loss with time.[1] Hedonic testing is a somewhat different approach that attempts to model the progressive loss of overall quality characteristics, using a graded hedonic scale.[1] The requirements on the sensory panel for uniformity, experience and size for hedonic testing are stricter than for the difference tests and often these requirements are not met, resulting in unreliable data. Another problem with this approach is the considerable difficulty in establishing a meaningful scale for each food product – an expert panel is not necessarily representative of consumers, let alone different consumer segments.[70]

A common approach to sensory testing is to assign the zero time value as 100% and the end of shelf-life value as 0% quality, and thus the times in between correspond proportionally to different levels of quality. This is based on the assumption that the sensory response is linear with time, which is often not true.[1] Typically, however, industry does not test to determine end of shelf-life. With the exception of rapid-decay foods – refrigerated products like milk – the consumer is not able to detect a sensory difference from day to day in shelf-stable items such as cereal or canned soup. Hence, typical sensory testing using hedonics or difference testing is not going to give a consistently accurate view of shelf-life for a given product.[1] The different statistical and graphical approaches for using sensory data in shelf-life testing were evaluated by Labuza and Schmidl[71] and Kilcast.[72] The maximum likelihood graphical procedure (Weibull method) was described as a good systematic approach to sensory testing.[13] The Weibull method is simple in that it asks only 'Is the product acceptable?'. The

Table 9.1 Examples of marker chemicals produced in food by insects and microbial activity which can be used as quality indices

Food product	Chemical	Spoilage problem (or off-flavor note)
Miscellaneous chemicals		
Prawn, lobster	*Bis*(methylthio)-methane trimethylarsine	Garlic
Prawn	Dimethyltrisulfide, indole	Rotten onion-like
Mushroom (canned), fish, water	2-Methyl isoborneol, geosmin	Earthy, muddy
French fries	*p*-Cresol, skatole, and indole	Pigsty-like
Cocoa beans (packaged)	Chlorinated anisole	Muddy
Meat (vacuum packed)	Indole, H_2S	Off-flavor
Fish	Trimethylamine	Off-flavor, loss of freshness
Meat, fish and cheese	Histamine, tyramine, cadavarine and putrescine	Past microbial activity, off-flavor, and health issues
Grain (corn, sorghum, soybean, and wheat)	1,4-Dimethoxybenzene and its 2-methyl, ethyl and methoxy derivatives	Off-odors
Volatile fatty acids		
Canned products, vegetables, fruit, meat, fish, and dairy products	*n*-Butyric acid	Swelling of low-acid canned foods
Canned meat	*n*-Valeric acid, *n*-butyric acid	Off-flavor
Paperboard	3-Methyl butyric, 2-methyl propionic and valeric acids	Transfer to food
Metabolites of sorbate		
Cheese, wine, and non-carbonated drinks	*Trans*-1,3-pentadiene	Hydrocarbon-, paint- and solvent-like
Wine	2-Ethoxyhexa-3,5-diene	Geranium defect
Metabolites of ferulic acid		
Lagers, beers and stouts	4-Vinylguaiacol	Phenolic taste
Presence of D-alanine		
Fruit juices	D-alanine (>1 ppm)	Quality indicator

Source: compiled from references 49–55 and 69.

intensity of testing is increased near the end of shelf-life so that a true shelf-life is determined. This method was successfully applied to determine the end of shelf-life of pasteurized milk and roasted and ground coffee.[73,74]

Besides the practical problems with regard to using sensory data in shelf-life modeling, further factors are the high cost involved with large testing panels and the problems connected with tasting spoiled or potentially hazardous samples. In some cases microbial growth or nutrient degradation could reach unacceptable levels while the food is still judged organoleptically acceptable. Sensory data are not 'objective' enough for regulatory purposes and in cases of legal action or dispute. Sometimes consumers can be 'trained' to accept lower-standard products by being exposed to products of gradually declining quality. That makes the need for alternative ways of assessing quality apparent.[1]

9.4 Conclusions and future trends

It is obvious from the above discussion that a wide range of methodologies have been employed for the purpose of shelf-life evaluation of foods. It is still not possible to describe any 'magic solutions' towards solving the puzzle of shelf-life testing, but we think that an integrated approach might work better. This will involve careful consideration of the product composition, processing parameters, packaging, environmental factors, chemical and biochemical reactions and type of microorganisms present, with the object of determining the food deterioration mechanisms. With this information in hand, the food scientist can choose proper chemical/microbial indices and identify the right tools and techniques to put in place for analysis of the product. The continued rapid progress in molecular techniques and analytical instrumentation technology is making it possible to detect and quantify spoilage using chemical and microbial indicators in minute quantities, indicating the onset of spoilage.

Better understanding of the spoilage mechanisms may help in developing new technologies, especially in the area of ingredients, processing and packaging, which will ensure the supply of safe, wholesome and high quality food products.

9.5 Sources of further information and advice

The authors would like to draw the reader's attention to some recently published excellent texts in the area of shelf-life of foods:

1. CADWALLADER K R and WEENEN, H (eds) (2003), *Freshness and Shelf Life of Foods*, ACS Symposium Series 836, Washington DC: American Chemical Society.
2. ESKIN N A M and ROBINSON D S (eds) (2001), *Food Shelf Life Stability: Chemical, Biochemical, and Microbiological Changes*, Boca Raton, FL: CRC Press.
3. KILCAST D and SUBRAMANIAM P (eds) (2000), *The Stability and Shelf Life of Foods*, Cambridge, UK: Woodhead Publishing.

4. MAN C M D and JONES, A. A. (eds) (1999), *Shelf life evaluation of foods*, Gaithersburg, MD: Aspen Publishers.

Additional information regarding water activity and shelf-life determination can also be obtained from the web page of Prof. T. P. Labuza at the Department of Food Science and Nutrition, University of Minnesota, St Paul, MN, USA (http://fscn.che.umn.edu/Ted_Labuza/tpl.html).

9.6 References

1. LABUZA T P (2000), 'The search for shelf life', *Food Testing and Analysis*, May, 26–36.
2. OFFICE OF TECHNOLOGY ASSESSMENT (1979), 'Open Shelf Life Dating of Food', Library of Congress, Cat. No. 79–600128.
3. IFT (1981), 'Open shelf life dating of foods', *Food Technol*, 35, 89–96.
4. CHEN S L and GUTMANIS, F. (1968), 'Auto-oxidation of extractable color pigments in chilli pepper with special reference to ethoxyquin treatment', *J. Food Sci*, 33, 274–280.
5. LABUZA T P (1971), 'Nutrient losses during and storage of dehydrated foods', *Crit Rev Food Sci* 2, 355.
6. HILLS C G and GRIEGER-BLOCK R A (1980), 'Kinetic data: generation, interpretation and use', *Food Technol*, 34, 56–66.
7. LABUZA T P (1984), 'Application of chemical kinetics to deterioration of foods', *J Chem Educ*, 61, 348–358.
8. LABUZA T P and KAMMAN J (1983), 'Reaction kinetics and accelerated tests simulation as function of temperature', in Saguy I, *Applications of Computers in Food Research*, New York: Marcel Dekker, 71–115.
9. SINGH R P (1999), 'Scientific principles of shelf life evaluation', in Man M D and Jones A A, *Shelf Life Evaluation of Food*, Gaithersburg, MD: Aspen Publications, 3–26.
10. TAOUKIS P and LABUZA T P (1995), 'An integrated approach to food chemistry: illustrative cases', in Fennema O R, *Food Chemistry*, 3rd edition, New York: Marcel Dekker, 1013–1042.
11. BLACKBURN C DE W (2000), 'Modeling shelf-life', in Kilcast D and Subramaniam P, *The stability and Shelf Life of Food,* Cambridge, UK: CRC Press – Woodhead Publishing, 54–78.
12. MIZRAHI S (2000), 'Accelerated shelf life testing', in Kilcast D and Subramaniam P, *The Stability and Shelf Life of Food,* Cambridge, UK: CRC Press – Woodhead Publishing, 107–128.
13. GACULA M C (1975), 'The design of experiments for shelf life study', *J Food Sci*, 40, 399–404.
14. SAGUY I and KAREL M (1980), 'Modeling of quality deterioration during food processing and storage', *Food Technol*, 34, 75–85.
15. TROLLER J A and CHRISTIAN J H B (1978), *Water Activity and Food*, New York: Academic Press.
16. LEISTNER L (1994), 'Further developments in the utilization of hurdle technology for food preservation', *J Food Eng*, 22, 421–425.

17. LABUZA T P and SCHMIDL M K (1985), 'Accelerated shelf life testing of foods', *Food Technol*, 39, 57–62, 64.
18. ELLIS, M. J. (1999), 'The methodology of shelf life determination', in Man M D and Jones A A, *Shelf Life Evaluation of Food*, Gaithersburg, Aspen Publications, 27–39.
19. ROBERTSON G L. (1993), Food Packaging, New York: Marcel Dekker.
20. ROOS Y H (2001), 'Water activity and plasticization', in Eskin N A M and Robinson D S, *Food Shelf Life Stability*, London: CRC Press, 3–36.
21. TAOUKIS P S, BREENE W B and LABUZA T P (1988), 'Intermediate moisture-foods', in Pomeranz Y, *Adv Cereal Sci Technol*, vol IX, 91–128.
22. MIZRAHI S, LABUZA T P and KAREL M (1970), 'Feasibility of accelerated tests for browning in dehydrated cabbage', *J Food Sci*, 35, 804–807.
23. AHN H-J, CHEORUN J, LEE J-W, KIM J-H, KIM K-H and BYUN M-W (2003), 'Irradiation and modified atmosphere packaging effects on residual nitrite, ascorbic acid, nitrosomyoglobin, and color in sausage', *J Agric Food Chem*, 51, 1249–1253.
24. SPENCER K C and HUMPHREYS D J (2003), 'Argon packaging and processing preserves and enhances flavor, freshness, and shelf life of foods', in Cadwallader K R and Weenen H, *Freshness and Shelf Life of Foods*, ACS Symposium Series 836, Washington, DC: American Chemical Society, 270–291.
25. TULIO A Z, YAMANAKA H, UEDA Y and IMAHORI Y (2002), 'Formation of methanethiol and dimethyl disulfide in crushed tissues of broccoli florets and their inhibition by freeze thawing', *J Agric Food Chem*, 50, 1502–1507.
26. OBENLAND D M, AUNG L H, BRIDGES D L and MACKEY B E (2003), 'Volatile emissions of navel oranges as predictors of freeze damage', *J Agric Food Chem*, 51, 3367–33.
27. ESKIN N A M and PRZYBYLSKI R (2001), 'Antioxidants and shelf life of foods', in Eskin N A M and Robinson D S, *Food Shelf Life Stability*, London: CRC Press, 175–210.
28. LABUZA T P, MCNALLY L, HAWKES J and HURTADO F J (1972), 'Stability of intermediate moisture food. 1. Lipid oxidation', *J Food Sci*, 37, 154–159.
29. ST ANGELO A J and SPANIER A M (1993), 'Lipid oxidation in meat', in Charalambous G, *Shelf Life Studies of Foods and Beverages: Chemical, Biological, Physical and Nutritional Aspects*, Amsterdam: Elsevier Science Publishers, 35–62.
30. JENSEN K, PETERSEN M A, POLL L and BROCKHOFF P B (1999), 'Influence of variety growing location on the development of off-flavor in precooked vacuum-packed potatoes', *J Agric Food Chem*, 47, 1145–1149.
31. KIM Y D and MORR C V (1996), 'Dynamic headspace analysis of light activated flavor in milk', *Int Dairy J*, 6, 185–193.
32. KIM G Y, LEE J-H and MIN D B (2003), 'Study of light induced volatile compounds in goat's milk cheese', *J Agric Food Chem*, 51, 1405–1409.
33. GROSCH W, MILO C and WIDDER S (1994), 'Identification and quantification of odorants causing off-flavors', in Maarse H and Van der Heij D G, *Trends in Flavor Research*, London: Elsevier Science, 409–415.
34. KARAGUL-YUCEER Y, DRAKE M A and CADWALLADER K R (2002), 'Volatile flavor components of stored nonfat dry milk', *J Agric Food Chem*, 50, 305–312.
35. KARAGUL-YUCEER Y, CADWALLADER K R and DRAKE M A (2003), 'Aroma characterization of fresh and stored-nonfat dry milk', in Cadwallader K R and Weenen H, *Freshness and Shelf Life of Foods*, ACS Symposium Series 836, Washington, DC: American Chemical Society, 108–123.

36. ULLRICH F and GROSCH W (1987), 'Identification of most intense volatile flavor compounds formed during autoxidation of linoleic acid', *Z Lebensm Unters Forsch*, 184, 277–282.
37. MARSILI R T (1999), 'SPME-MS-MVA as an electronic nose for the study of off-flavors in milk', *J Agric Food Chem*, 47, 648–654.
38. MARSILI R T (2000), 'Shelf-life prediction of processed milk by solid-phase microextraction, mass spectrometry, and multivariate analysis', *J Agric Food Chem*, 48, 3470–3475.
39. MARSILI R T and MILLER N (1998), 'Determination of the cause of off-flavors in milk by dynamic headspace GC/MS and multivariate data analysis', in Mussinan C, Contis E, Ho C T, Parliament T, Spanier A and Shahidi F, *Food Flavor Formation, Analysis, and Packaging Influences*, Amsterdam, Elsevier Science Publishers, 159–171.
40. SJOVALL O, VIRTALAINE T, LAPVTELAINEN A and KALLIO H (2000), 'Development of rancidity in wheat germ analyzed by headspace gas chromatography and sensory analysis', *J Agric Food Chem*, 48, 3522–3527.
41. LEE S-Y, GUINARD J-X and KROCHTA J M (2003), 'Relating sensory and instrumental data to conduct an accelerated shelf life testing of whey-protein-coated peanuts', in Cadwallader K R and Weenen H, *Freshness and Shelf Life of Foods*, ACS Symposium Series 836, Washington DC: American Chemical Society, 175–187.
42. MOTTRAM D S, SZAUMAN-SUMSKI C and DODSON A (1996), 'Interaction of thiol and disulfide flavor compounds with food components', *J Agric Food Chem*, 44, 2349–2351.
43. HOFMAN T, SCHIEBERLE P and GROSCH W (1996), 'Model studies on the oxidative stability of odor active thiols occurring in food flavors', *J Agric Food Chem*, 44, 251–255.
44. GUTH H, HOFMAN T, SCHIEBERLE P and GROSCH W (1995), 'Model reactions on the stability of disulfides in heated foods', *J Agric Food Chem*, 43, 2199–2203.
45. VAN SEEVENTER, P B, WEENEN H, WINKEL C and KERLE J. (2001), 'Stability of thiols in an aqueous process flavoring', *J Agric Food Chem*, 49, 4292–4295.
46. BEZMAN Y, ROUSEFF R L and NAIM M (2001), '2-Methyl-3-furanthiol and methional are possible off-flavors in stored orange juice: aroma-similarity, NIF/SNIF GC-O and GC analyses', *J Agric Food Chem*, 49, 5425–5432.
47. FERREIRA A C S, HOGG T and DE PINHO P G (2003), 'Identification of key odorants related to the typical aroma of oxidation-spoiled white wines', *J Agric Food Chem*, 51, 1377–1381.
48. GIJS L, PERPETE P, TIMMERMANS A and COLLINS S (2000), '3-Methylthiopropion-aldehyde as precursor of dimethyl trisulfide in aged beer', *J Agric Food Chem*, 48, 6196–6199.
49. ROBINSON D S (2001), 'The effect of oxidative enzymes in foods', in Eskin N A M and Robinson D S, *Food Shelf Life Stability*, London: CRC Press, 297–322.
50. BISHOP J R and WHITE C H (1986), 'Assessment of dairy product quality and potential shelf-life – a review', *J Food Prot*, 49, 739–753.
51. BOSSUYT R (1982), 'A 5–minute ATP platform test for judging the bacteriological quality of raw milk', *Neth Milk Dairy J*, 36, 355–364.
52. GRIFFITHS M W, PHILLIPS J D and MUIR D D (1981), 'Thermostability of proteases and lipases from a number of species of psychrotrophic bacteria of dairy origin', *J Appl Bacteriol*, 50, 289–303.
53. MUIR D D, PHILLIPS J D and DALGLEISH D G (1979), 'The lipolytic and proteolytic activity

of bacteria isolated from blended raw milk', *J Soc Dairy Technol*, 31, 137–144.
54. FUNG D Y C (1994), 'Rapid methods and automation in food microbiology: a review', *Food Rev Int*, 10, 357–375.
55. HUIS-INT VELD J H J (1996), 'Microbial and biochemical spoilage of foods: an overview', *Int J Food Microbiol*, 33, 1–18.
56. BALEIRAS-COUTO M M, HARTOG B J, HUIS-INT VELD J H J, HOFSTRA H and VAN DER VOSSEN J M B M (1996), 'Identification of spoilage yeast in food production chain by microsatellite polymerase chain reaction fingerprinting', *Food Microbiol*, 13, 59–67.
57. DAINTY R H (1996), 'Chemical/biochemical detection of spoilage', *Int J Food Microbiol*, 33, 19–33.
58. VALLEJO-CARDOBA B and NAKAI S (1994a), 'Keeping quality of pasteurized milk by multivariate analysis of dynamic headspace gas chromatographic data. 1. Shelf life prediction by principal component regression', *J Agric Food Chem*, 42, 989–993.
59. VALLEJO-CARDOBA B and NAKAI S (1994b), 'Keeping quality of pasteurized milk by multivariate analysis of dynamic headspace gas chromatographic data. 2. Flavor classification by linear discriminant analysis', *J Agric Food Chem*, 42, 994–999.
60. YU K, HAMILTON-KEMP T R, ARCHBOLD D D, COLLINS R W and NEWMAN M C (2000), 'Volatile compounds from *Escherichia coli* O157:H7 and their absorption by strawberry fruits', *J Agric Food Chem*, 48, 413–417.
61. HAMILTON-KEMP T R, ARCHBOLD D D, LOUGHRIN J H, COLLINS R W and BYERS M E (1996), 'Metabolism of natural volatile compounds by strawberry fruits', *J Agric Food Chem*, 44, 2802–2805.
62. WHITFIELD F B and TINDALE C R (1984), 'The role of microbial metabolites in food flavors', *Food Technol Aust*, 36, 204–213.
63. EYLES M J and ADAMS R F (1986), 'Detection of microbial metabolites by gas chromatography in the examination of spoiled canned foods and related products', *Int J Food Microbiol*, 3, 321–330.
64. ZIEGLEDER G, STOJACIC C and LUSTENBERGER M (1995), 'Causes and detection of off-odors in unprinted paperboard', *Pack Technol Sci*, 8, 219–228.
65. DALEY J D, LLOYD G T, RAMSHAW E H and STARK W (1986), 'Off-flavor related to the use of sorbic acid as a food preservative', *CSIRO Food Q*, 46, 59–63.
66. MADIGAN D and MCMURROUGH I (1994), 'Rapid determination of 4-vinyl Guaiacol ferulic acid in beers and worts by high performance liquid chromatography', *J Am Soc Brew Chem*, 52, 151–155.
67. GANDOLFI I, PALLA G, MARCHELLI R, DOSSENA A, PUELLI S and SALVADORI C (1994), 'D-alanine in fruit juices: a molecular marker for bacterial activity, heat treatments and shelf life', *J Food Sci*, 59, 152–154.
68. KILCAST D and SUBRAMANIAM P (2000), 'Introduction', in Kilcast D and Subramaniam P, *The Stability and Shelf Life of Food,* Cambridge, UK: CRC Press – Woodhead Publishing, 1–22.
69. SEITZ L M and RAM M S (2000), 'Volatile methoxybenzene compounds in grains with off-odors', *J Agric Food Chem*, 48, 4279–4289.
70. CADWALLADER K R (2000), 'Enzymes and seafood quality', in Haard N F and Simpson B K, *Seafood Enzymes: Utilization and Influence on Postharvest Seafood Quality*, New York: Marcel Dekker, 365–384.
71. LABUZA T P and SCHMIDL M K (1988), 'Use of sensory data in the shelf life testing of foods: principles and graphical methods for evaluation'. *Cereal Foods World*, 33, 193–205.

72. KILCAST D (2000), 'Sensory evaluation methods for shelf life assessment', in Kilcast D and Subramaniam P, *The Stability and Shelf Life of Food,* Cambridge, UK: CRC Press – Woodhead Publishing, 79–105.
73. DUYVESTEYN W S, SHIMONI, E and LABUZA, T P (2001), 'Determination of the end of shelf life for milk using Weibull Hazard analysis', *Lebensmittel-Wissenschaft und-Technologie*, 34, 143–148.
74. CARDELLI C and LABUZA T P (2001), 'Application of Weibull Hazard analysis to the determination of the shelf life of roasted coffee', *Lebensmittel-Wissenschaft und-Technologie*, 34, 273–278.

10

Verification and validation of food spoilage models

G. D. Betts and S. J. Walker, Campden and Chorleywood Food Research Association, UK

10.1 Introduction: the modelling process

The use of mathematical models to predict the response of microorganisms to their environmental conditions has increased steadily over the past few years, as modelling software has become more widely available. It is important, however, that users of models understand their strengths and limitations. They are not black boxes that give absolute answers. Instead, they require a certain amount of knowledge in their use and interpretation. The answers given can only be used as an indication of the likely microbial response, in the same way that data from traditional shelf-life studies or challenge tests provide an indication only.

The development of a predictive equation to describe microbial growth is not difficult, but developing models that are industrially robust has particular challenges. Many models are developed from microbiological data in broths and it is important to demonstrate that they are reliable not only when compared with laboratory data but more importantly when applied to real food situations. Assessing the reliability of a predictive microbiological model can be achieved by following the validation and verification stages described in this chapter.

There are many types of predictive model available which can be placed into different groups depending on the criteria used, e.g. the microbiological event studied, the modelling approach used or the variables considered (Whiting and Buchanan, 1994). Whiting and Buchanan (1993) proposed a three-tier system to describe the relative stages of the mathematical modelling process.

1. Primary models which describe the response (growth, death or survival) of an organism to a single set of environmental conditions in terms of, for example, growth rate.

2. Secondary models which use the answers from primary models obtained in a range of conditions to produce mathematical equations. These are then used to determine the microbial response under conditions not previously tested. It is possible to combine the primary and secondary modelling stages in a single-stage surface response surface routine (Jones, 1993), which may reduce the amount of data required for a model.
3. Tertiary level models which are the interface between the scientist and the end-user and consist of simple input screens where the user can enter a set of product formulation conditions and receive a prediction of growth parameters.

For reviews of the different types of models available see, for example, Skinner *et al.* (1994), McDonald and Sun (1999) and Whiting (1995). Generally, it is the secondary level models that require full validation and verification, although it is important to check the reliability of the individual curves in primary models.

There are different types of secondary model that describe different aspects of microbial response to the environment.

1. Kinetic growth models. These models are able to predict growth parameters of microorganisms such as lag time, growth rate and time taken to achieve a specific increase in numbers. These models can be used for situations where growth of, for example, a spoilage organism is acceptable but the rate of growth is important, as it can increase or decrease the shelf-life.
2. Time-to-growth models. These models are based on the time taken for specific microorganisms to initiate growth in a food. In the case of *Clostridium botulinum* such models often describe the time to toxin production (Genigeorgis *et al.*, 1991). These models are useful in the case of pathogens where reaching the end of lag phase and the start of growth would be unacceptable for any food formulation.
3. Growth/no-growth models or probabilistic models. These are similar to time-to-growth models but set out to define the boundary of environmental conditions allowing microbial growth or to determine the probability of growth occurring under any set of conditions in a given time period.
4. Inactivation/death models. These models are based on the death or inactivation kinetics of microorganisms as affected by heat or other physical process, e.g. irradiation, pulsed electric fields, or high pressure. They allow a prediction to be made on the likely decrease in numbers of microorganisms under different process conditions. These models have application in many products using heat treatments to reduce numbers of food poisoning or food spoilage organisms.
5. Survival models. These models predict the ability of microorganisms to survive under adverse environmental conditions (e.g. low pH or a_w values). They are particularly useful for organisms which have a low infective dose such as *Salmonella* or *Escherichia coli* O157. For these organisms, the ability to survive during storage of a food is of more interest than whether or not growth will occur, as the presence of a few cells can cause illness.

It is important that primary and secondary models are fully validated and that the latter are also verified for use under the intended conditions. This chapter is primarily concerned with model evaluation in respect to shelf-life predictions and therefore will focus on growth models. However, the procedures described below are applicable to all modelling types that give data on time, rate and changes in microbial numbers, and examples will be discussed where relevant.

10.2 Validation and verification: definitions and use

Validation is a term which has been widely applied to any assessment of the success or 'goodness-of-fit' of a model. Neumeyer *et al.* (1997) described validation as the process of comparing response times predicted by a model to the response time observed in food products, and this interpretation has been used by many others (e.g. Davey and Daughtry, 1995; Walls and Scott, 1996; Walls *et al.*, 1996; Wei *et al.*, 2001; Oscar, 2002). Mellefont *et al.* (2003) used the term 'performance evaluation' to describe their work on *Escherichia coli.* They suggested that 'evaluation', i.e. to assess, is more appropriate as a general term than 'validation', as many studies do not use confirmation studies of predictions in industrial situations when assessing models.

Another distinction which is sometimes made is whether the evaluation uses internal data, i.e. data which was used to create the model, or uses external or independent data, i.e. data not used within the model. The precise terminology used may not be important as long as it is clear what type of evaluation has been done.

In this chapter we have used the terms validation and verification. The two terms are similar but have slightly different meanings. The definitions in the context of this chapter are as follows (http://dictionary.reference.com, 2003).

validate: 'to establish the soundness of'

This term is used for any evaluation using internal data, where internal data means any broth studies carried out at the same laboratory using the same set of microorganisms as were used in creation of the model (see Section 10.2.1).

verify: 'to determine or test the truth of, by comparison, investigation or reference'

This term describes any evaluation using data generated from a different set of the target microorganisms, from different growth media or food, or by a different laboratory (see Section 10.2.2).

With data generated in inoculated food studies at the same laboratory using the same microorganisms as were used in the creation of the model, there is a high level of control, and better verification would be expected than from data obtained from naturally contaminated foods or literature studies which do not use the microorganisms or exactly reproduce conditions used in the creation of the model.

The aim of model validation and verification discussed within this chapter is to show that microbiological models can be successfully applied to the prediction of product shelf-life. A number of approaches have been suggested (e.g. McClure *et al.*, 1994; Whiting and Buchanan, 1994; Pin *et al.*, 1999) and these have been expanded below.

10.2.1 Validation

Assessment of a model against data used for its development should always be done. This can pick up, at an early stage, an underlying problem with the data or the model being used. A good agreement between predicted and observed data would be expected. With two-stage modelling techniques, both the primary and secondary fits to the data can be examined.

A model can be assessed against broth data not used in its creation in a cross-validation exercise. With this technique a model is created using only part (often 90%) of the data and the remaining data are used for validation purposes. Alternatively, data created in the laboratory at a different time using the same organisms, growth medium and conditions could be used instead. With the latter technique, the agreement between predicted and observed data may be less good due to variations in the preparation of the microorganism, growth medium or growth conditions employed.

10.2.2 Verification

Not all of the models available will necessarily have been verified for food use, and many will explicitly state this. When verification is to be done, a number of options exist on how to verify. The choice of option depends on the data available and the intended use of the model. Use of more than one option can increase the confidence in using the model or help identify the reasons why models fail in some food situations.

- Option 1: Comparison of model with food studies done in the same laboratory inoculated with the same microorganisms. Food validation is important as components present in the product formulation may not be present in laboratory media and could affect the extent to which the environmental factors will support or suppress growth (McDonald and Sun, 1999). This option is very useful as it adds the extra errors associated with the food matrix but uses the same organisms as used in the model. In some cases, sterile food is used so that there is no competition from the food's natural microflora. In other cases, the natural microflora remains present. The latter is a more robust verification, as it allows for microbial interactions, but is reliant on being able to selectively count the level of inoculated organisms in the presence of the natural flora. The agreement between predicted and observed data is likely to be less than for validation.
- Option 2: Comparison against naturally contaminated foods. If the model is to be used for predicting product shelf-life directly, then the most important

measure of the success of the model is how well it predicts microbial responses in new food situations. Use of naturally contaminated foods will show how well the model predicts the response of different strains of the target microorganism in foods of relevance to the model. It does, however, rely on the target microorganisms being naturally present in most samples; this may be unrealistic for many pathogens. If a model performs badly, then going through option 1 may help to identify whether the discrepancy lies with the food matrix or the microorganisms used.

- Option 3: Evaluation against independent data. This option can include data from inoculated broths or foods, uninoculated foods and other models published or produced completely independently from the data used to derive the model (McClure *et al.*, 1997). It provides a set of independent data, which will test the performance of the model over a wide range of conditions. Using literature data is often less costly than generating new microbiological data. The drawbacks to this are that experimental conditions will vary between studies, the quality of the microbial response data may be poor, and the data available may not adequately cover the model's domain of intended use. Also, some studies may not contain information of importance such as pH, acidulant or a_w of the food product, and a 'best guess' needs to be made based on similar product types. Where possible, literature data should be used for conditions where the experimental details are all known.

Many studies will compare predictions from in-house models to those which are publicly available such as the Pathogen Modelling Program (http://www.arserrc.gov/mfs/PATHOGEN.HTM). The models, however, should still be verified for use in a food situation as described previously. In addition, where an existing model has been expanded or data has been remodelled using a different approach, then it is useful to compare the predictions from the two models to see how they differ.

10.2.3 Conditions over which to evaluate models

When assessing the validity of a model it is important to assess the entire range of conditions or at least clearly define over what region the model has been tested.

All models will have an interpolation region. This is the region inside the boundaries of the experimental matrix under which data has been generated. This region is often termed the Minimum Convex Polyhedron (MCP) (Baranyi *et al.*, 1996) and this is shown in Fig. 10.1 for an experimental matrix comprising temperature, salt and pH. The MCP is usually smaller than the absolute ranges of the factors in a model. Outside the interpolation region, there is an extrapolation region. This can be defined as areas outside the boundary where there are no data points. Predictions should not be obtained in this region as they will not be reliable. Furthermore, as more factors used for predictions are included in models, the risks of extrapolation increase and it is important to

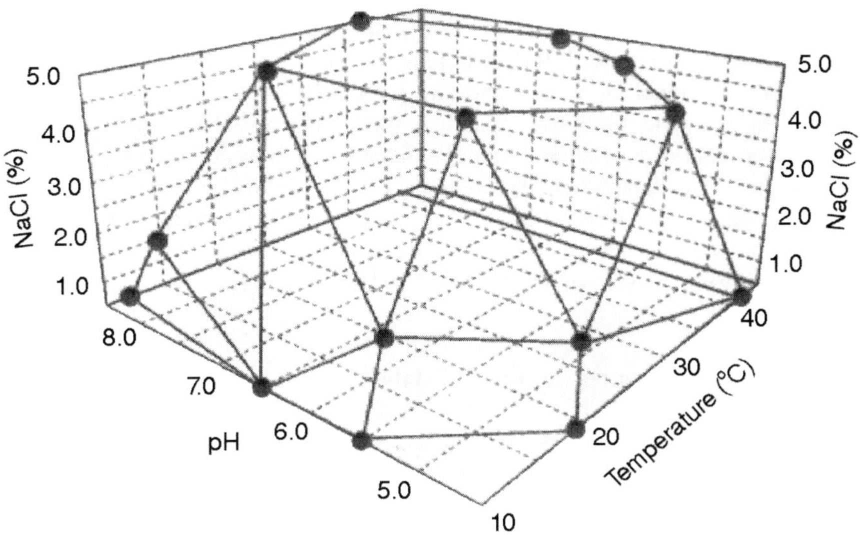

Fig. 10.1 Interpolation region (MCP) for three factor combinations (T, pH, NaCl). Reprinted from *Food Microbiology*, Vol. 17, Marcelo *et al.* © 2000 with permission from Elsevier.

ensure the combinations of factors are within the MCP (Masana and Baranyi, 2000a).

It is important that some comparison data are available over the entire region of the model. Much of the literature data is likely to be in the good growth region of the model and inoculated food studies at the boundaries of the model are therefore important. Care should also be taken in validating models that include data from extreme environmental conditions, given the variability in microbial response that may occur in such regions.

10.3 Evaluation techniques and data transformation

An important aspect of model evaluation is the use of the correct data transformation. Many graphical techniques, regression analysis and other measures of performance make an assumption that the data are normally distributed and have a variance which is independent of the mean values (Ratkowsky *et al.*, 1996). There are several publications which give consideration to variance-stabilising data transformations and discuss the significance of choosing the correct transformation (Alber and Schaffner, 1992; McMeekin *et al.*, 1993; Schaffner, 1998; Zwietering *et al.*, 1994). Many measures of microbial response are not normally distributed, and are routinely transformed into more normally distributed values. For time-based data, such as lag time or time to reach a target number, it is usual to use the $\log_{10}$ or natural log (ln) transformation (Buchanan *et al.*, 1993). With growth rate or death rate data, the reciprocal of the value, i.e. 1/time (Ross *et al.*, 2003), or the square root of time (Garcia-Gimeno *et al.*, 2003), are often used. If data transformation is not considered before any evaluations are done, the results and any associated errors may be misleading.

It is important to remember that if a model is created or validated using transformed data, the predicted values will need to be untransformed to be of any practical use. For example, a predicted lag time of $\log_{10}$ 1.8 h should be converted back to a lag time of 63 hours to have any practical meaning. The confidence intervals obtained from data that have been transformed and analysed will not be symmetrical when the data is retransformed.

10.3.1 Graphical representations of data

Plots of observed versus predicted data

One of the simplest measures of the goodness-of-fit of a model is to visually compare predicted and observed data (Bratchell *et al.*, 1990; Baranyi *et al.*, 1999). Figures 10.2(a)–(c) show the basic form of such graphical representation. Predicted values are plotted against observed values on appropriate axes and the closer the points fall to the line the better the agreement between predicted and observed values. Graphical representations can be made at any of the model evaluation stages above to show how well the predicted and observed data compare. The advantage of this technique is that it allows a crude assessment of any skew or bias in the data. Figure 10.2(b) shows a data set which is skewed or shows a proportional bias, whilst Fig. 10.2(c) shows a constant bias in the data set. These show some underlying errors in the data or the model, which may need to be investigated.

Any predicted value can be compared in this way, for example growth rate, generation time, lag phase duration, time to achieve a specific log increase or decrease and decimal reduction value (Jagannath *et al.*, 2003; Mellefont *et al.*, 2003; Augustin and Carlier, 2000; Zaika *et al.*, 1992). Plotting the data in this way allows an assessment to be made of whether the model is fail-safe or fail-dangerous (Fig. 10.2(a)). Fail-safe models tend to predict quicker growth than that seen in foods and so err on the side of safety; conversely, fail-dangerous models predict slower growth than is observed in foods, which could result in spoilage or safety problems (Table 10.1).

The advantage of graphical analysis is that such techniques are simple and do not require detailed statistical knowledge (Bratchell *et al.*, 1990). In addition, by plotting the experimental values differently (i.e. each temperature plotted with a different symbol/colour), an insight can sometimes be obtained into which areas of a model perform less well.

Residuals plots

Other useful plots of the data can be obtained using the statistical analyses, for example plots of the residuals, i.e. the difference between predicted and observed values. This will show whether there are any larger differences in any particular area of the model or whether the scatter is random.

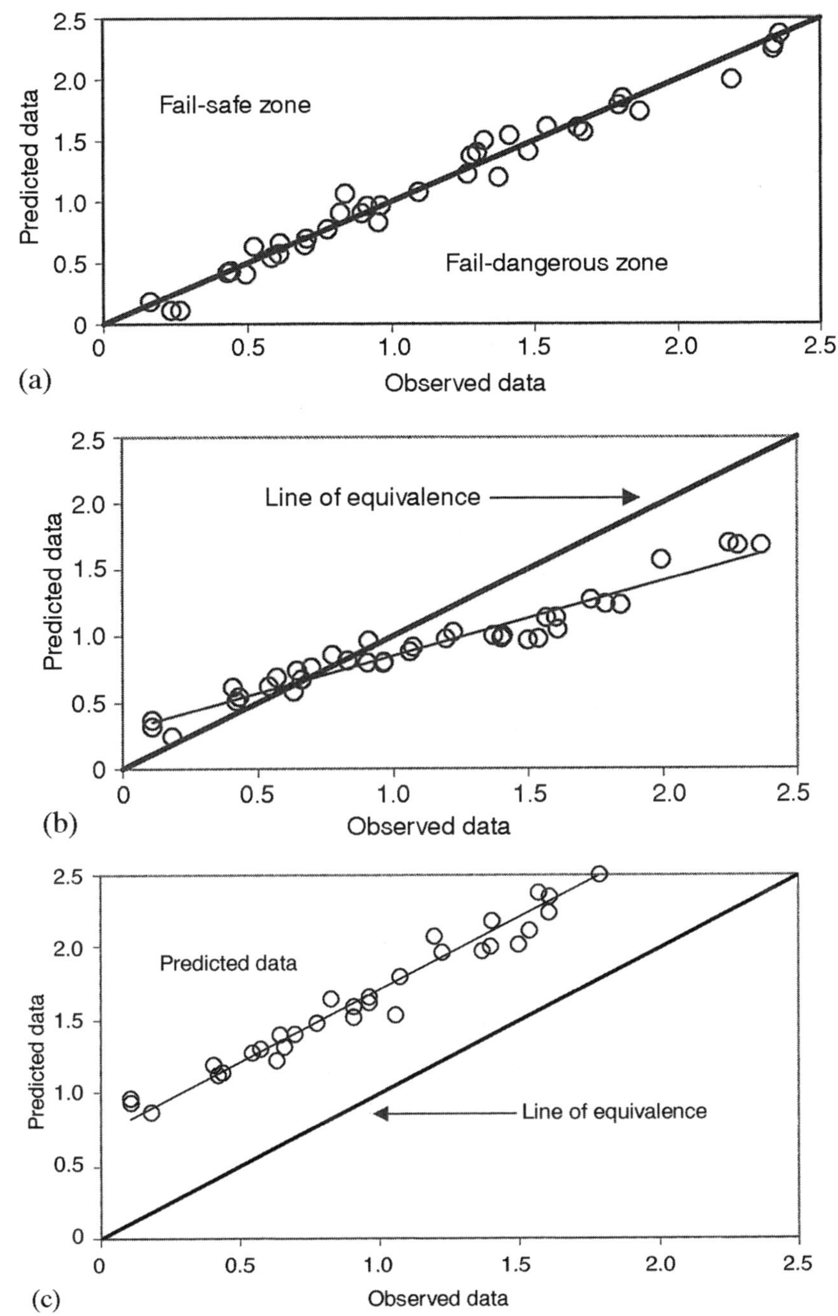

Fig. 10.2 (a) Typical plot of observed versus predicted growth data (o) showing good agreement between data sets along the line of equivalence (—); (b) Plot of observed versus predicted growth data (o) showing a proportional bias on predicted counts; (c) Plot of observed versus predicted growth data (o) showing a constant bias in predicted counts.

Table 10.1 Examples of fail-safe and fail-dangerous predictions for growth, survival and inactivation models

Prediction error	Kinetic growth model	Death/inactivation model	Survival model	Growth/no-growth model
Fail-safe:have a built-in safety margin	Predicted growth rate is quicker than observed in food; e.g. predicted shelf-life 5 days, observed shelf-life 7 days	Predicted death rate is slower than observed in foods; e.g. predicted *D* value 4 minutes, observed *D* value 3 minutes	Predicted survival time is longer than that observed in foods; e.g. predicted survival of *Salmonella* is 6 weeks, observed survival is 4 weeks	Growth predicted where none occurs
Fail-dangerous: may result in food safety or spoilage problems in foods	Predicted growth rate is slower than observed in food; e.g. predicted shelf-life 5 days, observed shelf-life 3 days	Predicted death rate is quicker than observed in foods; e.g. predicted *D* value 4 minutes, observed *D* value 5 minutes	Predicted survival time is shorter than that observed in foods; e.g. predicted survival of *Salmonella* is 6 weeks, observed survival is 8 weeks	Growth not predicted but does occur

10.3.2 Biological sense

Many of the techniques below rely on statistical measures of performance; however, an important aspect which should not be overlooked is whether the model makes biological sense. McClure *et al.* (1994) noted that poorly designed or over-parameterised models may give a good fit to the data but the predicted response may be erroneous. Zwietering *et al.* (1993) warned that 'non-sense' can be generated by models and computers and it is of great importance that a biological as well as a mathematical interpretation is given to the results. Graphical techniques can help in assessing whether the biological interpretation of the response is realistic.

Figure 10.3 shows the main effects of temperature, NaCl, pH and nitrite on the predicted growth responses of meat spoilage organisms (Betts *et al.*, 2001). The figures have been generated using Minitab Version 13.1 and show the mean response value at each level of the environmental factor. Figures 10.3(a) and 10.3(b) show the predicted growth rate and it can be seen that the response for environmental condition makes biological sense. The growth rate increases with temperature and decreases under more stressful conditions of NaCl or pH. With regard to predicted lag time (Fig. 10.3), whilst time response to NaCl and pH makes biological sense, the response with temperature does not. This suggests that closer examination of the observed data is necessary.

10.3.3 What parameters to compare

The choice of parameters for use in validation and verification will depend to some extent on the intended use(s) of the model. In many cases, secondary growth models are based on either maximum growth rate or lag time and so this will be the choice of comparison.

In relation to the shelf-life of perishable chilled foods, the time for a specific increase in numbers (e.g. T_{1000}: the time taken to achieve a 1000-fold increase in numbers) may be the most important factor, as the microbiological quality will be influenced by when a specific number of spoilage organisms has been reached. For pathogens or for shelf-stable products with a long shelf-life, the end of lag time may be the most important criterion, as this indicates the point at which growth has been initiated and could potentially lead to food safety or spoilage issues. The performance of models will be affected by the choice of parameter used in the comparison.

In general, lag time models have been shown to validate less well than growth rate models. Microbial lag time generally shows more variation than growth rate (Wei *et al.*, 2001), particularly when parameters are close to the conditions needed to prevent bacterial growth (Blackburn, 2000). Models for predicting lag time of *Listeria monocytogenes* were reported to be half as accurate as those for growth rate (Augustin and Carlier, 2000). Delignette-Muller (1995) evaluated the accuracy of several polynomial and square root models and found that the inaccuracy of lag time prediction ranged from 36% to 40%, whilst that for growth rate ranged from only 11% to 36%.

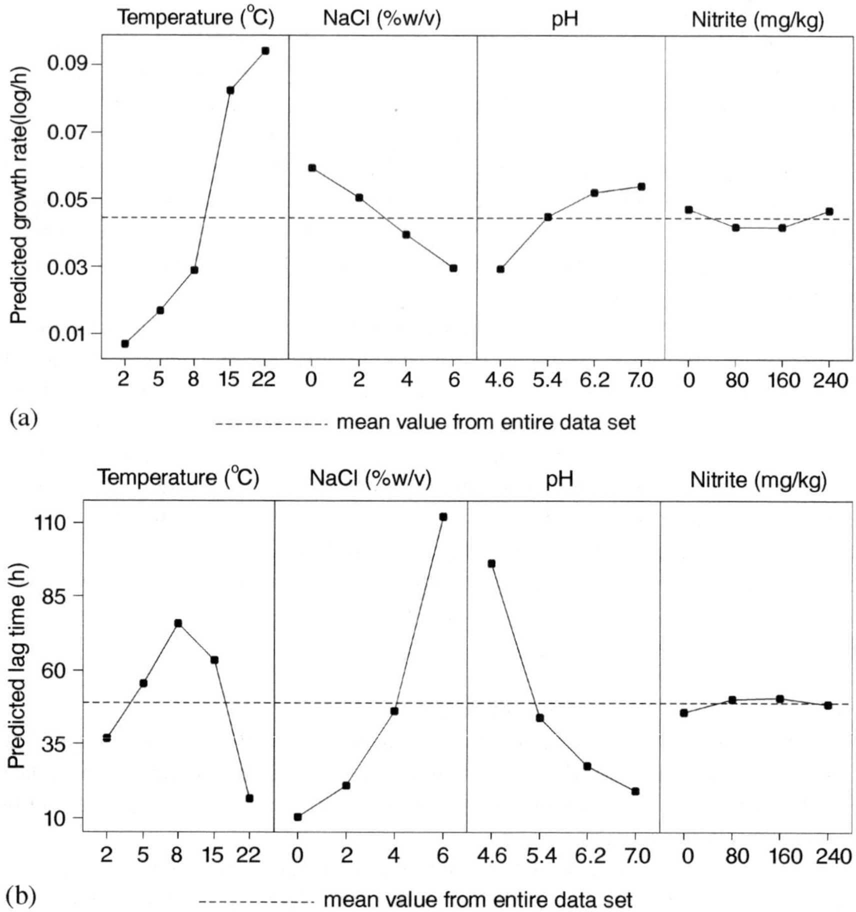

Fig. 10.3 (a) Main effects plot showing mean predicted growth rate time (log cfu/ml/h) at different levels of four environmental parameters; (b) main effects plot showing mean predicted lag time (h) at different levels of four environmental parameters.

If the model contains extreme growth conditions, the overall accuracy will decrease, even for growth rate predictions. Ross *et al.* (2003) also found larger errors in predictions for *E. coli* under conditions close to the boundary for growth, where predicted values of growth rate were 14 times longer than observed.

10.3.4 Statistical analysis techniques*

Multiple correlation

The multiple correlation (R^2) statistic is often used as an overall measure of the accuracy of a prediction (e.g. Devlieghere *et al.*, 2000). The higher the value, the

* For definition of terms see Table 10.2.

more accurate the predicted value (Wei *et al.*, 2001). The R^2 represents the proportion of the total variation in the data that is explained by fitting the model to the data, e.g. how well the changes in microbial response are accounted for by pH, a_w, temperature and other environmental factors. It is sometimes expressed as a percentage, but more usually as a value between 0 and 1, e.g. an R^2 of 0.95 is equivalent to 95% variation explained. The unexplained variation ($100 - R^2$) is due to other factors not known or not accounted for as well as natural variation in microbial growth. Systematic error in a model may not be shown up by R^2; although the predicted and observed values might be perfectly correlated, one may be consistently shifted or biased relative to the other.

The R^2 may not always be a good measure of model performance; the more parameters a model contains, the better is the fit to the data used to generate the model. Therefore the R^2 statistic itself cannot be used to decide whether a new model is an improvement (Baranyi and Roberts, 1995).

A similar measure of error attributed to the model fitting is the percent variance (%V) explained, but with an allowance for sample numbers. This uses the R^2 statistic in the calculation (Davey, 1994; Davey and Daughtry, 1995):

$$\%V = 1 - \frac{(1 - R^2)(n - 1)}{n - N - 1} \tag{10.1}$$

Therefore if a model with $N = 33$ terms (values used to characterise the model) and $n = 173$ data points had an R^2 of 0.949, the %V would be calculated thus:

$$\%V = 1 - \frac{(1 - 0.949)(172)}{139} = 1 - \frac{8.772}{139} = 1 - 0.0631 = 0.937(93.7\%) \tag{10.2}$$

For a given R^2 value, the %V will decrease as the number of terms in the model increases. The %V is also called the R^2-adjusted value and is calculated automatically by many modelling programs. The %V or R^2-adjusted value is more appropriate for models with relatively few data, as it takes into account the number of data points and the number of terms in the model (Davey, 1989).

Mean Square Error (MSE)

The MSE is calculated by the following equation (te Giffel and Zwietering, 1999).

$$\text{MSE} = \frac{\text{RSS}}{n} = \sum \frac{(Y_o - Y_p)^2}{n} \tag{10.3}$$

It describes the variability remaining in the fitted model, i.e. that which is not explained by factors such as temperature, pH, etc. The lower the MSE, the better is the model fit to the data.

Table 10.2 Definition of terms for equation-based statistical tests

R^2	Multiple correlation statistic
%V	Percent variance
%D	Percent discrepancy
n	Number of observations in the data set
N	Number of terms in the model
s	Number of fitted parameters (similar to N)
Y	Values of comparison e.g. lag time maximum growth rate, generation time
Y_{observed} or Y_o	Value observed in broths or foods
$Y_{\text{predicted}}$ or Y_p	Value predicted by the model
T_{1000}	Time to achieve a 1000-fold increase in numbers
RSS	Residual, sums of squares
S_{yx}	Standard deviation of the residuals
$\sum$	Sum of all relevant data
$\sum_{i=1}^{n}$	Sum of all data values (n) starting from the first value ($i = 1$)
$\lvert Y \rvert$	Absolute value of Y, i.e. no − or + signs attributed to the number

Standard deviation of the residuals

The standard deviation of the residuals (S_{yx}) or standard error of the measured count is sometimes calculated (Oscar, 2002) using equation [10.4]. It represents the amount of deviation that may be expected for any predicted value from the model. The smaller the S_{yx} the better.

$$S_{yx} = \sqrt{\sum (Y_o - Y_p)^2/(n - s)} \qquad [10.4]$$

Table 10.3 shows the statistical output from a single-stage response surface model with linear and cross-product terms (interactions), for a mixed group of Enterobacteriaceae under 16 combinations of pH, temperature, nitrite and a_w. It can be seen that the model had a high R^2-adjusted value (93.7%), which suggested a good fit to the data, and had an S_{yx} value of 0.737, which means the average error on each log count prediction was 0.737 log units.

Table 10.3 Statistical outputs from Enterobacteriaceae model

Source	DF	SS	MS
Regression	33	1.411E+03	42.8
Error	139	75.5	0.543
Total	172	1.487E+03	
$S_{yx} = 0.737$	$R^2 = 94.9\%$	$R^2(\text{adj}) = 93.7\%$	

DF, degrees of freedom; SS, sums of squares of errors; MS, mean squared error; S_{yx}, standard deviation of the residuals.

Bias and accuracy factors

Ross (1996) proposed two different indices for determination of model performance. The Bias factor (B_f) is a measure of the overall agreement between predicted and observed values. It will indicate whether, on average, the predictions lie above or below the line of equivalence, and by how much. Equal weighting is given to over-prediction, where the predicted value is greater than the observed, and under-prediction. The B_f is shown in equation [10.5]:

$$B_f = 10^{\left(\sum \log (Y_p/Y_o)/n\right)} \qquad [10.5]$$

Perfect agreement between observed and predicted values would give a B_f of 1. A value larger than 1 would indicate that, on average, the predictions were larger than the observed values and would represent a fail-dangerous model; e.g. a B_f of 1.2 would indicate that the predictions were 20% larger than observed values. A value less than 1 would indicate a fail-safe model where, on average, the predictions were shorter than observed. The B_f is now widely used as a measure of model performance and provides a common criterion for comparison of different models. In calculation of the B_f, the over (+) and under (−) predictions can cancel each other out. They do not therefore show the absolute errors in the model. This can be achieved by the use of the accuracy factor (A_f). This is based on a similar equation to [10.5] but disregards whether the difference between the predicted and observed value is positive or negative:

$$A_f = 10^{\left(\sum |\log (Y_p/Y_o)|/n\right)} \qquad [10.6]$$

As all differences are now positive, the A_f value will always be equal to (if there is perfect agreement) or greater than one. An accuracy of 1.5 means that, on average, the predicted value is 50% different (either smaller or larger) from the observed value. The accuracy and bias factors were originally developed to compare the model predictions to observations not used in the creation of the model, particularly with food data. Baranyi *et al.* (1999) refined these terms to also allow comparison with other models. The A_f measure was replaced by the percent discrepancy (%D) term.

$$\%D = (A_f - 1) \times 100 \qquad [10.7]$$

Relative error (RE)

Oscar (1999a,b) validated models against data not used in their development using equation [10.8] for each individual prediction. A value of zero would indicate perfect agreement between the observed and predicted data for each curve.

$$RE = \left[\frac{Y_p - Y_o}{Y_o}\right] \times 100 \qquad [10.8]$$

Median relative error (MRE)
The median relative error (MRE) has been used to assess the prediction bias of the model (Oscar, 2000). If the MRE of the entire data set were zero, this would represent a model which, on average, had no bias.

Mean absolute relative error (MARE)
The mean absolute relative error (MARE) is defined by Oscar (1999a,b) as

$$\mathrm{MARE} = \frac{1}{n}\sum_{i=1}^{n} |\mathrm{RE}|i \qquad [10.9]$$

It is similar to the A_f previously described and the answers are given as the percentage error in predictions. For example, Delignette-Muller (1995) reported MARE values of 36–40% for various lag time models.

10.3.5 Choice of measurement criteria

There are a number of different criteria to assess the performance of models during validation or verification stages. Many are based on similar principles and yield similar results. This is the case with measurements of model bias and accuracy. A worked example of how to calculate B_f, A_f, MRE and MARE is shown for some data of Gibson *et al.* (1988) on growth of *Salmonella* in Tryptone Soya Broth. Table 10.4a contains observed and predicted values for T_{1000} values under 20 different conditions of pH, salt and temperature. The full details are given in the original paper.

1. The B_f value is calculated by

   $$B_f = 10^{(\sum \log\,(Y_p/Y_o)/n)} = 10^{-0.24108/20} = 10^{-0.01205} = 0.972677$$

2. The A_f value is calculated by

   $$A_f = 10^{(\sum |\log\,(Y_p/Y_o)|/n)} = 10^{1.200253/20} = 10^{0.06001} = 1.15$$

3. For the MARE equation, the value is calculated by

   $$\mathrm{MARE} = \frac{1}{n}\sum_{i=1}^{n} |\mathrm{RE}|i = \frac{1}{20} \times 267 = 13\%\ \text{error}$$

It can be seen (Table 10.4b) that, for this data set, the A_f and the MARE give similar values of model accuracy and the B_f or MRE give similar values for model bias. It could be asked, therefore, which is the best measurement criterion to use. It is relatively easy using PC spreadsheet packages to calculate many different measures of model performance, but as can be seen, similar techniques give similar answers. Currently, it would appear that the A_f and B_f of Ross (1996) are the most widely used performance criteria for microbial growth models. As such they seem a good choice for indices of validation or verification of models as the results are easily comparable with other data. It is

Table 10.4a Predicted (Pr) and observed (Ob) data for T_{1000} values of *Salmonella* (data of Gibson *et al.*, 1988): example of calculations for various performance indices

ObT_{1000} (h)[1]	PrT_{1000} (h)[2]	Values for A_f and B_f calculations		Values for RE and MARE calculations	
		Log_{10} (PrT_{1000}/ ObT_{1000})	Absolute value Log_{10} (PrT_{1000}/ ObT_{1000})	RE	Absolute value of RE
176	180	0.00976	0.00976	2.2727	2.2727
237	171	−0.14175	0.14175	−27.8480	27.8480
249	188	−0.12204	0.12204	−24.4980	24.4980
288	236	−0.08648	0.08648	−18.0560	18.0560
394	372	−0.02495	0.02495	−5.5840	5.5840
37	47	0.10389	0.10389	27.0270	27.0270
41	45	0.04043	0.04043	9.7561	9.7561
47	52	0.04391	0.04391	10.6383	10.6383
85	70	−0.08432	0.08432	−17.6470	17.6470
98	99	0.00441	0.00441	1.0204	1.0204
14	17	0.08432	0.08432	21.4286	21.4286
19	17	−0.04830	0.04830	−10.5260	10.5260
20	19	−0.02228	0.02228	−5.0000	5.0000
36	38	0.02348	0.02348	5.5556	5.5556
8.5	9.4	0.04371	0.04371	10.5882	10.5882
10.7	9.2	−0.06560	0.06560	−14.0190	14.0190
16	17	0.02633	0.02633	6.2500	6.2500
20	21	0.02119	0.02119	5.0000	5.0000
9.6	7.2	−0.12494	0.12494	−25.0000	25.0000
14.2	17	0.07816	0.07816	19.7183	19.7183

[1] ObT_{1000}, observed time for 1000-fold increase.
[2] PrT_{1000}, predicted time for 1000-fold increase.

Table 10.4b Summary of model evaluation criteria[1] for the data of Gibson *et al.* (1998)

A_f	MARE	B_f	MRE
1.15 (15% error)	0.133 (13% error)	0.97 (3% error)	0.0164 (1.64% error)

[1] A_f, accuracy factor; MARE, mean absolute relative error; B_f, bias factor; MRE, median relative error.

recommended that a combination of graphical and statistical measurements be made. Use of a single technique may fail to reveal some forms of systematic error (te Giffel and Zwietering, 1999).

10.3.6 What performance criteria can be expected?

Microbiological data are naturally very variable and microbiologists tend to consider numbers of organisms in $\log_{10}$ values. Only differences between microbiological data of >0.5 log units are generally considered to be practically important. Smaller differences are within the normal limits of uncertainty of microbiological measurement and result from method variability and the heterogeneous distribution of microorganisms in foods. For example, recent studies have shown that for minced meat and meat carcasses, the measurement uncertainty is 0.3–0.4 logs for TVC and 0.4–0.5 logs for Enterobacteriaceae (Anon, 2003). Similar data were reported by Jarvis (2000) who considered that the error in any viable count method is approximately ±0.3 log units. Consequently, there will always be an inherent level of error with any microbiological data, including that for models. The developers and users of models should be aware of this and not expect a 'perfect' fit to data.

It could be asked which of the evaluation stages described is the most important. Obviously, if a model fails to validate, then there is little point in progressing with verification. If a model had a good A_f value when tested against internal broth data but was poor against data observed in foods, then the model could not be applied to industrial situations, as the issue is not how well the model fits the data but how well it predicts the microbial response in foods. The validation against the food group for which the model is intended to be used is therefore the most important verification stage. The uncertainty in model predictions increases as you work through the stages of model validation and verification. There should be good agreement of the model against the data used in its creation but as new broth conditions, food studies and literature data are included the agreement worsens. This is illustrated in Fig. 10.4 (Pin *et al.*, 1999) which showed that the error increased from 11.2% at the laboratory broth stage to 26.6% in inoculated foods to 53.5% in naturally spoiled foods.

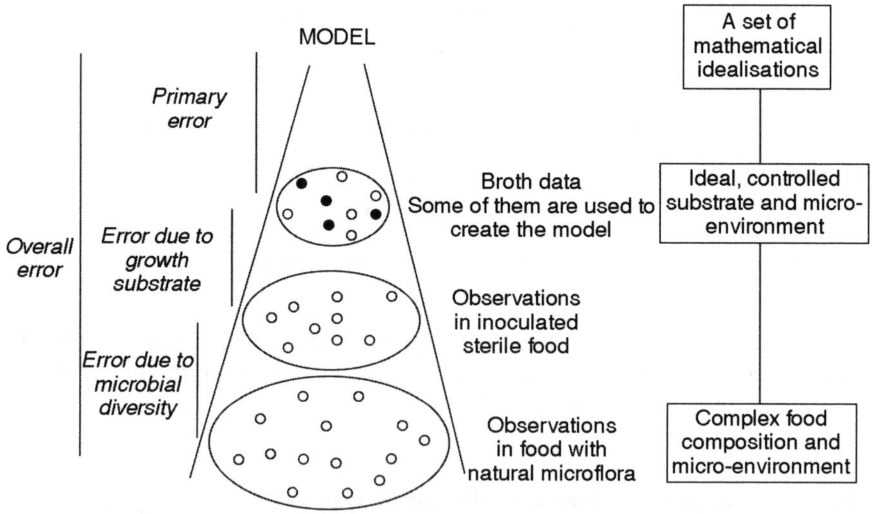

Fig. 10.4 Increasing errors with validation stage (Reprinted from *Journal of Applied Microbiology*, Vol. 87, Pin *et al.* © 1999 with permission from Blackwell Publishing).

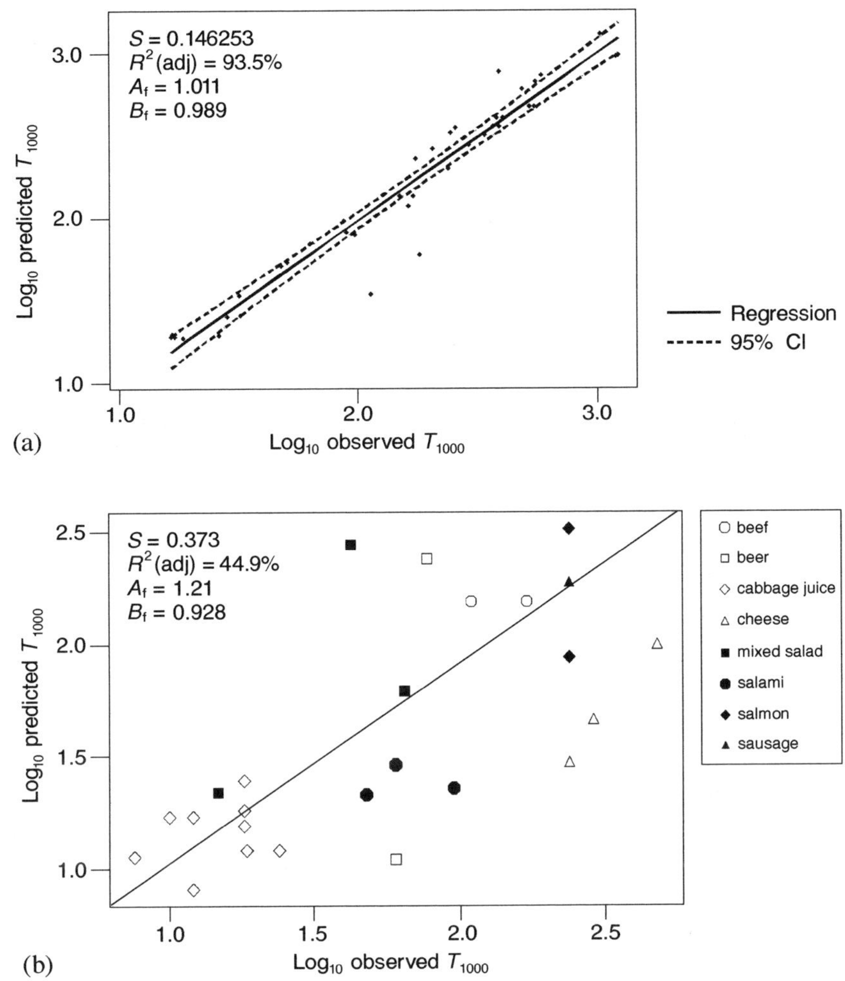

Fig. 10.5 (a) Validation of a model for lactic acid bacteria showing predicted T_{1000} values plotted against those obtained from the data used in the creation of the model; (b) verification of a model for lactic acid bacteria showing predicted T_{1000} values against those observed in foods.

Figure 10.5(a) shows the validation plot of observed versus predicted T_{1000} for lactic acid bacteria from a response surface model produced under different conditions of pH (2.9–5.8), salt (0.5–10% w/v) and temperature (2–30°C) (Betts and Linton, 1998). There was excellent agreement between the observed and predicted values, as may be expected, which shows that there were no underlying errors in the model or bias in the data. Figure 10.5(b) shows verification of the model against literature data. The model has a B_f of 0.93, which shows that it is slightly fail-safe, and an A_f of 1.2 or 20% error. Although worse than the validation against the broth data, this still shows good agreement of predictions against independent data. The value of plotting the data as well as simply calculating the B_f and A_f values can be seen by observation of Fig. 10.5(b). The poorest agreement was for data observed in cheese, where the predictions were very fail-safe. This may mean that there is something inhibitory in cheese that is affecting the accuracy of the predictions, or there could be something unusual about the experimental data in the cited literature that requires further investigation.

Other studies have quantified the level of error expected from model evaluation. Baranyi and Roberts (2000) found errors of 7–10% for growth rate models of *Listeria monocytogenes* compared with the data from which it was generated, and Ross *et al.* (2000) found the error in growth rate estimates to be about 10% per independent variable. As a rule of thumb, the acceptable accuracy factor would decrease by a further 10% for each independent variable (Mellefont *et al.*, 2003). Therefore a model with three variables, e.g. pH, temperature and a_w, would have an error of 30–40% or an A_f of 1.3 to 1.4.

With respect to B_f, care should be taken in using models with $B_f > 1$ for pathogens as they indicate a fail-dangerous situation. Four categories of B_f values are proposed. The 'good', 'acceptable' and 'unacceptable' model categories follow Ross (1999). The 'use with caution' category is new.

- Good model $B_f = 0.9–1.05$
- Acceptable model $B_f = 0.7–0.9$
- Use with caution $B_f = 1.06–1.15$
- Unacceptable model $B_f < 0.7$ and $B_f > 1.15$.

The 'Acceptable' range of the model includes those which are slightly fail-safe, but models which are slightly fail-dangerous are designated 'Use with caution'. Some A_f and B_f criteria from the literature for a range of pathogens and spoilage organisms are shown in Table 10.5.

There is limited guidance for other measures of model performance. For R^2, Seman *et al.* (2002) considered a value of 0.79 with a standard error of measurement of 1.08 to be good.

Fail-safe or fail-dangerous

As a general rule, models for pathogens should be fail-safe. If analysis of observed versus predicted graphs or calculations of B_f demonstrate that a model is fail-dangerous, then it should not be used for predicting the shelf-life of foods with respect to safety issues. Choice of modelled parameters or validation data should be investigated, and in areas of poor agreement more data should be obtained or the model domain curtailed and the data remodelled.

With respect to spoilage organisms, models do not necessarily have to be fail-safe, and if they are too fail-safe they may be of limited benefit. If a model has a tendency to be fail-dangerous ($B_f > 1$), but its accuracy is good, it may still be useful. A commercial decision would need to be made on whether any potential spoilage of a product could be allowed. Challenge test studies in the products of concern could be done to verify the model for that particular formulation.

Verification of dynamic growth models
Many microbial models are produced from data obtained under static growth conditions, i.e. constant temperature, water activity or pH. However, in many industrial situations there will be a fluctuating or dynamic environment where temperatures will rise and fall and pH may decline during product manufacture and storage. Models have been proposed for predicting microbial growth under dynamic temperatures for *Brochrothrix thermosphacta* (Baranyi *et al.*, 1995) and raw fish products (Koutsoumanis, 2001) and more recently under dynamic water activity conditions (Lebert *et al.*, 2003). It is important that the data used for verification of the model are collected under changing environments. Neumeyer *et al.* (1997) validated a model developed for *Pseudomonas* species under constant temperatures with broth data and industry data obtained under fluctuating conditions. The model had an A_f of 1.1 for laboratory data in foods at constant temperatures, an A_f of 1.3 for literature data on foods at constant temperature, and an A_f of 1.21 for industrial data in foods under fluctuating temperatures. The model developed under constant temperatures was therefore shown to give reasonable predictions of growth under fluctuating temperatures. The worst agreement was for the literature data, where the error between predicted and observed data was 30%. The authors concluded that use of literature data may prejudice the evaluation of model performance, and data obtained in controlled conditions may be preferable. This will depend on the intended purpose of the model. If it is to be used for a wide range of foods and industrial situations then comparison to a larger independent data set is useful, provided the errors are within the generally acceptable performance criteria for models.

Verification of growth/no-growth models
In some cases it is difficult to do the classical validation and verification stages as described above due to the lack of quantitative data. For example, in growth/no-growth models, the aim is to find the boundary between conditions which allow growth and those which do not in a given time period. In many of the experiments there will be no growth at all, and in others there will be long lag times followed by very slow growth which will not result in the typical growth curves used in conventional modelling approaches.

Here a different sort of validation exercise is needed. Figure 10.6 shows the fitted line of a survival model for the time-to-growth of yeasts as affected by alcohol, pH, sucrose, sorbate and temperature (Evans *et al.*, 2004) against the observed data in broths. Similar plots have been shown by McMeekin *et al.* (2000) for the growth/no-growth interface of *E. coli*. In this type of model, it is not the precise time to growth that is critical, but rather the ability to grow in a defined time period.

A useful validation or verification exercise is to categorise the data into broad bands of growth response (e.g. 1 = growth within 2 weeks, 2 = growth within 2 months) and calculate the percentage accuracy of predictions. This is demonstrated in Table 10.6 for the data of Evans *et al.* (2004) for the survival

Table 10.5 Accuracy (A_f) and bias (B_f) factors for literature data on model performance in foods

Organism	Model[1]	Medium	Growth parameter[2]	B_f	A_f	Reference
E.coli	In-house model[a]	Meat (49)	GT	1.05	1.11	Mellefont *et al.* (2003)
	FMM			0.93	1.15	
	PMP			0.72	1.38	
Lactic acid bacteria	In-house model[b]	Fish (34)	GR	1.14	1.22	Koutsoumanis *et al.* (2002)
	In-house model[c]			1.09	1.19	
Lactobacillus sake	In-house model[a]	Cooked meats (10)	Lag	1.26	1.26	Devlieghere *et al.* (1999)
			GR	0.96	1.26	
Listeria	FMM, PMP	Meats (92)	GR	0.75	1.73	te Giffel and Zwietering (1999)
L. monocytogenes	In-house model[d]	Cured meats (4)	GR	1.24	1.97	Seman *et al.* (2002)
L. monocytogenes	In-house model[a]	Prepared cooked meats	Lag	1.33	1.51	Devlieghere *et al.* (2001)
			GR	0.97	1.21	
L. monocytogenes	In-house model[e]	Prepared cooked meats	Lag	1.03	1.17	Devlieghere *et al.* (2001)
			GR	0.98	1.16	

Pseudomonas	In-house model[b]	Fish (33)	GR	0.93	1.22	Koutsoumanis *et al.* (2002)
	In-house model[c]			1.11	1.17	
Psychrotrophic pseudomonads	In-house model[a]	Chicken mince (8)	GR	0.92	1.29	Neumeyer *et al.* (1997)
Salmonella	In-house models[a,c,f]	Cooked chicken (21)	Lag	0.75–1.29	1.1–1.36	Oscar (2002)
S. aureus	FMM, PMP	Foods containing vegetables and chicken (22)	GR	0.37	2.67	Walls *et al.* (1996)
			Lag	2.0	2.0	
			L_3	2.2	2.32	
S. aureus	In-house models[a,e]	Ham	GR	1.56–6.85	1.64–7.12	Castilleio-Rodriguez *et al.* (2002)
		Turkey	GR	0.55–2.85	1.91–3.07	
		Chicken	GR	1.09–5.68	1.84–5.89	
S. aureus	FMM, PMP	Ham, turkey, chicken	GR	2.85–11.6	3.01–11.64	Castilleio-Rodriguez *et al.* (2002)

[1] FMM, Food MicroModel; PMP Pathogen Modelling Program
[2] GT, generation time; GR, growth rate; Lag, lag phase; L_3, 3 log increase in cell numbers.
In-house models: (a) square root or Ratkowsky model; (b) polynomial model; (c) Arrhenius model; (d) linear regression model; (e) response surface model; (f) hyperbolic model.
Medium: the number in brackets refers to the number of different types evaluated.

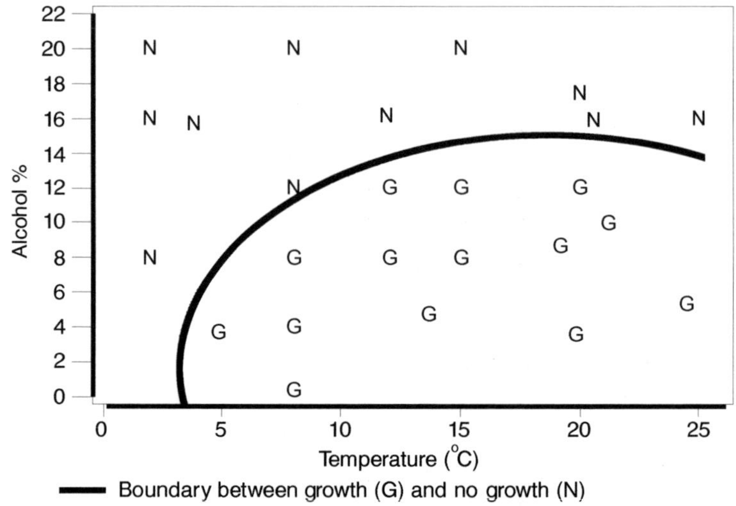

Fig. 10.6 Survival model of boundary between growth and no-growth of yeasts under different conditions of alcohol and temperature.

model for yeasts. Predicted time-to-growth in three broad categories has been compared to data observed in broths. There was agreement between the predicted and observed class on 798 occasions out of 900 trials, making an overall success rate of 88.6%.

Often, time-to-growth models are based on turbidity rather than plate count data. This can lead to issues with food validation as it may not be possible to observe turbidity due to the viscosity or colour of the foods. Other measures of growth may be needed, such as visible gas formation or fermented odour (Jenkins *et al.*, 2000). These, however, do not necessarily relate solely to microbial numbers, and so some degree of bias might be expected.

Table 10.6 Validation of a model for prediction of time-to-growth versus that observed in broth studies used in the creation of the model

Actual growth class[1]	Predicted growth class[1]			
	1	2	3	Total
1	194	43	0	237
2	3	30	7	40
3	4	45	574	623
Total	201	118	581	900

[1] Class 1 – growth within 30 days; Class 2 – growth between 30 and 149 days; Class 3 – no growth by 150 days.

10.4 Limitations of models

Predictive models are generally generated from well-controlled laboratory studies, using specific microorganisms in liquid media that support good growth of the target microorganism. They are then used for a wide range of foodstuffs, over a wide range of conditions to predict the growth of strains (or even species or genera for generic spoilage models) not included in the model development. It is therefore not surprising that all models 'fail' to some extent in their practical application. An understanding of the reasons for the discrepancy between the model and the observed data is essential for the correct application and interpretation of the model; to identify the likely cause of the difference and so try to correct the error so that best use can be made of the data already collected; and to develop future data collection and modelling techniques.

Whilst the limitations of models are widely discussed, many of the issues also apply to the other techniques used by industrial food microbiologists, e.g. storage trials and challenge tests. Such techniques are important and usually determine the microbial response directly in the food of concern and so include all of the factors affecting growth. The results, however, should only be applied to the specific conditions tested and are often based on relatively small studies. In practice, many industrial microbiologists will use predictive models to determine the likely growth response of a microorganism to its environment and then confirm the exact relationship with challenge tests.

There are several reasons why a model may 'fail' and these relate to either the validation or verification processes.

10.4.1 Validation issues

Models that do not validate well (i.e. do not reflect the data used to produce them) should not proceed to the verification stages. The reasons for poor validation involve the data generated and the modelling techniques used.

Data effects

The data should represent the range of conditions for the model. If an insufficient number of environmental conditions are examined, particularly in regions where the microbial response is greatly affected (i.e. near the boundaries for growth), the model may not reflect the observed response. For most kinetic growth models, the inoculum level should be well controlled, or the modelling technique may be used to account for the variation. An inoculum level of 10^2–10^4 cfu/ml or g is commonly used. This allows the numbers present in the initial phases of growth to be estimated well using traditional methods and provides a long period of exponential growth. If the inoculum level is too high, the microorganism may not fully enter the exponential phase of growth before it goes into the stationary phase and so the estimates of growth will be incorrect. If the inoculum level is very low, the lag phase may be underestimated as there will be too few cells present to allow for the range of variability expected in this

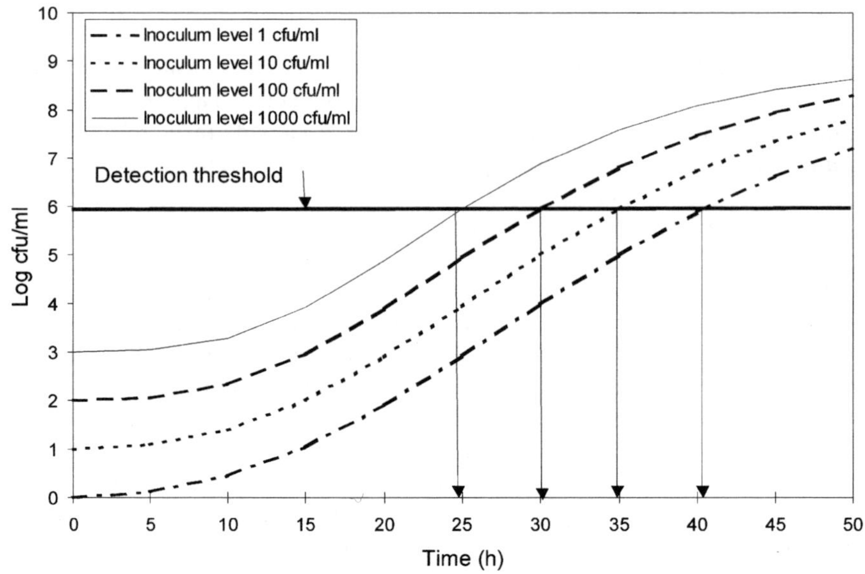

Fig. 10.7 Detection time for microbial growth as affected by inoculum size.

term (Ross *et al.*, 2003). Some models are based on the time to turbidity or to a specific optical density. In these cases, the time to reach the threshold value is affected not only by the environmental conditions, but also by the inoculum size; i.e. under identical environmental conditions, the detection time can vary with inoculum size (Fig. 10.7). Such models generally do not allow the user to adjust the initial inoculum concentration and so care is needed in their use. Walker and Jones (1993) provided some advice on the collection of data for modelling.

Modelling techniques

The choice of model can greatly affect the validation process. With many primary models the number and choice of data points for each environmental condition tested is important if the parameters of the growth curve are to be well defined. It is generally recognised that a growth curve should have at least eight data points and these should be positioned to identify the major points of inflection (i.e. end of lag, mid-exponential and end of exponential phases). With single-stage models, this is of less concern with an individual growth curve, provided there are enough points in the entire dataset. The choice of primary (e.g. Gompertz, Logistic) and secondary (polynomial, square root) models may affect the ability of the model to describe the data, as all are based on certain assumptions. Finally the techniques used to solve the models may be dependent on the starting values included for optimisation routines, which might result in an inappropriate solution for the model.

Failure in the validation phase of modelling can usually be picked up by the failure of a model to converge to a solution or a poor measure of fit to the data.

Even where a model appears to be a reasonable statistical fit to the data, it is recommended that the predicted values are graphically compared with the original data and that the effect of environmental factors makes biological sense, e.g. using 2D or 3D graphs.

10.4.2 Verification issues

Just because a model is a good reflection of the data used to produce it, it does not automatically follow that it will apply to other situations. There are several reasons why models may not verify under all conditions: these tend to relate to model data issues, verification data issues, or the presence of additional factors not present in the model.

Data effects

The data may be generated from a single microbial strain or a mixture of strains, species or even genera. The choice of the microorganism(s) is very important. If they are isolated from an environment very different from that normally expected in food, then their growth response may differ significantly. Conversely, if the strain(s) selected are chosen because of their resistance to the environmental conditions, then their response may allow significantly faster growth than normally expected. Models for pathogens are often based on the fail-safe approach, but this might not always be useful for spoilage models; it could result in the microbiological shelf-life of a food being unduly restricted. Although spoilage models may contain a variety of strains/species/genera, their choice should reflect the spoilage flora of concern. If the major spoilage organism is absent, e.g. *Shewenella putrefaciens* in fish spoilage, then the predictions are likely to fail.

It is generally accepted that the growth rate of a microorganism is dependent on the environmental conditions. However, the lag period or h_0 (the work to be done before growth: Pin *et al.*, 2002), is affected not only by the current environmental conditions, but also by the phase of growth when inoculated (Whiting and Bagi, 2002), the degree of cell injury (e.g. Smelt *et al.*, 2002) and the previous growth conditions (e.g. Walker *et al.*, 1990; Whiting and Bagi, 2002). Delignette-Muller *et al.* (2003) found that h_0 increased linearly with the degree of temperature shift between the original and current growth conditions. Consequently the verification of lag time is generally poor and frequently not reported. An attempt to include some effects of lag time in model verification has used the times for a specific increase in numbers (e.g. T_{1000}) rather than the lag time. Some tertiary modelling systems allow the user to obtain predictions without a lag time (which may be relevant where the target organism is well adapted to its growth environment) or to alter the duration of lag time based on experience.

The growth of microorganisms is more variable at extremes of conditions (McClure *et al.*, 1994; Blackburn, 2000). A model produced from such data may validate well, but verification with other data may be poor, simply due to the variability of the microbial response under such conditions even when repeated in the same laboratory under apparently identical conditions. The use of replicate data in model development, particularly towards the boundaries for growth, will help determine the significance of the issue. If poor replication is found to be the case, either from observed versus predicted plots or from analysis of the Relative Error for each condition, then it may be appropriate to remove the data at the boundaries and remodel the remaining data. This has the disadvantage that it will reduce the interpolation region or MCP, but will yield a model which is valid over a narrower range of conditions.

Verification data

The apparent failure of a model to verify may on occasion reflect the quality of the verification data. With data from the literature, not all the environmental conditions may have been measured and some may have to be estimated. In some published studies the microorganism may have been chosen because it was different from the strains usually found. With spoilage models, the microflora associated with particular foods may vary around the world depending on the raw materials and conditions. With complex foods, the microbial growth response may reflect the growth in only a part of the food conducive to growth, but the analysis of the environmental conditions (e.g. pH or a_w) may be for the whole food. Several models appear to be based on a_w, but were actually developed using a particular humectant, frequently NaCl, and this was transformed to an a_w value. If such models incorporating a_w are compared with verification data which used a different humectant, the results may not agree, particularly at lower a_w values.

Additional factor effects

On many occasions, models fail to verify for some food types because of factors not represented in the model. Foods are more complex than broth systems and potentially incorporate other factors affecting growth. These include the effects of food structure, additional antimicrobial systems (e.g. naturally present or added preservatives, different acids used in verification data, different storage atmospheres) and the presence of competitor microorganisms. In general, all these factors tend to make predictive models fail-safe. In addition, most models identify the range of conditions used to make the model. It is often assumed that all combinations of conditions are available for verification. In reality the interpolation region (MCP) of the model may be much reduced. Consequently the model fails to verify satisfactorily. The choice of acid is also important. Many models use hydrochloric acid, whilst foods use organic acids in their formulation. Organic acids have an additional effect above that caused by the pH level alone, which will not be accounted for in the model.

The verification process is often aimed at determining the limits within which the model can be used. In addition, the areas where verification is not good may be of importance in identifying future areas of work.

10.5 Future trends

10.5.1 Expanding conditions of models

Producing growth data for microbiological models is expensive. Therefore in order to make better use of currently available data there is a trend towards expanding current data sets to include new factors. Predictive models based on temperature, pH and a_w have been expanded to include nitrite (Buchanan and Bagi, 1994) and carbon dioxide (Sutherland *et al.*, 1997). The experimental design should ensure that there are enough conditions which are within the interpolation regions of the old and new data set to avoid extrapolation into regions not covered by the entire matrix. For more information on the risks of adding new data see Masana and Baranyi (2000b). In addition, the validation and verification stages will need to be repeated.

The greater use of models that predict microbial consequences under dynamic environmental conditions (i.e. changing temperature, pH, a_w) will require new approaches to ensure that they can be used with confidence.

10.5.2 Relationship of microbial predictions to other factors affecting shelf-life

Spoilage of food products can be caused by microbiological, biochemical and other sensory changes, e.g. loss of texture or colour. However, in many cases, microbial levels are used to determine the shelf-life; levels of 10^6–10^8 cfu/g for total count (Lyhs *et al.*, 2001), 10^6–10^8 cfu/g for *Pseudomonas* (Koutsoumanis and Nychas, 2000; Koutsoumanis *et al.*, 2002) and 10^7 cfu/g for *Shewanella putrefaciens* (Taoukis *et al.*, 1999) have been reported to equate to the onset of sensory spoilage. In other cases, however, the relationship between microbial numbers and sensory deterioration is not clear cut. Ahvenainen *et al.* (1989) found little correlation between the final numbers of microorganisms and shelf-life of modified-atmosphere and vacuum-packed ham at 5°C, and Leroi *et al.* (2001) found it was difficult to find a single rule for predicting shelf-life based on microbial levels. They showed prolonged storage of smoked fish at high microbial levels without sensory deterioration as well as unacceptable samples with low microbial numbers and concluded that chemical indices of shelf-life were more reliable than microbial levels. In the future, more models for predicting shelf-life will include sensory and chemical measures as well as microbial numbers. However, the performance of the models may well decrease as more factors are included and new acceptance criteria will need to be defined.

10.5.3 Increasing use of models by the food industry

Currently the percentage of the food industry regularly using predictive models is small but it is envisaged that this will increase over the next five years (Betts and Everis, 2002). This is particularly so since predictive models and microbiological response data are now more freely available (see below).

It is recommended that users of microbiological models develop their own in-house verification data. Predicted data on lag time, growth rate or T_{1000} values can be compared to those found in real foods during routine shelf-life trials. Accuracy and bias factors can be calculated using the techniques shown in this chapter and can give the users confidence in the validity of models for their foods.

10.6 Sources of further information and advice

A major achievement in predictive microbiology is the recent launch of Combase and Growth Predictor. Combase is a combined database of microbial responses to food environments (Baranyi and Tamplin, 2003) and Growth Predictor contains the pathogen models from the UK Government Food MicroModel program and is freely available (http://www.ifr.bbsrc.ac.uk/safety).

A useful overview of predictive microbiology can be found in *Predictive Microbiology. Theory and Application* by McMeekin *et al.* (1993). Although 10 years old, this book provides the most comprehensive description of predictive microbiology currently available.

10.7 References

AHVENAINEN R, SKYTTA E and KIVIKATAJA R L (1989), 'Factors affecting the shelf-life of gas-and-vacuum-packed cooked meat products Part 1: Sliced ham', *Lebensmittel-Wissenschaft und-Technologie*, 22, 391–398.

ALBER S A and SCHAFFNER D W (1992), 'Evaluation of data transformations used with the square root and schoolfield models for predicting bacterial growth rate', *Applied and Environmental Microbiology*, 58 (10), 3337–3342.

ANON (2003), *Progress made in the preparation of draft Commission Regulation on microbiological criteria for foodstuffs (SANCO/4198/2001, rev 5)*, Working document 2.6.2003/PM SANCO 420594/2003, European Commission.

AUGUSTIN J and CARLIER V (2000), 'Mathematical modelling of the growth rate and lag time for *Listeria monocytogenes*', *International Journal of Food Microbiology*, 56, 29–51.

BARANYI J and ROBERTS T A (1995), 'Mathematics of predictive food microbiology', *International Journal of Food Microbiology*, 26, 199–218.

BARANYI J and ROBERTS T A (2000), 'Principles and application of predictive microbiology of the effects of preservative factors on microorganisms', in Lund B, Baird-Parker T and Gould G, *The Microbiological Safety and Quality of Foods*, Gaithersburg, MD: Aspen Publishers, 342–358.

BARANYI J and TAMPLIN T (2003), 'A new international effort for the advance of Predictive Microbiology ComBase: combined database of microbial responses to food environment', in Van Impe J F M, Geeraerd A H, Leguerinel I and Mafart P, *Predictive Modelling in Foods, 4th International Conference*, ISBN 90–5682–400–7, 50–51.

BARANYI J, ROBINSON T P, KALOTI A and MACKEY B M (1995), 'Predicting growth of *Brochothrix thermosphacta* at changing temperatures', *International Journal of Food Microbiology*, 27, 61–75.

BARANYI J, ROSS B, MCMEEKIN T A and ROBERTS T A (1996), 'Effects of parameterization on the performance of empirical models used in predictive microbiology', *Food Microbiology,* 13, 83–91.

BARANYI J, PIN C and ROSS T (1999), 'Validating and comparing predictive models', *International Journal of Food Microbiology*, 48, 159–166.

BETTS G and EVERIS L (2002), 'Modelling systems and their impact in food microbiology' in *Emerging Technologies for the Food Industry,* CSIS, Madrid.

BETTS G D and LINTON P (1998), *Microbial Spoilage: Modelling growth of lactic acid bacteria in relation to chilled products,* CCFRA R&D Report No 61, CCFRA, UK.

BETTS G D, EVERIS L K and WHITE P (2001), *Microbial Spoilage: Modelling growth of meat spoilage bacteria*, CCFRA R&D Report No 138, CCFRA, UK.

BLACKBURN C DE W (2000), 'Modelling shelf-life' in Kilcast D and Subramanian P, *The Stability and Shelf-life of Food*, Cambridge, UK: Woodhead Publishing, 55–78.

BRATCHELL N, MCCLURE P J, KELLY T M and ROBERTS T A (1990), 'Predicting microbial growth: graphical methods for comparing models', *International Journal of Food Microbiology*, 11, 279–288.

BUCHANAN R L and BAGI L K (1994), 'Expansion of response surface models for the growth of *Escherichia coli* O157:H7 to include sodium nitrite as a variable', *International Journal of Food Microbiology*, 23, 317–332.

BUCHANAN R L, BAGI L K, GOINS R V and PHILIPS J G (1993), 'Response surface models for the growth kinetics of *Escherichia coli* O157:H7', *Food Microbiology,* 10, 303–315.

CASTILLEJO-RODRIGUEZ A M, GARCIA-GIMENO R M, ZURER COSANO G, BARCO ALCALA E and RODRIGUEZ PEREA M R (2002) 'Assessment of mathematical models for predicting *Staphylococcus aureus* growth in cooked meat products', *Journal of Food Protection*, 65 (4), 659–665.

DAVEY K R (1989), 'A predictive model for combined temperature and water activity on microbial growth during the growth phase', *Journal of Applied Bacteriology*, 67, 483–488.

DAVEY K R (1994), 'Review paper. Modelling the combined effect of temperature and pH on the rate coefficient for bacterial growth', *International Journal of Food Microbiology*, 23, 295–303.

DAVEY K R and DAUGHTRY B J (1995), 'Validation of a model for predicting the combined effect of three environmental factors on both exponential and lag phases of bacterial growth: temperature, salt concentration and pH', *Food Research International*, 28 (3), 233–237.

DELIGNETTE-MULLER M L (1995), 'Relation between the generation time and the lag time of bacterial growth kinetics', *International Journal of Food Microbiology*, 43, 97–104.

DELIGNETTE-MULLER M L, BATY F, CORNU M and BERGIS H (2003), 'The effect of temperature shift on the lag phase of *Listeria monocytogenes*', in J F M Van Impe, A H Gaeraerd, I Leguerinel and P Mafart (eds), *Predictive Modelling in Foods, 4th International Conference*, ISBN 90-5682-400-7, 209–211.

DEVLIEGHERE F, VANBELLE B and DEBEVERE J (1999), 'Shelf-life of modified atmosphere packed cooked meat products: a predictive model', *International Journal of Food Microbiology*, 46, 57–70.

DEVLIEGHERE F, GEERARD A H, VERSYK K J, BERNAERT H, VAN IMPE J F and DEBEVERE J (2000), 'Shelf-life of modified atmosphere packed cooked meat products: addition of Na-lactate as a fourth shelf-life determinative factor in a model and product validation', *International Journal of Food Microbiology*, 58, 93–106.

DEVLIEGHERE F, GEERARD A H, VERSYK K J, VANDEWATAERE, B VAN IMPE J F and DEBEVERE J (2001), 'Growth of *Listeria monocytogenes* in modified atmosphere packed cooked meat products: a predictive model', *Food Microbiology*, 18, 53–66.

EVANS D G, EVERIS L K and BETTS G D (2004), 'Use of survival analysis and classification and regression trees (CART) to model the growth/no growth boundary of spoilage yeasts as affected by alcohol, pH, sucrose, sorbate and temperature', *International Journal of Food Microbiology*, 92 (1), 55–67.

GARCIA-GIMENO R M, CASTILLEJO A M, RINCÓN F, ZURERA G and PEREZ-RODRIGUEZ F (2003), 'Response surface vs Davey model for the estimation of *St. aureus* growth and different conditions of temperature, pH, salt, nitrite and aerobic/anaerobic atmosphere', in Van Impe J F M, Geeraerd A H, Leguerinel I and Mafart P, *Predictive Modelling in Foods, 4th International Conference*, ISBN 90–5682–400–7, 132–134.

GENIGEORGIS C A, MENG J and BAKER D A (1991), 'Behaviour of non-proteolytic *Clostridium botulinum* Type B and E spores in cooked turkey and modelling lag phase and probability of toxigenesis', *Journal of Food Science*, 56 (2), 373–379.

GIBSON A M, BRATCHELL N and ROBERTS T A (1988), 'Predicting microbial growth: growth responses of Salmonellae in a laboratory medium as affected by pH, sodium chloride and storage temperature', *International Journal of Food Microbiology*, 6, 155–178.

JAGANNATH A, NAKAMURA I and TSUCHIDO T (2003), 'Modelling the combined effects of pH, temperature and sodium chloride stresses on the thermal inactivation of *Bacillus subtilis* spores in a buffer system', *Journal of Applied Microbiology*, 95, 135–141.

JARVIS B (2000), 'Sampling for microbiological analysis', in Lund B M, Baird-Parker T C and Gould G W, *The Microbiological Safety and Quality of Food, Volume II*, ISBN 0 8342 1323 0, 1691–1733.

JENKINS P, POULOS P G, COLE M B, VANDEVEN M H and LEGAN J D (2000), 'The boundary for growth of *Zygosaccharomyces bailii* in acidified products described by models for time to growth and probability of growth', *Journal of Food Protection*, 63 (2), 222–230.

JONES J E (1993), 'A real-time database/models base/expert system in predictive microbiology', *Journal of Industrial Microbiology*, 12, 268–272.

KOUTSOUMANIS K (2001), 'Modelling of the shelf-life of fish under nonisothermal conditions', *Applied and Environmental Microbiology*, 67 (4), 1821–1829.

KOUTSOUMANIS K and NYCHAS G E (2000), 'Application of a systematic procedure to develop a microbial model for rapid fish shelf-life predictions', *International Journal of Food Microbiology*, 60, 171–184.

KOUTSOUMANIS K, GIANNAKOUROU M C, TAOUKIS P S and NYCHAS G E (2002), 'Application of a shelf-life decision system (SLDS) to marine cultured quality', *International Journal of Food Microbiology*, 73, 375–382.

LEBERT I, BAUCOUR P, DAUDIN J D and LEBERT A (2003), 'Prediction of bacterial growth in dynamic water activity condition', in Van Impe J F M, Geeraerd A H, Leguerinel I and Mafart P, *Predictive Modelling in Foods, 4th International Conference*, ISBN 90-5682-400-7, 247–249.

LEROI F, JOFFRAUD J J, CHEVALIER F and CARDINAL M (2001), 'Research of quality indices for cold-smoked salmon using a step-wise multiple regression of microbiological counts and physico-chemical parameters', *Journal of Applied Microbiology*, 90, 578–587.

LYHS V, LAHTINEN J, FREDIKSSON-AHOMAA M, HYYTIA-TREES E, ELFING K and KORKEALA H (2001), 'Microbiological quality and shelf-life of vacuum-packaged gravad rainbow trout stored at 3 and 8°C', *International Journal of Food Microbiology*, 70, 221–230.

MASANA M O and BARANYI J (2000a), 'Adding new factors to predictive models: the effect on the risk of extrapolation', *Food Microbiology*, 17, 367–374.

MASANA M O and BARANYI J (2000b), 'Growth/no growth interface of *Brochothrix thermosphacta* as a function of pH and water activity', *Food Microbiology*, 17, 485–493.

McCLURE P J, BLACKBURN C DE W, COLE M B, CURTIS P S, JONES J E, LEGAN J D, OGDEN I D, PECK M W, ROBERTS T A, SUTHERLAND J P and WALKER S J (1994), 'Review paper. Modelling the growth, survival and death of microorganisms in foods: the UK FoodMicromodel approach', *International Journal of Food Microbiology*, 23, 265–275.

McCLURE P J, BEAUMONT A L, SUTHERLAND J P and ROBERTS T A (1997), 'Predictive modelling of growth of *Listeria monocytogenes*. The effects on growth of NaCl, pH, storage temperature and $NaNO_2$', *International Journal of Food Microbiology*, 34, 221–232.

McDONALD K and SUN D-W (1999), 'Predictive food microbiology for the meat industry: a review', *International Journal of Food Microbiology*, 52, 1–27.

McMEEKIN T A, OLLEY J N, ROSS T and RATKOWSKY D A (1993), *Predictive Microbiology. Theory and Application*, Taunton, Research Studies Press Ltd.

McMEEKIN T A, PRESSER K, RATKOWSKY D, ROSS T, SALTER M and TIENUNGOON S (2000), 'Quantifying the hurdle concept by modelling the bacterial growth/no growth interface', *International Journal of Food Microbiology*, 55, 93–98.

MELLEFONT L A, McMEEKIN T A and ROSS T (2003), 'Performance evaluation of a model describing the effects of temperature, water activity, pH and lactic acid concentration on the growth of *Escherichia coli*', *International Journal of Food Microbiology*, 82, 45–58.

NEUMEYER K, ROSS T, THOMSON G and McMEEKIN T A (1997), 'Validation of a model describing the effects of temperature and water activity on the growth of psychrotrophic pseudomonads', *International Journal of Food Microbiology*, 38, 55–63.

OSCAR T P (1999a), 'Response surface models for effects of temperature, pH, and previous growth pH on growth kinetics of *Salmonella typhimurium* in brain heart infusion broth', *Journal of Food Protection*, 62 (2), 106–111.

OSCAR T P (1999b), 'Response surface models for effects of temperature and previous temperature on lag time and specific growth rate of *Salmonella typhimurium* on cooked ground chicken breast', *Journal of Food Protection*, 62 (10), 1111–1114.

OSCAR T P (2000), 'Variation of lag time and specific growth rate among 11 strains of *Salmonella* inoculated onto sterile ground chicken breast burgers and incubated at 25°C', *Journal of Food Safety*, 20, 225–236.

OSCAR T P (2002), 'Development and validation of a tertiary simulation model for predicting the potential growth of *Salmonella typhimurium* on cooked chicken', *International Journal of Food Microbiology*, 76, 177–190.

PIN C, SUTHERLAND J P and BARANYI J (1999), 'Validating predictive models of food spoilage organisms', *Journal of Applied Microbiology*, 87, 491–499.

PIN C, GARCÍA DE FERNANDO G D, ORDÓÑEX J A and BARANYI J (2002), 'Analysing the lag-growth rate relationship of *Yersinia enterocolitica*', *International Journal of Food Microbiology*, 73, 197–201.

RATKOWSKY D A, ROSS T, MACARIO N, DAMMETT T W and KAMPERMAN L (1996), 'Choosing probability distributions for modelling generation time variability', *Journal of Applied Bacteriology*, 80, 131–137.

ROSS T (1996), 'Indices for performance evaluation of predictive models in food microbiology', *Journal of Applied Bacteriology*, 81, 501–508.

ROSS T (1999), *Predictive Food Microbiology in the Meat Industry*, MSRC 003, Meat and Livestock Australia.

ROSS T, BARANYI J and McMEEKIN T A (2000), 'Predictive microbiology and food safety', in Robinson R K, Batt C A and Patel P D, *Encyclopaedia of Food Microbiology*, London: Academic Press, 1679–1710.

ROSS T, RATKOWSKY D A, MELLEFONT L A and MCMEEKIN T A (2003), 'Modelling the effects of temperature, water activity, pH and lactic acid concentration on the growth of *Escherichia coli*', *International Journal of Food Microbiology*, 82, 33–43.

SCHAFFNER D W (1998), 'Predictive modelling Gedanken experiment. Why do microbial growth data require a transformation?', *Food Microbiology*, 15, 185–189.

SEMAN D L, BORGER A C, MEYER J D, HALL P A and MILKOWSKI A L (2002), 'Modelling the growth of *Listeria monocytogenes* in cured ready-to-eat processed meat products by manipulation of sodium chloride, sodium diacetate, potassium lactate and product moisture content', *Journal of Food Protection*, 65 (4), 651–658.

SKINNER G E, LARKIN J W and RHODEHAMEL E J (1994), 'Mathematical modelling of microbial growth: a review', *Journal of Food Safety*, 24, 175–217.

SMELT J P P M, OTTEN G D and BOSS A P (2002), 'Modelling the effect of sublethal injury on the distribution of lag times of individual cells of *Lactobacillus plantarum*', *International Journal of Food Microbiology*, 73, 207–212.

SUTHERLAND J P, BAYLISS A T, BRAXTON D S and BEAUMONT A L (1997), 'Predictive modelling of *Escherichia coli* O157: H7. Inclusion of carbon dioxide as a fourth factor in a pre-existing model', *International Journal of Food Microbiology*, 37, 113–120.

TAOUKIS P S, KOUTSOUMANIS K and NYCHAS G J E (1999), 'Use of time–temperature integrators and predictive modelling for shelf-life control of chilled fish under dynamic storage conditions', *International Journal of Food Microbiology*, 53, 21–31.

TE GIFFEL M C and ZWIETERING M H (1999), 'Validation of predictive models describing the growth of *Listeria monocytogenes*', *International Journal of Food Microbiology*, 46, 135–149.

WALKER S J and JONES J E (1993), 'Protocols for data generation for predictive modelling', *Journal of Industrial Microbiology*, 12, 273–276.

WALKER S J, ARCHER P and BANKS J G (1990), 'Growth of *Listeria monocytogenes* at refrigeration temperatures', *Journal of Applied Bacteriology*, 68, 157–161.

WALLS I and SCOTT V N (1996), 'Validation of predictive mathematical models describing the growth of *Escherichia coli* O157:H7 in raw ground beef', *Journal of Food Protection*, 59 (12), 1331–1335.

WALLS I, SCOTT V N and BERNARD D T (1996), 'Validation of predictive mathematical models describing growth of *Staphylococcus aureus*', *Journal of Food Protection*, 59 (1), 11–15.

WEI Q K, FANG T J and CHEN W C (2001), 'Development and validation of growth model for *Yersinia enterocolitica* in cooked chicken meats packaged under various atmosphere packaging and stored at different temperatures', *Journal of Food Protection*, 64 (7), 987–993.

WHITING R C (1995), 'Microbial modelling in foods', *Critical Reviews in Food Science and Nutrition*, 35 (6), 467–494.

WHITING R C and BAGI L K (2002), 'Modelling the lag phase of *Listeria monocytogenes*', *International Journal of Food Microbiology*, 73, 291–295.

WHITING R C and BUCHANAN R L (1993), 'A classification of models for predictive microbiology', *Food Microbiology*, 10, 175.

WHITING R C and BUCHANAN R L (1994), 'Microbial modelling. Scientific status report', *Food Technology*, 48 (6), 113–120.

ZAIKA L L, PHILLIPS J G and BUCHANAN R L (1992), 'Model for aerobic growth of *Shigella flexneri* under various conditions of temperature, pH, sodium chloride and sodium nitrite concentrations', *Journal of Food Protection*, 35 (7), 509–513.

ZWIETERING M H, ROMBOUTS F M and VAN'T RIET K (1993), 'Some aspects of modelling microbial quality of food', *Food Control,* 4 (2), 89–95.

ZWIETERING M H, CUPPERS H G A M, DEWIT J C and VAN'T RIET K (1994), 'Evaluation of data transformations and validation of a model for the effect on bacterial growth', *Applied and Environmental Microbiology*, 60 (1), 195–203.

11

Measuring and modelling the glass transition temperature

I. A. Farhat, University of Nottingham, UK

11.1 Introduction

With the exception of lipid oxidation processes, the stability of foods and pharmaceuticals and their main ingredients usually decreases dramatically when the system undergoes a transition from the glassy to the rubbery state. Examples of this are illustrated for a range of physical and chemical deterioration processes in the other chapters.

The appreciation of the paramount importance of the glass transition concept in understanding the performance of food systems in terms of processing, stability, perception, etc. led to the development of a crucial aspect of modern food science, namely 'food polymer science' or 'food material science'. This discipline interprets the behaviour of food components in relation to their molecular mobility or relaxation time, in a manner adopted from organic and inorganic materials (e.g. synthetic polymers, ceramics, metals, etc.). Molecular mobility and other directly or indirectly related physical properties such as specific volume and density, viscosity, modulus and other rheological properties, are usually described in relation to a 'scaled' (T_g/T) or 'shifted' ($T - T_g$) temperature, where T is the temperature of interest, e.g. processing temperature or storage temperature in shelf-life studies, and T_g is the temperature at which the reversible transition from glass to rubber or from rubber to glass occurs (e.g. Slade and Levine, 1991). The experimental determination of T_g and its dependence on composition are therefore the most important starting points in understanding the processing of a food and predicting its stability (e.g. Slade and Levine, 1991; Roos, 1995; Fan *et al.*, 1994; Farhat *et al.*, 2000). It is important to remember that the glass–rubber transition occurs over a wide range of temperatures which for some systems such as starch can cover several tens of degrees.

While the crystalline state lends itself to a range of analytical tools such as X-ray diffraction, microscopy, solid-state NMR, etc., which enable full three-dimensional characterisation, studying the glassy state often relies on monitoring the changes of physical properties (see above) that occur as the glassy material enters the rubbery state at T_g. The choice of a measurement technique depends on several fundamental and practical issues. The glass transition is a kinetic transition and thus the temperature at which it occurs is not a physical constant of the system but is dependent on the time scale probed. Typically (Sperling, 1986), a decrease by a factor of 10 of the time scale of a measurement (or an increase in frequency) is predicted to lead to an increase of T_g by ~3°C according to the Williams–Landel–Ferry equation (Williams *et al.*, 1955). It is therefore very important to qualify each measurement technique in terms of this characteristic or attempt to extrapolate to a common time scale.

From a practical point of view, the nature of the sample often dictates the choice of analytical method. The main issues to consider are sampling (amount and shape of sample required), sensitivity, control of water gain or loss during measurement, sample preparation, expertise, time requirements, cost, etc. Most of these issues are discussed below for each method. In this chapter, only the most commonly used techniques for measuring the T_g of foods and food ingredients are described.

11.2 Measuring the glass transition temperature

11.2.1 Dilatometry

The increase in the specific volume V (volume per unit mass, m^3kg^{-1}) of a glassy system with increasing temperature exhibits, as it approaches its glass transition temperature range, a detectable change in slope reflecting the difference in the volumetric thermal expansion coefficient α ($m^3kg^{-1}K^{-1}$), between the glassy and rubbery states (Fig. 11.1). This phenomenon was among the first to be used to determine the T_g of polymers (Sperling, 1986).

Dilatometry, where the expansion of the glassy system immersed in a non-plasticising fluid (e.g. mercury or silicon oil) is measured over a range of temperatures, is one of the methods used to measure T_g (Sperling, 1986). This method is simple and could be performed using a relatively inexpensive glass dilatometer immersed in a temperature-controlled transparent fluid, the choice of which depends on the temperature range of interest. This approach is, however, prone to artefacts often due to trapped air. Another drawback is the limitation in precise control of heating/cooling rates and the limited range of these rates available – typically low heating rates of less than 1 $K\,min^{-1}$.

A typical example of the use of a simple borosilicate glass dilatometer (Merck UK Ltd) to measure the glass transition of a model starch-based food system is shown in Fig. 11.2. Sophistication of dilatometry evolved mostly around the range of ways the change in sample dimensions with increasing temperature is sensed, for example through the change in capacitance when a

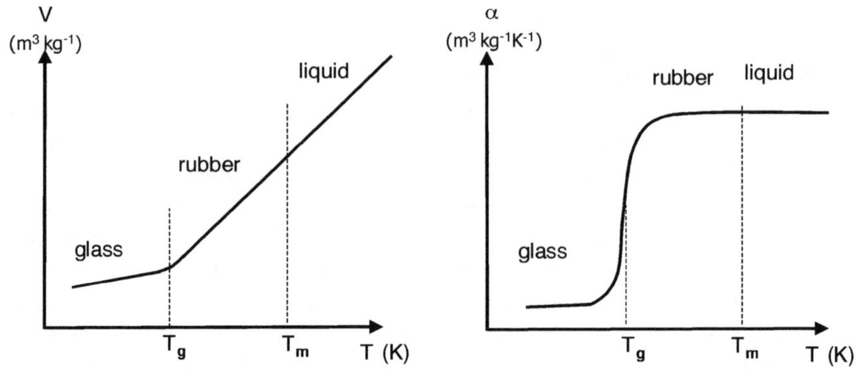

Fig. 11.1 The changes of the specific volume (V) and volumetric thermal expansion coefficient (α) around T_g (adapted from Sperling, 1986).

sample is confined between two electrodes (Capacitance Scanning Dilatometry or CSD, e.g. Bauer *et al.*, 2000) or the use of electronic transducers as in Thermo Mechanical Analysis (TMA). TMA is used for determining the linear thermal expansion coefficient by measuring the dimensional changes in a sample as a function of temperature or as a function of time at a given temperature, usually by employing a probe applied to the sample surface (see Fig. 11.3). Moisture loss during heating is a major drawback of the technique.

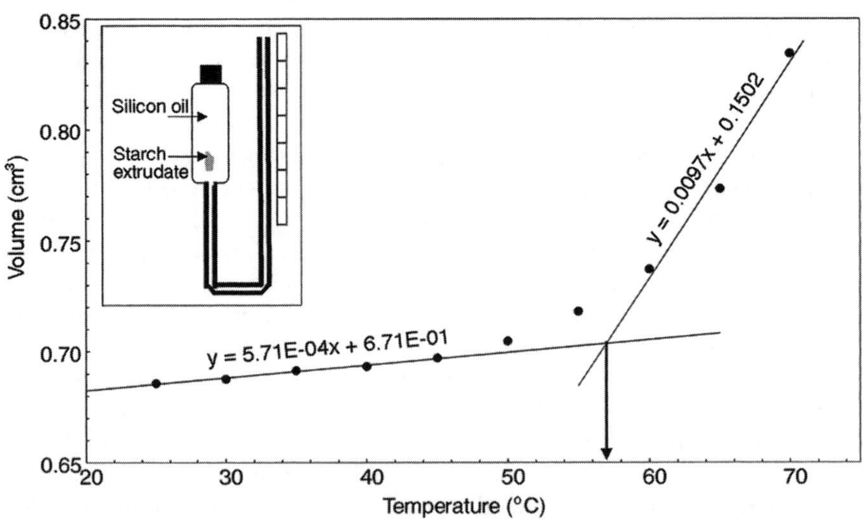

Fig. 11.2 Dilatometric measurement of the glass–rubber transition of a model food system (adapted from Khalid and Farhat, unpublished). The sample, an extruded wheat starch–sugar blend (80:20) containing 15% water (wet basis), was immersed in silicon oil.

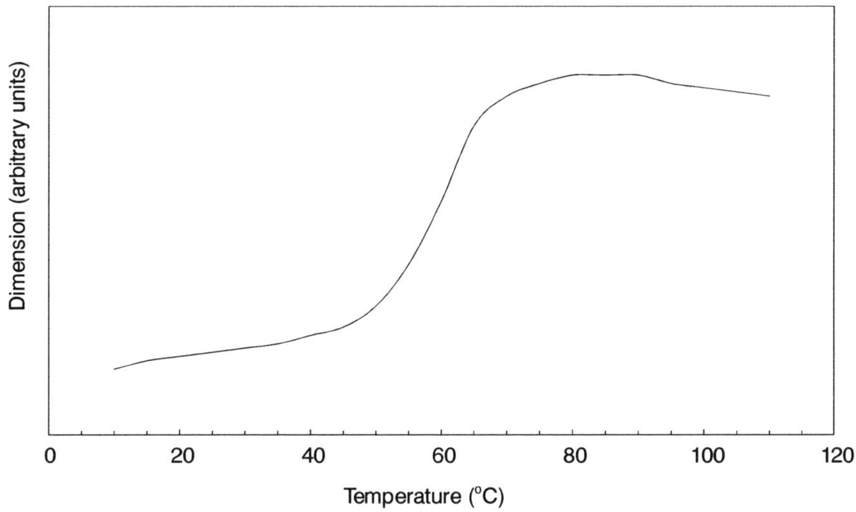

Fig. 11.3 TMA analysis of the glass transition of brown rice kernel containing 14.4% water (adapted from Sun *et al.*, 2002).

11.2.2 Calorimetry

As a material is heated from the glassy state, a change in enthalpy H (J kg^{-1}), similar to that described above for the specific volume, occurs at T_g, resulting in a step change in the specific heat capacity C (J kg^{-1}K^{-1}), also referred to as C_p since, in most food-related studies, constant pressure or moderate pressure change conditions prevail (Fig. 11.4). These changes are very often used to monitor the glass–rubber transition using calorimetric methods such as Differential Scanning Calorimetry (DSC), which is arguably the most widely used technique employed for the determination of the glass transition temperature (Laye, 2002). The sample, typically a few milligrams, is sealed in a special pan,

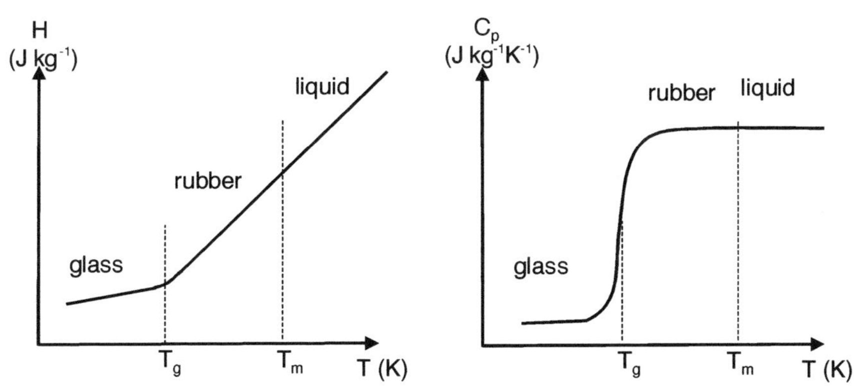

Fig. 11.4 The changes of enthalpy (H) and specific heat capacity (C_p) around T_g (adapted from Sperling, 1986).

usually made from aluminium or stainless steel. Two main types of differential scanning calorimeters can be found. When using a heat flux calorimeter, the sample and reference (usually an empty DSC pan) are placed in the same furnace and the temperature difference between the sample and the reference during steady-state heat flow from/to the furnace is measured and is proportional to heat flux (the proportionality constant is determined through calibration). In a power-compensated calorimeter, the difference in electric power required to heat the sample at the same rate as a reference is recorded. The sample and reference are placed in two different furnaces and subjected to the same time–temperature profile (including isothermal steps if required) (Mathot, 1994).

DSC thermograms are often represented as a plot of heat flux (dq/dt W) versus temperature. The use of appropriate calibrations of the heat flux data in the steady-state region using a material of known specific heat capacity such as sapphire (Hohne, 1991; Cammenga *et al.*, 1993) allows the plot of absolute specific heat capacity versus temperature. A typical specific heat capacity versus temperature thermogram for a commercial breakfast cereal containing approximately 17% sugar and 5% water (wwb) is shown in Fig. 11.5.

As mentioned above, the measured T_g value is dependent on the measurement time scale, which in the case of DSC can be varied through the heating rate. Most DSC glass transition determinations are usually carried out at heating rates of 10–20°C min^{-1} and usually T_g increases linearly on heating rate (e.g. Kalichevsky *et al.*, 1993; Mizuno *et al.*, 1998).

Very often the first heating scan for glassy systems, depending on their T_g and thermal history, exhibits an endotherm occurring between the temperature at which

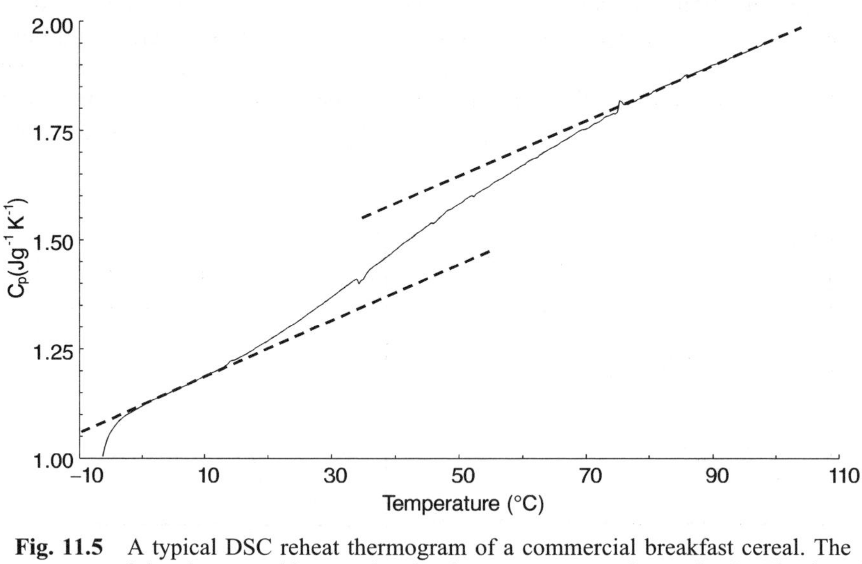

Fig. 11.5 A typical DSC reheat thermogram of a commercial breakfast cereal. The occurrence of the glass transition can be clearly seen as a step change in C_p. The dotted lines are added as a guide for the eye representing the specific heat of the glass and rubber/liquid states (adapted from Martinet and Farhat, unpublished).

the sample was stored and that at which the glass–rubber transition is complete. This endotherm is the enthalpy relaxation associated with the physical ageing of glassy systems (Lourdin *et al.*, 2002). Because enthalpy relaxation is usually not reversible within the DSC experimental time scale, the glass transition is best characterised from the second heating run. To a large extent, the use of temperature-modulated DSC techniques can overcome this issue of overlap between so-called reversing (e.g. the glass transition) and non-reversing (e.g. enthalpy relaxation) thermal events (see the series of papers by M. Reading and co-workers over the last two decades (e.g. Jones *et al.*, 1997; Reading, 2001).

11.2.3 Rheology

The mechanical properties of a glassy system change dramatically as the temperature increases above the glass–rubber transition temperature. Viscosity and modulus decrease by several orders of magnitude as temperature exceeds T_g.

Dynamic mechanical analysis (DMA) is one of the most commonly used rheological techniques to determine the T_g of polymers through the change in their viscoelastic properties (α-relaxation) (Wetton, 1986; Price, 2002). Measurement of the elastic and loss Young moduli (E′ and E″ or G′ and G″) over a frequency range typically between 0.01 and 100 Hz can be carried out in bending, tension, shear or compression modes, with the first being the most commonly used. The glass transition temperature is usually taken as the temperature of the onset of the decrease in elastic modulus or, more often, as the temperature where a peak in the loss tangent, tan δ is reached (tan $\delta = E''/E'$). Kalichevsky *et al.* (1992) described a good correlation between DMTA and DSC, with the T_g value measured by DSC (midpoint) being intermediate between those measured from the E′ drop and the tan δ peak. The tan δ is for some systems very broad and thus the determination of T_g can be subjected to significant uncertainties.

One of the main disadvantages of this method is the requirement for the sample to have a defined geometry (typically a 30 mm × 10 mm × 2 mm slab) and thus, for most foods, the sample requires pressing and/or cutting which could interfere with any thermal history-related information. Recently, developments in sampling devices made the determination of state transitions of powders possible (see Triton Technology website). Also, MacInnes (1993) developed a sample holder enabling the study of liquids using DMTA. A typical example of DMA on starch is shown in Fig. 11.6.

The frequency (f, Hz) dependence of the measured glass transition temperature can be used to determine the activation energy (E_a, kJ mol^{-1}) of the transition from the slope of the Arrhenius plot of $\ln f$ versus $1/T_g$ (e.g. Kalichevsky *et al.*, 1993).

The phase transition analyser

An equipment based on a closed-chamber capillary rheometer referred to as the Phase Transition Analyser™ (PTA), where the compressibility of powdered materials is measured as a function of temperature under a constant pressure

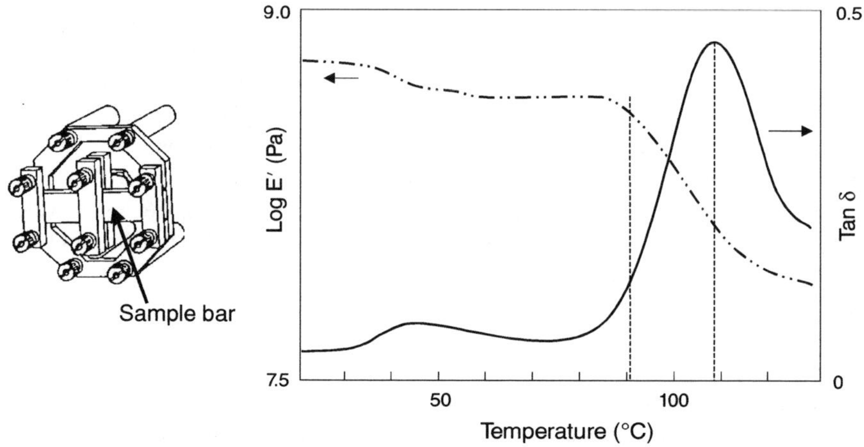

Fig. 11.6 Measurement of the glass transition of starch using DMA. A typical bending mode dual cantilever sampling setup is shown (Wetton, 1986). The sample was pregelatinised waxy maize starch powder, hydrated 20–25% (w/w), pressed (20 tonnes, 90–100°C, 5–10 min) and then re-equilibrated at room temperature and RH = 43% prior to analysis (adapted from Kalichevsky *et al.*, 1992). The elastic modulus (E′) and the loss tangent (tan δ) results are shown.

load (Strahm *et al.*, 2000), is available (Wenger Manufacturing, Inc.) for the determination of the so-called 'controlling T_g'. The device monitors the displacement of the piston applying the pressure as a function of temperature. The onset of the glass–rubber transition is characterised by a pronounced displacement as the glassy particles become more compressible at T_g.

11.2.4 Other techniques

The techniques described above are the most routinely used ones, but many other techniques have been employed to study the glass–rubber transition, particularly through its effect on molecular relaxation and mobility. The most common of these techniques are briefly discussed below.

Dielectrics

Dielectric probes are used to investigate molecular relaxation processes occurring as the material undergoes its glass–rubber transition by monitoring the ability of ions, dipolar molecules and dipolar groups in molecules to reorient. The technique measures the dielectric properties (dielectric conductivity ϵ', dielectric loss ϵ'' and the loss tangent tan δ) of a sample, usually a thin film, as a function of temperature. The results are similar to DMA in appearance but dielectric spectroscopy affords a much wider range of frequency (typically between 10^{-3} and 10^6 Hz).

Dielectric Thermal Analysis (DETA) has been used to study the relaxation processes of many polymers, particularly thin films (Price, 2002). Most of the work on food-related systems has been limited to model systems with low water contents (Noel *et al.*, 1992, 1996) due to concerns relating to the dominating effect of ions and water in defining the dielectric characteristics of a food compared to the contribution of the matrix molecules (usually the polymers) which is often sought in glass transition studies.

Nuclear magnetic resonance

Solid state NMR can be employed to monitor the molecular mobility of the components of a food over a range of temperatures encompassing T_g. Proton relaxometry has been extensively used to study the glass transition of food-related systems (e.g. Kalichevsky *et al.*, 1992; Van den Dries *et al.*, 1998a; Kumagai *et al.*, 2002). The most common approach is monitoring the effect of temperature (or water content) on the mobility of a solid food matrix through the second moment (M_2 s^{-2}) of its spin–spin relaxation decay. M_2 measures dipolar interactions in the solid and decreases with increasing mobility as dipolar interactions are averaged out in mobile systems. A typical determination of the change in molecular mobility of a model glassy confectionery system as it is heated through its glass transition temperature range is shown in Fig. 11.7.

Other forms of NMR, including ^{13}C solid state Magic Angle Spinning spectroscopy and relaxometry, have also been deployed to study the change in molecular mobility across the glassy and rubbery states. These techniques afford a chemical shift resolved information allowing, for example, the monitoring of

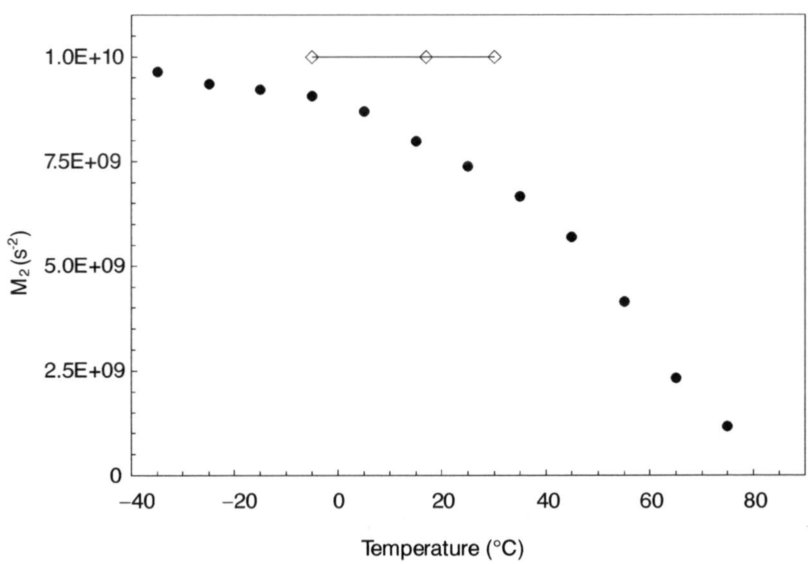

Fig. 11.7 Increase in the mobility of the solid-like component of glucose syrup containing 0.9% κ-carrageenan, 0.14% KCl and 8.4% water as the material undergoes a glass–rubber transition. The open symbols show, from left to right, the onset, midpoint and end of the glass transition range measured by DSC on the same sample (adapted from Kumagai *et al.*, 2002).

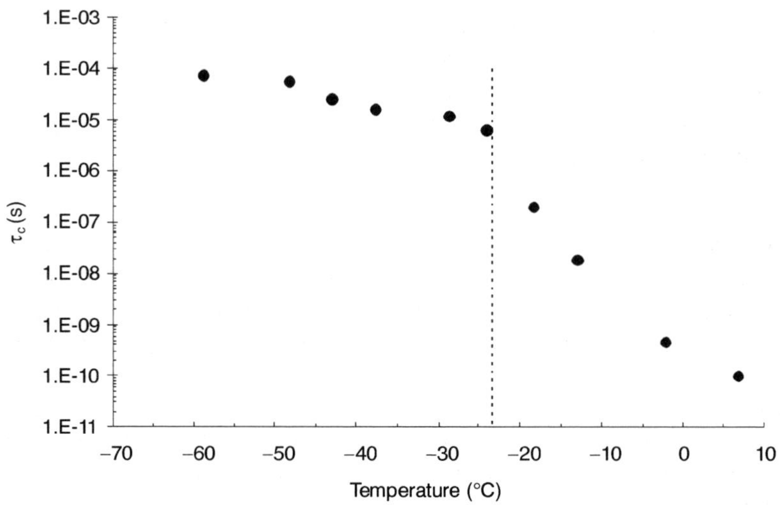

Fig. 11.8 Rotational correlation time of TEMPOL as a function of a 20% solution of maltotriose in water (adapted from Roozen *et al.*, 1991). The dotted line shows the DSC-measured $T_g{}'$ for this mixture as reported by Levine and Slade (1988).

the change in the mobility of individual carbons of a mono- or disaccharide (MacNaughton *et al.*, unpublished).

Electron spin resonance

The dependence of the lineshape of the Electron Spin Resonance spectrum of free radicals on their rotational mobility is often used to monitor the mobility of spin probes, which are stable free radicals, dispersed in a range of matrices such as synthetic polymers and food-related systems such as carbohydrate–water mixtures (Hemminga *et al.*, 1993; Roozen *et al.*, 1991; Van den Dries *et al.*, 1998a, 1998b). Typically, the rotational correlation time (τ_c s) of a spin probe, usually a nitroxide-based free radical such as TEMPOL (4-hydroxy, 2,2,6,6-tetramethyl-piperidino-oxyl), is defined from the ESR spectrum lineshape. A typical example of the dramatic difference in rotational correlation time of a spin probe between the freeze-concentrated glassy state ($\tau_c \geq 10^{-5}$ s) and the rubbery state (τ_c decreasing to $\sim 10^{-10}$ s) is shown in Fig. 11.8.

11.3 Modelling the glass transition temperature

Since most foods and food ingredients contain more than one component, the prediction of the glass transition temperature of mixtures based on the knowledge of the T_g of the individual constituents is an important step in the understanding of processing and stability.

11.3.1 Effect of molecular weight on T_g

Fox and Flory (1950) formalised the dependence of T_g of a polymer on its molecular weight through the following equation:

$$T_g = T_{g_\infty} - \frac{K}{M} \qquad [11.1]$$

where $T_{g\infty}$ is the limiting value for T_g and K is a constant that depends on the type of polymer. Roos (1995) suggested that this equation applies for a range of synthetic and natural polymers, and for polydisperse systems, M would refer to the number average molecular weight. Figure 11.9 shows the relationship between T_g and molecular weight (MW) for a range of carbohydrates.

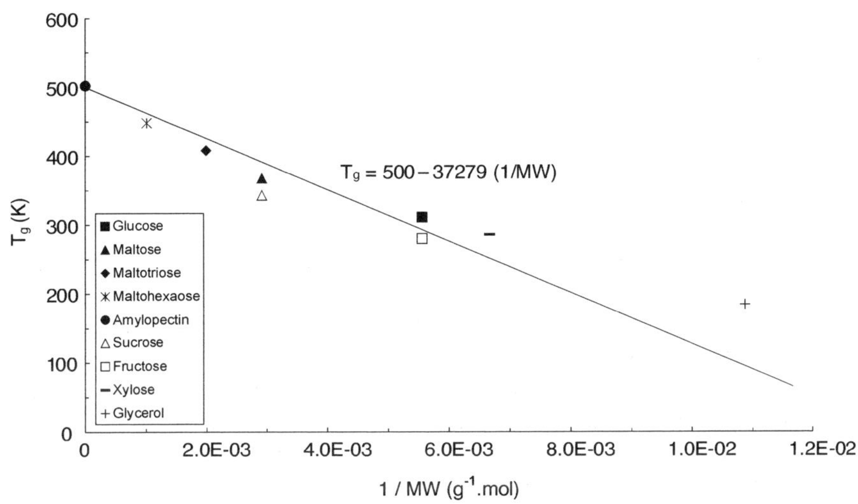

Fig. 11.9 Relationship between T_g and molecular weight (MW) for a range of carbohydrates (T_g values obtained from a range of literature sources including Orford *et al.*, 1989).

11.3.2 T_g of mixed systems and plasticisation

The T_g of a homogeneously mixed multicomponent system is therefore defined by its average molecular weight. This is the most direct and useful way to consider the plasticisation of biopolymer by compatible low molecular weight additives such as water, glycerol, sugars, etc. Using this definition, lipids would not plasticise most food biopolymers, which are usually hydrophilic.

Several equations have been used to rationalise the measured T_g of mixtures and to predict the T_g of systems where a measured value is not available. The most commonly used equations are as follows.

- The Gordon–Taylor (1952) equation which can be extended to multi-component systems as follows:

$$T_g = \frac{w_w T_{g_w} + \sum k_i w_i T_{g_i}}{w_w + \sum k_i w_i} \qquad [11.2]$$

where w_w and T_{g_w} refer to the weight fraction of water and its glass transition temperature (134 K) respectively, while w_i and T_{g_i} are the weight fractions and T_g values for each of the remaining components. The constants k_i depend on the nature of each of the components.

- The Ten Brinke *et al.* (1983) equation based on the well-known Couchman–Karasz (1978) equation, which again can be extended to multicomponent systems as follows:

$$T_g = \frac{\sum_i w_i \Delta C_{p_i} T_{g_i}}{\sum_i w_i \Delta C_{p_i}} \qquad [11.3]$$

where ΔC_p is the difference in specific heat capacity (C_p) between the rubbery/liquid and the glassy states at T_g (Figs 11.4 and 11.5).

It is worth noting that equations [11.2] and [11.3] can be written with, for each component:

$$k_i = \frac{\Delta C_{p_i}}{\Delta C_{p_w}} \qquad [11.4]$$

Table 11.1 lists T_g and ΔC_p values for water, glycerol and a range of carbohydrates.

Because of the difficulty in obtaining 'pure' components (particularly in terms of obtaining totally anhydrous glasses) to measure T_{g_i} and k_i (or ΔC_{p_i}) for substances for which these characteristics are not available, values are usually obtained by fitting equations [11.2] or [11.3] to experimental values of T_g of samples of different compositions (Mousia *et al.*, 2000), for example using the solver function of MS EXCEL™.

Table 11.1 T_g and ΔC_p values for water, glycerol and a range of carbohydrates (from a range of literature sources including Orford *et al.*, 1989; Kalichevsky *et al.*, 1992; Roos, 1995)

Component	ΔC_p (J g^{-1} K^{-1})	T_g (K)
Water	1.94	134
Glycerol	N/A	184
Xylose	0.94	286
Fructose	0.83	280
Glucose	0.88	311
Maltose	0.79	368
Sucrose	0.73	343
Maltotriose	0.53	408
Maltohexaose	0.49	448
Amylopectin	0.41	502

This approach has been adopted for many model and real food systems where the exact composition is not always known. For these, as long as the unknown constituents are expected to mix homogeneously with the rest of the system, one set of T_g and k (or ΔC_p) would be assigned and fitted to experimental T_g values measured, for example, over a range of water contents.

One of the main limitations of this approach is that it relies on homogeneous mixing and compatibility between components. This is, however, not always the case as foods are inherently heterogeneous and often their desired sensory attributes result not only from their chemical composition but from their physical structures, such as phase separation and domain size, emulsification, droplet dimensions, multilayering, etc. In many instances, however, knowledge of the composition and structure of the food can be used to produce the spatial distribution of T_g and predict its behaviour from that of its domains. For example, Farhat and co-workers (e.g. Mousia *et al.*, 2000; Farhat *et al.*, 2001) showed that this approach for predicting the T_g of heterogeneous biopolymer blends could be successful if one accounts for the unequal partitioning of water between the components.

11.4 Conclusion and recommendations

The importance of T_g in understanding and predicting the stability and processing performance of foods and their ingredients is widely acknowledged. There are now many experimental tools for the measurement of T_g; the choice of technique is guided by a range of fundamental and practical requirements. The kinetic character of the glass transition must always be considered when comparing values obtained by different techniques or by different protocols on the same technique. The broad range of temperatures over which the glass transition occurs should not be overlooked. Furthermore, one must never forget that water is by far the most effective plasticiser for foods and thus artefacts due to changes in water content during T_g measurements should not be underestimated.

The calculation of T_g of foods by adapting equations developed for synthetic polymers can be a useful tool, at least for predicting general processing and stability trends, but the heterogeneous nature of the food must be taken into account.

11.5 Sources of further information and advice

BLANSHARD J M V and LILLFORD P J (eds) (1993), *The Glassy State in Foods*, Nottingham University Press.

HAINES P J (ed.) (2002), *Principles of Thermal Analysis and Calorimetry*, Royal Society of Chemistry.

SPERLING L H J (1986), *Introduction to Physical Polymer Science*, Wiley, New York.

11.6 References

BAUER C, RICHERT R, BÖHMER R and CHRISTENSEN T (2000), Dynamic thermal expansivity near the glass transition, *Journal of Non-Crystalline Solids*, 262, 276–281.

CAMMENGA H K, EYSEL W, GMELIN E, HEMMINGER W, HOHNE G W H and SARGE S M (1993), The temperature calibration of scanning calorimeters 2: calibration substances, *Thermochimica Acta*, 219, 333–342.

COUCHMAN P R AND KARASZ F E (1978), A classical thermodynamic discussion of the effect of composition on glass-transition temperatures, *Macromolecules*, 11, 117–119.

FAN J T, MITCHELL J R and BLANSHARD J M V (1994), A computer-simulation of the dynamics of bubble-growth and shrinkage during extrudates expansion, *Journal of Food Engineering*, 23, 337–356.

FARHAT I A, BLANSHARD J M V and MITCHELL J R (2000), The retrogradation of waxy maize starch extrudates: effects of storage temperature and water content, *Biopolymer*, 53, 411–422.

FARHAT I A, MOUSIA Z and MITCHELL J R (2001), Water redistribution during the recrystallisation of amylopectin in extruded amylopectin-gelatin blends, *Polymer*, 42, 4763–4766.

FOX T G and FLORY P J (1950), Second-order transition temperatures and related properties of polystyrene. I. Influence of molecular weight, *Journal of Applied Physics*, 21, 581–591.

GORDON M and TAYLOR J S (1952), Ideal copolymers and the second order transitions of synthetic rubbers. I. Non-crystalline copolymers, *Journal of Applied Chemistry*, 2, 593–600.

HEMMINGA M A, ROOZEN M J G W and WALSTRA P (1993), Molecular motions and the glassy state, in Blanshard J M V and Lillford P J (eds), *The Glassy State in Foods*, Nottingham University Press, 157–171.

HOHNE G W H (1991), Remarks on the calibration of differential scanning calorimeters, *Journal of Thermal Analysis*, 37, 1987–2000.

JONES K J, KINSHOTT I, READING M, LACEY A A, NIKOLOPOULOS C and POLLOCK H M (1997), The origin and interpretation of the signals of MTDSC, *Thermochimica Acta*, 305, 187–199.

KALICHEVSKY M T, JAROSZKIEWICZ E M, ABLETT S, BLANSHARD J M V and LILLFORD P J (1992), The glass transition of amylopectin measured by DSC, DMTA and NMR. *Carbohydrate Polymers*, 18, 77–88.

KALICHEVSKY M T, BLANSHARD J M V and MARSH R D L (1993), Applications of mechanical spectroscopy to the study of glassy biopolymers and related systems, in Blanshard J M V and Lillford P J (eds), *The Glassy State in Foods*, Nottingham University Press, 133–156.

KUMAGAI H, MACNAUGHTAN W, FARHAT I A and MITCHELL J R (2002), The Influence of carrageenan on the molecular mobility in low moisture amorphous sugars, *Carbohydrate Polymers*, 48, 341–349.

LAYE P G (2002), Differential Thermal Analysis and Differential Scanning Calorimetry, in Haines PJ *Principles of Thermal Analysis and Calorimetry*, Royal Society of Chemistry, 55–93.

LEVINE H and SLADE L (1988), Principles of 'cryostabilization' from structure/property relationships of carbohydrate/water systems – a review. *Cryoletters*, 9, 21–63.

LOURDIN D, COLONNA P, BROWNSEY G J, NOEL T R and RING S G (2002), Structural relaxation and physical ageing of starchy materials, *Carbohydrate Research*, 337, 827–833.

MACINNES W M (1993), Dynamic mechanical thermal analysis of sucrose solutions, in Blanshard J M V and Lillford P J (eds), *The Glassy State in Foods*, Nottingham University Press, 223–248.

MATHOT V B F (1994), in *Calorimetry and Thermal Analysis of Polymers*, Hanser Publishers, 18–45.

MIZUNO A, MITSUIKI M and MOTOKI M (1998), Effect of crystallinity on the glass transition temperature of starch, *Journal of Agricultural and Food Chemistry*, 46, 98–103.

MOUSIA Z, FARHAT I A, BLACHOT J F and MITCHELL J.R (2000), Effect of water partitioning on the glass-transition behaviour of phase separated amylopectin-gelatin mixtures, *Polymer*, 41, 1841–1848.

NOEL T R, RING S G and WHITTAM M A (1992), Dielectric relaxations of small carbohydrate molecules in the liquid and glassy states. *Journal of Physical Chemistry*, 96, 5562–5567.

NOEL, T R, PARKER R and RING S G (1996), A comparative study of the dielectric relaxation behaviour of glucose, maltose, and their mixtures with water in the liquid and glassy states, *Carbohydrate Research*, 282, 193–206.

ORFORD P D, PARKER R, RING S G and SMITH A C (1989), Effect of water as a diluent on the glass transition behaviour of malto-oligosaccharides, amylose and amylopectin, *International Journal of Biological Macromolecules*, 11, 91–96.

PRICE D M (2002), Thermomechanical, Dynamic Mechanical and Dielectric Methods, in Haines P J, *Principles of Thermal Analysis and Calorimetry*, Royal Society of Chemistry, 94–128.

READING M (2001), The use of modulated temperature programs in thermal methods. *Journal of Thermal Analysis and Calorimetry*, 64, 7–14.

ROOS Y (1995), *Phase transitions in foods*, Academic Press, Inc, San Diego.

ROOZEN M J G W, HEMMINGA M A and WALSTRA P (1991), Molecular motion in glassy water-maltooligosaccharide (maltodextrin) mixtures studied by conventional and saturation-transfer spin-probe ESR spectroscopy, *Carbohydrate Research*, 215, 229–237.

SLADE L and LEVINE H (1991), in *CRC Critical Reviews in Food Science and Nutrition*, 30, 115.

SPERLING L H (1986), in *Introduction to Physical Polymer Science*, Wiley, New York.

STRAHM B, PLATTNER B, HUBER G and ROKEY G (2000), Application of Food Polymer Science and Capillary Rheometry in Evaluating Complex Extruded Products. *Cereal Foods World*, 45, 300–302.

SUN Z H, YANG W D, SIEBENMORGEN T, STELWAGEN A and CNOSSEN A (2002), Thermomechanical transitions of rice kernels, *Cereal Chemistry*. 79, 349–353.

TEN-BRINKE G, KARASZ F E and ELLIS T S (1983), Depression of the glass transition temperatures of polymer networks by diluents, *Macromolecules*, 16, 244–249.

VAN DEN DRIES I J, VAN DUSSCHOTEN D and HEMMINGA M A (1998a). Mobility in maltose–water glasses studies with ^{1}H NMR. *Journal of Physical Chemistry B*, 102, 10483–10489.

VAN DEN DRIES I J, DE JAGER P A and HEMMINGA M A (1998b). Sensitivity of saturation transfer electron spin resonance extended to extremely slow mobility in glassy materials. *Journal of Magnetic Resonance*, 131, 241–247.

WETTON R E (1986), Dynamic mechanical thermal analysis of polymers and related systems, in Dawkins J V, *Developments in Polymer Characterisation*, Elsevier Applied Science Publishers, London, 179–221.

WILLIAMS M L, LANDEL R F and FERRY J D (1955), The temperature dependence of relaxation mechanisms in amorphous polymers and other glass-forming liquids, *Journal of the American Chemical Society*, 77, 3701–3707.

12

Detecting spoilage yeasts

V. Loureiro and M. Malfeito-Ferreira, Instituto Superior de Agronomia and A. Carreira, STAB Vida, Portugal

12.1 Introduction: food spoilage yeasts

The detection and enumeration of spoilage yeasts in foods and beverages is essential to the quality assurance of the products susceptible to their detrimental effects. The term 'spoilage yeasts' relates to yeasts capable of inducing visible or detectable, by odour and/or taste, changes of physical and sensorial properties of food as a result of their metabolic activity. The species regarded as having spoilage ability represent a minority of food contamination species. The group of spoilage yeasts is even shorter when confined to those species able to cause deterioration in foods which have been processed and packaged according to the standards of good manufacturing practices (GMPs) (Pitt and Hocking, 1985, 1999), and regarded as *sensu stricto* spoilage yeasts. If these practices are not adhered to, many other adventitious, or innocent, yeast contaminants can develop in a product (Pitt and Hocking, 1985, 1999). Species like *Zygosaccharomyces bailii*, *Saccharomyces cerevisiae*, *Pichia membranifaciens* and *Dekkera bruxellensis* are among the most feared contaminants of several food commodities. The main spoilage effects are haze formation, cloudiness, refermentation, product discoloration and off-flavour production, depending on the type of food commodity (Loureiro and Malfeito-Ferreira, 2003a).

Spoilage yeasts are characterised by having a high tolerance to stress conditions, such as low pH, low a_w and high levels of preservatives, which favour their competitiveness over bacterial contaminants, especially in preserved, processed and refrigerated foods (Fung and Liang, 1990; Deak and Beuchat, 1996). Tudor and Board (1993) also defined a 'second-division' group of spoilage species (e.g. *Candida lypolytica*, *C. parapsilosis*, *Hansenula anomala*, *Kluyveromyces marxianus*) which are associated with deterioration of foods that do not impose extreme conditions for microbial enrichment allowing the growth of bacteria, mainly Gram-negative ones having simple nutritional requirements but marked nutritional versatility. In these foods the factor(s) favouring the growth of yeasts over bacteria cannot be identified, unlike in *sensu stricto* spoilage species which usually appear in foods presenting the above-mentioned extreme abiotic stress factors (Tudor and Board, 1993). The strictly aerobic nature of moulds tends to favour yeast activity in environments with low oxygen tension or in liquid foods (Pitt and Hocking, 1985).

Although there are few truly spoilage species, their technological significance is not related to their taxonomic nomenclature. In fact, strains of the same species may have different spoilage abilities. Furthermore, the spoilage activity also depends on the food, or beverage product, and on the processing step where contamination occurs. For instance, *S. cerevisiae* is responsible for wine fermentation, but the same strain, or other strains not used in fermentation, may be dangerous spoilers when contaminating the bottled product (Malfeito-Ferreira *et al.*, 1997). Another species, *D. bruxellensis*, is regarded as one of the most dangerous spoilage yeasts at all steps of wine production (Loureiro and Malfeito-Ferreira, 2003b), but in lambic and gueze beers it is part of the desired fermenting microflora (van Oevelen *et al.*, 1977). Furthermore, in many cases microbial spoilage is not easily defined, particularly in fermented foods and beverages, where the produced metabolites contribute to the flavour, aroma, and taste of the final products. In many cases, as in cheese-making and meat processing, there is only a slight line between what is perceived as either spoilage or beneficial activity (Fleet, 1992). All these factors contribute to the different approaches used by technologists and taxonomists to characterise spoilage yeasts. Unlike the latter, technologists tend to be more concerned with the spoilage ability of a particular strain than to identify the problem species.

The frequent changes in yeast nomenclature also hinder communication between technologists and taxonomists, especially when the former nomenclature distinguished particularly significant technological characteristics. Two examples may be illustrated by the changes in *Saccharomyces* spp. nomenclature. The former *S. diastaticus*, now *S. cerevisiae*, is one of the most feared contaminants in bottled beer due to its ability to hydrolyse dextrins, so brewing technologists clearly prefer to distinguish it from other species (Moreira-da-Silva *et al.*, 1994). The present species, *S. bayanus*, was for some time included in *S. cerevisiae* (Pretorius *et al.*, 1999). For wine technologists this distinction is significant because *S. bayanus* is the starter most used in the fermentation of sparkling wine in the bottle and is also a dangerous wine spoiler (Malfeito-Ferreira *et al.*, 1989a).

The deterioration of foods and beverages by yeasts has gained increasing importance in food technology (Thomas, 1993) and only recently have edited handbooks been dedicated fully (Deak and Beuchat, 1996) or to a large extent (Boekhout and Robert, 2003) to this subject. Among the reasons that have contributed to this increasing importance are the use of modern technologies in food processing, the great variety of new formulations of foods and beverages, the tendency to reduce the use of preservatives, particularly those effective against yeasts (e.g. sulphur dioxide and benzoic acid), and the less severe processing (Fleet, 1999; Loureiro and Querol, 1999). Simultaneously, although foodborne yeasts are recognised as non-pathogenic, certain foods of animal origin (e.g. cheeses, sausages) are frequently contaminated with yeast species regarded as opportunistic pathogens to certain consumer risk groups (e.g. HIV carriers, transplant and cancer patients, pregnant and elderly people) (Hazen, 1995; Murphy and Kavanagh, 1999). In spite of these observations, the techniques used to control and monitor such yeasts in the food industry remained practically unchanged for decades and did not follow the recognition of their increasing importance and the scientific developments in this field.

12.2 Detection and enumeration: viable and direct count techniques

Several techniques are used to detect and/or count foodborne yeasts in foods, and these can be classified in different groups. A distinction will be made here between techniques providing yeast counts by cultivation, by microscopic observation and by instrumental methods.

12.2.1 Pre-inoculation techniques

Classical enumeration techniques used in the food industry are plate counting of colonies growing on culture media, after spreading (preferably for yeasts), incorporation or membrane filtration, and Most Probable Number (MPN) counts (Harrigan, 1998). Before inoculation in culture media several steps may be performed, including sampling, sample preparation and treatment, dilution and, occasionally, enrichment.

12.2.2 Sampling

Sampling plans based on statistics are required to estimate the microbial load of food commodities. These plans consider, at least, the nature of the food (liquid or solid), lot size, sample size and sampling rate. However, they are conditioned by their inherent costs and the time for their execution, and it is necessary to establish whether their use in industy is worthwhile. They are especially relevant for foods susceptible to pathogenic bacteria. On the contrary, in industries processing foods not susceptible to health hazards, the most frequent situation concerning yeast control is that sampling plans have no sound statistical background and are based on empirical experience. The discussion of sampling criteria is beyond the scope of this chapter and detailed information may be found elsewhere (Anon, 1982; Jarvis, 2000).

12.2.3 Sample preparation and treatment

This operation is much more important in solid or liquid foods with suspended solids than in clear liquids (e.g. wine, beer), because yeast cells may adhere, with different intensity, to the solid surfaces. A more complicated situation is the entrapment of cells in the reticulate structures of certain foods, which usually are immobilised and localised in high densities (Fleet, 1999). For the isolation of yeast and moulds, maceration/blending procedures may consist of manually shaking of the sample, previously grounded if necessary, with a known volume of diluent, mixing with a diluent in a blade blender, or pummelling in a Stomacher™ (Colworth) or Pulsifier™ (Kalyx) (Boer and Beumer, 1999). Diluents commonly used range from distilled water, saline and phosphate buffer to, most commonly, 0.1% (w/v) peptone water (Deak, 2003). Contact time ranges from less than one minute to several minutes (generally 5–10 min).

Stronger methods such as vigorous shaking, jet-streaming, and sonication of samples, should be used to recover yeasts from natural habitats (Martini *et al.*, 1980) or from the surface of frozen foods (Diriye *et al.*, 1993). These pre-isolation treatments allow the recovery of a much higher number of species and of counts when compared to the traditional enrichment procedures. However, it seems that vigorous shaking with Vortex mixer is more effective than water jet or sonication (Deak, 2003).

12.2.4 Sample dilution

To many microbiologists sample dilution is a routine operation, perfectly defined and innocuous or harmless for yeast cells. There is even the conviction that yeasts are more resistant to osmotic shock than bacteria and so the diluent composition is not relevant. However, holding periods of one hour (Beuchat *et al.*, 2002) or two hours (Mian *et al.*, 1997) have been reported as causing significant reductions in yeast populations, regardless of the type of diluent. The reductions mentioned were also observed in 0.1% peptone water which yielded the highest recoveries (Mian *et al.,* 1997). These periods may seem long to a bench microbiologist, but are common in industry, mainly when performing swabs of equipment and plant surfaces before media inoculation in the laboratory. In addition, settling periods longer than one minute may significantly reduce population size (Deak, 2003). Sterile peptone water (0.1% w/v) is the recommended diluent for preparing samples to be plated on general-purpose enumeration media (Samson *et al.*, 1992), to which may be added Tween 80 as an aid to cell dispersion in the diluent (Beuchat, 1993). However, given the diversity of food-intrinsic characteristics and yeast biodiversity, there is no ideal diluent. For instance, xerophilic yeasts presented higher recovery percentages when diluted in 30% (w/w) glycerol than on 0.1% (w/w) peptone (Andrews *et al.*, 1997) and dilution of fruit concentrates should be made with 30% (w/w) glucose to obtain higher yields of *Z. rouxii* (van Esch, 1992). The International Commission on Food Mycology recognises that there are not yet specific protocols for the different types of foods, concerning either food nature or sample contact time, making it more difficult to compare results from different food industry laboratories. In addition, according to Fleet (1999), based on an international collaborative study under the auspices of the above-mentioned Commission, it is possible to conclude that, apart from diluent composition and timing between dilution and plating, other factors, such as stage of cell life cycle, cell stress prior to dilution, degree of cell clumping and aggregation, shear forces during shaking, presence of contaminating metal ions, pH and temperature could all impact on the survival of the yeast cells during the dilution operation. Furthermore, it should not be forgotten that the most important objective is to recover spoilage yeasts and so dilution conditions should favour their isolation. Contrary to what is generally believed in food industry, dilution may not be an inoffensive operation to yeast recovery.

12.2.5 Sample enrichment

Enrichment procedures are commonly used in food bacteriology, namely to detect pathogenic species and other minority species present in foods. For yeasts such procedures are not common, since it is not yet clear whether they bring any advantage at all. As mentioned before, the results of Martini *et al.* (1980) suggested that vigorous and disruptive sample treatments of natural substrates give better results than enrichment cultures. However, the same authors suggest the use of enrichment cultures to elicit the presence of fermenting species. Taking into account that most of the spoilage yeasts are fermenting species and are present in foods and natural substrates at very low levels, it is conceivable that this technical step may improve their detection. In addition, the recovery of cells sublethally injured by heat, cold, osmotic and acid shocks may determine the utilisation of specific repair steps (Fleet, 1992; Deak and Beuchat, 1996), which may also be designed to select the spoiling microflora (Thomas and Ackerman, 1988).

12.2.6 Culture media

General purpose media

Basically, the general purpose isolation and enumeration media for foodborne yeasts are complex and nutritionally rich, containing sugar as energy source (e.g. glucose, fructose, sucrose), a digested protein as nitrogen source (e.g. peptone, tryptone, casitone), and a complex vitamin supplement (e.g. yeast extract, malt extract). Additionally, they may or may not contain one or more antibiotics against bacteria (e.g. oxytetracycline, chloramphenicol), a compound to inhibit the most rapidly spreading moulds (e.g. rose bengal, diphenyl, dichloran, sodium propionate, the antibiotic oligomycin), sometimes a pH indicator (e.g. bromocresol green, bromophenol blue), and agar, depending on their use as solid or broth media (Loureiro and Querol, 1999). Many studies have concluded that these media are generally better for recovering yeasts than the earlier acidified media with organic or inorganic acids to get a pH value around 3.5

(Beuchat, 1993). Unfortunately, all of these media are specially designed to recover the biggest number of yeast cells present in foods while inhibiting bacterial growth and reducing fungal spreading (Deak, 2003), instead of targeting only spoilage yeasts. This limitation is greater than apparently thought, because most innocent yeasts are fast growers, inhibiting the growth of slow-growing yeasts, such as some of the most dangerous spoilage yeasts (e.g. *Zygosaccharomyces* spp., *Dekkera* spp.). The usual media for enumeration of total viable yeast counts may even not support growth of *B. anomalus* (Deak *et al.*, 2001; Beuchat *et al.*, 2001). Additionally, dichloran rose bengal chloramphenicol agar (DRBC), widely used for detection of yeasts in the presence of moulds, may affect yeast growth, in particular of spoilage species such as *Z. rouxii* (Andrews *et al.*, 1997). In conclusion, the usual culture media used in food mycology may be inappropriate to give a 'real image' of the food ecosystem in terms of quality control. Ideally, the best medium to enumerate yeast cells on foods should prevent the growth of all innocent yeasts and promote the growth of all spoilage yeasts. As this is virtually impossible, other strategies must be used, as described below.

Selective and differential media

The classical development of selective media relies on choosing formulations and incubation conditions favourable to particular groups of yeasts without taxonomic significance, but with technological significance, and used in a wide range of food commodities, such as 'acid-resistant yeasts', 'xerotolerant/osmophilic yeasts' or 'psychrotroph yeasts' (Davenport, 1980; Fleet, 1992; Deak and Beuchat, 1996; Loureiro and Querol, 1999). The introduction of stress factors is directed to the selection of a few or a single species but may leave undetected strains of the target species with lower resistance to such stress factors, or may detect highly resistant strains of species assumed as sensitive. This situation may be irrelevant for a particular food, but may be a strong limitation for widespread use in the food industry. For instance, concerning media developed for 'acid-resistant yeasts', the *Zygosaccharomyces bailii* Agar (ZBA) medium has proved to be efficient in the detection of *Z. bailii* in acidified ingredients, mainly due to the addition of acetic and sorbic acids (Erickson, 1993). However, when it was tested with other strains and in other food commodities, it was judged less efficient to recover *Z. bailii* than another general-purpose medium added to acetic acid (Tryptone Glucose Yeast Extract Agar, TGYA) (Hocking, 1996), particularly for heat-stressed cells in acidic foods with reduced a_w (Makdesi and Beuchat, 1996a) or acidic beverages (Makdesi and Beuchat, 1996b). In the case of *Z. bailii*, these acidified media recovered numbers similar to those recovered on antibiotic-supplemented media, but normally the use of antimicrobials tends to increase incubation time. This drawback may be compensated if long incubation periods may be an additional differential characteristic, as described below. For enumerating yeasts from low a_w foods ($a_w < 0.85$) it may not be necessary to use a selective medium because the dominant species must be xerophilic, but from intermediate moisture foods it is desirable to use a selective medium to suppress the competition of osmo-sensitive yeasts (Andrews *et al.,* 1997).

Several media have been proposed presenting high concentrations of sugars (Cava and Hernández, 1994; Casas *et al.*, 1999; Andrews *et al.*, 1997), glycerol (Hocking and Pitt, 1980) or salt (Andrews *et al.*, 1997). Glycerol (18%) supplemented media are indicated for moderately xerotolerant yeasts, while glucose (up to 60% w/w) may be used to detect and count moderate to extreme xerophiles (Beuchat, 1993). Dichloran 18% Glycerol agar (DG18) was found to be best for recovering xerophilic yeasts in the presence of xerophilic moulds (Braendlin, 1996) but it is not adequate as a general enumeration medium (Deak, 2003). These reduced a_w media were shown better than the current general purpose media to recover cells sublethally injured by osmotic shock (Andrews *et al.*, 1997; Casas *et al.*, 1999), despite their ability to have incubation periods as long as 10 (Beuchat, 1993) or 28 days (Tilbury, 1980). Most spoilage yeasts are mesophilic and so current incubation temperatures range from room temperature (20–22°C) to 30°C (Deak, 2003). The decrease to 5–7°C, in order to detect and enumerate 'psychrotrophic yeasts', also determines the convenience of incubation periods longer than 14 days (Beuchat, 1993). In addition, incubation should be carried out with plates in the upright position (Samson *et al.*, 1992).

Another approach exploits particular enzymatic features which are common to a variable range of species (e.g. 'lipolytic yeasts' or 'proteolytic yeasts' (Davenport, 1980; Deak and Beuchat, 1996)) or restricted to a single species, in conjugation or otherwise, with stress factors. The medium of Chaskes and Tyndall (1975) may be regarded as the precursor of this sort of media. This medium allows a clear-cut identification of the pathogen yeast *Cryptococcus neoformans* by containing 3,4-dihydroxyphenylalanine (DOPA), which is converted by this species to black pigments. Following this strategy, but exploiting other physiological features as well, several media have been proposed specifically for particular spoilage species (Table 12.1). The degrees of efficiency of the media mentioned are variable. For instance, *Yarrowia lypolytica* Medium (YLM) is fully differential for *Y. lypolytica* due to its unique ability to produce brown pigments from tyrosine within 24 hours (Carreira and Loureiro, 1998). *Zygosaccharomyces* Differential Medium (ZDM) relies on the ability of *Z. bailli* and *Z. bisporus* to assimilate glucose and formic acid simultaneously with medium alkalinisation (evidenced by bromocresol green added to the medium) and provided excellent results in wines (Schuller *et al.*, 2000). However, its differentiating ability is lower when testing a wider range of species from other sources (*D. hansenii*, *C. tropicalis*, *C. parapsilosis*, *P. guilliermondii*) (unpublished results).

Dekkera/Brettanomyces Differential Medium (DBDM) selects *D. bruxellensis*, and other species, owing to the utilisation of ethanol (6% v/v) as single carbon and energy source and resistance to cycloheximide. The differential ability is conferred by the use of a pH indicator and *p*-coumaric acid, the precursor of 4-ethylphenol, which is easily detected by imparting a phenolic smell to medium plates. This medium proved its high efficiency to recover *D. bruxellensis* from

Table 12.1 Culture media addressed to the enumeration of specific foodborne yeasts

Medium	Target species	Differential characteristics	Target food	Reference
YLM	*Yarrowia lypolytica*	Brown discoloration of the agar medium containing tyrosine	Cheese	Carreira and Loureiro (1998)
KDM	*Kluyveromyces marxianus*, *K. lactis*	Blue colonies indicating presence of β-galactosidase in the absence of lactose, in X-gal containing medium	Dairy products	Valderrama *et al.* (1999)
ZDM	*Zygosaccharomyces bailii*, *Z. bisporus*	Blue colonies growing on glucose and formic acid, and agar colour changes from green to blue	Wine	Schuller *et al.* (2000)
–	*Debaryomyces hansenii*, *K. marxianus*	Salmon and dark blue colonies, respectively, in Salmon-Gluc and X-gal containing medium	Intermediate moisture foods	Silóniz *et al.* (2000)
DBDM	*Dekkera/Brettanomyces* spp.	Pin-point yellow to green colonies, phenolic odour, slow growth, agar colour changes from blue to yellow	Wine, soft drinks	Rodrigues *et al.* (2001)

wine samples (Rodrigues *et al.*, 2001). However, its use in a wider range of samples (grapes, insects, grape juice) showed the presence of *P. guilliermondii*, which was for the first time recognised as a strong 4-ethylphenol producer (Dias *et al.*, 2003). To distinguish these two species, another feature must be considered: colonies of *D. bruxellensis* take at least 6–7 days to become visible, while those of *P. guilliermondii* appear within 2–3 days. This feature exemplifies another limitation of current enumeration protocols which usually consider that 48–72 h of incubation is enough to detect foodborne yeasts. This is not so, particularly for slow-growing spoilage species such as *Z. bailii* or *D. bruxellensis*, where incubation periods as long as 14 days may be necessary (Millet and Lonvaud-Funel, 2000; Rodrigues *et al.*, 2001; Dias *et al.*, 2003), or, at least, of 10 days for preservative-stressed cells (van Esch, 1992).

Concerning specific industries, many culture media have been developed based on different enzymatic and/or stress resistances, particularly in the brewing and wine industries. In brewing, it is relevant to discriminate between *S. cerevisiae* fermenting ale or lager yeasts and 'wild yeasts', which may be undesirable *S. cerevisiae* strains, other *Saccharomyces* species, or non-*Saccharomyces* species (Campbell, 1987). Many culture media have been developed with that goal (Lysine agar, Lin's medium, Schwarz Differential Medium, Crystal Violet Medium, Copper Sulphate Medium, Starch Medium, etc.) with different efficiencies (Campbell, 1987; Deak, 2003). A recent report claimed that Copper Sulphate Medium was the best to discriminate wild yeasts (including wild *S. cerevisiae*) from fermenting yeasts in lager beers (Kuhle and Jespersen, 1998). Increasing incubation temperature from the usual 25°C to 37°C also promotes growth of wild yeasts and *S. cerevisiae* ale yeasts (Lawrence, 1983) or wild *S. cerevisiae* strains (Kuhle and Jespersen, 1998). In wines, ethanol (11.4% v/v) has been successfully used as a selective agent in a medium developed to detect 'wine-spoiling yeast' (Thomas and Ackerman, 1988). Heard and Fleet (1986) used Lysine agar to detect non-*Saccharomyces* species and preferred the generic medium Malt Extract Agar to count *S. cerevisiae* instead of Ethanol Sulphite Agar (containing 12% v/v ethanol and 150 mg/l total sulphite). This medium was developed by Kish *et al.* (1983) to differentiate wine yeast in the presence of excessive numbers of apiculate yeasts. The Cadaverin Lysine Ethylamine Nitrate (CLEN) agar directed to non-*Saccharomyces* in beer (Martin and Siebert, 1992) did not seem appropriate for wines because it elicited the growth of *S. cerevisiae* wine strains (Fernández *et al.*, 2000). Fugelsang (1997) described media developed for the detection of *Dekkera* spp. and of *Z. bailii* in wines, based on their particular stress resistances. The above-mentioned ZDM and DBDM were developed primarily for the wine industry (see Table 12.1).

In many of the above-mentioned types of media the colonies of the targeted species are differentiated by colony morphology, as in general-purpose media, and so results are regarded as presumptive. The differential properties of a medium may also be enhanced by the addition of certain dyes which may be chromogenic or fluorogenic substrates (for a review on the mode of action of dyes see Fung and Liang, 1990). The two latter compounds induce brightly coloured or fluorescent products when reacting with specific enzymes or metabolites present in cell biomass. Among the dyes, the inclusion of pH indicators may give additional differentiating characteristics by changing colony and/or medium colour. That is the case of Molybdate agar supplemented with 0.125% propionate which was found adequate to distinguish several yeast species isolated from tropical fruits (Rale and Vakil, 1984), or of WLN agar for brewing (Lawrence, 1983) and wine-related species (Pallman *et al.*, 2001). Silóniz *et al.* (1999) devised a scheme based on growth on three selective/ differential media followed by detection of specific enzymatic activities, directed to yeasts recovered from high sugar products (marzipan, fruit concentrates and syrups). A basal media (Yeast Malt Agar) was added to (1) eosin and methylene blue, (2) acetic acid, and (3) acetic acid and potassium tellurite, and three enzymatic activities were determined in an APIZym kit. The overall procedure enabled the determination of *S. cerevisiae*, *D. hansenii*, *I. orientalis*, *Z. rouxii* and *Z. bailii*. These authors reported that the chromogenic substrate triphenyltetrazolium, used to differentiate clinical species of *Candida* spp., did not permit a clear distinction of the tested species because of intraspecific variation. In fact, dyes are mostly used in clinical diagnosis but yeast response to many of them is not yet well established (Fung and Liang, 1990; Deak, 2003), although several fluorophores have been proposed to characterise strains of the genera *Candida*, *Hansenula*, *Kluyveromyces*, *Pichia*, *Rhodotorula* and *Saccharomyces* (Geyer *et al.*, 1991).

The above-mentioned media were addressed to plating, with or without previous membrane filtration. However, they may also be used as broths. The classical technique of the Most Probable Number (MPN) (Harrigan, 1998) is broadly used in bacteriology but seems to be overlooked in estimating yeast counts, perhaps because it is regarded as disadvantageous when compared with plating (Deak, 2003). However, it is quite appropriate when samples cannot be membrane filtered, as in liquid samples with suspended solids, such as fortified wines (Vaz-Oliveira *et al.*, 1995). In addition, this technique was also found to be essential to recover *D. bruxellensis* present in proportions much less than 0.1% of total microbial flora, and/or in the presence of moulds, which prevented its detection in standard counting plates (Rodrigues *et al.*, 2001). We believe that the MPN technique, using selective media, is especially adequate for the recovery and estimation of spoilage or fermenting minority species. The broths may also act as enrichment or repair media and prevent the growth of moulds, avoiding the use of antibiotics which may also affect yeast growth. Its relatively low utilisation by beverage industries, laboratories, when compared to filter membrane plating, is perhaps due to higher labour requirements and to the absence of commercial kits. However, a commercial system (MicroCount^{TM}, Millipore) uses MPN to calculate automatically cell counts after growth in 24 wells of microplates.

Commercial media

In spite of the large number of selective/differential media referred in the literature, they do not seem to interest companies producing diagnostic media (Table 12.2). Most commercial media are of general purpose, only inhibiting or retarding bacterial or mould growth, or aimed at clinical specimens. To the best of our knowledge, only two 'specific' media are commercially available and both target the detection of *Dekkera* sp.: *Brettanomyces* Specific Broth (BSM), Millipore, and *Dekkera Brettanomyces* Differential Medium (DBDM), STAB Vida. BSM is a selective medium that contains cycloheximide and 20 g/l glucose. According to our experience, a culture medium with 20 g/l of glucose is not appropriate to detect *Brettanomyces* cells, because the sugar favours fast-growing species (e.g. *K. apiculata*, *C. tropicalis*, *P. guilliermondii*) (Rodrigues *et al.*, 2001). It is not, then, a medium specific for *Brettanomyces*, but merely a medium for species resistant to cycloheximide, mainly fast-growing species. Confirming our expectations, practical results demonstrate the recovery on BSM of colonies of non-*Dekkera* species characterised by small spheroid cells which may be confused with *Dekkera* by technicians without experience (Stender *et al.*, 2001). DBDM is partially selective (cycloheximide) and totally differential for *Dekkera* sp. and does not contain glucose. In this medium, the growth of this species is differentiated by four characteristic features: slow growth (more than 5 days), yellow colour of the colonies that darken with time, change in the colour of the medium (from blue to yellow), and formation of a phenolic smell. By combining selectivity and differentiability, this medium allows a clear-cut identification of *Dekkera* sp., even in samples contaminated with cycloheximide resistant species (Dias *et al.*, 2003; Rodrigues *et al.*, 2001). DBDM is presently under patent protection.

In the field of clinical microbiology the media addressed to the selective and differential recovery of pathogenic yeasts (see Table 12.2) are based on different enzymatic features (Freydiere *et al.*, 2001), but due to the targeted species (mostly *Candida* spp.), they do not seem of great utility for the most dangerous food spoilers. However, their utility evaluation has just provided the first results concerning CHROMagar Candida and it remains to be seen whether the different colony morphologies reported for some food spoilage species (Deak, 2003) are also observed under practical situations and for a wider range of strains.

12.2.7 Plating techniques and their variations

Microbial counts may be determined by techniques other than the standard plate counts (with or without using filter membranes), developed mainly to facilitate analytical procedures. The spiral plater adds the sample to the plate without the need for serial dilutions (Boer and Beumer, 1999). The culture media may also be deposited in ready-to-use structures facilitating inoculation procedures, like the systems Petrifilm™, Redigel™ and SimPlate™, which yield counting results comparable to conventional plating (Beuchat *et al.*, 1990, 1998). Another

Table 12.2 Commercially available culture media for yeasts

Name	Companies[1]	Purpose/principle	Improvements/adjustments
General media for yeasts			
Czapek Dox Agar[2]	Biomedics, Difco, EMSL, Fluka, HiMedia, Lab.Conda, Merck, Oxoid	Chemically defined medium for the cultivation and maintenance of moulds and yeasts. Contains sucrose as sole carbon source and nitrate as sole source of nitrogen. Only non-fastidious bacteria are able to grow in these conditions.	These media can be rendered selective for yeasts and moulds either by the addition of antibiotic to suppress bacterial growth (e.g. chloramphenicol, streptomycin, gentamycin, etc.) or by acidifying the medium (e.g. pH 3.5).
Malt Extract Agar[3]	BBL, Difco, EMSL, Fluka, HiMedia, Lab.Conda, Merck, Oxoid	General purpose media for the detection, enumeration and cultivation of yeasts and moulds. The composition particularly favours the growth of these organisms.	
Mycological Agar	Difco		
Potato Dextrose/Glucose Agar[4]	Biomedics, BBL, Difco, EMSL, Fluka, HiMedia, Lab.Conda, Merck, Oxoid		
Yeast Extract Agar	HiMedia, Lab.Conda, Mast, Merck, Oxoid		
Yeast and Mould Agar	bioMérieux, Difco, Oxoid		
Cooke Rose Bengal Agar	Difco, HiMedia	General purpose medium for the detection, enumeration and cultivation of yeasts and moulds; Rose Bengal restricts the size and growth of spreading moulds.	

WLN Wallerstein Laboratories Nutrient Agar	BBL, Difco, Fluka, HiMedia, Merck, Oxoid	A good general medium that supports the growth of yeast and bacteria. It contains an indicator dye (bromocresol green) that is taken up differentially by different yeast strains. Mixed cultures will therefore sometimes demonstrate mixed colony morphology (different shades of green and white). Its pH of 5.5 is optimal for the enumeration of brewers' yeast.	• If bakers' or distillers' yeast is to be examined, the pH should be adjusted to 6.5 (better yields). • When cultivating microorganisms from an alcoholic mash, tomato juice should be added to the medium. • Addition of antibiotics to suppress bacterial growth.
Selective media for yeasts			
Chloramphenicol (Glucose/ Dextrose) Agar	Biokar, Lab.Conda	Selective agar for detection and enumeration of yeasts and moulds. Contains chloramphenicol to inhibit bacterial growth.	For heavily loaded materials, it is advisable to add other antibiotics (e.g. gentamycin, oxytetracycline, etc.).
Dichloran Rose Bengal Chloramphenicol Agar (DRBCA)	Difco, Fluka, HiMedia, Lab.Conda, Merck, Oxoid	Selective agar for detection and enumeration of yeasts and moulds. Contains chloramphenicol to inhibit bacterial growth and Rose Bengal (RB) to restrict the size and growth of spreading moulds. RB also assists enumeration as the colour is taken up by the colonies. Dichloran inhibits the rapid spreading of mucoraceous fungi and also restricts colony sizes of other genera, easing the colony count.	This medium is one of the most commonly used for the enumeration and isolation of yeasts from foods, particularly those containing high numbers of bacteria and moulds. Addition of other antibiotic might be necessary to further inhibit bacterial growth.
Oxytetracycline Glucose Yeast (extract) Agar (OGYA)[5]	Difco, Fluka, HiMedia, Merck, Oxoid	Selective agar for detection and enumeration of yeasts and moulds. Contains oxytetracycline to inhibit bacterial growth.	Recommended for all types of samples (foodstuffs, clinical specimens, etc.). For proteinaceous foods, or other foods heavily loaded with bacteria, the addition of gentamycin is recommended.

Table 12.2 Continued

Name	Companies[1]	Purpose/principle	Improvements/adjustments
Rose Bengal Chloramphenicol Agar (RBCA)[6]	Biokar, Difco, EMSL, HiMedia, Lab.Conda, Merck, Oxoid	Selective agar for detection and enumeration of yeasts and moulds. Contains chloramphenicol to inhibit bacterial growth and Rose Bengal (RB) to restrict the size and growth of spreading moulds. RB also assists enumeration as the colour is taken up by the colonies.	This medium is particularly recommended for fresh proteinaceous foods whose bacterial flora is mainly Gram-negative. But for heavily loaded materials, it is advisable to add other antibiotics (e.g. gentamycin, oxytetracycline, etc.)
Sabouraud Dextrose Agar (with or without antibiotics)[7]	BBL, Biokar, Biomedics, bioMérieux, Difco, EMSL, HiMedia, Int.Microbio, Lab.Conda, Mast, Merck, Oxoid	Medium of choice of most pharmacopeias for the enumeration of yeasts and moulds in non-sterile products. Contains a high glucose content to optimise fungal growth, and the acid pH and the inclusion of other antibiotics increases the selectivity of the medium.	The addition of other inhibiting substances (chloramphenicol, cycloheximide, gentamycin) depends on the accompanying microflora: • Mixture of cycloheximide, penicillin, and streptomycin (to inhibit the most common contaminant when analysing dermatophytes) • Chloramphenicol and cycloheximide (for the same purpose) • Cycloheximide, colistin, and novobiocin to isolate *C. albicans* • TTC (triphenyltetrazolium chloride) can be used to differentiate *Candida albicans*: these are unpigmented or pale pink, whilst other *Candida* species and other fungi form deeper pink or red colonies.

Yeast Glucose Chloramphenicol Agar (YGCA)[8]	BioMérieux, Difco, Fluka, Merck	Selective agar for isolating and counting yeasts and moulds in milk and milk products. Contains chloramphenicol to suppress bacterial growth.	For yeast counts in samples heavily loaded with moulds (e.g. mould-ripened cheeses), it is advisable to use oligomycin to inhibit mould growth.
Selective media for groups of yeasts			
Dichloran 18% Glycerol Agar (DG18)	EMSL, Merck, Oxoid	Selective agar with low water activity (a_w) for the enumeration and isolation of xerotolerant (osmophilic) yeasts and moulds in dried and semi-dried foods, such as dried fruits, meat and fish products, spices, confectionery, cereals, nuts, grains and flours, as well as from high sugar foods, such as fruit concentrates, syrups and beverage bases. The inclusion of dichloran inhibits the rapid spreading of mucoraceous fungi and restricts colony sizes of other genera, easing the colony count. The addition of chloramphenicol and the reduced a_w prevent the growth of bacteria.	It has also proved to be a useful general purpose medium for counting yeasts and moulds in a wide range of foodstuffs. Addition of Triton-X to DG18 agar increases the inhibition of vigorously spreading fungi.
Lin's Wild Yeast Medium (LWYM)[9]	Siebel I.T.	For detection and quantitative determination of wild yeast populations in brewing culture yeast. The growth of culture yeast is suppressed, while wild yeasts grow as large distinct colonies. This medium is specially formulated to encourage the growth of *Saccharomyces* wild yeast, and contains crystal violet, which inhibits brewery yeast. A number of non-*Saccharomyces* yeasts may also grow on this medium.	If wild yeast populations are mainly non-*Saccharomyces,* an alternative to LWYM is Lin's Cupric Sulfate Medium (LCSM, Siebel I.T.), which was specially designed to encourage the growth of non-*Saccharomyces* wild yeasts.

Table 12.2 Continued

Name	Companies[1]	Purpose/principle	Improvements/adjustments
Lysine Medium[9]	Difco, HiMedia, Oxoid	A complex medium for the isolation and enumeration of wild yeasts in pitching yeast (brewing). It is based on the use of lysine: strains of *S. cerevisiae* and *S. carlsbergensis* do not use it, while many other yeasts, including wild yeasts, do so.	
Schwarz Differential Agar	Siebel I.T.	A nutrient medium that will detect most organisms commonly encountered in a brewery. Acid-producing organisms are identified by the development of a clear zone around the colonies. Further identification is facilitated by characteristic colour reactions.	Cycloheximide can be added to the medium to suppress the growth of culture yeast.
WLD Wallerstein Laboratories Differential Agar	Difco, HiMedia	WLD has the same formula as WLN, but with added cycloheximide (actidione) (CHX) to suppress the growth of CHX-sensitive yeasts, which include *Saccharomyces* spp. For *Saccharomyces*, if anything grows on this medium, the culture is considered contaminated.	
Wort Agar	BBL, Biokar, Difco, Fluka, Merck, Oxoid	General purpose mycological medium, particularly suitable for the cultivation, isolation and enumeration of osmophilic yeasts, especially in butter, syrups, lemonades, and more generally, in sweet drinks. The medium, which duplicates the composition of natural wort, is of an acidity that is optimal for many yeasts but inhibitory to most bacteria.	• pH 3.5 or 4.0, instead of 4.8, to increase selectivity towards bacteria. • For the examination of osmophilic yeast in sugar products, these components shoul be dissolved in a syrup containing 35 parts w/w of sucrose and 10 parts w/w of glucose (osmophilic agar).

Medium	Supplier	Description
Selective/differential media for specific yeasts		
Albicans ID	bioMérieux	For the isolation of yeasts and immediate identification of *C. albicans* (blue colonies), based on a chromogenic substrate that detects β-galactosaminidase activity.
Bismuth Glycine Glucose Yeast Agar (BiGGY Agar)	BBL, Biokar, Biomedics, Difco, HiMedia, Lab.Conda, Merck, Oxoid	Recommended for the detection, isolation and presumptive identification of *Candida* species. Bismuth sulphite inhibits bacterial growth, and is reduced by *Candida* to sulphide, which colours the colonies brown. Typical colonies are: • *C. albicans*: brown to black colonies with no pigment diffusion and no sheen • *C. krusei*: shiny, wrinkled, brown to black colonies with black centres, black pigment diffusion and a sheen • *C. pseudotropicalis*: flat, shiny red to brown colonies with no pigment diffusion • *C. parakrusei*: flat, shiny, wrinkled, dark reddish-brown colonies with light reddish-brown peripheries and a yellow fringe • *C. stellatoidea*: flat, dark brown colonies with a light fringe. Addition of other antibiotics (e.g. neomycin) is advisable in order to better inhibit the growth of the accompanying bacterial flora.
Brettanomyces Selective Broth	Millipore	For the detection/enumeration of *Brettanomyces*. It is a selective medium: contains inhibiting factors for other yeast species (cycloheximide), or bacteria (antibiotic).
Candichrom®II	Int.Microbio	Substrate incorporation ID medium for isolation and enumeration of yeasts, and direct identification of *C. albicans*.
Candida BCGAgar	Difco	For use with added neomycin for the primary isolation of *Candida* sp.
Candida ID Agar	bioMérieux	Contains a chromogenic indolyl glucosaminide substrate, which is hydrolysed by *C. albicans* to give a turquoise or blue insoluble product. *C. tropicalis, C. lusitaniae*, and *C. guilliermondii* appear pink, and other species of *Candida* appear white.
Candida Ident Agar	Fluka	For the selective isolation and identification of *C. albicans* from clinical material. Contains chloramphenicol and a chromogenic mixture.
Candida Isolation Agar	Difco	For primary isolation and differentiation of *C. albicans*.
CandiSelect	Sanofi Diagnostics Pasteur	Chromogenic medium based on a chromogenic substrate for the detection of β-galactosaminidase. The colonies of *C. albicans* are differentiated by their blue colour.

Table 12.2 Continued

Name	Companies[1]	Purpose/principle	Improvements/adjustments
CHROMagar™ Candida	Becton Dickson	This medium allows not only the growth and detection of yeasts, like traditional media, but in addition contains a mixture of chromogenic substrates that instantly allow one to differentiate various *Candida* species just by the colour of the colony: *C. albicans* (green), *C. tropicalis* (steel blue), and *C. krusei* (rose, fuzzy).	
CHROMOGEN ALBICANS	Biomedics	For the presumptive identification of *Candida albicans.*	
Fluoroplate Candida Agar	Merck	For the detection and presumptive identification of *C. albicans*, based on a fluogenic substrate to detect β-galactosaminidase activity. In this medium, the colonies of *C. albicans* are fluorescent.	
Dekkara Brettanomyces Differential Medium (DBDM)	STAB Vida	For detection/enumeration of *Brettanomyces*. Contains inhibiting factors for other yeast species (cycloheximide) and bacteria (chloramphenicol). It is also a differential medium, in which the growth of these yeasts is followed by a change in the colour of the medium (from blue to yellow), and by the formation of a phenolic smell. The colonies are yellowish and darken throughout incubation.	
MAST ID™ – CHROMagar®	Mast	Identification medium for the simultaneous detection and identification of *C. albicans, C. tropicalis* and other *Candida* species, by the incorporation of species specific chromogenic substrates which enable the individual colonies to be identified by the coloration acquired during growth as well as colony morphology.	

[1] Data obtained from commercial support documents: BBL/Becton Dickson (New Jersey, USA); Biokar Diagnostics (Beauvais, France); Biomedics (Madrid, Spain); bioMérieux (Marcy l'Etoile, France); Laboratorios Conda, S.A. (Madrid, Spain); Difco Laboratories, USA (Detroit, MI); EMSL Analytics Inc. (New Jersey, USA); Fluka (Buchs, Switzerland); International Microbio (Signes, France); MAST Group Ltd. (Mereyside, UK); Millipore (Bedford, USA); MERCK (Darmstadt, Germany); Oxoid (Basingstoke, UK); Sanofi Diagnostics Pasteur Inc. (Sanofi-Synthélabo, France); SIEBEL Institute of Technology (Chicago, USA); STAB Vida (Oeiras, Portugal). Standard Methods Reference: [2] SMWW; [3] AOAC; [4] BAM, COMPF; [5] AFNOR, COMPF, DIN, ISO, SMD; [6] COMPF, SMD; [7] COMPF, BAM, EP; [8] AFNOR, BS, DIN, IDF, ISSO, [9] ASBC (**AFNOR:** Association Française de Normalisation; **AOAC:** Association of Official Analytical Chemists; **ASBC:** American Society of Brewing Chemists; **BAM:** Bacteriological Analytical Manual; **BS:** British Standards; **COMPF:** Compendium of Methods for the Microbiological Examination of Foods; **DIN:** Deutsches Institut für Normung; **EP:** European Pharmacopeia; **IDF:** International Dairy Federation; **ISO:** International Standards Organization; **SMD:** Standard Methods for the Examination of Dairy; **SMWW:** Standard Methods for the Examination of Water and Wastewater).

system, Compactdry™, consists of inoculating the sample in the centre of a self-diffusible medium, amended with a cold soluble gelling agent, placed in a disposable plastic dish (Mizuochi and Kodaka, 2000). The hydrophobic grid membrane filter technique enables one to decrease the number of dilutions to count highly contaminated samples (Boer and Beumer, 1999). Microscopical observation of stained microcolonies by optical or fluorescent dyes, growing on filter membranes treated with optical brighteners, provides faster detection of contaminants (Andrews, 1982; Navarro *et al.*, 1987; Campbell, 1987; Parker, 1989), especially for occasional highly contaminated samples, like those related to unexpected rupture of sterile filters prior to wine bottling. Image analysis techniques may also be used to facilitate counting colonies (Parker, 1989; Boer and Beumer, 1999).

Surface analysis is commonly performed by swabbing a defined area and suspending in a diluent, followed by inoculation in culture media. Other techniques include replica plating, molten agar, sterile sponge, etc. (see review of Lee and Fung, 1986). Fung *et al.* (2000) devised a hands-free 'Pop-up' adhesive tape method for microbial sampling of meat surfaces. The tape is used to obtain the sample and is transferred to an agar surface which, after 15 s of contact, is incubated, and colonies are counted in due time.

12.2.8 Direct count techniques

The above-mentioned methods are time consuming because they depend on cultivation, but other techniques avoid this step and provide cell counts, such as direct microscopy with or without a counting chamber and epifluorescence microscopy. In optical light microscopy several stains may be used to evaluate cell viability, such as methylene blue and Ponceau S (Kunkee and Neradt, 1974). Toluidine blue has also been reported as an adequate stain for *in situ* detection of microorganisms in foods (Dodd and Waites, 1991). Fluorochromic stainings may provide easier detection than the former visible stainings and may also be adapted to *in situ* techniques (for a review see Dodd and Waites, 1992). Classical examples are fluorescein diacetate (Paton and Jones, 1975), aniline blue (Koch *et al.*, 1986) and acridine orange used in direct epifluorescent filter technique (DEFT) (Day, 1987). DEFT techniques taking less than 30 minutes were applied to raw milk after lysis of somatic cells (Pettipher *et al.*, 1980), to suspensions of solid foods with different degrees of correlation with standard plate counts (Pettipher and Rodrigues, 1982), and to count osmophilic yeasts with a 24-hour pre-incubation period (Pettipher, 1987). Problems associated with interpretation of viability and photofading may be circumvented by using the fading retardant Citifluor AF2 (Navarro *et al.*, 1987) or by staining with berberine sulphate followed by acridine orange (Peladan and Leitz, 1991). Today, other fluorescent stainings may be used, such as SYTO 9, oxonol, propidium iodide, FUN 1, SYBR Green II and carboxyfluorescein diacetate (Prudêncio *et al.*, 1998; Couto *et al.*, 1999; van Zandycke *et al.*, 2003; Yamaguchi *et al.*, 2003), which may also be applied in the flow cytometric analysis described below. In addition, microorganisms may be counted directly on the surface of adhesive sheets by fluorescence microscopy (Yamaguchi *et al.*, 2003).

The results obtained with DEFT are not always reliable (Rodrigues and Kroll, 1986; Meidell, 1987) but provide fast results when cells are concentrated (Meidell, 1987), as mentioned before regarding counting of microcolonies. Therefore these direct techniques might be very useful for real-time monitoring when contamination is suspected, as in the verification of beer pasteurisation (Peladan and Leitz, 1991). In addition, microscope-based techniques are not adequate for daily routine use because of operators' fatigue, but automated and image analysis systems may overcome this drawback (Fung, 2002). These techniques could also be included in the next section regarding rapid or instrumental methods.

12.3 Detection and enumeration: instrumental techniques

Instrumental techniques have been developed mainly to reduce test time and labour involved in food and beverage analyses, and may also be designated as rapid methods. However, the speed of analyses is dependent on the fundamentals of the determination. If there is a culture-mediated process, the analysis time is longer than when cells or molecules are directly detected. Examples of culture-dependent methods are those based on turbidity, electrometry and media optical (colour) changes. Methods which may be applied without cultivation steps include chemiluminescence, flow cytometry, immunology and molecular methods (for recent critical reviews on instrumental techniques see for instance Boer and Beumer, 1999; Veal *et al.*, 2000; Fung, 2002). This last group of methods are preferentially used for yeast identification and typing and so will be discussed in Section 12.4. Immunological methods are also included here although they may not always be regarded as instrumental methods. The above-mentioned distinction between culture-dependent or independent methods is not always clear-cut, as microbial cultivation may precede the utilisation of chemiluminescence, flow cytometry, immunology or molecular methods. Furthermore, these techniques may be applied together and so their distinction is not always clear.

12.3.1 Electrometry

Electrometric methods estimate the number of microorganisms based on changes in electrical characteristics (impedance, capacitance or conductance) in a culture medium, which are correlated with microbial counts. The use of impedance has been developed since the beginning of the 1970s, primarily addressed to clinical bacteria (Ur and Brown, 1975, and references cited therein). For counting yeasts capacitance was shown to provide faster results than impedance (Fleischer *et al.*, 1984) or conductance (Fleischer *et al.*, 1984; Schaertel *et al.*, 1987). As it is a culture-dependent technique the speed of analysis depends on the initial contamination of the sample and on microbial growth rate (Campbell, 1987). Thus it is especially quick (less than 24 hours) for highly contaminated samples and for bacteria which have mean growth rates higher than yeasts (Weihe *et al.*, 1984). A report on wine spoilage yeasts referred to their enumeration, with detection times ranging from 43 h for 10 viable cells/ml, to 16 h for 10^5 viable cells/ml (Henschke and Thomas, 1988). The selection of culture medium influences the electrical signal (Fleischer *et al.*, 1984; Schaertel *et al.*, 1987) and the type of growing microorganisms (Henschke and Thomas, 1988). In beers, yeasts caused an increase in impedance while lactic bacteria caused a decrease (Evans, 1982), making it difficult to interpret results of mixed populations. More recently, indirect conductance methodologies, by measuring the release of CO_2, appear to be specially appropriate to beverages (see review of Deak and Beuchat, 1994). These can be used in developing systems to predict food shelf-life, because of the shorter detection times due to lower detection thresholds (*ca.* 10^4–10^5 yeast cells/ml) when compared with direct conductimetry (Deak and Beuchat, 1993a, 1994, 1995).

12.3.2 Optical measurements

There are several techniques that employ optical measurements to detect and enumerate microorganisms. Optical instruments measure the changes in colour of growth medium as a result of carbon dioxide production or changes in pH and redox potential due to microbial metabolism (Fung, 2002). Firstenberg-Eden *et al.* (2002) described an optical instrument primarily developed for bacterial counts (2–11 h time response) and also adequate for counting yeasts in yoghurts and soft cheeses after 40 h of incubation. The less than one day time response and the ability to test sanitation swabs allow the incorporation of this system in Hazard Analysis and Critical Control Point (HACCP) programs (Russel and Bailey, 2000). Other colorimeters (e.g. BacT/Alert Microbial Detection System, Omnispec Bioactivity Monitor System) are preferentially used in clinical and pathogen detection (Fung, 2002).

Another sophisticated technique involves the entrapment of microorganisms in gel microdroplets (10 to 100 μm in diameter), being surrounded by a hydrophobic fluid and counted by colorimetric or fluorescence pH indicators (Williams *et al.*, 1990). These authors applied the technique to clinical bacteria and claim that it provides results much more quickly than plating methods. Cell counting may also be performed by flow cytometry (Katsuragi *et al.*, 2000; Zengler *et al.*, 2002) while uncultured microorganisms showed the ability to grow in these microdroplets (Zengler *et al.*, 2002).

12.3.3 Bioluminescence

Bioluminescence is based on the quantification of ATP of microbial cells using the luciferase enzyme and the luciferin cofactor. The light produced by the hydrolysis of ATP is measured in a luminometer and so light intensity is proportional to the amount of ATP in the sample. Hence, detected ATP may be due to microbial or other sources (e.g. food processing residues). Very low levels of microbial contamination may be detected with filterable samples (e.g. <10 yeasts/250 ml) but former successful applications in several industries (Graumlich, 1985; Littel and LaRocco, 1985, 1986; Miller and Galston, 1989; Simpson *et al.*, 1989; Thomas and Ackerman, 1988) were not followed by extensive practical utilisation (Deak, 2003). Nowadays bioluminescence is specially adapted to hygiene monitoring (Poulis *et al.*, 1993) where there is no need to distinguish the source of measured ATP, or for sterile products without organic residues. The speed of the technique (a few minutes) renders it a valuable tool to HACCP programs because sanitation processes may be monitored practically in real time (Colquhoun *et al.*, 1998). If ATP levels are higher than acceptable then it is possible to repeat the sanitation operation. However, the type and concentration of the chemical cleaner or sanitiser in the sample may affect ATP measurements (Green *et al.*, 1999) and commercially available systems have different sensitivities and reproducibilities (Colquhoun *et al.*, 1998).

12.3.4 Flow cytometry

Flow cytometry estimates the number of microbial cells passing by a beam of laser light. Its main advantages are its sensitivity and speed but many food matrixes interfere with the extent of light scattering and affect the results. Earlier reports were concerned with counting yeasts in soft drinks (Pettipher, 1991) and beers (Jespersen *et al.*, 1993). Today, flow cytometry, adapted to detect cells stained with fluorochromes, is a rather sensitive technique and may provide real-time information with an expected bright future (for a review see Veal *et al.*, 2000). In the wine industry, fluorocytometry yielded real-time results for a threshold of 5×10^4 yeasts/ml in grape juices (Bouix *et al.*, 1999) and of 10^3 yeasts/ml in wines (Malacrino *et al.*, 2001). In fruit yoghurts, this technique enabled the routine detection of spoilage yeasts at levels of one yeast per pot of finished product within 24 h of sampling (Groote *et al.*, 1995). In addition, these authors developed a protocol for shelf-life evaluation within 48 h which is much faster than the usual seven-day accelerated keeping quality tests. Sensitivity levels of 100 viable yeasts per gram of prepared salads within 45 min, or one viable yeast per 20 grams of processed fruit products after incubation during 48 h, have been reported (Roy and Mulard, 1996). Due to the rapid response time flow cytometry was proposed to be used in proactive quality management systems (Bouix *et al.*, 1999; Groote *et al.*, 1995) and may be also used in conjunction with nucleic acid-based fluorescent probes (Porter and Pickup, 2000).

Solid phase cytometry is a novel technique adapted to count single cells. After sample filtration, the microorganisms are fluorescently labelled on the membrane and counted by a laser scanning device (D'Haese and Nelis, 2002). Epifluorescence microscopy may be used to observe occasional fluorescent spots by means of a computer-driven moving stage (D'Haese *et al.*, 2001). The application of this technique in clinical yeasts of *Cryptococcus neoformans* labelled with immunofluorescent dyes yielded detection limits of 3–6 cells per ml within 30 min (Bauters *et al.*, 2003).

12.3.5 Immunology

Immunological methods may also be used to detect food spoilage yeasts. An immunofluorescence-based method has been adopted by advisory institutions of brewing industries (Campbell, 1987) and was successfully applied in practical conditions to detect yeast contaminants within two and half hours (Legrand and Ramette, 1986). The quantification of fluorescence response of brewing wild yeasts detected by immunofluorescence may be done by flow cytometry (Hutter *et al.*, 1979). An enzyme-linked immunosorbent assay (ELISA) has been described to detect *Brettanomyces* spp. directly from wines, the results being scored both visually and spectrophotometrically (Kuniyuki *et al.*, 1984). The sensitivity of the assay was 91.5×10^{-12} g of protein which was calculated as being equivalent to the detection of 34 total cells per ml of wine. Despite these promising reports, immunological techniques are not widely used in industry for detecting yeasts, perhaps because of the lack of commercially available antigens (Deak, 2003). Hence, the recent advances and future prospects concerned with immunodetection and identification of foodborne pathogens (Fung, 2002) seem to be far from being applicable to food spoilage yeasts.

12.3.6 Instrument utilisation in the food and beverage industries

Several methods formerly described for either enumeration or identification/typing purposes (see former reviews of Dziezak, 1987, and Gutteridge, 1987) were hardly used in industrial routine and the instruments presently available are the result of improvements aimed at increasing their utilisation by industries. For instance, turbidimetric measurements using plate readers which have been reported as feasible for industrial use (Thomas and Ackerman, 1988) are practically restricted to research laboratories where they are a powerful tool to obtain simultaneously a large number of growth-dependent parameters. Instruments based on radiometry, infrared spectrophometry and microcalorimetry were also commercially available (Goldschmidt and Fung, 1978; Dziezak, 1987) and radiometers were somewhat frequent in the American citrus industry (Weihe *et al.*, 1984). However, to the best of our knowledge, they were not well succeeded for microbiological control of foodborne yeasts and are not mentioned among the instruments presently available (Fung, 2002). The occasional disappearance of commercial instruments (e.g. the colorimeter Omnispec Bioactivity Monitor System recently mentioned by Fung (2002) for pathogen detection has been discontinued and not replaced by an improved instrument) and the reliability of technical assistance also contribute to the decision by food technologists to adopt instrumental techniques.

Fung (2002) estimated, for the USA, that about 70% of total microbial tests were done using manual or conventional methods, while 30% used 'rapid methods'. Prospects for 2005 were 50% for each, while for pathogens 60–70% of the tests will be some form of 'rapid tests'. Numbers for yeasts were not provided but are surely much lower for these latter techniques and, at least for market reasons, industry utilisation of instrumental methods addressed to yeasts will depend on the development of equipment for clinical and pathogen indicator applications. This may be the case for the rapid quantification of total viable counts in meats by Fourier Transform Infrared Spectroscopy (FT-IR) followed by machine learning for data treatment (Ellis *et al.*, 2002). These authors demonstrated that at levels of 10^7 bacteria per gram the main biochemical indicator of spoilage was the onset of proteolysis. The speed of this non-invasive technique (reported time response of 60 s) renders it a valuable tool for HACCP programs.

The results of an inquiry made by us and mailed to several instrument manufacturers are summarised in Table 12.3 for the few companies answering to our request. Although absolute numbers of industrial users were not always provided by the companies, we believe that only a small proportion of food industries are potential clients of routine instruments for microbiological control. At the initial stage of their development instrumental methods were probably seen with too much optimism concerning their utilisation by industry (Sharpe, 1979; Dziezak, 1987), although difficulties were not underestimated (Goldschmidt and Fung, 1978), in particular by technological research institutions (Griffiths *et al.*, 1984; Betts and Banks, 1988). Their potential use remains intact or is even greater today because of the improvement in instrumentation and selection of the most appropriate ones. However, the high cost of most equipment and the need for skilled labour limit its utilisation to high-capital non-food industries (e.g. cosmetics, pharmaceuticals), to industries processing foods susceptible to pathogenic microorganisms or to large production plants (e.g. dairies, breweries). The increase in the consumption of ready-to-eat meals, which have shelf-life periods of less than one month, should also stimulate the utilisation of rapid methods due to the need for short response times. Many food industries, particularly those processing food and beverages unaffected by health hazards, are a relative small market for instrument manufacturers (Fung, 2002) and do not have the financial capacity or do not think it profitable to invest significantly in microbiological control. The increase in the number of microbiological tests expected in the future in the fruit and vegetable industries due to recent foodborne outbreaks related to these food commodities (Fung, 2002) may change this attitude. Then, the versatility of instruments to detect both pathogens and yeast spoilers will be a major factor determining yeast monitoring by these techniques. In addition, many instrumental methods depend on previous cultivation and so suffer from the limitations discussed above concerning the lack of culture media specific for spoilage yeasts. On the contrary, bioluminescence, due to its increasing operating simplicity and decreasing cost for portable apparatus, will be common equipment for hygiene assessment. It also induces an improvement in production efficiency by means of a 'psychological effect', because staff do not enjoy having to reclean equipment and, therefore, there is a strong incentive to get it right first time (Ogden, 1994).

Table 12.3 Commercially available instruments to detect and enumerate yeasts in foods and beverages (data supplied by manufacturers)

Manufacturer	Equipment	Brand	Industrial users (number or percentage in brackets)	Approximate cost (euro)	Approximate cost per test (euro)	Number of simultaneous tests	Time to result	Observations
Chemunex	Flow cytometer	D-*Count*®	Cosmetics (11), dairy (22), fruit juices (9), fruit preparation (3), processed food (soup, salad) (4), ingredients (1)	137 000 (includes sample loader)	4–5 (cosmetics), 2–4 (food)	50 samples per batch	48 samples/h	For filterable and non-filterable products, sensitivity: 50–1000 cells/ml or g of product
	Solid phase cytometer	Chem*Scan*®	Pharmaceuticals (58), drinking water (18), biotechnology (10), semi-conductor (6), food (7), other (10)	154 000	7 (total viable counts and fungi), 20 (coliforms)	One sample at a time	Total viable counts (90 min), fungi (4 h), coliforms (3.5 h)	For filterable products, sensitivity: one cell per volume filtered
BD Biosciences	Flow cytometer	FACSCalibur™ FACSCan™ (discontinued)	In Europe: cosmetics (4), dairy (8), brewery (3), neutraceuticals (2), other food (13)	Starting at 60 000 (without sample loader)	2.85	40 samples per batch	Sample preparation (5 min), analysis (1–3 min)	Counting of live and dead cells stained with thiazole orange and propidium iodide

Table 12.3 Coontinued

Manufacturer	Equipment	Brand	Industrial users (number or percentage in brackets)	Approximate cost (euro)	Approximate cost per test (euro)	Number of simultaneous tests	Time to result	Observations
FOSS Electric	Optical detector	MF32® MF128®	Dairy (61%), meat (17%), general (12%), others (10%)	Not supplied	Not supplied	32 or 128 samples per batch	Yeast (48 h), total viable count (12–14 h), coliform (14–17 h)	Used for food samples or equipment swabs
Charm Sciences	Luminometer	LUM-96®/ Sterylite®	Milk, juice, soy	Not supplied	Not supplied	96 samples per batch	25 min	Microplate reader for testing 'sterile' samples
	Luminometer	Micro-Q®	Wine, water, clear juice	Not supplied	Not supplied	One sample at a time	3 min	Portable reader, for total microbial load of filterable products without organic debris
Celsis International	Luminometer	Advance™	Cosmetics, pharmaceuticals, dairy, fruit juices, water, wine	40 000–66 500	0.40–5.00	164 samples per batch		Pre-incubation for 24 h or 48 h

		Advance™ Coupe				30 samples per batch		
		CellScan™ M596				96 samples per batch	20 min	Microplate reader, pre-incubation 48 h
Millipore	Luminometer	MicroStar®	Pharmaceuticals, water	88 000	Not supplied	One sample at a time	Few minutes	With or without pre-incubation up to 24 h, readings by image analysis after filtration through hydrophobic membranes
		SteriScreen™	Pharmaceuticals, water	88 000	Not supplied	164 samples per batch	Few minutes	Pre-incubation for 24 h or 48 h
		MicroCount™	Pharmaceuticals, water	88 000	Not supplied	4 samples per batch	8 min	Microplate reader, pre-incubation for 24 h or 48 h, counting with MPN in 24 wells for each sample, sensitivity: 10^2–10^3 cfu/ml

12.4 Methods of identifying and characterising foodborne yeasts

Besides yeast detection and counting, microbiological control may be further aimed at the identification and characterisation or typing of foodborne yeasts. Classical identification is based on physiological, biochemical or sexual characteristics, but the recent advances in yeast taxonomy are based on molecular methods (Kurztman *et al.*, 2003). Yeast characterisation or typing is related to the utilisation of methods based on the chemical and molecular composition of cells and may, or may not, provide information at subspecies level.

12.4.1 Classical methods and their simplifications

Classical methods of identification cannot be routinely utilised in the food industry due to time and skilled labour implications. As a consequence, various miniaturised and simplified identification methods have been developed. However, they are based on the same approach to the classical yeast identification methods, being time consuming, even when procedures are automated and computerised, and often lead to false or equivocal identifications (Fung and Liang, 1990; Fleet, 1992; Deak and Beuchat, 1996). Presently, physiological and biochemical tests are still being developed and used but demand somewhat skilled data interpretation (Robert, 2003). On the contrary, identification at a practical level is easier with dichotomic keys. Pitt and Hocking (1999) presented a simplified key to identify the most dangerous food spoilage species. The Simplified Identification Method (SIM) updated and published by Deak and Beuchat (1996) includes a dichotomic key that has been currently applied to fruit juice industrial isolates (Sancho *et al.*, 2000). Velásquez *et al.* (2001) also presented a phenotypic system to identify foodborne yeasts, while Middelhoven (2002) published a dichotomic key to identify yeasts recovered from sour fermented foods and fodders. Once more, clinical specimens are the main target of miniaturised systems, like Vitek, API 32C, API 20C AUX, MicroScan, Yeast Star, Auxacolor and RapID Yeast Plus (Fung and Liang, 1990; St-Germain and Beauchesne, 1991; Land *et al.*, 1991; Deak and Beuchat, 1996; Robert, 2003), with low applicability to the food industry unless other tests are devised. For instance, the systems RapID Yeast Plus and API 20C AUX correctly identified only 35% and 13% of orange fruit isolates, respectively (Arias *et al.*, 2002). Further improvements in commercial systems directed to foodborne yeast species depend on the demand by industrial laboratories. Until then, it is possible to combine the responses from the miniaturised systems with the computer programs used in classical methodologies (Fleet, 1992) or with Deak and Beuchat's SIM database and obtain higher percentages of correct identifications (Deak and Beuchat, 1993b). Conventional identification methodologies may also include chemotaxonomic determinations, like the ubiquinone and cell wall monosaccharide compositions (Kurtzman and Fell, 1998), which may occasionally sort out the identity of strains having conflicting results between genotypic and phenotypic data (Prillinger *et al.*, 1999).

12.4.2 Chemical methods

To overcome the drawbacks of classical methodologies, alternative faster typing methods have been developed, based, among others, on analysis of total proteins (van Vuuren and van der Meer, 1987, 1988), of isoenzymes (Duarte *et al.*, 1999), and of total long-chain fatty acids (Augustyn *et al.*, 1992; Moreira-da-Silva *et al.*, 1994; Malfeito-Ferreira *et al.*, 1989b, 1997; Saldanha-da-Gama *et al.*, 1997; Sancho *et al.*, 2000; Dias *et al.*, 2003). These methods are not intended to replace classical identification but provide additional characterisation features, although they are seldom used in industry. They have been questioned because of their dependence on the physiological state of yeast cells (Querol *et al.*, 2003; van der Vossen *et al.*, 2003) but this is only relevant for extreme environmental conditions which are not used for screening food spoilage yeasts (Loureiro, 2000). One system based on fatty acid profiling is commercially available (MIDI, Newark, USA) and widely used to identify bacteria (Buyer, 2003) and for clinical and environmental yeasts grown in agar media (see http://www.midi-inc.com). The development of zymological indicators may also be based on yeast long-chain fatty acid compositions and is claimed to have a technological significance unlike molecular biological methods (see Section 12.5.1).

Other promising techniques dependent on complex instruments have also been developed for microbial fingerprinting with utilisation limited to research centres and highly trained operators. Examples are the cases of circular intensity differential scattering (CIDS) and of pyrolysis mass spectrometry (see review of Gutteridge, 1987). The first was overtaken by flow cytometry and the second led to a much simpler and faster spectrometric technique – matrix assisted laser desorption time of flight mass spectrometry (MALDI-TOF-MS). This has been applied for direct profiling of proteins from mammalian cells (Chaurand and Caprioli, 2002), for rapid differentiation of intact cells of clinical bacteria (Claydon *et al.*, 1996; Evanson *et al.*, 2001) and for direct surface analysis of fungal species (Valentine *et al.*, 2002). To the best of our knowledge this technique has not been reported for foodborne yeasts.

12.4.3 Nucleic acid based typing

Molecular identification techniques rely mostly on sequencing the 600–650 nucleotide D1/D2 domain of large subunit (26S) ribosomal DNA because of its taxonomic significance, but sequencing is not a suitable tool for industrial utilisation despite the availability of gene sequence databases (see review by Kurtzman *et al.*, 2003). At the beginning of the last decade molecular biology techniques were still regarded as promising for spoilage yeasts with few known applications (Fleet, 1992). Recent progress has contributed to the development of powerful typing techniques which are summarised in Table 12.4. For descriptions of techniques and reviews on the bases of molecular characterisation applied to foodborne yeasts see, for instance, Loureiro and Querol (1999), Querol *et al.* (2003) and van der Vossen *et al.* (2003) and references cited therein. Many of the methodologies presented in Table 12.4 were applied to few

Table 12.4 Recent molecular biological techniques for typing foodborne yeasts

Technique	Target species	References	Observations
RFLP–mtDNA restriction analysis with several restriction enzymes	*Saccharomyces* spp. and several other species	Querol *et al.* (1992a,b), Vezinhet *et al.* (1992), Ibeas *et al.* (1996), Romano *et al.* (1996), Constantí *et al.* (1998), Guillamón *et al.* (1994), Epifanio *et al.* (1999), Esteve-Zarzoso *et al.* (2000), Comi *et al.* (2000), Gutiérrez *et al.* (2001), Fernández-Espinar *et al.* (2001), Torija *et al.* (2002)	Biomass from broth cultures or plate colonies
RFLP–mtDNA restriction analysis with *Hinf*Ie *Dde*I	*S. cerevisiae* and several wine-related species	López *et al.* (2001)	Improvement of Querol *et al.* (1992a,b) technique; time: 25.5 h
PCR-RFLP of the 5.8S, 18S or 28S ITS regions of the rRNA gene, followed by restriction with several enzymes	Several yeast species	Constantí *et al.* (1998), Guillamón *et al.* (1998), Ramos *et al.* (1998), Esteve-Zarzoso *et al.* (1999), Dlauchy *et al.* (1999), Granchi *et al.* (1999), Fernández *et al.* (2000), Nguyen *et al.* (2000), Egli and Henick-Kling (2001), Gutiérrez *et al.* (2001), Redzepovic *et al.* (2002), Dias *et al.* (2003)	Biomass from broth cultures or plate colonies
PCR amplification of DNA δ sequences	*Saccharomyces* spp.	Ness *et al.* (1993), Fernández-Espinar *et al.* (2001)	Biomass from plate colonies
DNA and PCR fingerprinting with several primers	Several yeast species	Lieckfeldt *et al.* (1993)	Biomass from broth cultures, DNA test: 5 days time, PCR test: 14 h time
PCR fingerprinting with $(GAC)_5$ and $(GTG)_5$ primers and NTS region amplification and restriction with *Hae*III and *Msp*I	Several yeast species	Baleiras-Couto *et al.* (1995, 1996), Caruso *et al.* (2002)	Biomass from broth cultures

PCR-RAPD of total DNA with several primers	Several species	Baleiras-Couto *et al.* (1995), Quesada and Cenis (1995), Romano *et al.* (1996), Mitrakul *et al.* (1999), Prillinger *et al.* (1999), Andrighetto *et al.* (2000), van der Westhuizen *et al.* (2000a,b)	Biomass from plate colonies or broth cultures
Electrophoretic karyotyping of chromosomal DNA by PFGE	*S. cerevisiae*	Yamamoto *et al.* (1991), Vezinhet *et al.* (1992)	Biomass from broth cultures
Electrophoretic karyotyping of chromosomal DNA by CHEF	Several species	Schütz and Gafner (1993), Ibeas *et al.* (1996), Egli *et al.* (1998), Mitrakul *et al.* (1999), van der Westhuizen *et al.* (2000a,b), Fernández-Espinar *et al.* (2001)	Biomass from plate colonies
PCR-RFLP of rDNA, with several restriction enzymes	*Dekkera/Brettanomyces* spp.	Molina *et al.* (1993)	Biomass from broth cultures
PCR of ITS region of rDNA	*Saccharomyces* spp.	Valente *et al.* (1996)	Biomass from broth cultures
	Several species	Sancho *et al.* (2000)	Biomass from plate culture, time: 4 h
Nested PCR of DNA	*Dekkera/Brettanomyces* spp.	Ibeas *et al.* (1996), Alguacil *et al.* (1998)	Plate colonies or broth cultures, time: <10 h. Direct analysis from wine, threshold: >10^4 cell/ml.
AFLP selective PCR amplification of restriction fragments of total DNA	Several species	Barros-Lopes *et al.* (1999)	Biomass from broth cultures
Staircase electrophoresis of low molecular weight RNA profiles	Several species	Velásquez *et al.* (2001)	
PCR amplification of DNA SSRs	*S. cerevisiae* starters	Techera *et al.* (2001)	Biomass from broth cultures, time: 14 h
PNA FISH targeting the D1–D2 region of 26S rDNA	*D. bruxellensis*	Stender *et al.* (2001), Dias *et al.* (2003)	Biomass from plate colonies, time: 3 h

Table 12.4 Continued

Technique	Target species	References	Observations
PNA CISH targeting targeting the D1–D2 region of 26S rRNA	*D. bruxellensis*	Connell *et al.* (2002)	Biomass from plate micro-colonies
PCR of introns in the mitochondrial gene COX 1	*S. cerevisiae* starters	López *et al.* (2002)	Direct analysis from grape juice, time: 8 h
DGGE of PCR amplified 26S rRNA genes	Several species	Cocolin *et al.* (2000, 2001), Mills *et al.* (2002)	Direct analysis from grape juice, threshold: $>10^3$ cell/ml, sample volume: 100 ml, time: 1 day
		Cocolin *et al.* (2002)	Direct analysis from grape juice, time: 8 h

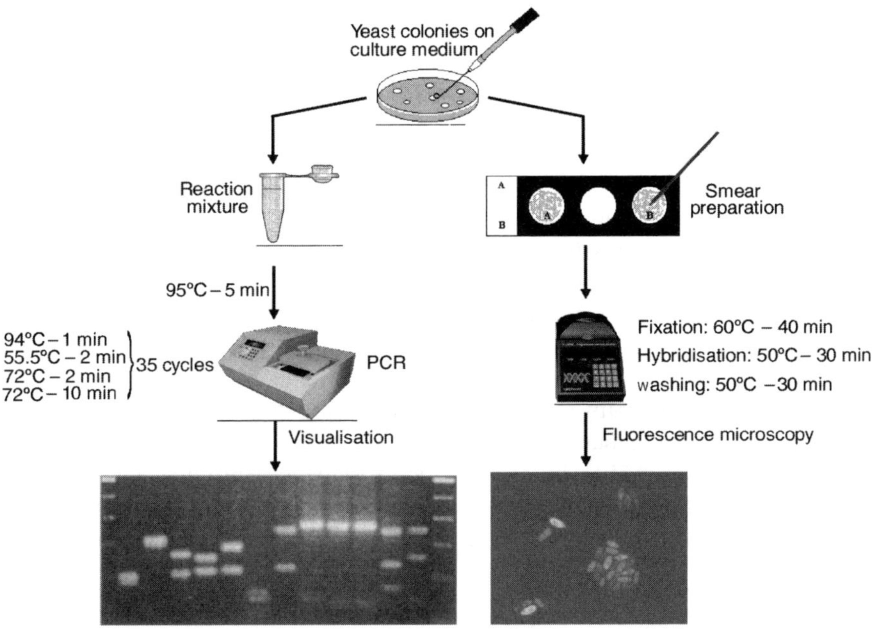

Fig. 12.1 Schematic representations of 5.8S-ITS profiling and of PNA-*FISH* techniques, according to Esteve-Zarosso *et al.* (1999) and Stender *et al.* (2001) respectively.

species or genera and, in spite of their efficiency, were not followed by confirmatory research with a larger number of species or in a wider variety of food industries.

Polymerase chain reaction (PCR) is now the basis of most rapid typing techniques (van der Vossen *et al.*, 2003) and from Table 12.4 emerges the analysis of restriction fragment length polymorphim by PCR (RFLP-PCR) of the 5.8S-ITS (internal transcribed spacer) region of rRNA which has been presented as the first available molecular method for yeast species identification at practical level (Querol *et al.*, 2003) (Fig. 12.1). The database of 5.8S-ITS profiles produced by Querol's group includes 300 yeast species and is available on the website http://motor.edinfo.es/iata (Querol *et al.*, 2003). Arias *et al.* (2002) judged it as the best tool for rapid and accurate identification of yeasts isolated from fruit products. In addition, these authors recommended the use of the taxonomically relevant sequence of the 26S rRNA gene and classical methodologies when corroboration of results is necessary. Heras-Vazquez *et al.* (2003) advised the use of sequence analysis of the ITS region until the above-mentioned database by restriction analysis is completed.

Concerning strain typing at an infraspecific level, the mitochondrial DNA (mtDNA) restriction analysis has been largely used to characterise mainly strains of *S. cerevisiae*, using different restriction enzymes (Table 12.4). The number and variety of restriction enzymes depend on the strains under study and are determined, frequently, after failure of the first tried enzymes to discriminate those strains. This also holds true for the restriction enzymes used in 5.8S-ITS profiling, and for the selection of primers in random amplified polymorphic (RAPD) based techniques, making their utilisation in routine industry laboratories difficult. PCR-fingerprinting using the $(GTG)_5$ simple repeat primers provides information at the subspecies level making it applicable for tracing routes of contamination in processing lines together with RFLP-PCR of 18S rDNA and ITS (van der Vossen *et al.*, 2003). The research laboratory of these authors developed a database of PCR fingerprint patterns from 650 yeast species to cope with industrial needs. Amplified fragment length polymorphism (AFLP) is another PCR-based technique applicable to foodborne yeasts and its good reproducibility is an advantage over RAPD and PCR fingerprinting (van der Vossen *et al.*, 2003). In many of these electrophoretic dependent techniques, the interpretation of a large range of gel lanes requires the utilisation of specific software (van der Vossen *et al.*, 2003) with additional costs and need for skilled labour.

Oligonucleotide probes based on unique sequences provide a fast typing technique at different taxonomic levels, and advances in molecular array technologies may enable the handling of a large number of probes (Kurtzman *et al.*, 2003). The utilisation of peptidonucleic acid (PNA) molecular probes appears to have a promising future in the rapid identification of yeast species (see review of Stender *et al.*, 2002), particularly for industrial users, due to its simplicity. After strain isolation, the results take about two hours to obtain, hybridisation occurs in microscope slides and positive results are checked by microscopy (Fig. 12.1), which is easier to interpret than profiles of gel bands characteristic of the above-mentioned RFLP, RAPD or AFLP techniques. Our experience with a fluorescent assay using a PNA probe specific to *D. bruxellensis* stands for the success of routine application of this technique (Dias *et al.*, 2003). On the contrary, another PNA probe directed to *Z. bailii* (Perry-O'Keefe *et al.*, 2000) was found to not be completely specific, requiring further improvement (unpublished observations). At equipment level, the main flaw is the necessity for a costly fluorescence microscope. However, in spite of their great potential, PNA probes for food spoilage yeasts are not yet commercially available. Other recent *in situ* techniques cited in Table 12.4 (e.g. Direct Gel Gradient Electrophoresis (DGGE), COX1 introns, real-time PCR), which avoid the culturing step, are also promising. These techniques are much faster than the culture-dependent ones, because yeast growth and strain purification are avoided. However, they do not give information on cell viability and, if their sensitivity is low, only dominant strains are detected. Minority populations, as occurs frequently with dangerous spoilage yeasts, may be underestimated by the use of low-sensitive *in situ* techniques. In addition, instruments necessary to make these analyses are expensive for industrial use (e.g. real-time PCR apparatus costs about €40.000) and still depend on a certain degree of technical expertise.

Despite the huge efforts in rendering molecular techniques more user friendly, they have not yet reached the bench of the industry microbiologists. It is likely that these continuous improvements are an obstacle to their application in industry routine because the practical microbiologist has no basis for a critical assessment of appropriate methodologies. Perhaps, for the industry, it is wiser to wait and see which method resists continuous changes and remains as reliable as in the work published by research laboratories. Furthermore, polyphasic approaches usually reveal that the best discriminatory results are obtained by combination of two or more techniques (Querol and Ramon, 1996; Prillinger *et al.*, 1999; Querol *et al.*, 2003; van der Vossen *et al.*, 2003), which contributes to preventing adoption by industry. The gap between industry and science will be reduced when a new method clearly proves its efficiency at practical level and at comparable cost. Then it will be rapidly spread over industrial laboratories that are eager to have efficient routine diagnostic tools for spoilage yeasts at hand. At the present stage, particularly concerning small and medium enterprises, molecular techniques are more appropriate to research laboratories, certification institutions or other support laboratories, which may provide epidemiological studies or occasional consultancy to industry demands.

12.5 The use of microbiological indicators to monitor food quality and spoilage

In the modern food industry, with a well-assimilated HACCP system, the evaluation of the microbiological quality of foods is not limited to the retrospective analysis of the final product. It also includes the estimation of the microbiological quality of raw materials, ingredients, sanitation procedures, processing operations and of the product shelf-life. At the same time, international trade is evolving towards the evaluation of microbiological quality of foods and beverages according to standardised methods and analytical parameters – microbiological or chemical – accepted by all parties involved. In this context, microbiological indicators are essential tools, either for production control, or for quality assessment and food trade regulation.

12.5.1 Zymological indicators

In foods where yeasts are the main dominant microflora and pathogens do not grow or survive, the presence of spoilage yeasts is the main microbiological concern. Two occasional exceptions are the yeasts affecting particular groups of ill individuals (Hazen, 1995; Murphy and Kavanagh, 1999) and the explosion of bottles due to CO_2 production by refermentation (van Esch, 1992; Thomas, 1993). The evaluation of food zymological (zymo = yeast) quality is, therefore, specially addressed to preventing the hazard of final product deterioration and is done, as a rule, through yeast plate counting, using a general culture medium. This enumeration of 'total' viable yeasts (broadly known as 'yeasts and moulds'), like the indicator 'total viable count' used in food bacteriology, provides rather limited information, which is clearly insufficient from the food stability point of view. Moreover, it gathers in one determination two groups of microorganisms with different significances. But, most food industries accept 'what is currently done' (Mossel and Struijk, 1992), instead of choosing target organisms (or methods) on the basis of their technological significance. This limitation comes from the low ability to detect/enumerate spoilage yeasts, because general-purpose media clearly favour fast-growing yeasts, giving an underestimation of most dangerous yeasts – slow-growers – or inhibiting them completely. Of course this situation does not apply to the case where spoilers are the only microorganisms present. However, this indicator is quite useful to evaluate, in a retrospective way, the efficiency of sanitation operations, enabling the detection of contamination sources in the production chain. Other indicators are available to evaluate the presence or activity of spoilage yeasts, but only a few are used in industry, unlike bacterial indicators, revealing the low importance given by food microbiologists and technologists to this subject. These indicators may be grouped into three categories which may or may not be used together, as described below.

Indicators based on cell cultivation

Yeast enumeration on selective and/or differential media may be used as a zymological indicator. For instance, 'acid-resistant yeasts', enumerated on ZBA (Erickson, 1993) or TGYA (Makdesi and Beuchat, 1996a,b), may be particularly useful for assessing the zymological quality of acidified food and beverage products. 'Xerotolerant/osmophilic yeasts' may be enumerated on general-purpose media added to high concentrations of sugars (Cava and Hernández, 1994; Casas *et al.*, 1999), glycerol (Hocking and Pitt, 1980) and sodium chloride (Andrews *et al.*, 1997), being useful for fruit concentrates, confectionery and brines. ZDM (Schuller *et al.*, 2000) may be used in foods and beverages susceptible to *Z. bailii* and *Z. bisporus.* Lysine agar may be directed to the detection of 'non-*Saccharomyces* species' (Heard and Fleet, 1986), which may be regarded, under certain conditions, as a hygiene indicator. Another important zymological indicator to assess the quality of red wines, particularly those aged in oak barrels, is the '4-ethylphenol producing yeasts', obtained by growth on DBDM medium (Rodrigues *et al.*, 2001). The ESA medium (Kish *et al.*, 1983) was not found appropriate to detect *S. cerevisiae* (Heard and Fleet, 1986). Alternatively, a comparison between counts in general-purpose medium and Lysine medium (Heard and Fleet, 1986), or a general-purpose medium with 4 ppm of cycloheximide, may be used to estimate the population of *Saccharomyces* spp. This approach is a slight approximation to that used in bacteriology where a wide set of bacterial indicators may be used to characterise the contaminants of different foodstuffs. The results obtained by these media should be regarded as presumptive for identification of the targeted species and further confirmatory tests, using identification or typing techniques, may be performed additionally.

Chemical and organoleptic indicators

An alternative approach to zymological indicators based on microbiological cultures, which are time consuming, is to examine food samples for chemical (or sensorial) evidence of past microbial activity. The use of metabolites as indicators of spoilage (e.g. volatile acidity as a measure of acetic bacteria activity in wines) is often more convenient and faster than using microbiological counts and some of the before-mentioned instrumental methods measure, in fact, variations in chemical/physical characteristics. However, instrumental characteristics are supposed to be well correlated with microbial counts, which is not necessarily true for the evaluation of chemical markers. Only very few chemical compounds have received acceptance as a means of assessing the degree of yeast spoilage in foods. Ethanol and acetoin levels provide reliable indexes of the quality of the fruit on arrival at the factory and of hygiene in the processing plant, respectively (Mossel *et al.*, 1995); analysis of carbon dioxide in the headspace of sealed culture vials was proposed for rapid enumeration of fermentative yeasts in food, using a selective medium and gas-chromatographic analysis (Guerzoni *et al.*, 1985), and for shelf-life prediction (Gardini *et al.*, 1988); 1,3–pentadiene was reported as indicator of the activity of osmophilic spoilage yeasts in marzipan (Casas *et al.*, 1999); and 4-ethylphenol can be used as a sensorial or chemical marker to spot wines infected by *Dekkera/ Brettanomyces* sp. (Rodrigues *et al.*, 2001). These chemical markers may be regarded as 'spoilage predictors', because they may be detected at low levels before the food commodity deteriorates. This is not the case with ethyl acetate, which could not be used as a chemical marker to evaluate the spoiling activity of dangerous yeasts (e.g. *P. anomala*) during pre-fermentative maceration and white grape juice settling, because its production is very fast (Plata *et al.*, 2003), and when results are obtained the juice may have already deteriorated.

Indicators based on biomarkers

Another approach to establish zymological indicators is based on the long-chain fatty acid composition of contamination yeasts (Malfeito-Ferreira *et al.*, 1989b, 1997; Augustyn *et al.*, 1992; Sancho *et al.*, 2000). The rationale of this approach has been given elsewhere (Loureiro and Querol, 1999; Loureiro, 2000). The fatty acid characterisation of yeast species is presented in Table 12.5, gathering data obtained by us and from literature sources. This observation is of great importance because it proves that the different culture conditions used in the literature do not influence significantly the biomass fatty acid compositions as generally believed (Deak and Beuchat, 1996; Querol *et al.*, 2003; van der Vossen *et al.*, 2003). Based on the presence or absence of polyunsaturated C18 fatty acids, it was possible to separate the foodborne yeasts present into three broad groups (Table 12.5) with different technological significances.

The most dangerous yeasts for some food industries are located in group II: *Dekkera* spp. and *Zygosaccharomyces* spp. for the wine industry; *Zygosaccharomyces* spp. and *T. delbrueckii* for fruit concentrate and the juice industry; *Y. lipolytica* for the cheese industry; *Zygosaccharomyces* spp. for the industries of mayonnaise, salad dressings and marzipan. The species of this group which are not dangerous, such as *S. dairensis*, are extremely rare and have ecological niches quite well defined. Under these conditions, the group

Table 12.5 Separation of the foodborne yeast species based on their polyunsaturated C18 fatty acids composition

Yeast species	Yeast species
Group I [C18:2(−); C18:3(−)]	Group III [C18:2(+); C18:3(+)]
Hanseniaspora uvarum	*Candida cantenulata*
Hanseniaspora valbyensis	*Candida diddiense*
Hanseniaspora vineoe	*Candida famata*
Saccharomyces aceti	*Candida guilliermondii*
Saccharomyces bayanus	*Candida haemulonii*
Saccharomyces capensis	*Candida humilis*
Saccharomyces carlbergensis	*Candida krusei*
Saccharomyces cerevisiae	*Candida norvegica*
Saccharomyces chevalieri	*Candida parapsilosis*
Saccharomyces diastaticus	*Candida sake*
Saccharomyces exiguus	*Candida solanii*
Saccharomyces fructuum	*Candida stellata*
Saccharomyces globusus	*Candida tropicalis*
Saccharomyces oxidans	*Candida utilis*
Saccharomyces paradoxus	*Candida valida*
Saccharomyces pastorianus	*Citeromyces matritensis*
Saccharomyces steineri	*Cryptococcus albidus*
Saccharomyces unisporus	*Cryptococcus flavus*
Saccharomyces willianus	*Cryptococcus humicolus*
Saccharomycodes ludwigii	*Cryptococcus laurentii*
Schizosaccharomyces malidevorans	*Cryptococcus terreus*
Schizosaccharomyces octoporus	*Debaryomyces hansenii*
Schizosaccharomyces pombe	*Debaryomyces polymorphus*
Wickerhamiella domercqiae	*Issatchenkia orientalis*
	Issatchenkia terricola
Group II [C18:2(+); C18:3(−)]	*Kluyveromyces lactis*
Dekkera anomala	*Kluyveromyces marxianus*
Dekkera bruxellensis	*Kluyveromyces thermotolerans*
Dekkera intermedia	*Lodderomyces elongisporus*
Dekkera naardenensis	*Metschnikowia pulcherrima*
Endomyces fibuliger	*Pichia anomala*
Pichia etchellsii	*Pichia fermentans*
Saccharomyces dairensis	*Pichia guilliermondii*
Torulaspora delbrueckii	*Pichia jardini*
Torulaspora globosa	*Pichia membranaefaciens*
Torulaspora pretoriensis	*Pichia norvergensis*
Yarrowia lipolytica	*Rhodotorula* sp.
Zygosaccharomyces bailii	*Rhodotorula glutinis*
Zygosaccharomyces microellipsoides	*Rhodotorula mucilaginosa*
Zygosaccharomyces rouxii	*Saccharomyces kluyveri*

Source: adapted from Loureiro (2000).

displaying C18:2 (+) and C18:3 (−) can be seen as a fair zymological indicator for all industries in which the genera *Zygosaccharomyces*, *Dekkera*, *Torulaspora* and *Yarrowia* are feared. The yeasts of group I may be, in certain situations, dangerous to the wine, concentrates, fruit juices and soft drinks industries. This group, characterised by the absence of C18:2 and C18:3, includes species of *Saccharomyces* spp., the genus *Hanseniaspora* spp., which is not a dangerous spoilage agent, and the genera *Saccharomycodes* and *Schizosaccharomyces*. These last two genera may be quite dangerous for the wine industry, but are quite rare and their cellular morphology is easily recognised.

In group III, a large number of species are gathered with some characteristics in common, like their predominant oxidative metabolism and, sometimes, their ability to form a pellicle on a liquid surface. Generally, the species in this group are not very dangerous as food spoilage agents. In fact, they can be associated with the lack of good manufacturing practices of some industries, such as deficient hygiene and product contact with oxygen. Under these situations (e.g. in the wine and beer industries) the group of yeasts containing C18:2 and C18:3 can be regarded as a zymological indicator of lack of good manufacturing practices. However, the species *I. orientalis* may be problematic for the mayonnaise and salad dressings industries, and *D. hansenii* for the marzipan industry. In the dry-cured pork meat and olive brines industries the most significant pair of yeasts – *Cryptococcus* spp. and *D. hansenii*, and *P. membranifaciens* and *D. hansenii*, respectively – belong also to group III. However, these species are not regarded as having a particularly dangerous spoiling ability.

The above-mentioned 'broad tuning' separation into three groups may be followed by a 'fine tuning' discrimination using multivariate statistical analysis, which allocates strains to distinct clusters of spoilage species (Malfeito-Ferreira *et al.*, 1997; Sancho *et al.*, 2000; Dias *et al.*, 2003). Nucleic acid based methodologies so far have no technological significance and may be used after a first screening by fatty acid profiling to confirm species identity or to provide further intraspecific information. The negative tests using molecular probes or primers do not provide information on the probable identity and so their use after fatty acid profiling reduces the number of probes/primers to be tested. In conclusion, the most significant advantage of fatty acid profiling does not lie in its specificity – molecular techniques are more taxonomically significant – but in gathering the species in broad groups characterised by having different technological significance. Cases of this approach have been reported for spotting *Z. bailii* in wine bottling facilities (Malfeito-Ferreira *et al.*, 1997) and fruit concentrates (Sancho *et al.*, 2000) and *D. bruxellensis* in wines (Dias *et al.*, 2003). However, its widespread use in industry is strongly limited by the lack of readily available databases and by the difficulty of interpreting fatty acid profiles under industry conditions, a drawback also common to the evaluation of electrophoretic gel bands from molecular methods. The utilisation of fatty acid profiles may be avoided if a differential medium restricts the diversity of flora to a single or a few species. If positive presumptive colonies are detected, then a single or a few specific molecular probes or primers may be used to confirm identity (Dias *et al.*, 2003), for which DNA array techniques at affordable prices are welcomed.

Accelerated incubation tests

Accelerated incubation tests are common in some industries, like the beer and canning industries, although they are rarely used in others. Essentially, they can be used in three ways: (1) directly on the food, giving favourable incubation conditions to contaminating microflora (e.g. incubation at 25–30ºC); (2) directly on the food, giving favourable incubation conditions to a particular spoilage strain inoculated in a food sample; or (3) inoculating product samples in an appropriate culture medium – significance broth – with slightly more favourable conditions to microbial growth than in the food itself. The first case is essentially identical to the industry quarantine regime to which packaged products with high contamination levels are subjected, being released to market only when, after a long incubation period, they are microbiologically stable (van Esch, 1992; Andrews, 1992; Loureiro and Malfeito-Ferreira, 2003b). In brewing, this 'forcing test' includes beer incubation at elevated temperature for up to six weeks and dates back to the beginning of the twentieth century (Evans, 1985). The second case has been described by Groote *et al.* (1995) to predict the shelf-life of fruit yoghurt within 48 h. The last case may be exemplified by the work of Thomas and Ackerman (1988), in wines, using relatively short incubation periods (72 h) and growth detection by bioluminescence.

12.5.2 Acceptable levels of yeasts and spoilage prediction in foods

The establishment of acceptable levels of microorganisms in the final product is a concern common to many food industries. The aim of the technologist is to comply with levels that are attainable under current industrial conditions and ensure product stability during its shelf-life. Scientific and technical literature on acceptable levels and spoilage prediction of foods and beverages caused by yeasts is surprisingly scarce, revealing the low importance given to it by food microbiologists and technologists.

One viable cell of a virulent strain in each packaged unit of product may cause spoilage and so rather stringent acceptable levels should be established accordingly, as has been advised for *Z. bailii* in wines (Davenport, 1986; Thomas, 1993) and carbonated soft drinks (Pitt and Hocking, 1999), and for *S. cerevisiae* var. *diastaticus* and *Dekkera/Brettanomyces* spp. in beers (Andrews, 1992). Van Esch (1992) stated that a useful guideline as to the risk of spoilage of preserved carbonated drinks is the presence of more than 100 yeasts per litre, and that *Z. bailii/Z. bisporus* should not be detected in 10 g of fruit concentrate.

Concerning spoilage prediction, the work of Delle, made in Odessa, at the beginning of the twentieth century (cited by Amerine and Kunkee, 1965) is a remarkable exception to the few works in the field, although it was only developed for dessert wines. According to his work, the biological stability of wine is reached when the sum (Delle units) of the sugar content (% by weight) and six times the ethanol content (% by weight) must equal at least 78. Although with different values of Delle units to ensure stability, this approach was used in Californian musts fortified at various steps of fermentation by Amerine and Kunkee (1965), and in sweet and dry table wines by Whiteley (1979). More recently, a commercial package – Food MicroModel, version 2 (Anon, 1996) – presented several predictive models (mostly for pathogenic bacteria) including one for wines and another for acidic and low a_w ingredients. Unfortunately, predictive models are seldom used by industry, which still uses, as a rule, the classical determination of 'total viable counts'. It remains to be seen whether such an attitude is due to model deficiency or simply to industry resistance to changing routines.

Acceptable levels of yeasts in the food industry and in wholesalers/retailers
In a survey made by Andrews (1992) covering several Australian beer, wine and fruit juice industries, the listed acceptable levels for yeasts were relatively low. A similar inquiry was performed by us regarding wine companies and wholesalers/retailers in some countries (Loureiro and Malfeito-Ferreira, 2003b). The industry tends to play on the safe side and so very low levels of contamination are reported as acceptable, being frequently less than one cell per 100 ml as a maximum acceptable level, even if these low values have little statistical basis. As a rule, the quantification of contaminant flora is obtained after growth on general media and so results reflect the 'total' flora and not the spoiling one. When yeast levels are higher than acceptable, most food industries hold the product for enough time to meet specifications or to re-process the affected product. This procedure gives a picture of the contaminant flora because if counts increase, the food is likely to be contaminated with spoiling yeasts, but it takes time! Most industries that monitor yeast contamination of final product also monitor the efficiency of sanitation. Furthermore, some of their commercial clients (wholesalers/retailers) demand and monitor GMPs. Therefore, there are microbiological criteria to distinguish an acceptable product from an unacceptable one. Microbiological criteria for acceptable levels are also applied to sanitation and processing procedures, as enforced by HACCP and ISO 9000. In this way, the analysis of final product should be regarded as a final check of all production process and leads to low acceptable levels in foods.

To define microbiological criteria it is advisable to standardise the analytical procedures – sampling, sample volume, diluents, culture media and incubation conditions. Presently, the methodologies are somewhat variable, which makes comparison of results difficult. Commercial contracts established between producers and wholesalers also include microbiological criteria which are accepted by industry, or industry proposes an attainable specification which is accepted and checked by wholesalers. It would not be reasonable to establish commercial specifications more severe than those used in the food industries.

12.6 Future trends

Evolution in food zymology usually follows the advances acquired in medical microbiology and in food bacteriology which are more closely related to human health and, consequently, provide a wider market for the commercialisation of modern detection and identification techniques. It is also possible that the increasing availability of affordable rapid field techniques dealing with bioterrorism threats (Anon, 2003) may lead to subsequent adaptations to food microbiology, as happened before with the impetus given by the NASA space programme to the development of instrumental methods to detect microorganisms (Day, 1987). Most of the near-future trends listed by Fung (2002) for rapid detection and automation techniques are not applicable to industrial zymological control but may provide an orientation for today's research. If we compare the present state of available commercial tools to monitor yeasts in foods with Fung's ages of development in rapid and automated microbiology (1965–75: miniaturisation and diagnostic kits; 1975–85: immunological test kits; 1985–95: genetic probes, molecular testing systems, PCR applications; 1995 to date: biosensors, computer chip technology, microarray systems, proteomics, etc.), it is easy to conclude that much has to be done to fill the gap. In fact, the zymological scientific research has not yet been able to produce the knowledge needed to solve the questions raised by technologists concerning monitoring spoilage yeasts. In our opinion, future efforts should be aimed at the following:

- Establishing analytical procedures, after development of a set of culture media, to fully characterise foodborne yeast microbiota according to its technological significance and for each type of food commodity.
- Improving diagnostic kits for identification/typing targeting food spoilage yeasts, such as PCR-based methods, DNA chips or biosensors addressed to virulent strains (resistant to preservatives or other stress conditions).
- Adjusting the current rapid instrumental methods of yeast detection and typing to the needs of small and medium enterprises (SMEs), which increasingly need to comply with HACCP requirements, mainly by reducing equipment prices and skilled labour requirements.
- Making analytical detection and enumeration procedures consistent in order to establish standardised zymological indicators.
- Quantifying food susceptibility to spoilage yeast colonisation in order to assess processing spoilage risks and to evaluate the effect of spoilage yeasts on product shelf-life.

The problem of technology transfer from research to industry also depends on technologists who, besides being aware of current findings, should demand specific tools to monitor spoilage yeasts in order to interest diagnostic companies in their commercialisation.

12.7 Sources of further information

The books edited by Skinner *et al.* (1980), Rose and Harrison (1993), Deak and Beuchat (1996), Pitt and Hocking (1999) and Boekhout and Robert (2003) include chapters covering most relevant aspects of food contamination and spoilage by yeasts. Microbiological methods are described in the monographs edited by King *et al.* (1986) and Samson *et al.* (1992) covering the efforts of many research groups in the development of appropriate techniques for yeast detection and enumeration. Rapid methods and automation methods were thoroughly reviewed by Fung (2002) and this work is freely available at the website of the Institute of Food Technologists (http//www.ift.org). Yeast taxonomic and identification treatises used at present are those of Kurtzman and Fell (1998) and Barnett *et al.* (2000). In addition, Boekhout *et al.* (2002) edited a CD-Rom describing the characteristics of 768 species of yeasts. Most of the yeast research groups assemble annually in the International Specialised Symposium on Yeasts which is in its 23rd edition for 2003. A newsletter edited by A. Lachance under the auspices of the International Commission on Yeasts of the International Union of Microbiological Societies (IUMS) summarises 'in real time' the research produced in this field. Major professional and regulatory bodies include institutions concerned with specific industries like brewing (European Brewery Convention, American Society of Brewing Chemists), dairy (International Dairy Federation) and wine (Office International de la Vigne et du Vin), and with the standardisation of analytical methods (see footnote to Table 12.2).

12.8 Acknowledgements

We are indebted to the companies answering our inquiries and to A. Barata for preparing Fig. 12.1.

12.9 References

ALGUACIL, M., FIDALGO, M., JIMÉNEZ, J., LOZANO, J.I., NEVA, M.A. and PERDIGONES, F. (1998), Detección de *Brettanomyces/Dekkera* en instalaciones de vendimia mediante PCR. *Alimentacion, Equipos y Tecnología* 10, 81–85.

AMERINE, M.A. and KUNKEE, R.E. (1965), Yeast stability tests on dessert wines. *Vitis* 5, 187–194.

ANDREWS, D. (1982), A note on a microcolony method for the rapid detection of yeasts using Euchrysine 2GNX. *J. Appl. Bacteriol.* 52, 117–118.

ANDREWS, S. (1992), Specifications for yeasts in Australian beer, wine and fruit juice products. In: Samson, R.A., Hocking, A.D., Pitt, J.I. and King, A.D. (eds), *Modern Methods in Food Mycology*. Amsterdam, Elsevier, 111–118.

ANDREWS, S., DE GRAAF, H. and STAMATIN, H. (1997), Optimisation of methodology for enumeration of xerophilic yeasts. *Int. J. Food Microbiol.* 35, 109–116.

ANDRIGHETTO, C., PSOMAS, E., TZANETAKIS, N., SUZZI, G. and LOMBARDI, A. (2000), Randomly amplified polymorphic DNA (RAPD) PCR for the identification of yeasts isolated from dairy products. *Lett. Appl. Microbiol.* 30, 5–9.

ANON. (1982), *Microorganisms in Foods 2, Sampling for Microbiological Analysis: Principles and Applications*. International Commission on Microbiological Specification of Foods (ICMSF) of the International Association of Microbiological Societies. Toronto, University of Toronto Press.

ANON. (1996), *Food MicroModel Software* (version 2). Leatherhead, UK, Food MicroModel Ltd.

ANON. (2003), *Biodetection Technologies, Advances and Applications in Detecting Biological Agents*. Abstracts of the Knowledge Foundation's 3rd International Symposium, 1–3 June, Arlington, VA, USA.

ARIAS, C.R., BURNS, J.K., FRIEDRICH, L.M., GOODRICH, R.M. and PARISH, M.E. (2002), Yeast species associated with orange Juice: Evaluation of different identification methods. *Appl. Environm. Microbiol.* 68, 1955–1961.

AUGUSTYN, O., KOCK, J. and FERREIRA, D. (1992), Differentiation between yeast species, and strains within a species, by cellular fatty acid analysis 5. A feasible technique? *System. Appl. Microbiol.* 15, 105–115.

BALEIRAS-COUTO, M.M., VOGELS, J.T.W.E., HOFSTRA, H., HUIS IN'T VELD, J.H.J.. and VAN DER VOSSEN, J.M.B.M. (1995), Random amplified polymorphic DNA and restriction enzyme analysis of PCR amplified rDNA in taxonomy: two identification techniques for food-borne yeasts. *J. Appl. Bacteriol.* 79, 525–535.

BALEIRAS-COUTO, M.M., HARTOG, B.J., HUIS IN'T VELD, J.H.J., HOFSTRA, H. and VAN DER VOSSEN, J.M.B.M. (1996), Identification of spoilage yeasts in a food-production chain by microsatellite polymerase chain reaction fingerprinting. *Food Microbiol.* 13, 59–67.

BARNETT, J., PAYNE, R. and YARROW, D. (2000), *Yeasts: Characteristics and Identification*, 3rd edition. Cambridge, Cambridge University Press.

BARROS-LOPES, M., RAINIERI, S., HENSCHKE, P.A. and LANDRIDGE, P. (1999), AFLP fingerprinting for analysis of yeast genetic variation. *Int. J. System. Bacteriol.* 49, 915–924.

BAUTERS, T., SWINNE, D., STOVE, V. and NELIS, H. (2003), Detection of single cells of *Cryptococcus neoformans* in clinic samples by solid-phase cytometry. *J. Clin. Microbiol.* 41, 1736–1737.

BETTS, R. and BANKS, P. (1988), *An evaluation of the Autotrak – a rapid method for the enumeration of microorganisms*. Technical Memorandum No. 491. The Campden Food and Drink Research Association, Chipping Campden, England.

BEUCHAT, L.R. (1993), Selective media for detecting and enumerating foodborne yeasts. *Int. J. Food Microbiol.* 19, 1–14.

BEUCHAT, L.R., NAIL, B.V., BRACKETT, R.E. and FOX, T.L. (1990), Evaluation of a Culture Film (Petrifilm TM YM) Method for enumerating yeasts and molds in selected dairy and high-acid foods. *J. Food Protect.* 53, 864, 869–874.

BEUCHAT, L.R., COPELAND, F., CURIALE, M.S., DANISAVICH, T., GANGAR, V., KING, B.W., LAWLIS, T.L., LIKIN, R.O., OKWUSOA, J., SMITH, C.F. and TOWNSEND, D.E. (1998), Comparison of the SimPlate total plate count method with Petrifilm, Redigel, and conventional pour-plate methods for enumerating aerobic microorganisms in foods. *J. Food Prot.* 61, 14–18.

BEUCHAT, L., FRANDBERG, E., DEAK, T., ALZAMORA, S., CHEN, J., GUERRERO, S., LÓPEZ-MALO, A.,

OHLSSON, I., OLSEN, M., PEINADO, J., SCHNURER, J., SILONIZ, M. and TORNAI-LEHOCZKI (2001), Performance of mycological media in enumerating dessicated food spoilage yeasts: an interlaboratory study. *Int. J. Food Microbiol.* 70, 89–96.

BEUCHAT, L.R., SCOUTEN, A.J. and JABLONSKA, J. (2002), Influence of composition of diluent on populations of yeasts and moulds recovered from raw fruits. *Lett. Appl. Microbiol.* 35, 399–402.

BOEKHOUT, T. and ROBERT, V. (eds) (2003), *Yeasts in Food, Beneficial and Detrimental Aspects*. Hamburg, Behr's Verlag.

BOEKHOUT, T., ROBERT, V., SMITH, M., STALPERS, J., YARROW, D., BOER, P., GIJSWIJT, G., KURTZMAN, C., FELL, J., GUÉHO, E., GUILLOT, J. and ROBERTS, I. (2002), *Yeasts of the World – Morphology, physiology, sequences and identification*. CD-ROM from ETI Information Services Ltd, Wokingham, UK.

BOER, E. and BEUMER, R. (1999), Methodology for detection and typing of foodborne microorganisms. *Int. J. Food Microbiol.* 50, 119–130.

BOUIX, M., GRABOWSKI, A., CHARPENTIER, M., LEVEAU, J.Y. and DUTEURTRE, B. (1999), Évaluation rapide de la contamination des moûts par cytrométrie en flux. *J. Int. Sci. Vigne Vin* 33, 25–32.

BRAENDLIN, N. (1996), Enumeration of xerophilic yeasts in the presence of xerophilic moulds: a collaborative study. *Int. J. Food Microbiol.* 29, 185–192.

BUYER, J. (2003), Improved fast gas chromatography for FAME analysis of bacteria. *J. Microbiol. Meth.* 54, 117–120.

CAMPBELL, I. (1987), Wild yeasts in brewing and distilling. In: Priest, F. and Campbell, I. (eds), *Brewing Microbiology*. London, Elsevier Applied Science, 187–206.

CARREIRA, A. and LOUREIRO, V. (1998), A differential medium to detect *Yarrowia lipolytica* in 24 hours. *J. Food Mycol.* 1, 3–12.

CARUSO M., CAPECE, A., SALZANO, G. and ROMANO, P. (2002), Typing of *Saccharomyces cerevisiae* and *Kloeckera apiculata* strains from Aglianico wine. *Lett. Appl. Microbiol.* 34, 323–328.

CASAS, E., VALDERRAMA, M. and PEINADO, J. (1999), Sorbate detoxification by spoilage yeasts isolated from marzipan products. *Food Technol. Biotechnol.* 27, 87–91.

CAVA, R. and HERNÁNDEZ, P. (1994), Comparison of media for enumerating osmotolerant yeasts in orange juice concentrates. *Int. J. Food Microbiol.* 22, 291–295.

CHASKES, S. and TYNDALL, R.L. (1975), Pigment production by *Cryptococcus neoformans* from *para-* and *ortho-*diphenols: effect of the nitrogen source. *J. Clin. Microbiol.* 1, 509–514.

CHAURAND, P. and CAPRIOLI, R. (2002), Direct profiling and imaging of peptides and proteins from mammalian cells and tissue sections by mass spectrometry. *Electrophoresis* 23, 3125–3135.

CLAYDON, M., DAVEY, S., EDWARDS-JONES, V. and GORDON, D. (1996), The rapid identification of intact microorganisms using mass spectrometry. *Nat. Biotechnol.* 14, 1584–1586.

COCOLIN, L., BISSON, L.F. and MILLS, D.A. (2000), Direct profiling of the yeast dynamics in wine fermentations. *FEMS Microbiol. Lett.* 189, 81–87.

COCOLIN, L., HEISEI, A. and MILLS, D. A. (2001), Direct identification of the indigenous yeasts in commercial wine fermentations. *Am. J. Enol. Vitic.* 52, 49–53.

COCOLIN, L., MANZANO, M., REBECCA, S. and COMI, G. (2002), Monitoring of yeast population changes during a continuous wine fermentation by molecular methods. *Am. J. Enol. Vitic.* 53, 24–27.

COLQUHOUN, K., TIMMS, S. and FRICKER, C. (1998), A simple method for the comparison of commercially available ATP hygiene-monitoring systems. *J. Food Prot.* 61, 499–501.

COMI, G., MAIFRENI, M., MANZANO, M., LAGAZIO, C. and COCOLIN, L. (2000), Mitochondrial DNA restriction enzyme analysis and evaluation of the enological characteristics of *Saccharomyces cerevisiae* strains isolated from grapes of the wine-producing area of Collio (Italy). *Int. J. Food Microbiol.* 58, 117–121.

CONNELL, L., STENDER, H. and EDWARDS, C.G. (2002), Rapid detection and identification of *Brettanomyces* from winery air samples based on peptide nucleic acid analysis. *Am. J. Enol. Vitic.* 53, 322–324.

CONSTANTÍ, M., REGUANT, C., POBLET, M., ZAMORA, F. and MAS, A. (1998), Molecular population dynamics: effect of sulphur dioxide and inoculum on must fermentation. *Int. J. Food Microbiol.* 41, 169–175.

COUTO, J., GRA, A., SOARES-FRANCO, J. and HOGG, T. (1999), A novel fluorochromic system for the rapid detection and estimation of wine lactic acid bacteria by DEFT application to quality control in a bottling line. *Proceedings of the 6th International Oenology Symposium in Bordeaux*. Paris, Tec and Doc Editions, 407–410.

DAVENPORT, R.R. (1980), An outline guide to media and methods for studying yeasts and yeast-like organisms. In: Skinner, F., Passmore, S. and Davenport, R.R. (eds), *Biology and Activities of Yeasts*. London, Academic Press, 261–278.

DAVENPORT, R.R. (1986), Unacceptable levels for yeasts. In: King, A., Pitt, J. Beuchat, L. and Corry, J. (eds), *Methods for the Mycological Examination of Food*. New York, Plenum Press, 214–215.

DAY, A. (1987), Rapid detection of microbial spoilage. In: Priest, F. and Campbell, I. (eds), *Brewing Microbiology*. London, Elsevier Applied Science, 207–232.

DEAK, T. (2003), Detection, enumeration and isolation of yeasts. In: Boekhout, T. and Robert, V. (eds), *Yeasts in Food, Beneficial and Detrimental Aspects*. Hamburg, Behr's Verlag, 39–68.

DEAK, T. and BEUCHAT, L.R. (1993a), Evaluation of the indirect conductance method for the detection of yeasts in laboratory media and apple juice. *Food Microbiol.* 10, 255–262.

DEAK, T. and BEUCHAT, L.R. (1993b), Comparison of the SIM, API 20C, and ID 32C systems for identification of yeasts isolated from fruit juice concentrates and beverages. *J. Food Prot.* 56, 585–592.

DEAK, T. and BEUCHAT, L.R. (1994), Use of indirect conductimetry to predict the growth of spoilage yeasts, with special consideration of *Zygosaccharomyces bailii*, *Int. J. Food Microbiol.* 23, 405–417.

DEAK, T. and BEUCHAT, L.R. (1995), Modified indirect conductimetric technique for detecting low populations of yeasts in beverage concentrates and carbonated beverages. *Food Microbiol.* 12, 165–172.

DEAK, T. and BEUCHAT, L.R. (1996), *Handbook of Food Spoilage Yeasts*. Boca Raton, FL, USA, CRC Press.

DEAK, T., CHEN, J., GOLDEN, D., TAPIA, M., TORNAI-LEHOCZKI, J., VILJOEN, B., WYDER, M. and BEUCHAT, L. (2001), Comparison of dichloran 18% (DG18) agar with general purpose mycological media for enumerating food spoilage yeasts. *Int. J. Food Microbiol.* 67, 49–53.

D'HAESE, E. and NELIS, H. (2002), Rapid detection of single cell bacteria as a novel approach in microbiology. *J. AOAC Int.* 85, 979–983.

D'HAESE, E., NELIS, H. and REYBROECK, W. (2001), Determination of somatic cells in milk by solid phase cytometry. *J. Dairy Res.* 68, 9–14.

DIAS, L., DIAS, S., SANCHO, T., STENDER, H., QUEROL, A., MALFEITO-FERREIRA, M. and LOUREIRO, V. (2003), Identification of yeasts originated from wine related environments and capable of producing 4–ethylphenol. *Food Microbiol.*, in press.

DIRIYE, F., SCORZETTI, G. and MARTINI, A. (1993), Methods for the separation of yeast cells from the surfaces of processed, frozen foods. *Int. J. Food Microbiol.* 19, 27–37.

DLAUCHY, D., TORNAI-LEHOCZLI, J. and PÉTER, G. (1999), Restriction enzyme analysis of PCR amplified rDNA as a taxonomic tool in yeast identification. *System. Appl. Microbiol.* 22, 445–453.

DODD, C. and WAITES, W. (1991), The use of toluidine blue for *in situ* – detection of micro-organisms in foods. *Lett. Appl. Microbiol.* 13, 220–223.

DODD, C. and WAITES, W. (1992), Development of microscopic techniques for examining microbial growth *in situ* in food ecosystems. *J. Appl. Bacteriol.* (Symp. Suppl.) 73, 148S-154S.

DUARTE, F., PAIS, C., SPENCER-MARTINS, I. and LEÃO, C. (1999), Distinctive electrophoretic isoenzyme profiles in *Saccharomyces sensu stricto*. *Int. J. System. Bacteriol.* 49, 1907–1913.

DZIEZAK, J.D. (1987), Rapid methods for microbiological analysis of food. *Food Technol.* 7, 56–73.

EGLI, C.M. and HENICK-KLING, T. (2001), Identification of *Brettanomyces/Dekkera* species based on polymorphism in the rRNA internal transcribed spacer region. *Am. J. Enol. Vitic.* 52, 241–247.

EGLI, C.M., EDINGER, W., MITRAKUL, C. and HENICK-KLING, T. (1998), Dynamics of indigenous and inoculated yeast populations and their effect on the sensory character of Riesling and Chardonnay wines. *J. Appl. Microbiol.* 85, 779–789.

ELLIS, D., BROADHURST, D., KELL, D., ROWLAND, J. and GOODACRE, R. (2002), Rapid and quantitative detection of the microbial spoilage of meat by Fourier transform infrared spectroscopy and machine learning. *Appl. Environ. Microbiol.* 68, 2822–2828.

EPIFANIO S., GUTIERREZ, A. R., SANTAMARÍA, M. P. and LÓPEZ, R. (1999), The influence of enological practices on the selection of wild yeast strains in spontaneous fermentation. *Am. J. Enol. Vitic.* 50, 219–224.

ERICKSON, J. (1993), Hydrophobic membrane filtration method for the selective recovery and differentiation of *Z. bailii* in acidified ingredients. *J. Food Prot.* 563, 234–238.

ESTEVE-ZARZOSO, B., BELLOCH, C., URUBURU, F. and QUEROL, A. (1999), Identification of yeasts by RFLP analysis of the 5.8S rRNA gene and the two ribosomal internal transcribed spacers. *Int. J. Syst. Bacteriol.* 49, 329–337.

ESTEVE-ZARZOSO, B., GOSTÍNCAR, A., BOBET, R., URUBURU, F. and QUEROL, A. (2000), Selection and molecular characterization of wine yeasts isolated from the 'El Penedès' area (Spain). *Food Microbiol.* 17, 553–562.

EVANS, H.A.V. (1982), A note on two uses for impedimetry in brewing microbiology, *J. Appl. Bacteriol.* 53, 423–426.

EVANS, H.A.V. (1985), New approaches to the detection and evaluation of beer spoilage micro-organisms. *Proceedings of the 20th Congress of the European Brewery Convention*. Helsinki, 451–458.

EVANSON, D., CLAYDON, M. and GORDON, D. (2001), Exploring the limits of bacterial identification by intact cell-mass spectrometry. *J. Am. Soc. Mass Spectrom.* 12, 49–54.

FERNÁNDEZ, M., ÚBEDA, J.F. and BRIONES, A. I. (2000), Typing of non-*Saccharomyces* yeasts with enzymatic activities of interest in wine-making. *Int. J. Food Microbiol.* 59, 29–36.

FERNÁNDEZ-ESPINAR, M.T., LÓPEZ, V., RAMÓN, D., BARTRA, E. and QUEROL, A. (2001), Study of the authenticity of commercial wine yeast strains by molecular techniques. *Int. J. Food Microbiol.* 70, 1–10.

FIRSTENBERG-EDEN, R., FOTI, D., MCDOUGAL, S. and BAKER, J. (2002), Optical instrument for the rapid detection of microorganisms in dairy products. *Int. Dairy J.* 12, 225–232.

FLEET, G.H. (1992), Spoilage Yeasts. *Crit. Rev. Biotechnol.* 12, 1–44.

FLEET, G.H. (1999), Microorganisms in food ecosystems. *Int. J. Food Microbiol.* 50, 101–117.

FLEISCHER, M., SHAPTON, N. and COOPER, P. (1984), Estimation of yeast numbers in fruit mix for yogurt. *J. Soc. Dairy Technol.* 37, 63–65.

FREYDIERE, A.-M., GUINET, R. and BOIRON, P. (2001), Yeast identification in the clinical microbiology laboratory: phenotypical methods. *Medic. Mycol.* 39, 9–33.

FUGELSANG, K. (1997), *Wine Microbiology*. New York, Chapman and Hall.

FUNG, D. (2002), Rapid methods and automation in Microbiology. *Comp. Revs. Food Sci. Food Safety* 1, 3–22.

FUNG, D. and LIANG, C. (1990), Critical review of isolation, detection and identification of yeasts from meat products. *Crit. Revs. Food Sci. Nutr.* 29, 341–379.

FUNG, D., THOMSON, L., CROZIER-DODSON, B. and KASTNER, C. (2000), Hands-free 'Pop-up' adhesive type method for microbial sampling of meat surfaces. *J. Rapid Method Automat. Microbiol.* 8, 209–217.

GARDINI, F., GUERZONI, M. and LERICI, C. (1988), CO_2 determination with the GC method for assessing shelf-life of fruit system in relation to a_w and storage temperature. *Lebensm.-Wiss. u.-Technol.* 21, 137–143.

GEYER, W., BRUEGGEMANN, L., FLEMMING, I. and NAGEL, B. (1991), Characterization of yeast strains of the genera *Candida*, *Hansenula*, *Kluyveromyces*, *Pichia*, *Rhodotorula*, and *Saccharomyces* by mixed-dye fluorometry. *Int. J. System. Bacteriol.* 41, 249–254.

GOLDSCHMIDT, M.C. and FUNG, D. (1978), New methods for microbiological analysis of food. *J. Food Prot.* 41, 201–219.

GRANCHI, L., BOSCO, M., MESSINI, A. and VINCENZINI, M. (1999), Rapid detection and quantification of yeast species during spontaneous wine fermentation by PCR-RFLP analysis of the rDNA ITS region. *J. Appl. Microbiol.* 87, 949–956.

GRAUMLICH, T.R. (1985), Estimation of microbial populations in orange juice by Bioluminescence. *J. Food Sci.* 50, 115–124.

GREEN, T., RUSSELL, S. and FLETCHER, D. (1999), Effect of chemical cleaning agents and commercial sanitizers on ATP bioluminescence measurements. *J. Food Protec.* 62, 86–90.

GRIFFITHS, M., PHILLIPS, J. and MUIR, D. (1984), *Rapid detection of post-pasteurization contamination*. Bulletin No. 10. Hannah Research Institute, Ayr, Scotland.

GROOTE, J., VROEGROP, L. and BRAILSFORD, M. (1995), A proactive tool for the control of yeast spoilage in fermented milk products. *Microbiol. Eur.* 3, 24–27.

GUERZONI, M., PIVA, M. and GARDINI, F. (1985), Proposal of a rapid HS-GLC method for microbiological control of food. *Lebensm.-Wiss. u.-Technol.* 18, 128–132.

GUILLAMÓN, J.M., BARRIO, E., HUERTA, T. and QUEROL, A. (1994), Rapid characterization of four species of the *Saccharomyces* sensu stricto complex according to mitochondrial DNA Patterns. *Int. J. System. Bacteriol.* 44, 708–714.

GUILLAMÓN, J.M., SABATÉ, J., BARRIO, E., CANO, J. and QUEROL, A. (1998), Rapid identification of wine yeasts species based on RFLP analysis of the ribosomal internal transcribed spacer (ITS) region. *Arch Microbiol.* 169, 387–392.

GUTIÉRREZ, A., EPIFANIO, S., GARIJO, P., LÓPEZ, R. and SANTAMARÍA, P. (2001), Killer yeasts: incidence in the ecology of spontaneous fermentation. *Am. J. Enol. Vitic.* 52, 352–356.

GUTTERIDGE, C. (1987), Methods for the rapid identification of micro-organisms. In Priest, F. and Campbell, I. (eds), *Brewing Microbiology.* London, Elsevier Applied Science, 233–268.

HARRIGAN, W. (1998), Laboratory methods in food microbiology. San Diego, CA, Academic Press.

HAZEN, K.C. (1995), New and emerging yeast pathogens. *Clin. Microbiol. Rev.* 8, 462–478.

HEARD, G.M. and FLEET, G.H. (1986), Evaluation of selective media for enumeration of yeasts during wine fermentation. *J. Appl. Bacteriol.* 60, 477–481.

HENSCHKE, P.A. and THOMAS, D.S. (1988), Detection of wine-spoiling yeasts by electronic methods. *J. Appl. Bacteriol.* 64, 123–133.

HERAS-VAZQUEZ, F., MINGORANCE-CAZORLA, L., CLEMENTE-JIMENEZ, J.M. and RODRIGUEZ-VICO, F. (2003), Identification of yeast species from orange fruit and juice by RFLP and sequence analysis of the 5.8S rRNA gene and the two internal transcribed spacers. *FEMS Yeast Res.* 3, 3–9.

HOCKING, A.D. (1996), Media for preservative resistant yeasts: a collaborative study. *Int. J. Food Microbiol.* 29, 167–175.

HOCKING, A. and PITT, J. (1980), Dichloran-glycerol medium for enumeration of xerophilic fungi from low moisture foods. *Appl. Environm. Microbiol.* 39, 488–492.

HUTTER, K.J., PUNESSEN, U. and EIPEL, H.E. (1979), Rapid determination of the purity of yeast cultures by immunofluorescence and flow cytometry. *J. Inst. Brew.* 85, 21–22.

IBEAS, J.I., LOZANO, I., PERDIGONES, F. and JIMÉNEZ, J. (1996), Detection of *Dekkera-Brettanomyces* strains in sherry by a nested PCR method. *Appl. Environm. Microbiol.* 62, 998–1003.

JARVIS, B. (2000), Sampling for microbiological analysis. In: Lund, B. Baird-Parker, T. and Gould, G. (eds), *The Microbiological Safety and Quality of Food.* Vol. II. Gaithersburg, MD, USA, Aspen Publishers, 1691–1733.

JESPERSEN, L., LARSEN, S. and JAKOBSEN, M. (1993), Flow cytometric detection of wild yeast in lager breweries. *Int. J. Food Microbiol.* 17, 321–328.

KATSURAGI, T., TANAKA, S., NAGAHIRO, S. and TANI, Y. (2000), Gel microdroplet technique leaving microorganisms alive for sorting by flow cytometry. *J. Microbiol. Methods* 42, 81–86.

KING, A.D., PITT, J.I., BEUCHAT, L.R. and CORRY, J.E. (eds) (1986), *Methods for the Mycological Examination of Food.* New York, Plenum Press.

KISH, S., SHARF, R. and MARGALITH, P. (1983), A note on a selective medium for wine yeasts. *J. Appl. Bacteriol.* 55, 177–179.

KOCH, H.A., BANDLER, R. and GIBSON, R. (1986), Fluorescence microscopy procedure for quantitation of yeasts in beverages. *Appl. Environm. Microbiol.* 52, 599–601.

KUHLE, A. and JESPERSEN, L. (1998), Detection and identification of wild yeasts in lager breweries. *Int. J. Food Microbiol.* 43, 205–213.

KUNIYUKI, A.H., ROUS, C. and SANSERSON, J.L. (1984), Enzyme-linked immunosorbent assay (ELISA) Detection of *Brettanomyces* contaminants in wine production. *Am. J. Enol. Vitic.* 35, 143–145.

KUNKEE, R. and NERADT, F. (1974), A rapid method for detection of viable yeast in bottled wines. *Wines & Vines* 12, 36–39.

KURTZMAN, C.P. and FELL, J.W. (eds) (1998), *The Yeasts, A Taxonomic Study*. 4th edition. Amsterdam, Elsevier.

KURTZMAN, C.P., BOEKHOUT, T., ROBERT, V., FELL, J. and DEAK, T. (2003), Methods to identify yeasts. In: Boekhout, T. and Robert, V. (eds), *Yeasts in Food, Beneficial and Detrimental Aspects*. Hamburg, Behr's Verlag, 69–121.

LAND, G.A., SALKIN, I.F., EL-ZAATARI, M., MCGINNIS, M.R. and HASHEM, G. (1991), Evaluation of the Baxter-MicroScan 4–hour enzyme based yeast identification system. *J. Clinic. Microbiol.* 29, 718–722.

LAWRENCE, D.R. (1983), Yeast differentiation and identification. *Proceedings of the 19th Congress of the European Brewery Convention*. London, 449–456.

LEE, J. and FUNG, D. (1986), Surface sampling technique for bacteriology. *J. Environm. Health* 48, 200–205.

LEGRAND, M. and RAMETTE, P. (1986), La technique de l'immunofluorescence appliquée a la détection rapide des contaminants dans les levains de brasserie. *Cerevisia* 4, 173–183.

LIECKFELDT, E., MEYER, W. and BÖERNER, T. (1993), Rapid identification and differentiation of yeasts by DNA and PCR fingerprinting. *J. Basic Microbiol.* 33, 413–426.

LITTEL, K.J. and LAROCCO, K.A. (1985), Bioluminescent standard curves for quantitive determination of yeast contaminants in carbonated beverages. *J. Food Prot.* 48, 1022–1024.

LITTEL, K.J. and LAROCCO, K.A. (1986), ATP screening method for presumptive detection of microbiologically contaminated carbonated beverages. *J. Food Sci.* 51, 474–476.

LÓPEZ, V., QUEROL, A., RAMÓN, D. and FERNÁNDEZ-ESPINAR, M. T. (2001), A simplified procedure to analyse mitochondrial DNA from industrial yeasts. *Int. J. Food Microbiol.* 68, 75–81.

LÓPEZ, V., FERNÁNDEZ-ESPINAR, M. T., BARRIO, E., RAMÓN, D. and QUEROL, A. (2002), A new PCR-based method for monitoring inoculated wine fermentations. *Int. J. Food Microbiol.* 81, 63–71.

LOUREIRO, V. (2000), Spoilage yeasts in food and beverages: characterisation and ecology for improved diagnosis and control. *Food Res. Int.* 33, 247–256.

LOUREIRO, V. and MALFEITO-FERREIRA, M. (2003a), Yeasts in spoilage. In: Caballero, B. Trugo, L. and Finglas. P. (eds), *Encyclopedia of Food Sciences and Nutrition* (2nd edition). London, Academic Press, 5530–5537.

LOUREIRO, V. and MALFEITO-FERREIRA, M. (2003b), Spoilage yeasts in the wine industry. *Int. J. Food Microbiol.*, in press.

LOUREIRO, V. and QUEROL, A. (1999), The prevalence and control of spoilage yeasts in foods and beverages. *Trends Food Sci. Technol.* 10–11, 356–365.

MAKDESI, A.K. and BEUCHAT, L.R. (1996a), Evaluation of media enumerating heat-stressed, benzoate-resistant *Zygosaccharomyces bailii*. *Int. J. Food Microbiol.* 33, 169–181.

MAKDESI, A.K. and BEUCHAT, L.R. (1996b), Performance of selective media for enumerating *Zygosaccharomyces bailii* in acidic foods and beverages. *J. Food Prot.* 59, 652–656.

MALACRINO, P., ZAPPAROLI, G., TORRIANI, S. and DELLAGLIO, F. (2001), Rapid detection of viable yeasts and bacteria in wine by flow cytometry. *J. Microbiol. Meth.* 45, 127–134.

MALFEITO-FERREIRA, M., LOPES, J. and LOUREIRO, V. (1989a), Infecting yeasts in Portuguese bottled white-wines. *Proceedings of the XIIIth International Symposium on Yeasts.*

Leuven, Belgium.

MALFEITO-FERREIRA, M., ST AUBYN, A. and LOUREIRO, V. (1989b), The long-chain fatty acid composition as a tool for differentiating spoilage wine yeasts. *Mycotaxon* 36, 35–42.

MALFEITO-FERREIRA, M., TARECO, M. and LOUREIRO, V. (1997), Fatty-acid profiling: a feasible typing system to trace yeast contamination in bottling plants. *Int. J. Food Microbiol.* 38, 143–155.

MARTIN, C. and SIEBERT, K. (1992), Evaluation of multi-nitrogen source media for wild yeast detection in brewing culture yeast. *J. Amer. Soc. Brew. Chem.* 50, 134–138.

MARTINI, A., FEDERICI, F. and ROSINI, G. (1980), A new approach to the study of yeast ecology of natural substrates. *Can. J. Microbiol.* 26, 856–859.

MARTINI, A., CIANI, M. and SCORZETTI, G. (1996), Direct enumeration and isolation of wine yeasts from grape surfaces. *Am. J. Enol. Vitic.* 47, 435–440.

MEIDELL, J. (1987), Unsuitability of fluorescence microscopy for the rapid detection of small numbers of yeast cells on a membrane filter. *Am. J. Enol. Vitic.* 38, 159–160.

MIAN, M.A., FLEET, G.H. and HOCKING, D. (1997), Effect of diluent type on viability of yeasts enumerated from foods or pure culture. *Int. J. Food Microbiol.* 35, 103–107.

MIDDELHOVEN, W.J. (2002), Identification of yeasts present in sour fermented foods and fodders. *Molec. Biotechnol.* 21, 279–292.

MILLER, R. and GALSTON, G. (1989), Rapid methods for the detection of yeast and *Lactobacillus* by ATP bioluminescence. *J. Inst. Brew.* 95, 317–319.

MILLET, V. and LONVAUD-FUNEL, A. (2000), The viable but non-culturable state of wine micro-organisms during storage. *Lett. Appl. Microbiol.* 30, 136–141.

MILLS, D.A., ERIC, A.J. and COCOLIN, L. (2002), Yeast diversity and persistence in Botrytis-affected wine fermentations. *Appl. Environm. Microbiol.* 68, 4884–4893.

MITRAKUL, C. M., HENICK-KLING, T. and EGLI, C. M. (1999), Discrimination of *Brettanomyces/Dekkera* yeast isolates from wine by using various DNA fingerprinting methods. *Food Microbiol.* 16, 3–14.

MIZUOCHI, S. and KODAKA, H. (2000), Evaluation of dry sheet medium culture plate (Compactdry) method for determining numbers of bacteria in food samples. *J. Food Prot.* 63, 665–667.

MOLINA, F.I., SHEN, P. and JONG, S. (1993), Validation of the species concept in the genus *Dekkera* by restriction analysis of genes coding for rRNA. *Int. J. Syst. Bacteriol.* 43, 32–35.

MOREIRA-DA-SILVA, M., MALFEITO-FERREIRA, M., ST AUBYN, A. and LOUREIRO, V. (1994), Long-chain fatty acid composition as a criterion for yeast distinction in the brewing industry. *J. Inst. Brew.* 100, 17–22.

MOSSEL, D.A. and STRUIJK, C.B. (1992), The contribution of microbial ecology to management and monitoring of the safety, quality and acceptability (SQA) of foods. *J. Appl. Bacteriol.* (Symp. Suppl.) 73, 1S–22S.

MOSSEL, D.A., CORRY, J.E., STRUIJK, C.B. and BAIRD, R.M. (1995), *Essentials of the Microbiology of Foods*. Chichester, John Wiley & Sons.

MURPHY, A. and KAVANAGH, K. (1999), Emergence of *Saccharomyces cerevisiae* as a human pathogen, implications for biotechnology. *Enz. Microbiol. Technol.* 25, 551–557.

NAVARRO, A., MORENO, P., CARBONELL, J.V. and PINAGA, F. (1987), Practical approaches to the use of epifluorescence microscopy as a rapid method for microbiological quality control in brewing. *Proceedings of the 21st Congress of the European Brewery Convention*. Madrid, 465–472.

NESS, F., LAVALÉE, F., DUBOURDIEU, D., AIGLE, M. and DULAU, L. (1993), Identification of yeast strains using the polymerase chain reaction. *J. Sci. Food Agric.* 62, 89–94.

NGUYEN, H., PULVIRENTI, A. and GAILLARDIN, C. (2000), Rapid differentiation of the closely related *Kluyveromyces lactis* and *Kluyveromyces marxianus* strains isolated from dairy products using selective media and PCR/RFLP of the rDNA non transcribed spacer2. *Can. J. Microbiol.* 46, 1115–1122.

OGDEN, K. (1994), Using ATP-bioluminescence to control brewery plant hygiene. *Proceedings of the 24th Congress of the European Brewery Convention*. Nutfield, 134–147.

PALLMAN, C.L., BROWN, J.A., OLINEKA, T. L., COCOLIN, L., MILLS, D.A. and BISSON, L. (2001), Use of WL medium to profile native flora fermentations. *Am. J. Enol. Vitic.* 52, 198–203.

PARKER, M.J. (1989), The application of automated detection and enumeration of microcolonies using optical brighteners and image analysis to brewery microbiological control. *Proceedings of the 22nd Congress of the European Brewery Convention*. Zurich, 545–552.

PATON, A.M. and JONES, S.M. (1975), The observation and enumeration of micro-organisms in fluids using membrane filtration and incident fluorescence microscopy. *J. Appl. Bacteriol.* 38, 199–200.

PELADAN, F. and LEITZ, R. (1991), New method for the differential staining of dead and living cells of yeasts and bacteria. *Proceedings of the 23rd Congress of the European Brewery Convention*. Lisbon, 481–487.

PERRY-O'KEEFE, H, STENDER, H., BROOMER, A., OLIVEIRA, K., COULL, J. and HYLDIG-NIELSEN, J.J. (2000), Filter-based PNA in situ hybridization for rapid detection, identification and enumeration of specific micro-organisms. *J. Appl. Microbiol.* 90, 180–189.

PETTIPHER, G.L. (1987), Detection of low numbers of osmophilic yeasts in creme fondant within 25h using a pre-incubated DEFT count. *Lett. Appl. Microbiol.* 4, 95–98.

PETTIPHER, G.L. (1991), Preliminary evaluation of flow cytometry for the detection of yeasts in soft drinks. *Lett. Appl. Microbiol.* 12, 109–112.

PETTIPHER, G.L. and RODRIGUES, U.M. (1982), Rapid enumeration of microorganisms in foods by the direct epifluorescent filter technique. *Appl. Environm. Microbiol.* 44, 809–813.

PETTIPHER, G.L., MANSELL, R., MCKINNON, C. and COUSINS, C. (1980), Rapid membrane filtration-epifluorescent microscopy technique for direct enumeration of bacteria in raw milk. *Appl. Environm. Microbiol.* 39, 423–429.

PITT, J.I. and HOCKING, A.D. (1985), *Fungi and Food Spoilage*. Sydney, Academic Press.

PITT, J.I. and HOCKING, A.D. (1999), *Fungi and Food Spoilage*, 2nd edition. Gaithersburg, MD, USA, Aspen Publishers.

PLATA, C., MILLÁN, C., MAURICIO, J. C. and ORTEGA, J. M. (2003), Formation of ethyl acetate and isoamyl acetate by various species of wine yeasts. *Food Microbiol.* 20, 217–224.

PORTER, J. and PICKUP, R. (2000), Nucleic acid-based fluorescent probes in microbial ecology: application of flow cytometry. *J. Microbiol. Meth.* 42, 75–79.

POULIS, J.A., DE PIJPER, M., MOSSEL, D.A. and DEKKERS, P. (1993), Assessment of cleaning and disinfection in the food industry with the rapid ATP-bioluminescence technique combined with the tissue fluid contamination test and a conventional microbiological method. *Int. J. Food Microbiol.* 20, 109–116.

PRETORIUS, I., VAN DER WESTHUIZEN, T. and AUGUSTYN, O. (1999), Yeast biodiversity in vineyards and wineries and its importance to the South African wine industry: a review. *S. Afr. J. Enol. Vitic.* 20, 61–74.

PRILLINGER, H., MONÁR, O., ELISKASES-LECHNER, F and LOPANDIC, K. (1999), Phenotypic and genotypic identification of yeasts from cheese. *Antonie van Leeuwenhoek* 75, 267–283.

PRUDÊNCIO, C., SANSONETTY, F. and CÔRTE-REAL, M. (1998), Flow cytometric assessment of cell structural and functional changes induced by acetic acid in the yeasts *Zygosaccharomyces bailii* and *Saccharomyces cerevisiae*. *Cytomet.* 31, 307–313.

QUEROL, A. and RAMON, P. (1996), The application of molecular techniques in wine microbiology. *Trends Food Sci. Technol.* 7, 73–78.

QUEROL, A., BARRIO, E., HUERTA, T. and RAMÓN, D. (1992a), Molecular monitoring of wine fermentations conduced by active dry yeast strains. *Appl. Environm. Microbiol.* 58, 2948–2953.

QUEROL, A., BARRIO, E. and RAMÓN, E. (1992b), A comparative study of different methods of yeast strains of *Saccharomyces cerevisiae* characterization. *Syst. Appl. Microbiol.* 58, 2948–2953.

QUEROL, A., FERNÁNDEZ-ESPINAR, M.T. and BARRIO, E. (2003), Genetic variability of Yeast in Wine Fermentation. In: Aroa D.K. (ed.), Bridge, P.D. and Bhatnagar, D. (assoc. eds), *Handbook of Fungal Biotechnology*, 2nd edition, ch. 38. In press.

QUESADA, M.P. and CENIS, J.L. (1995), Use of random amplified polymorphic DNA (RAPD-PCR) in the characterization of wine yeasts. *Am. J. Enol. Vitic.* 46, 204–208.

RALE, V.B. and VAKIL, J.R. (1984), A note on an improved molybdate agar for the selective isolation of yeasts from tropical fruits. *J. Appl. Bacteriol.* 56, 409–413.

RAMOS, J.P., VALENTE, P., HAGLER, A.N. and LEONCINI, O. (1998), Restriction analysis of the ITS region for characterization of the *Debaryomyces* species. *J. Gen. Appl. Microbiol.* 44, 399–404.

REDZEPOVIC, S., ORLIC, S., SIKORA, S., MAJDAK, A. and PRETORIUS, I.S. (2002), Identification and characterization of *Saccharomyces cerevisiae* and *Saccharomyces paradoxus* strains isolated from Croation vineyards. *Lett. Appl. Microbiol.* 35, 1–6.

ROBERT, V. (2003), Data processing. In: Boekhout, T. and Robert, V. (eds), *Yeasts in Food, Beneficial and Detrimental Aspects*. Hamburg, Behr's Verlag, 139–169.

RODRIGUES, N., GONÇALVES, G., MALFEITO-FERREIRA, M. and LOUREIRO, V. (2001), Development and use of a differential medium to detect yeasts of the genera *Dekkera/Brettanomyces*. *Int. J. Food Microbiol.* 90, 588–599.

RODRIGUES, U.M. and KROLL, R.G. (1986), Use of the direct epifluorescent filter technique for the enumeration of yeasts. *J. Appl. Bacteriol.* 61, 139–144.

ROMANO, A., CASAREGOLA, S., TORRE, P. and GAILLARDIN, C. (1996), Use of RAPD and mitochondrial DNA RFLP for typing of *Candida zeylanoides* and *Debaromyces hansenii* yeast strains isolated from cheese. *System. Appl. Microbiol.* 19, 255–264.

ROSE, A.H. and HARRISON, J.S. (eds) (1993), *The Yeasts*, vol. 5, 2nd edition. London, Academic Press.

ROY, D. and MULARD, Y. (1996), Rapid microbiology using a laser flow cytometer. *Food Tech. Europe* 3/4, 36–40.

RUSSEL, S. and BAILEY, J. (2000), Plug'n play HACCP. *WATT Poultry USA* 6, 30–34.

ST-GERMAIN, G. and BEAUCHESNE, D. (1991), Evaluation of the MicroScan rapid yeast identification panel. *J. Clin. Microbiol.* 29, 2296–2299.

SALDANHA-DA-GAMA, A., MALFEITO-FERREIRA, M. and LOUREIRO, V. (1997), Characterization of yeasts associated with Portuguese pork-based products. *Int. J. Food Microbiol.* 37, 201–207.

SAMSON, R.A., HOCKING, A.D., PITT, J.I. and KING, A.D. (eds) (1992), *Modern Methods in Food Mycology*. London, Academic Press.

SANCHO, T., GIMÉNEZ-JURADO, G., MALFEITO-FERREIRA, M. and LOUREIRO, V. (2000), Zymological indicators: a new concept applied to the detection of potential spoilage yeast species associated with fruit pulps and concentrates. *Food Microbiol.* 17, 613–624.

SCHAERTEL, B.J., TSANG, N. and FIRSTENBERG-EDEN, R. (1987), Impedimetric detection of yeast and mold. *Food Microbiol.* 4, 155–163.

SCHULLER, D., CÔRTE-REAL, M. and LEÃO, C. (2000), A differential medium for the enumeration of the spoilage yeast *Zygosaccharomyces bailii* in Wine. *J. Food Protec.* 63, 1570–1575.

SCHÜTZ, M. and GAFNER, J. (1993), Analysis of yeast diversity during spontaneous and induced alcoholic fermentations. *J. Appl. Bacteriol.* 75, 551–558.

SHARPE, A.N. (1979), An alternative approach to food microbiology for the future. *Food Technol.* 33, 71–74.

SILÓNIZ, M.I., VALDERRAMA, M.J., PAYO, E. and PEINADO, J.M. (1999), Advances in the development of a methodology to identify common yeast contaminants of high sugar food products. *Food Technol. Biotechnol.* 37, 277–280.

SILÓNIZ, M.I, VALDERRAMA, M.J. and PEINADO, J.M. (2000), A chromogenic medium for the detection of yeasts with β-galactosidase and β-glucosidase activities from intermediate moisture foods. *J. Food Prot.* 63, 651–654.

SIMPSON, W.J., HAMMOND, J.R.M., THRUSTON, P.A. and KYRIAKIDES, A.L. (1989), Brewery process control and the role of 'instant' microbiological techniques. *Proceedings of the 22nd Congress of the European Brewery Convention*. Zürich, 663–674.

SKINNER, F., PASSMORE, S. and DAVENPORT, R. (eds) (1980), *Biology and Activities of Yeasts*. London, Academic Press.

STENDER, H., KURTZMAN, C., HYLDIG-NIELSEN, J., SORENSEN, D., BROOMER, A., OLIVEIRA, K., PERRY-O'KEEFE, H., SAGE, A., YOUNG, B. and COULL, J. (2001), Identification of *Dekkera bruxellensis* (*Brettanomyces*) from wine by fluorescence in situ hybridization using peptide nucleic acid probes. *Appl. Environ. Microbiol.* 67, 938–941.

STENDER, H., FIANDACA, M., HYLDIG-NIELSEN, J.J. and COULL, J. (2002), PNA for rapid microbiology. *J. Microbiol. Meth.* 48, 1–17.

TECHERA, A.G., JUBANY, S., CARRAU, F. M. and GAGGERO, C. (2001), Differentiation of industrial wine yeast strains using microsatellite markers. *Lett. Appl. Microbiol.* 33, 71–75.

THOMAS, D.S. (1993), Yeasts as Spoilage Organisms in Beverages. In: Rose A. H. and Harrison J. S. (eds), *The Yeasts*, vol. 5, 2nd edition. London, Academic Press, 517–561.

THOMAS, D.S. and ACKERMAN, J.C. (1988), A selective medium for detecting yeasts capable of spoiling wine. *J. Appl. Bacteriol.* 65, 299–308.

TILBURY, R. (1980), Xerotolerant (osmophilic) yeasts. In: Skinner, F., Passmore, S. and Davenport, R.R. (eds), *Biology and Activities of Yeasts*. London, Academic Press, 153–179.

TORIJA, M.J., ROZÈS, N., POBLET, M., GUILLAMÓN, J.M. and MAS, A. (2002), Effects of fermentation on the strain population of *Saccharomyces cerevisiae*. *Int. J. Food Microbiol.* 80, 47–53.

TUDOR, E.A. and BOARD, R.G. (1993), Food spoilage yeasts. In: Rose, A.H. and Harrison, J.S. (eds), *The Yeasts*, vol. 5, 2nd edition. London, Academic Press, 435–516.

UR, A. and BROWN, D. (1975), Impedance monitoring of bacterial activity. *J. Med. Microbiol.* 8, 19–28.

VALDERRAMA, M.J., SILÓNIZ, M.I., GONZALO, P. and PEINADO, J.M. (1999), A differential medium for the isolation of *Kluyveromyces marxianus* and *Kluyveromyces lactis* from dairy products. *J. Food Prot.* 62, 189–193.

VALENTE, P., GOUVEIA, F.C., LEMOS, G.A., PIMENTEL, D., VAN ELSAS, J.D., MENDONÇA-HAGLER, L.C. and HAGLER A.N. (1996), PCR amplification of the rDNA internal transcribed spacer region for differentiation of *Saccharomyces* cultures. *FEMS Microbiol. Lett.* 137, 253–256.

VALENTINE, N., WAHL, J., KINGSLEY, M. and WAHL, K. (2002), Direct surface analysis of fungal species by matrix-assisted laser desortion/ionization mass spectrometry. *Rapid Commun. Mass Spectrom.* 16, 1352–1357.

VAN DER VOSSEN, J.M., RAHAOUI, H., DE NIJS, M. and HARTOG, B. (2003), PCR methods for tracing and detection of yeasts in the food chain. In: Boekhout, T. and Robert, V. (eds), *Yeasts in Food, Beneficial and Detrimental Aspects*. Hamburg, Behr's Verlag, 123–138.

VAN DER WESTHUIZEN, T., AUGUSTYN, O., KHAN, W. and PRETORIUS, I.S. (2000a), Geographical distribution of indigenous *Saccharomyces cerevisiae* strains isolated from vineyard in the coastal region of the Western Cape in South Africa. *S. Afr. J. Enol. Vitic.* 21, 3–9.

VAN DER WESTHUIZEN, T., AUGUSTYN, O., KHAN, W. and PRETORIUS, I.S. (2000b), Seasonal variation of indigenous *Saccharomyces cerevisiae* strains isolated from vineyard of the Western Cape in South Africa. *S. Afr. J. Enol. Vitic.* 21, 10–15.

VAN ESCH, F. (1992), Yeast in soft drinks and concentrated fruit juices. *Brygmesteren* 4, 9–20.

VAN OEVELEN, D., SPAEPEN, M., TIMMERMANS, P. and VERACHTERT, H. (1977), Microbiological aspects of spontaneous wort fermentation in the production of Lambic and Gueuze. *J. Inst. Brew.* 83, 356–360.

VAN VUUREN, H.J.J. and VAN DER MEER, L.J. (1987), Fingerprinting of yeasts by protein electrophoresis. *Am. J. Enol. Vitic.* 38, 49–53.

VAN VUUREN, H.J.J. and VAN DER MEER, L.J. (1988), Characterization of brewing yeast strains by numerical analysis of total soluble cell protein patterns. *J. Inst. Brew.* 94, 245–248.

VAN ZANDYCKE, S., SIMAL, O., GUALDONI, S. and SMART, K.A. (2003), Determination of yeast viability using fluorophores. *J. Am. Soc. Brew. Chem.* 61, 15–22.

VAZ-OLIVEIRA, M., BARROS, P. and LOUREIRO, V. (1995), Analyse microbiologique du vin. Techniques des tubes multiples pour l'énumeration de microorganisms dans les vins. *Feuilles Verts O.I.V.* 987, Paris.

VEAL, D.A., DEERE, D., FERRARI, B., PIPER, J. and ATTFIELD, P.V. (2000), Fluorescence staining and flow cytometry for monitoring microbial cells. *J. Immunol. Meth.* 243, 191–210.

VELÁSQUEZ, E., CRUZ-SÁNCHEZ, J. M., RIVAS-PALÁ, T., ZURDO-PINEIRO, J. L., MATEOS, P. F., MONTE, E., MARTÍNEZ-MOLINA, E. and CHORDI, A. (2001), YeastIdent-Food/ ProleFood, a new system for the identification of food yeasts based on physiological and biochemical tests. *Food Microbiol.* 18, 637–646.

VEZINHET, F., HALLET, J., VALADE, M. and POULARD, A. (1992), Ecological survey of wine yeasts strains by molecular methods of identification. *Am. J. Enol. Vitic.* 43, 83–86.

WEIHE, J.L., SEIBT, S.L. and HATCHER JR, W.S. (1984), Estimation of microbial populations in frozen concentrated orange juice using automated impedance measurements. *J.*

Food Sci. 49, 243–245.

WHITELEY, L. (1979), *The Microbiological Spoilage of White Wines*. Second report to the Consortium of Wine Companies. Long Ashton, Bristol, UK.

WILLIAMS, G.B., WEAVER, J.C. and DEMAIN, A.L. (1990), Rapid microbial detection and enumeration using gel microdroplets and colorimetric or fluorescence indicator systems. *J. Clin. Microbiol.* 28, 1002–1008.

YAMAGUCHI, N., ISHIDOSHIRO, A., YOSHIDA, Y., SAIKA, T., SENDA, S. and NASU, M. (2003), Development of an adhesive sheet for direct counting of bacteria on solid surfaces. *J. Microbiol. Methods* 53, 405–410.

YAMAMOTO, N., AMEMIYA, H., YOKOMORI, Y., SHIMIZU, K. and TOTSUKA, A. (1991), Electrophoretic karyotypes of wine yeasts. *Am. J. Enol. Vitic.* 42, 358–363.

ZENGLER, K., TOLEDO, G., RAPPÉ, M., ELKINS, J., MATHUR, J., SHORT, J. and KELLER, M. (2002), Cultivating the uncultured. *Proc. Natl. Acad. Sci. USA* 99, 15861–15686.

13

Measuring lipid oxidation

J. W. Irwin and N. Hedges, Unilever R & D, Sharnbrook, UK

13.1 Introduction: lipid oxidation

Lipid oxidation is one of the key problems associated with the loss of quality of foods. It is the process by which oxygen reacts with unsaturated lipids present in the foodstuff. Although the process of lipid oxidation is highly favourable thermodynamically, the direct reaction between oxygen and even highly unsaturated lipids is kinetically hindered.[1] Hence an activating reaction is necessary to initiate free radical chain reactions. It has been proposed that lipid oxidation can be initiated and/or promoted by a number of mechanisms. These include the production of singlet oxygen, enzymatic and non-enzymatic generation of partially reduced or free radical oxygen species (i.e., hydrogen peroxide, hydroxy radicals), active oxygen iron complexes and thermal or iron-mediated homolytic cleavage of hydroperoxides. The exact details of these mechanisms will not be covered here but have been reviewed extensively in other articles.[2,3] They have also been described in other chapters of this book. However, the complexity of the oxidation process means that a whole range of possible reaction products can be produced, and the type and concentration of molecules will depend upon a number of key factors. These include oxygen concentration in the local environment, surface area exposed to oxygen, fatty acid composition of lipids, levels of endogenous antioxidants and/or endogenous oxidative catalysts, and storage temperature.

The net result of the action of oxygen on lipids is the production of lipid hydroperoxides which react further to generate low molecular weight volatile components, such as aldehydes and ketones. It is the taste and smell of these volatile molecules that renders a foodstuff unpleasant and of unacceptable quality to the consumer. This unpleasant organoleptic experience is described as rancidity. Rancidity is, of course, a subjective measure of food quality. One person may describe a product as unpalatable or 'off', but for another the flavour may still be acceptable. These differences in perception may reflect culture, individual genetic variation, age, etc.[4] Therefore, to be able to measure 'rancidity' by an analytical method offers some obvious benefits, particularly in the removal of subjectivity from the assessment. It may be argued that a trained sensory panel can also assess rancid off-flavours in an objective manner, and that appropriate acuity tests can be used to screen out insensitive tasters. These statements are of course true; however, analytical measurements can give an additional insight into the mechanistic origin of the rancidity. In addition, there is likely to be a threshold below which people do not perceive off-flavours even though lipid oxidation products are present. Therefore, if analytical measures can detect rancidity or the precursors to rancidity at an earlier stage then this will have advantages. For example, in tests on the effectiveness of antioxidants, shorter shelf-life experiments would be possible by reliable early detection of key oxidation products. Finally, the use of a sensory panel does require a large number of samples, and a reasonably large number of trained panellists. To use sensory testing to screen a wide variety of different treatments or product types, especially when it is not clear whether any of the samples offer an improvement, is likely to be both time consuming and expensive. It might be argued, therefore, that this would not be the best use of a sensory facility.

It is the aim of this chapter to describe many of the analytical tools that can be used to measure products of lipid oxidation. The potential pitfalls in employing each measurement technique will be discussed and a comparison of each method's relative merits and weaknesses will be made. Also, the relationship between analytical measurement and sensory evaluation will be discussed, particularly the relationship between the presence of key oxidation products and the perception of rancidity. Establishing this link is a key part in determining the usefulness of any of the measurement techniques for reporting on food quality.

It has also been suggested that lipid oxidation products can affect the colour and nutritional value of food.[5] Nutritionally important unsaturated fats, such as n-3 fatty acids, may become degraded in purified fish oils.[6] However, the degree to which these molecules are degraded in foods such as fish fillets is still an area of some uncertainty.[7] The impact of lipid oxidation and lipid oxidation products on the colour and nutritional quality of foods will not be covered in this chapter. However, they too are important factors that influence a consumer's perception of the quality of a food product.

13.2 Chemical methods of measuring lipid oxidation

13.2.1 Free fatty acids

Free fatty acids (FFA) are formed in foods as a result of hydrolytic rather than oxidative rancidity. However, their measurement is frequently carried out alongside products of lipid oxidation and is particularly important in frying oils.

Also, they may be used as an indicator of storage stability in fried foods, such as potato chips.[8] FFAs can of course contribute to off-flavour, particularly in dairy products, and have been shown to correlate with off-flavour, e.g. harsh odour note in palm oil.[9] They may be formed in raw food ingredients by the process of lipolysis (see Chapter 8). The FFAs then act as the substrate for enzymes responsible for oxidative deterioration, e.g. lipoxygenase.

Although there are a variety of techniques that can be used to determine FFA content, titration with sodium or potassium hydroxide remains the most widely used method. Typically, a 1 gram portion of the extracted oil or fat is dissolved in solvent, usually ethanol or equal volumes of ethanol and diethyl ether, then titrated with 0.1M solution of alkali using either phenolphthalein or Alkali Blue 6B as indicator. For highly coloured oils where the slight pink colouration of the endpoint would be difficult to detect, an autotitrator may be used to perform a potentiometric titration or with an endpoint set at around pH 10.8. Results are expressed either as % FFA content or as the 'acid value', defined as the weight in mg of potassium hydroxide required to neutralise the acidity of 1 gram of oil or fat.

13.2.2 Peroxide value

As hydroperoxides are formed in the early stages of lipid oxidation, the determination of peroxide value (PV) may serve to give an early indication of rancidity as, for example, in lard.[10] PV is readily determined by titration of liberated iodine, produced by the reaction of the hydroperoxides with potassium iodide. To do this a portion of fat or oil is first dissolved in a 60:40 mixture of acetic acid with chloroform or, more commonly now, iso-octane. Potassium iodide is added in the form of a saturated solution, then after a set time period the reaction is stopped by the addition of water. Agitation is achieved throughout the assay by bubbling with nitrogen or carbon dioxide, which also serves to remove dissolved oxygen. The aqueous layer is then titrated with 0.002M sodium thiosulphate using starch as indicator, or by potentiometric titration with the use of an autotitrator. PV is calculated as milli-equivalents (or sometimes millimoles) of active oxygen per kilogram of fat.

As hydroperoxides readily decompose at elevated temperature, it is important that only a cold extraction procedure is used to obtain the oil or fat. Considerable care should also be taken during the solvent evaporation stage to ensure losses do not occur. To avoid evaporation, extraction of the fat into chloroform or iso-octane for dry foods, or into chloroform–methanol for wet foods, may be carried out. Separate aliquots of the solvent extract can then be taken, one used for PV determination whilst the other is evaporated to obtain the weight of fat.

As PV is a relatively simple test to perform it is frequently used to investigate shelf-life or to examine the antioxidant properties of various additives. For instance, in the investigation of the frying oil and packaging used for potato chip manufacture, PV was used in conjunction with FFA and sensory data to demonstrate that a shelf-life of 90 days could be obtained with sunflower oil and aluminium foil packaging.[11] In a study of cooked, comminuted flesh of herring, it was shown that the use of chitosans reduced PV, as well as TBARS (thiobarbituric acid reactive substances) and volatile aldehyde levels.[12] However, hydroperoxides may degrade during storage as propagation reactions take off, therefore PV cannot be relied upon to give a complete picture of the oxidative status. A trend for PV values to increase to a maximum and then decrease was observed during storage of edible oil blends.[13]

13.2.3 Anisidine value

Anisidine value (AV) is a measure of the aldehyde levels in an oil or fat, in particular those that are unsaturated (and principally the 2–alkenals). To determine AV, a solution of the oil or fat in iso-octane is reacted with *p*-anisidine in glacial acetic acid to form yellowish reaction products. The AV is then determined from the absorbance measured at 350 nm, both before and after reaction. By convention, AV is defined as 100 times the increase in absorbance measured at 350 nm in a 1 cm cell of a solution resulting from the reaction of 1 g oil or fat in 100 ml solvent with anisidine reagent. The method is not suitable for highly coloured oils, particularly those that have a high absorbance at this wavelength.

As a measure of secondary oxidation products, AV is used instead of, or together with, PV to assess thermally stressed oils. It may be used, for instance, in studies carried out to maximise the life of frying oils.[14] AV may also be used to support sensory data, as in an investigation of the storage of tortilla chips fried in various oils.[15] In another study, using accelerated storage tests to investigate commercial fat samples, it was found to be a good predictor of storage stability.[16]

13.2.4 Totox value

The anisidine value (AV) is sometimes measured in conjunction with peroxide value for determination of the totox value (TV), where TV = 2PV + AV. This allows both the oxidative history of the oil and the potential for further deterioration to be taken into account. As a general guide, an oil considered to be of acceptable quality would be expected to have a TV of less than 10.

During the initial phase of lipid oxidation, hydroperoxides may steadily increase to a maximum, then start to decrease. At the same time, as hydroperoxides react further, aldehydes begin to form. In some studies, it has been found that TV increases linearly during storage, for instance in accelerated storage study of grapeseed oil.[17] In an assessment of frozen beef quality, TV was found to be the best indicator of fat oxidation whilst certain other tests proved less reliable.[18]

13.2.5 TBARS

The thiobarbituric acid (TBA) test was previously thought to be a measure predominantly of malondialdehyde (MDA); however, it is now known that TBA reacts with a number of different compounds and so is referred to as TBARS (thiobarbituric acid reactive substances). The red pigment, measured at 532 nm, produced by reaction with 2-thiobarbituric acid can be formed by 2-alkenals and 2,4-alkadienals as well as by MDA. Measurement of a yellow pigment at 450 nm (TBARS 450) has also been used by a number of workers, such as in the study of freeze-dried meat.[19]

TBARS is commonly used as an indicator of lipid oxidation, particularly in meat and fish products. The method is usually carried out either by direct extraction or by a distillation procedure. The direct extraction method, which is applicable to raw meat and fish, involves maceration of the sample in a 7.5% solution of trichloroacetic acid (TCA). After filtration of the extract, an aliquot is reacted with TBA by placing in a boiling water bath for 35 minutes. After cooling, the solutions are read against the reagent blank and compared with standards prepared from 1,1,3,3 tetraethoxypropane. A plot of μg malondialdehyde against absorbance is used to calculate TBARS, which is expressed as mg MDA/kg.

Cooked meats or samples with high fat content where turbidity can be a problem may be analysed by a distillation procedure. Following extraction in TCA, and then centrifugation, an aliquot is distilled at a controlled rate until a predetermined volume has been collected. A portion of the distillate is then taken for reaction with TBA. The recovery of MDA with this method will not be complete, therefore it is important that the standards are also distilled. The distillation procedure can also be used for oil and fats and on lipid extracts from meat or fish samples. Methodology for TBARS determination has been reviewed by Fernandez *et al.*[20] Variations on the method include the use of a microplate technique for large numbers of samples and the use of continuous flow analysis in the study of brewery materials.[21,22]

TBARS can be a good indicator of rancidity, particularly for meat and fish products. However, it should be remembered that other substances present in foods, such as sugars, acids, esters, amino acids and oxidised proteins, may also react with TBA.[23] As the fatty acid composition will have an influence on the actual level of TBARS that may form in a food sample, comparison of results from different sample types should be treated with some caution.

13.2.6 Kreis test

One of the first commercially available tests developed for lipid oxidation was the Kreis test and was used as an early indicator of rancidity. It involves a colorimetric assay that measures the red colour produced by the reaction of phloroglucinol with epihydrin aldehyde and other products of lipid oxidation. The red colour is measured in a colour comparator such as the Lovibond tintometer model F or PFX995 (The Tintometer Ltd, Salisbury, UK), and reported in red units on the Lovibond scale. The Kreis test has been studied extensively by Narasimhan-Shanthi, to examine the kinetics of the method, determine optimal experimental conditions and investigate correlations with sensory data.[24,25,9]

13.3 Physical methods of measuring lipid oxidation

13.3.1 Conjugated dienes

Hydroperoxides of linoleic and linolenic acid contain a conjugated diene, which absorbs ultraviolet light at around 234 nm. Certain secondary oxidation products such as diketones, e.g. 2,3-butadione and conjugated trienes, have a principal absorption band at around 268 nm. Therefore, the measurement of the absorbance at one or both of these wavelengths will provide an indication of the degree of lipid oxidation. A dilute solution of the oil is made in a high grade solvent such as cyclohexane or iso-octane, then the absorbance is measured against a 'blank' of, e.g., methyl stearate.

When using this method it should be remembered that non-oxidised fatty acids may naturally contain conjugated dienes and that hydroperoxides of mono-enes such as oleic acid will not. Even though factors can be applied to the absorbance readings to give a measure of specific compounds, it must be remembered that some interference from co-extracted material may occur.[26] Therefore some consideration should be given to the type of sample to which this method is applied, and care taken in interpretation of the results.

13.3.2 Dielectric constant

Dielectric constant (DC) may be used to measure the degree of thermal or oxidative breakdown that occurs when an oil or fat is subjected to prolonged heating, as encountered during frying. Consequently, DC is normally used to monitor the condition of frying oils rather than food products themselves. However, a certain amount of oil is absorbed into a fried food and so the condition of the frying oil will have a direct impact on the quality of the product. Although not often used in storage studies, DC has been used, for instance, to assess the oil extracted from instant noodles following two months' storage at 37°C.[27]

The Food Oil Sensor (FOS), produced by Northern Instruments Co., Minnesota, USA, has been widely used to measure DC of frying oil and was shown to be a useful technique for monitoring oil condition, and to be both rapid and reliable.[28,29] Unfortunately, though, the FOS is no longer commercially available. It has been shown that the DC correlates with the levels of total polar compounds in frying oils for which there are legal limits in many countries.[28] The main use of DC, therefore, is as a quick and simple method to indicate the 'discard point' of a frying oil.

13.3.3 Induction period tests

Induction period (IP) tests are a form of accelerated test that is carried out at relatively high temperature, to measure the resistance of an oil or fat to oxidative rancidity. They are a means of predicting shelf-life and avoiding the long storage periods that would otherwise be necessary to determine shelf-life. Instruments that can be used to perform this test include the Rancimat, produced by Metrohm, Basle, Switzerland; the Oxidograph, produced by Mikrolab Aarhus A/S, Hojbjerg, Denmark; and the Oxidative Stability Instrument (OSI), produced by Omnion, Inc., Rockland, MA, USA.

The Rancimat test is widely used in the edible oil industry and is an automated version of the Swift test or active oxygen method (AOM), which was based on PV measurement. The Rancimat apparatus maintains the oil sample at a predetermined temperature and provides a flow of air as a source of oxygen. The volatile carboxylic acids produced by the oil as it becomes rancid are swept away by the air and collected in a reservoir of distilled water to be measured by electrical conductivity. The latest version, the 743 Rancimat, can analyse up to eight samples simultaneously and performs 'temperature extrapolation' to convert the results to the storage temperature. The Omnion OSI is a similar instrument that can analyse up to 24 samples simultaneously. Both of these instruments use disposable reaction vessels, which reduces cleaning time and helps to improve accuracy and reliability of results.

The Oxidograph is an instrument that measures the reduction in pressure as oxygen is used up by a sample of oil in a sealed vessel. In the original Sylvester test, on which the instrument is based, the oil was heated to 100°C in a closed vessel and shaken continuously until the IP had been determined. The Oxidograph, which can anlayse up to six samples, works on a similar principle but with the oil sample being stirred rather than shaken. The FIRA–Astell apparatus, which was developed by the Leatherhead Food RA, also worked on a similar principle, though it is no longer commercially available.

IP tests are well established in the food industry and are included in various international standards such as ISO 6886. It must be remembered, however, that they give only a prediction of shelf-life rather than an actual value. In particular, some caution should be exercised if using either the Rancimat or OSI to study certain antioxidants such as BHA or BHT. As they are relatively volatile, they will gradually be removed from the oil by the flow of air, resulting in underestimation of IP and hence shelf-life. With the Oxidograph or FIRA–Astell apparatus, however, there should not be a problem as the test is carried out in a sealed vessel.

13.4 Chromatographic methods of measuring lipid oxidation

13.4.1 Total polar compounds by liquid chromatography

During deep fat frying, the quality of the frying oil deteriorates due to the combined effects of hydrolysis, lipid oxidation and polymerisation. These processes lead to the formation of lipid breakdown products such as FFA, mono- and diglyceride and oxidised or polymerised triglyceride, all of which are measured as total polar compounds (TPC). Frying oils are considered unfit for human consumption when the level of TPC exceeds 25–30%, with many countries having either legal or unofficial limits somewhere between these two figures. Although measurement of the dielectric constant, e.g. using the FOS, will give an indication of TPC, the official method (ISO 8420) is carried out by liquid chromatography on silica gel.

To perform the chromatographic separation, a glass column is packed with silica gel of mesh size 70 to 230, and with a water content of 5%. After applying 1 gram of the oil sample to the top of the column the non-polar fraction, which consists mainly of non-oxidised triglycerides, is eluted from the column with light petroleum ether–diethyl ether (87:13). After rotary evaporation of the solvent, the non-polar content is determined gravimetrically, then TPC is calculated by difference. TPC may be eluted from the column with 100% diethyl ether, but recovery is insufficient for accurate determination. However, to assess column efficiency, thin layer chromatography can be used to examine the non-polar and polar fractions.

As the official chromatographic method is very time consuming, other techniques have frequently been used to provide an indication of TPC. These include differential scanning colorimetry and various 'quick tests' based on dielectric constant, viscosity or spectrophotometry.[30] A viscosity measurement is made when using the Fri-check available from MirOil, Allentown, PA, USA, whilst a spectrophotometric measurement is the basis of the VERI-FRY PRO quick test from Libra Technologies, Inc., Metuchen, NJ, USA. All of these tests have shown some correlation with TPC and their performance has been compared by a number of authors.[31,32]

13.4.2 Lipid classes by HPLC

FFA, mono- and diglycerides, as the products of hydrolytic breakdown of triglyceride, may be determined by HPLC using normal phase chromatography and a simple binary gradient. In our laboratory, we have used a 10 cm by 4.6 mm, 5 μm Lichrosorb Si60 column and a gradient based on toluene, ethyl acetate and formic acid (Table 13.1). The lipid classes were detected using a PL-EMD 950 evaporative light scattering detector from Polymer Laboratories, Church Stretton, Shropshire, UK and quantified against external standards using a multi-level calibration.

Table 13.1 HPLC gradient for the separation of 'neutral' lipid classes

Step	Time (min)	Solvent %A	Solvent %B	Curve
0	0.4	100	0	–
1	1.5	100	0	–
2	10.5	50	50	1
3	2	50	50	–
4	16	100	0	0

Solvent A: toluene–ethyl acetate–formic acid, 400:5:4. Solvent B: ethyl acetate–formic acid, 99.5:0.5. Flowrate: 1.5 ml/min.

Table 13.2 HPLC gradient for the separation of neutral and polar lipid classes

Step	Time (min)	Solvent %A	%B	%C	%D	Curve
0	0.4	100	0	0	0	0
1	1.5	100	0	0	0	0
2	10.5	50	0	0	50	1.0
3	24	0	95	5	0	1.5
4	12	0	100	0	0	0
5	22	100	0	0	0	0

Solvent A: toluene–ethyl acetate–formic acid, 400:6:4. Solvent B: methanol–chloroform–formic acid, 600:100:35. Solvent C: water–formic acid, 99.5:0.5. Solvent D: ethyl acetate–formic acid, 99.5:0.5. Flowrate: 1 ml/min.

If the food product under storage contains uncooked meat, fish or vegetables, then the activity of enzymes such as lipase and lipoxygenase may be an issue. The action of lipase or phospholipase results in the formation of free fatty acids, which act as the favoured substrate for lipoxygenase. Monitoring the lipolysis of both storage lipid (triglyceride) and certain membrane lipids (phospholipids and galactolipids) may therefore be important in the study of lipid oxidation (see Chapter 8). As the chromatographic separation of both neutral and polar lipid classes is quite difficult to achieve, it may be necessary to use a quaternary gradient to effect complete separation. In our laboratory, we have used a Perkin Elmer series 410 HPLC pump to run a multi-step, non-linear gradient (Table 13.2) with a 10 cm by 4.6 mm, 5 μm Lichrosorb Si60 column, and using the PL-

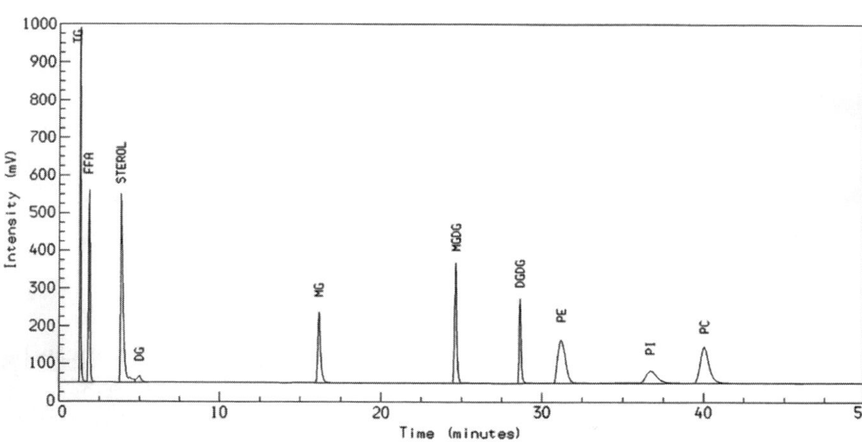

Fig. 13.1 HPLC separation of neutral and polar lipid classes: standard mix. TG = triglyceride, FFA = free fatty acid, DG = diglyceride, MG = monoglyceride, MGDG = monogalactosyl diglyceride, DGDG = digalactosyl diglyceride, PE = phosphatidylethanolamine, PI = phosphatidylinositol, PC = phosphatidylcholine.

EMD 950 detector. This method, which is a modification of one published by Hammond, has proved to give a reliable and reproducible separation of neutral lipids, phospholipids and galactolipids (Fig. 13.1).[33]

13.4.3 Hydroperoxides by HPLC

HPLC of individual hydroperoxides has been used both in the study of rancidity development and in the study of lipoxygenase activity in uncooked meat, fish and vegetables. HPLC with UV detection at 234 nm was used to analyse the products of linoleic acid oxidation catalysed by pea seed lipoxygenases, with GC–MS used for identification.[34] In this study, the feasibility of analysing hydroperoxide, hydroxide and keto fatty acids in a single chromatographic run was demonstrated. For HPLC analysis, hydroperoxides may be reduced to their more stable hydroxy derivatives by reaction with sodium borohydride. However, they have also been analysed without derivatisation, for example in the characterisation of hydroperoxides formed in oxidised linoleic and linolenic acids, using MS detection.[35]

The development of two methods, based on post-column luminescence, using cytochrome c and luminol was reported by Christensen and Holmer.[36] These methods were used to study the formation of hydroperoxides in butter and dairy spreads during storage. The methodology was reported to be rapid and sensitive and in this study showed a linear relationship with PV.

13.4.4 Volatiles by gas chromatography

The later stages of lipid oxidation result in the production of a wide range of volatile compounds including alcohols, aldehydes, ketones, alkanes and organic acids. These volatiles, many of which are directly responsible for off-flavour, can be measured quite readily by gas chromatography (GC). When this technique is coupled with mass spectrometry (GC–MS), the volatile species can also be easily identified. Probably the biggest problem faced by the analyst is in selection and optimisation of the appropriate sampling technique. Those available include simultaneous distillation and extraction (SDE), static headspace, dynamic headspace, solid phase micro-extraction (SPME) and stir bar sorptive extraction (SBSE).

13.4.5 Simultaneous distillation extraction (SDE)

SDE is a relatively simple sampling technique to perform, although it can be quite time consuming. It is frequently performed using the Likens–Nickerson apparatus, which allows simultaneous steam distillation with volatile extraction into solvent. Typically, 20 g of sample is boiled under reflux for 90 minutes with a low boiling solvent such as pentane or diethyl ether boiled in another flask. Then in the condenser, where both sets of vapours meet, the volatiles transfer to the solvent vapour, condense and collect in the flask. Gentle evaporation under a stream of nitrogen is then used to concentrate the extract to low volume for direct injection into the GC. As relatively large sample amounts can be used, the technique provides good sensitivity and is particularly useful for recovering the higher molecular weight, volatile compounds.

Potential problems with SDE can be minimised by taking certain precautions such as thorough cleaning of glassware to avoid carry-over and contamination from cleaning agents. Contamination can even come from perfume or after-shave worn by the analyst. Even with the use of high purity solvents, it is sensible to perform a blank extraction as any impurities present in the solvent will be concentrated by evaporation. With the relatively high temperatures involved in SDE, artefact production can also be a problem. An investigation of the flavour volatiles in cod indicated that artefacts were produced by oxidation of the lipids, but could be suppressed by the use of antioxidants or by removal of oxygen.[37] Despite these drawbacks, SDE has been used in various food storage studies, for example in a study of fragrant rice.[38] It was shown that, after 3 months' storage at 30ºC and 84% relative humidity, total volatile levels in the rice increased, due mainly to the formation of aldehydes and ketones.

13.4.6 Static headspace

One very simple technique to perform is static headspace sampling. With this, the sample is heated in a sealed vial for a period of time to allow the volatile compounds to partition into the headspace above the sample and reach equilibrium. A volume of the headspace, typically 1–5 ml, is then injected into the GC for analysis. Static headspace is simple and rapid and serves to provide a sample of the volatiles that is representative of the food in terms of sensory perception. However, despite the injection of large volumes, the technique suffers from poor sensitivity and only the more abundant compounds may be detected. The other main drawback is that sample size is limited by the vial capacity, making it difficult, at least with certain foodstuffs, for a representative sample to be taken.

13.4.7 Dynamic headspace

The limitations of static headspace can be overcome by the use of dynamic headspace sampling, otherwise known as 'purge and trap'. With this technique, the volatiles are purged from the sample using a flow of inert gas, which is directed onto the surface of the sample or perhaps bubbled through if it is liquid. The flow of gas, typically nitrogen or helium, is then directed through a trap containing an absorbent material such as Tenax or Chromosorb. The volatiles are normally transferred from the trap to the GC by thermal desorption, although solvent elution followed by direct injection is also possible.[39]

Instruments are commercially available that allow the whole procedure to be automated, for example the Tekmar LSC 2000 (now replaced by a new instrument: Velocity XPT, from Teledyne Tekmar, Los Angeles). With this system, the sample is placed in the bottom of a u-tube, fitted with a thermostatically controlled heating jacket, and then connected to the instrument. A flow of helium is used to purge the sample and to transfer the volatiles to a Tenax trap. A 'dry purge' is performed by passing helium only through the trap, which helps in the removal of excess water. Next, thermal desorption is carried out to transfer the volatiles to the top of the GC column where they are cryofocused using liquid nitrogen. Finally, the top of the column is rapidly heated and the chromatographic run commenced. The main disadvantage with this type of system is that the whole process including the chromatographic run must be completed before starting the next sample. This severely limits the number of samples that can be analysed in any one day.

An alternative procedure is to perform the purge and trap separately to that of the thermal desorption and GC analysis. This may be carried out using an automated thermal desorption (ATD) system such as the ATD400 from PerkinElmer, Beaconsfield, Buckinghamshire, UK. With this system, the sampling 'trap' consists of a stainless steel desorption tube, which is packed with the appropriate adsorbent. For sample purging, the tube may be connected either directly or via a short piece of tubing to a user-selected purge vessel. This allows a great deal of flexibility in terms of the size and geometry of the purge vessel and hence the type and amount of food which can be sampled. With the ATD400, the desorption tubes are placed on a carousel, which allows the contents of each tube to be analysed in turn. The volatiles are thermally desorbed from the tube in a flow of helium and transferred, via an inlet split, to the cold trap. The cold trap is then rapidly heated to transfer the volatiles, via an outlet split, to the GC column. The use of both inlet and outlet splits allows the flow of helium to be fully optimised for each stage of the process to give efficient transfer of the volatiles and removal of excess water.

With this type of system, greater flexibility is possible with the purge and trap procedure and an increase in sample throughput can be gained. The disadvantages are, first of all, that 'breakthrough' of the most volatile compounds may occur if the duration, temperature or flowrate of the purge and trap process is excessive. However, although there may be a significant loss of a compound due to breakthrough, the amount trapped is normally proportional to the actual amount present in the sample. Secondly, a high percentage of the volatiles, which have been concentrated on the desorption tube, are lost through use of the splits. Whichever type of system is used, the purge and trap technique provides much greater sensitivity than that of static headspace and thereby allows a large number of compounds to be detected.

13.4.8 Solid phase microextraction (SPME)

SPME is another technique that is relatively easy to perform and can be used with almost any GC. The apparatus consists of a fibre, which is a short length of fused silica with an absorbent coating, held for protection within a needle, and mounted on a syringe-like device. The fibre is coated with a thin layer of an absorbent or adsorbent such as divinylbenzene or polydimethylsiloxane

(PDMS). To sample the volatiles, the SPME fibre is pushed out of the needle and exposed to the headspace above the sample, typically contained in a sealed vial. As with normal static headspace, sufficient time is allowed for equilibrium to be reached before the fibre is retracted back into the needle. For GC analysis, the fibre is then exposed in the heated injector port, to thermally desorb the volatiles onto the GC column. The process can easily be automated, for use with either GC or HPLC instruments.

There are a number of parameters that need to be considered for efficient extraction of a volatile onto a SPME fibre, including sampling time and temperature, sample amount (or concentration), headspace volume and type of sorbent used. For any given conditions, the amount of a particular compound absorbed will be determined by the partition coefficient between sample and headspace and between the headspace and the fibre coating. To maximise sensitivity, the fibre with the appropriate type and thickness of coating should be selected, based on the polarity and the volatility of the compound.

Perhaps one of the main drawbacks of the SPME technique is that it may be necessary to devote some time and effort optimising the conditions before satisfactory results can be obtained. This is true particularly for the determination of a range of volatile compounds, for which the fibre coatings will have varying selectivity. The main advantages of the SPME headspace technique are that it is simple, relatively fast and low cost, and may offer some improvement in sensitivity over normal static headspace. Examples of its use include the measurement of hexanal and pentanal in cooked turkey, the determination of flavour threshold for acetaldehyde in milk and water and the analysis of free fatty acids in whey products.[40–42] In one investigation of lipid oxidation products in milk, it was found that SPME was as sensitive as purge and trap for the detection of pentanal and hexanal.[43]

13.4.9 Stir bar sorptive extraction (SBSE)

SBSE has been developed quite recently and may be considered as a variation of the SPME technique. As its name implies, SBSE employs a stir bar, which is coated with the sorbent PDMS. Volatile sampling is carried out by placing the stir bar either in an aqueous sample or in the headspace above a solid sample. After stirring for a period of 30 to 60 minutes, the stir bar is placed in a glass tube and loaded onto a thermal desorption unit fitted to the GC.

In a review covering the theory and principles of the technique it was claimed that SBSE offers up to a 500-fold increase in sensitivity over that of SPME.[44] But as yet, there has not been a great deal of published literature regarding the use of this technique for studying lipid oxidation and shelf-life. However, one example in the literature described its use in the determination of stale flavour, carbonyl compounds in beer.[45]

13.4.10 Detection of volatiles

There are a number of detector types that can be used for the detection of volatile compounds. The flame ionisation detector (FID) is considered a 'universal' detector as it has good sensitivity for virtually all organic compounds together with a very wide linear range. It is also inexpensive, simple in design and easy to maintain, and consequently its use has always been very popular. Other detectors can be used for specific compounds such as the flame photometric detector (FPD) which can be set up for compounds containing either sulphur or phosphorus. The nitrogen phosphorus detector (NPD) detects both nitrogen- and phosphorus-containing compounds.

The mass spectrometer (MS), although a technique in itself, can be considered a 'universal' detector when coupled to either GC (GC–MS) or HPLC (LC–MS). With the availability of relatively inexpensive, bench-top instruments, it is now in common use. The use of MS offers a number of advantages over FID, such as improved sensitivity, identification of the compounds, and the ability to 'separate' co-eluting compounds. The use of a spectral library, an invaluable tool for the analyst, makes for rapid identification, although isomeric compounds may not easily be distinguished. For this to be carried out it is necessary to collect data in full scan mode so that all the fragment masses and thus a spectrum can be obtained. However, to monitor a number of samples for known compounds, data can be collected by selected ion monitoring (SIM), which gives much increased sensitivity over that of full scan. When two or more compounds co-elute, the MS software can be used to disentangle the spectra of the two compounds, and then to quantify them through their respective characteristic ion chromatograms. The combination of purge and trap and GC–MS has allowed, for example, the identification and measurement of 40 volatile compounds in milk samples that had been stored for 4 months at 25°C.[46]

One type of detector that has not been mentioned so far is the human nose, which of course is used in the technique GC–olfactometry. Alternatively known as 'GC–sniff', this technique can be used to obtain odour descriptors and to determine odour potency, or to help identify which volatiles are responsible for a particular odour or off-flavour. To perform GC–sniff, the outlet from the GC column is split between a detector e.g. FID, and a nose cone with humidified air added for the 'panellist' to sniff. When a smell is detected, the panellist records an odour descriptor either against the analyte peak on a chart recorder or 'blind' so that they are not influenced by the appearance of the peak. If an indication of odour potency is required, the panellist is required to give some indication of the strength of the odour. Alternatively, aroma extraction and dilution analysis (AEDA) may be performed, by running a series of dilutions of the sample extract until the odour can no longer be detected.

With the availability of equipment such as the olfactory detector outlet ODO 11 from SGE, Inc., Austin, Texas, USA, it is now relatively straightforward to split the outlet of the GC between a sniff port and MS. This then allows the odour potent compounds present in a food to be identified, measured and sensorially described. The main drawback with the GC–sniff technique is the time required for panellist training as well as in performing the analysis.

13.4.11 Other chromatographic methods

In the context of lipid oxidation, the main use of GC is in the study of volatile compounds. Another use, however, is in the determination of fatty acid composition. For example, the loss of polyunsaturated fatty acids (PUFA) during photo-oxidation of fish lipids has been studied.[47] A linear relationship was observed between PUFA losses and PV, though a significant loss occurred only when PV was very high. GC has also been used to determine the profile of FFA, for example in the study of Manchego cheese during ripening.[48]

As already discussed, HPLC can be used to examine the lipid class composition or to separate individual hydroperoxides. It has also been used as an alternative to the spectrophotometric assay in the determination of MDA in foods. For instance, reverse phase HPLC with fluorescence detection was found to give a sensitive and reproducible method for MDA concentration in various foods, such as butter, margarine, oil, meat and fish.[49] In another study, an HPLC method, which employed cetrimide as an ion-pairing agent, gave results that correlated highly with the chemical method for refrigerated, ground pork.[50] Recently, HPLC has been used in conjunction with MS to analyse non-volatile products of lipid oxidation in vegetable oils.[51] In this study, an LC–MS method was used to separate and identify different triglyceride (TAG) products such as epoxy-TAG, oxo-TAG, hydroperoxy-TAG, hydroxy-TAG and 2.5 glycerides.

13.5 Measurement issues

The main issues that need to be considered are the treatment of the food sample prior to analysis, extraction of the lipid from the food and, of course, selection of the appropriate method(s). Ideally, a sample would be analysed immediately on completion of a storage period. However, if a number of time points are involved in the study then it may well be more practical to analyse all of the samples at the very end of the storage trial. In this case, the samples should be stored at the lowest temperature possible and ideally below their glassy state temperature, so that further oxidative deterioration is minimised. This tends to be easier for frozen samples. But for frozen storage trials in particular, a further possible complication may occur. If during the storage trial, oxidative reactions are initiated then they may still proceed, albeit at a slower rate, in low temperature storage. This is especially true if the temperature used to arrest further reaction is not particularly low.

To eliminate the possibility of additional free radical reactions occurring between the end of the storage period for a given sample and the time when the storage trial is complete and the samples can be analysed, reverse storage trials should be employed. For example, if the effect of storage at −9°C was to be investigated, and the take-off points were 3, 6, 9 and 12 weeks, then the trials would be conducted in the following manner. It will be assumed that a −80°C freezer is being used to arrest any changes on frozen storage. Initially all samples would be equilibrated in the −80°C freezer. The 12-week samples would then be transferred to the −9°C freezer for the trials. After 3 weeks the 9-week samples would be transferred to −9°C, and after a further 3 weeks the 6-week samples would be transferred, and so on. In this way at the end of the 12 weeks all samples would have received their allocated storage at −9°C but the need to store the 3-week samples for 9 weeks at −80°C, after any oxidative reactions had been initiated by storage at −9°C, would be removed.

Upon removal of the samples from storage, analysis or lipid extraction, if required, should be commenced as soon as possible. In the case of 'biological' material such as raw vegetables, meat or fish (with PUFA and/or active enzymes present) lipid extraction should be performed on the still frozen, or at least semi-frozen, sample. Otherwise, cellular damage from ice crystal formation will give rise to rapid hydrolytic and oxidative deterioration. Also, to avoid further oxidation, the use of heat in the extraction process, such as with Soxhlet or other reflux extraction, should be avoided wherever possible. Other factors to consider are exclusion of air and lighting conditions, i.e. the use of tungsten lighting rather than fluorescent to avoid UV. Various procedures have been used for lipid extraction and some comparisons have been made by a number of different workers.[52–54]

The relative merits of each of the methods have been discussed earlier to some extent. The choice of method(s) employed will be based on the type of foodstuff under investigation and the anticipated extent of lipid oxidation, and also on whether a full understanding of the deterioration process is needed, or just an indication of its extent. Other factors that need to be considered include simplicity of the method, availability of instrumentation and numbers of samples to be tested. It may be that a prediction of shelf-life is required, in which case an IP test or the measurement of hydroperoxides may be appropriate. On the other hand, a measurement of specific or total volatile compounds may well relate more closely to sensory evaluation and consumer perception of rancidity.

13.6 Correlating analytical measurements with sensory evaluation

A key part of establishing the usefulness of any measurement technique, as a way of measuring food flavour quality, is to determine how well results correlate with the sensory quality of a foodstuff. The measurement may be used to predict the immediate flavour/flavour quality, or it may be a method by which future, or remaining, shelf-life can be determined. It is the aim of this section to describe some of the papers that have attempted to correlate analytical measures with a trained sensory assessment. Because of the size of the topic, this section will focus primarily on meat and fish products, but will refer to other foodstuffs where appropriate.

13.6.1 Free fatty acids (FFA)

Although FFAs are not oxidation products, they can contribute to the perception of 'rancidity' in two ways, either directly or by their subsequent oxidation. They are formed by the action of lipases or phospholipases on lipids or phospholipids respectively. For example, it has been reported that free linoleic acid and free linolenic acid had the highest impact on off-flavour development, particularly bitterness problems associated with soy lecithins used in emulsions.[55] Similarly, rancid and goat-like flavours in milk products have been attributed to the presence of C4 and C6 FFAs, whilst soapy and bitter flavours were correlated with C10 and C12 FFAs.[56] Conversely, these flavour notes can be positive, with lipases used to improve the ripening rate and the flavour of, for example, Manchego-type cheeses.[57] In this study it was recommended that the process be carefully controlled so that the preferred FFAs were produced. Similarly the generation of FFAs in butter by lipolytic bacteria has been reported as a route to speeding up flavour development.[58] Thus, it is clear that the presence of FFAs can have a direct impact on the flavour perception of products, and that this may be positive or negative depending on the types of molecule that are liberated, and the type of product in question.

13.6.2 Peroxide values (PV)

The PV is determined by titration of iodine released from the reaction between a lipid extract and potassium iodide solution. Problems may occur due to lack of sensitivity, from interference from coloured, lipid soluble components, or from additional oxidation occurring during extraction of the lipids. Despite these limitations, PVs have been correlated with rancidity development in oils.[59,60] However, not all authors have found a good correlation between PV and sensory assessment of quality of oils.[61] Similarly, poor correlations have also been found between sensory assessment of rancidity and PV in solid foodstuffs, such as fried snack foods and mackerel.[62,63] Thus, it is clear that the correlation achieved between PV and sensory evaluation is variable, and is likely to be dependent upon the food under investigation.

13.6.3 TBARS

A number of authors have reported correlation between TBA or TBARS results with sensory assessment. The degree to which the test results correlate seems to be quite variable. In some cases no or poor correlation has been reported.[64,65] Other studies suggest much better correlation.[66–68] Some of the variability of the test results may be due to the presence of interfering compounds, not associated with lipid oxidation.

A further complication has been expressed by Younathan *et al.*[69] In their study they found a high positive correlation between TBA results for ground, cooked beef and panel scores for rancidity. However, panellists gave much lower scores for rancid odour in turkey samples despite the fact that the TBA levels were high. This result was explained by a lack of sensitivity to rancid odours in turkey meat by the panellists. These observations raise a number of questions. Firstly, was the TBA test really reporting on the key volatile or volatiles associated with the unpleasant odour? Secondly, what are the important volatile components? Thirdly, did the turkey have any other masking aromas that may have reduced the perception of rancidity?

An example of how volatile components that are not generated by oxidative degradation of lipids may be influential in the perception of quality comes from a study of the effects of ice storage on hake.[70] TBA, total volatile base nitrogen (TVB-N) and trimethylamine nitrogen (TMA-N) levels were correlated with visual inspection and sensory assessment of spoilage during ice storage. From the results it appeared that the most suitable indices of spoilage in this case were TVB-N and TMA-N levels. The TBA index, which remained at a low level, did not correlate with the assessments of spoilage, but was consistent with the sensory assessment of low rancidity. Thus, in complex food systems many reactions may occur that can generate flavour and aroma compounds. Exogenous factors such as microbial contamination may be important, as too may be endogenous factors such as naturally occurring enzyme reactions. Therefore, a more complete analysis of possible flavour/aroma molecules may often need to be undertaken in order to obtain a clearer picture of what changes have occurred in the sample.

13.6.4 GC, GC–MS, GC–sniff

Gas chromatography (GC) is becoming increasingly used to measure oxidative rancidity in oils and foods. Volatile oxidation products can be measured directly by headspace GC methods or as more stable derivative products. The headspace method determines the volatile components in the headspace above the food. This method will give an accurate profile of the volatile components, and this is of value since some of these volatile molecules will contribute to the rancid odour and/or flavour. Furthermore, this methodology reduces the amount of sample preparation that is required. A number of papers have cited aldehydes[71–73] and particularly hexanal[73,74] as key marker molecules for oxidative degradation of lipids. Good correlations between the presence of increasing concentration of hexanal and the perception of rancid odour and flavour by sensory assessment have been reported.[68,75–77] For example, the development of hexanal in frying fats has been negatively correlated with aroma quality, and positively with grassy, rancid, painty and chemical aromas.[78] The production of other volatile molecules has been reported in meat and fish products as correlating with sensory assessment. These include 1-pentene-3-ol, 2,3-pentanedione and nonane, and 3-hydroxy-2-butanone.[79–81] However, many of these volatiles also correlate with the production of hexanal. Therefore, hexanal is probably the most commonly reported marker molecule for rancid off-flavour. However, it should be remembered that in some cases a poor correlation of hexanal concentration with sensory assessment has also been noted.[82]

Some studies have used hexanal as an indicator of secondary lipid oxidation products. One study using election spin resonance (ESR) measured the production of free radicals and found a good correlation between ESR signal, hexanal production and sensory evaluation.[83] Similarly Zhang and Lee used hexanal as an example of a volatile molecule in their feasibility study using silica gel adsorption/near-infrared reflectance spectroscopy.[84]

The use of GC–sniff in combination with GC–MS allows the identification of a compound together with its odour description. Therefore it is possible to relate sensory descriptors of aroma and flavour directly to a molecule. In a study of rehydrated vegetables, it was found that 2-methylpropanol was associated with a chocolate aroma whilst 2,3-butadione was associated with a caramel/fatty aroma.[85] Overall it was concluded that the volatile components provided a significant contribution to the flavour of rehydrated vegetables. In this way a picture can be built up which relates the volatile molecules that are produced to their aroma. From this is may be possible to then relate the aroma molecules with key sensory descriptors for flavour and/or aroma attributes of a product.

13.6.5 Electronic nose

Related to the GC volatile methods is the technique of employing an electronic nose. The essential feature of this technique is the use of a gas sensor array to analyse the volatile lipid oxidation products that may be released during storage. This method has been used to detect rancidity in virgin olive oil and also lipid oxidation in herring.[86–88] The results from the studies on herring indicated that this technique could be used as a rapid and non-destructive method for determining the development of rancidity. Having a reliable, easy to use, hand-held device to 'measure rancidity' would offer clear benefits, and have the real potential to simplify the assessment process.

13.6.6 Fluorescence techniques

The presence of auto-fluorescence has been correlated with the development of aldehydes and the sensory assessment of rancidity.[89] The results indicated that the increase in fluorescence which was occurring during storage was due to the reaction between the aldehydes produced by lipid oxidation and the protein matrix in the meat. Multivariate regression was used to correlate fluorescence data with rancidity, as assessed by a sensory panel, volatile compounds (GC–MS), and TBARS. Results verified that the increase in fluorescence during storage originated from aldehydes reacting with the meat matrix. The results produced indicated that front face fluorescence was a useful tool for non-destructive analysis of oxidation progress in meat products.

13.7 Measurement techniques and shelf-life improvement

Some of the key benefits of measuring the development of rancidity fall under a number of broad headings. Firstly, by monitoring the production of lipid oxidation products it is possible to understand better the mechanism by which an off-flavour has developed. For example, is it the production of FFAs by hydrolysis of lipids that is the direct cause of flavour change, or is it due to oxidation of lipids, or is another mechanism involved? By gaining a better understanding of the mechanism, better strategies to prevent flavour change can be devised.

Another key benefit is the identification of markers for oxidative damage. If a marker molecule can be identified and measured which correlates with the sensory perception of rancid off-flavour and more especially with the consumer rejection of a product, then a number of benefits will accrue. For example, it may be possible to develop much better shelf-life indicators of oxidative damage. Although this would not improve product stability as such, it would be useful in identifying potential weak points in the supply chain, which could ultimately lead to improvements in product quality. Also, by having good markers for oxidative damage, the study of antioxidant function, or indeed any stability improvement strategy, could be made much easier. If rancidity could be reliably detected at an earlier stage in a storage trial then this would speed up the testing procedure. This would be especially useful for frozen foods. With these, the development of sufficiently high concentrations of volatile molecules to allow reliable detection by trained sensory panellists may take many months. Finally, markers for oxidative damage can help corroborate, and possibly assist in interpreting results from a sensory panel.

Although this chapter has focused on the measurement of the products of lipid oxidation, the quantification of the amounts of natural or added antioxidants may also prove to be a useful marker in predicting remaining shelf-life of products. If the kinetics of the depletion of antioxidants is known, then this too could be used in developing, for instance, time–temperature indicators (TTI).

13.8 Conclusions and future trends

13.8.1 Summary

We hope that this chapter has given you, the reader, an insight into the myriad of techniques that have been employed to monitor oxidative damage. In particular, it should have given an overview of commonly used techniques together with their applicability, advantages and disadvantages. In many cases there is often a trade-off between simplicity of use and the degree of insight that the test results can give the researcher. Also, some techniques are better suited to a particular foodstuff than others.

Even without a direct correlation with a sensory assessment, having a reliable marker for oxidative damage is still useful. Its measurement can be used to determine whether or not lipid oxidation has occurred, and thereby establish whether it is the cause of an off-flavour. It can also be used to evaluate whether a particular treatment has eliminated lipid oxidation. Difficulties are likely to arise, however, when only a limited degree of oxidation has been determined. In this case, it may be more difficult to establish whether or not this, in sensory terms particularly, is significant.

Generally methods which analyse the composition of the volatile components are more likely to give a more complete picture of what is happening at a mechanistic level. Clearly, evidence in the literature does suggest that good correlations do exist between key volatile components and the sensory perception of rancidity. In particular, the combination of GC–MS and GC–sniff techniques with the sensory descriptors of key flavour/rancid notes has the real potential to offer valuable insights into the likely origins of perceived rancidity or off-flavours.

However, a good correlation between the measurement and the trained sensory panellists is only part of the story. Why the correlation exists needs to be known also. This is important in disentangling cause and effect, and without this important link extrapolation outside the established experimental conditions may prove to be unreliable. Finally, and perhaps most difficult to determine, is how the analytical measurement ultimately relates to consumer reaction to and rejection of a product. If this final piece of the puzzle can be found then it will allow the development of, for example, better markers for quality as well as better TTIs and prediction of shelf-life.

13.8.2 Future trends

The consumer trend towards natural, 'unadulterated' foods seems likely to impact upon the type of molecules that the food scientist would want to use to maintain flavour quality. Thus, the challenge will be to select the most natural options (i.e. options that the consumer recognises as relating to their own experiences in food preparation) to offer products with the quality expected by increasingly discerning consumers. The control of off-flavour by natural means is likely to set real challenges for the researches in this area. Because the development of rancidity in foods is a perennial and complex problem, there is unlikely to be any generic solution. Thus, each food system will need to be treated as a separate problem.

The use of sensory panelling is an important aspect of shelf-life investigations and is always likely to be the 'gold standard'. However, to screen new antioxidant molecules in a variety of systems solely by trained sensory assessors is likely to prove extremely time consuming and costly. Thus instrumental measurement still has an important role to play.

Apart from the established methods that have been discussed in this chapter, there are some relatively new techniques whose use may become more common in the future. One of these is ESR spectroscopy, used for the determination of oxidative stability or in the assessment of antioxidant activity. Various aspects of this technique, including the relationship between free radical concentration and ESR signal, have been reviewed by Anderson and Skibsted.[90] Another technique, nuclear magnetic resonance (NMR), was used by Shahidi and Spurvey to study the oxidative stability of dark and light muscle of mackerel.[91] The results obtained were reported to compare well with those from established methods. Oxygen radical absorbance capacity (ORAC) may be used to measure antioxidant activity and so has some potential for the determination of oxidative stability, at least in certain samples. Hence, in a study of 33 virgin olive oil samples, ORAC values were found to correlate well with total phenols and to a lesser extent with IP values determined by Rancimat.[92]

The ELISA technique has been used as a measure of lipid oxidation, for instance to quantify hexanal–protein adducts in a meat model system.[93] In this study, results from monoclonal and polyclonal antibody-based ELISA showed strong correlation with volatile and TBARS methods. The electronic nose (EN) which uses a sensor array, the electronic tongue (ET) based on lipid membranes, and chemical sensors based on MS all have some potential for the detection of off-flavour and have been reviewed by Kress-Rogers and Brimelow.[94] To produce and interpret the results from these techniques, multivariate data analysis or chemometrics are often used. Chemometrics may also be used, of course, to process and compare the large amount of data obtained from, for example, the volatile and sensory analysis of a number of samples in a storage trial.

13.9 Sources of further information

AKOH C C and MIN D B (2002), *Food Lipids: Chemistry, Nutrition and Biotechnology*, New York, Marcel Dekker.

ALLEN J C and HAMILTON R J (1999), *Rancidity in Foods*, Gaithersburg, MD, Aspen Publishers.

BAIGRIE B (2003), *Taints and Off-flavours in Food*, Cambridge, Woodhead Publishing.

BREMNER H A (2002), *Safety and Quality Issues in Fish Processing*, Cambridge, Woodhead Publishing.

ERICKSON M C and HUNG Y C (1997), *Quality in Frozen Food*, London, Chapman and Hall.

FRANKEL E N (1998), *Lipid Oxidation*, Dundee, The Oily Press.

O'BRIEN R D, FARR W E and WAN P J (2000), *Introduction to Fats and Oils Technology*, Champaign, IL, AOCS Press.

PEARSON A M and DUTSON T R (1994), *Quality Attributes and Their Measurement in Meat, Poultry and Fish Products*, Glasgow, Blackie.

ROUSEFF R L and CADWALLADER K R (2001), *Headspace Analysis of Foods and Flavours: Theory and Practice*, New York, Kluwer Academic/Plenum Publishers.

SIKORSKI Z E and KOLAKOWSKA A (2002), *Chemical and Functional Properties of Food Lipids*, Boca Raton, FL, CRC Press.

13.10 References

1 LABUZA T P (1971), 'Kinetics of lipid oxidation in foods', *CRC Critical Reviews in Food Technology*, **2**(3) 355–405.

2 ALLEN J C and HAMILTON R J (1999), *Rancidity in Foods*, Gaithersburg, MD, Aspen Publishers.

3 ERICKSON M C and HUNG Y C (1997), *Quality in Frozen Foods*, New York, International Thomson Publishing.

4 BRYHNI E A, BYRNE D V, RODBOTTEN M, CLAUDI-MAGNUSSEN C, AGERHEM H, JOHANSSON M, LEA P and MARTEN M (2002), 'Consumer perceptions of pork in Denmark, Norway and Sweden', *Food Qual Pref*, **12**(5) 257–266.

5 TIMMERMAN F and MEGREMIS C (1996), 'More to it than meets the eye', *Food Ingred Anal Int*, **18**(3) 41–43.

6 MENDEZ E, SANHUEZA J, SPEISKY H and VALENZUELA A (1996), 'Validation of the Rancimat test for the assessment of the relative stability of fish oils', *J Am Oil Chem Soc* **73**(8) 1033–1037.

7 POLVI S M, ACKMAN R G, LALL S P and SAUNDERS R L (1991), 'Stability of lipids and omega-3 fatty acids during frozen storage of Atlantic salmon', *J Food Proc Pres*, **15**(3) 167–181.

8 PETUKHOV I, MALCOLMSON L J, PRZYBYLSKI R and ARMSTRONG L (1999), 'Storage stability of potato chips fried in genetically modified canola oils', *J Am Oil Chem Soc*, **76**(8) 889–896.

9 NARASIMHAN S, RAJALAKSHMI D, CHAND N, MAHADEVIAH B and INDIRAMMA A R (2001), 'Palm oil quality in different packaging materials – sensory and physiochemical parameters', *J Am Oil Chem Soc*, **78**(3) 257–264.

10 KOCH S and KLEIN D (2000), 'Investigations on the pull date of lard', *Fleischwirtschaft*, **80**(6) 99–102.

11 SANDHU K S, BAL A and AHLUWALIA P (2002), 'Studies on suitability of cultivar, frying medium and packaging for potato chips', *J Food Sci Agric*, **39**(4) 394–402.

12 KAMIL J Y V A, YOU-JIN-JEON and SHAHIDI F (2002), 'Antioxidative activity of chitosans of different viscosity in cooked comminuted flesh of herring (*Clupea harengus*)', *Food Chem*, **79**(1) 69–77.

13 MURTHI T N, SHARMA M, DEVDHARA V D, CHATTERJEE S and CHAKRABORTY B K (1987), 'Storage stability of edible oils and their blends', *J Food Sci and Technol – India*, **24**(2) 84–87.

14 ANON (2001), 'Maximising the life of frying oil', *Asia Pacific Food Industry*, **13**(6) 28–30.

15 HAWRYSH Z J, ERIN M K, KIM S S and HARDIN R T (1995), 'Sensory and chemical stability of tortilla chips fried in canola oil, corn oil and partially hydrogenated soybean oil', *J Am Oil Chem Soc*, **72**(10) 1123–1130.

16 GROMOPHONE M A (1991), 'Anisidine value as a measure of latent deterioration of fats', *Grasas-y-Aceites*, **42**(1) 8–13.

17 CASTRO M E, MASSON S L and VALENZUELA G F (1986), 'Kinetics of accelerated deterioration of grapeseed oil', *Alimentos*, **11**(4) 5–8.

18 CATTANEO P and CANTONI C (1995), 'Chemical indices for evaluation of the quality of frozen beef', *Ingegneria Alimentare le Conserve Animali*, **11**(4) 16–20.

19 HAVENS A L, FAUSTMAN C, SENECAL A and RIESEN J W (1996), 'Use of thiobarbituric acid reactive substances at 450 nm to measure oxidation in freeze-dried meats', *IFT Annual Meeting: Book of Abstracts.*

20 FERNANDEZ J, PEREZ-ALVAREZ J A and FERNANDEZ-LOPEZ J A (1997), 'Thiobarbituric acid test for monitoring lipid oxidation in meat', *Food Chem*, **59**(3) 345–353.

21 GAEBLER D M, JONES S J and MANDIGO R W (2002), 'Adaptation of microplate reader for measuring oxidative rancidity in meat products', *Meat Sci*, **62**(2) 193–198.

22 BAIER B, HARMS J and KRUEGER E (1994), 'Application of continuous flow analysis technology for quality control in the brewery industry. V. Determination of the thiobarbituric acid count in malt, brewing wort and beer', *Monatsschrift-fuer-Brauwissenschaft*, **47**(9) 278–283.

23 GUILLEN-SANS R and GUZMAN-CHOZAS M (1998), 'The thiobarbituric acid (TBA) reaction in foods: a review', *Crit Rev Food Sci Nutr*, **38**(4) 315–330.

24 NARASIMHAN S, SARVA-MANGALA G K, VASANTH-KUMAR A K, CHAND N and RAJALAKSHMI-D (1994), 'Kinetics of the Kreis test through response surface methodology', *J Am Oil Chem Soc*, **71**(11) 1267–1271.

25 NARASIMHAN S, VASANTH-KUMAR A K, RAVVI R and CHAND N (1999), 'Optimisation of Kreis test for edible oils', *J Food Lipids*, **6**(2) 107–115.

26 WHEATLEY R A (2000), 'Some recent trends in the analytical chemistry of lipid peroxidation', *Trends in Analytical Chemistry*, **19**(10) 617–628.

27 YEN G C and LAI S H (1989), 'Oxidative stability of instant noodles fried with sesame oil–vegetable oil blends', *Journal of the Chinese Agricultural Chemical Society*, **27**(2) 196–201.

28 WU P-F and NAWAR W W (1986), 'A technique for monitoring the quality of used frying oils', *J Am Oil Chem Soc*, **63**(10) 1363–1367.

29 DU PLESSIS L M and MEREDITH A J (1999), 'Palm olein quality parameter changes during industrial production of potato chips', *J Am Oil Chem Soc*, **76**(6) 731–738.

30 TAN C P and CHE-MAN Y B (1999), 'Quantitative differential scanning calorimetric analysis for determining total polar compounds in heated oils', *J Am Oil Chem Soc*, **76**(9) 1047–1057.

31 CHU Y H (1991), 'A comparative study of analytical methods for evaluation of soybean oil quality', *J Am Oil Chem Soc*, **68**(6) 379–384.

32 GERTZ C (2001), 'Routine analysis of deep-frying fats and oils', *Lipid Technology*, **13**(2) 44–47.

33 HAMMOND E W (1993), *Chromatography for the Analysis of Lipids*, London, CRC Press.

34 WU Z, ROBINSON D S, DOMONEY C and CASEY R (1995), 'High-performance liquid chromatographic analysis of the products of linoleic acid oxidation catalysed by pea (*Pisum sativum*) seed lipoxygenases', *J Agric Food Chem*, **43**(2) 337–342.

35 NUNEZ A, FOGLIA T A and PIAZZA G J (2001), 'Characterisation of lipoxygenase oxidation products by high-performance liquid chromatography with electron impact–mass spectrometric detection', *Lipids*, **36**(8) 851–856.

36 CHRISTENSEN T C and HOLMER G (1996), 'Lipid oxidation determination in butter and dairy spreads by HPLC', *J Food Sci*, **61**(3) 486–489.

37 MCGILL A S and HARDY R (1977), 'Artefact production in the LikensNickerson apparatus when used to extract the volatile flavorous components of cod', *J Sci Food Agric*, **28**(1) 89–92.

38 WIDJAJA R, CRASKE J D and WOOTTON M (1996), 'Changes in volatile components of paddy, brown and white fragrant rice during storage', *J Sci Food Agric*, **71**(2) 218–224.

39 KARAHADIAN C and JOHNSON K A (1993), 'Analysis of headspace volatiles and sensory characteristics of fresh corn tortillas made from fresh masa dough and

spray-dried masa flour', *J Agric Food Chem*, **41**(5) 791–799.

40 BRUNTON N P, CRONIN D A, MONAHAN F J and DURCAN R (2000), 'A comparison of solid-phase microextraction (SPME) fibres for measurement of hexanal and pentanal in cooked turkey', *Food Chem*, **68**(3) 339–345.

41 VAN AARDT M, DUNCAN S E, BOURNE D, MARCY J E, LONG T E, HACKNEY C R and HEISEY C (2001), 'Flavour threshold for acetaldehyde in milk, chocolate milk, and spring water using solid phase microextraction gas chromatography for quantification', *J Agric Food Chem*, **49**(3) 1377–1381.

42 TOMAINO R M, PARKER J D and LARICK D K (2001), 'Analysis of free fatty acids in whey products by solid-phase microextraction', *J Agric Food Chem*, **49**(8) 3993–3998.

43 MARSILI R T (1999), 'Comparison of solid-phase microextraction and dynamic headspace methods for the gas chromatographic-mass spectrometric analysis of light-induced lipid oxidation products in milk', *J Chromatogr Sci*, **37**(1) 17–23.

44 BALTUSSEN E, SANDRA P, DAVID F and CRAMERS C (1999), 'Stir bar sorptive extraction (SBSE), a novel extraction technique for aqueous samples: theory and principles', *J Microcolumn Sep*, **11**(10) 737–747.

45 OCHIAI N, SASAMOTO K, DAISHIMA S, HEIDEN A C and HOFFMAN A (2003), 'Determination of stale-flavour carbonyl compounds in beer by stir bar sorptive extraction with in-situ derivatization and thermal-desorption–gas chromatography–mass spectrometry', *J Chromatogr A*, **986**(1) 101–110.

46 VALERO E, VILLAMIEL M, MIRALLES B, SANZ J and MARTINEZ-CASTRO I (2001), 'Changes in flavour and volatile components during storage of whole and skimmed UHT milk', *Food Chem*, **72**(1) 51–58.

47 KOLAKOWSKA A, SZCYGIELSKI M and GLOWACKA D (1998), 'Losses in n-3 PUFAs during oxidation of fish lipids', *Polish Journal of Food and Nutrition Sciences*, **7/48**(1) 41–49.

48 POVEDA J M, PEREZ-COELLO M S and CABEZAS L (1999), 'Evolution of the free fatty acid fraction in Manchego cheese during ripening', *Milchwissenschaft*, **54**(12) 685–687.

49 BERGAMO P, FEDELE E, BALESTRIERI M, ABRESCIA P and FERRARA L (1998), 'Measurement of malondialdehyde levels in food by high-performance liquid chromatography with fluorometric detection', *J Agric Food Chem*, **46**(6) 2171–2176.

50 KEY-WHANG (1999), 'High performance liquid chromatography (HPLC) detection of malonaldehyde–thiobarbituric acid (MA–TBA) complex in ground pork', *Food Sci Nutr*, **4**(3) 171–174.

51 STEENHORST-SLIKKERVEER L, LOUTER A, JANSSEN H G and BAUER-PLANK C (2000), 'Analysis of nonvolatile lipid oxidation products in vegetable oils by normal-phase high-performance liquid chromatography with mass spectrometric detection', *J Am Oil Chem Soc*, **77**(8) 837–845.

52 UNDELAND I, HARROD M and LINGNERT H (1998), 'Comparison between methods using low-toxicity solvents for the extraction of lipids from herring (*Clupea harengus*)', *Food Chem*, **61**(3) 355–365.

53 CROWE T D, CROWE T W, JOHNSON L A and WHITE P J (2002), 'Impact of extraction method on yield of lipid oxidation products from oxidised and unoxidised walnuts', *J Am Oil Chem Soc*, **79**(5) 453–456.

54 BOSELLI E, BONOLI M, CABONI M F and LERCKER G (2002), 'Determination of cholesterol oxidation products in the supercritical carbon dioxide extract of egg

yolk powder: comparison with conventional liquid solvent extraction methods', *European Food Res Technol*, **215**(1) 72–75.

55 STEPHAN A and STEINHART H (2000), 'Bitter taste of unsaturated free fatty acids in emulsions: contribution to the off-flavour of soyabean lecithins', *European Food Res Tech*, **212**(1) 17–25.

56 WOO A H Y (1983), 'Characterisation of the relationship between free fatty acids and dairy flavours', *Diss Abs Int-B*, **44**(3) 742.

57 FERNANDEZ-GARCIA E, LOPEZ-FANDINO R, ALONSO L and RAMOS M (1994), 'The use of lipolytic and proteolytic enzymes in the manufacture of Manchego type cheese from ovine and bovine milk', *J Dairy Sci*, **77**(8) 2139–2149.

58 FAID M, CHABARD J L, BERGER J A and LARPENT J P (1989), 'Experimental processing of Moroccan smen: application of lipolytic microorganisms', *Revue Française Corps Gras*, **36**(5) 221–225.

59 CHUN H N and KIM Z U (1991), 'Evaluation of soyabean oil rancidity by pentanal and hexanal determination', *J Korean Agric Chem Soc*, **34**(2) 149–153.

60 MALCOLMSON L H, VAISEY-GENSER M, PRZYBYLSKI R, RYLAND D, ESKIN N A M and ARMSTRONG L (1997), 'Characterisation of stored regular and low-linolenic canola oils at different levels of consumer acceptance', *J Am Oil Chem Soc*, **73**(9) 1153–1160.

61 NOR-AINI-IDRIS, ABDULLAH A and HALIM A H (1992), 'Evaluation of palm oil quality: correlating sensory with chemical analysis', *J Am Oil Chem Soc*, **69**(3) 272–275.

62 PADDMAJA R J, RAMESH V B, SUDERSHAN R V and NADAMUNI N A (2001), 'Suitability of chemical parameters in setting quality standards for deep-fried snacks', *Food Qual Pref*, **12**(4) 223–228.

63 HARDY R and SMITH J G M (1976), 'The storage of mackerel (*Scomber scombrus*). Development of histamine and rancidity', *J Sci Food Agric*, **27**(7) 595–599.

64 KOLKOWSKA A and DEUTRY J (1983), 'Some comment on the usefulness of 2-thiobarbituric acid test (TBA) for the evaluation of rancidity in frozen fish', *Nahrung*, **27**(5) 513–518.

65 STOICK S M, GRAY J I, BOOREN A M and BUCKLEY D J (1991), 'Oxidative stability of restructured beef steaks processed with oleoresin rosemary, tertiary butylhydroquinone, and sodium tripolyphosphate', *J Food Sci*, **56**(3) 597–600.

66 MARCUSE R and POKORNY J (1994), 'Higher correlation with sensory evaluation of oxidative rancidity by modified TBA test', *Fett Wissenschaft Technologie-Fat Sci Tech*, **96**(5) 185–187.

67 BOU R, GUARDIOLA F, GRAU A, GRIMPA S, MANICH A, BARROETA A and CODONY R (2001), 'Influence of dietary fat source, alpha tocopherol and ascorbic acid supplementation on sensory quality of dark chicken meat', *Poultry Sci*, **80**(6) 800–807.

68 MIELNIK M B, AABY K, ROLFSEN K, ELLEKJAER M R and NILSOSON A (2002), 'Quality of comminuted sausages formulated from mechanically deboned poultry meat', *Meat Sci*, **61**(1) 73–84.

69 YOUNATHAN M T, MARJAN Z M and ARSHAD F B (1980), 'Oxidative rancidity in stored ground turkey and beef', *J Food Sci*, **45**(2) 274–278.

70 RUIZ-CAPILLAS C and MORAL A (2001), 'Correlation between biochemical and sensory quality indices in hake stored in ice', *Food Res Int*, **34**(5) 441–447.

71 DEWINNE A and DIRINCK P (1996), 'Studies on vitamin E and meat quality. 2. Effect of feeding high vitamin E levels on chicken meat quality', *J Agric Food Chem*, **44**(7) 1691–1696.

72 FREEMAN D W and SHANNON C W (1994), 'Rancidity in frozen catfish fillets as influenced by antioxidant injection and storage', *J Aquatic Food Prod Tech*, **3**(1) 65–76.

73 TIMON A I B, CAVA R, GARCIA R J, CARMONA J M, SOARES M and BOTE C J L (1999), 'Dietary alpha tocopherol acetate supplementation modifies volatile aldehyde and sensory properties of dry cured ham', *Irish J Agric Food Res*, **38**(1) 137–142.

74 CRAIG J A, BOWERS J A, WANG X Y and SEIB P A (1996), 'Inhibition of lipid oxidation in meats by inorganic phosphate and ascorbate salts', *J Food Sci*, **61**(5) 1062–1067.

75 FRITSCH C W and GALE J A (1977), 'Hexanal as a measure of rancidity in low fat foods', *J Am Oil Chem Soc*, **54**(6) 225–228.

76 CABRAL A C D (1979), 'Determination of hexanal in fat-rich foods by GLC', *Coletanea do Instituto de Tecnologia de Alimentos*, **10** 73–97.

77 ST ANGELO A J, VERCELLOTTI J R, LEGRTENDRE M G, VINNETT C H, KUAN J W, JAMES C and DUPUY H P (1989), 'Chemical and instrumental analysis of warmed-over flavour in beef', *J Food Sci*, **52**(5) 1163–1168.

78 BREWER M S, VEGA J D and PERKINS E G (1999), 'Volatile compounds and sensory characteristics of frying fats', *J Food Lipids*, **6**(1) 47–61.

79 HUMAN J and KHAYAT A (1981), 'Quality evaluation of raw tuna by gas chromatography and sensory methods', *J Food Sci*, **46**(3) 868–873.

80 BUSCAILHON S, BERDAGUE J L, BOUSSET J, CORNET M, GANDEMER G, TOURAILLE C and MONIN G (1994), 'Relations between compositional traits and sensory qualities of French dry-cured ham', *Meat Sci*, **37**(2) 229–243.

81 BREWER M S, IKINS W G and HARBERS C A Z (1992), 'TBA values, sensory characteristics and volatiles in ground pork during long term frozen storage – effects of packaging', *J Food Sci*, **57**(3) 558–563, 580.

82 CRACKEL R L, GRAY J I, BOOREN A M, PEARSON A M and BICKLEY D J (1988), 'Effect of antioxidants on lipid stability in restructured beef steaks', *J Food Sci*, **53**(2) 656–657.

83 NISSEN L R, MANSSON L, BERTELSEN G, HUYNH-BA T and SKIBSTED L H (2000), 'Protection of dehydrated chicken meat by natural antioxidants as evaluated by electron spin resonance spectrometry', *J Agric Food Chem*, **48**(11) 5548–5556.

84 ZHANG H Z and LEE T C (1997), 'A novel gel adsorption/near infrared spectroscopic method for the determination of hexanal as an example of a volatile compound', *J Agric Food Chem*, **45**(8) 3083–3087.

85 RUTH S M, ROOZEN J P, COZIJSEN J L and POSTHUMUS M A (1995), 'Volatile compounds of rehydrated French beans, bell peppers and leeks. II Gas chromatography/sniffing port analysis and sensory evaluation', *Food Chem*, **54**(1) 1–7.

86 APARICIO R, ROCHA S M, DELGADILLO I and MORALES M T (2000), 'Detection of rancid defect in virgin olive oil by the electronic nose', *J Agric Food Chem*, **48**(3) 853–860.

87 GARCIA-GONZALEZ D L and APARICIO R (2002), 'Detection of defective olive oils by metal oxide sensors', *European Food Res Tech*, **215**(2) 118–123.

88 HAUGEN J E and UNDERLAND I (2003), 'Lipid oxidation in herring fillets (*Clupea harengus*) during ice storage measured by a commercial hybrid gas-sensor array system', *J Agric Food Chem*, **51**(3) 52–759.

89 WOLD J P, MIELNIK M K, PETTERSEN K A and BAARDSETH P (2002), 'Rapid assessment of rancidity in complex meat products by front face fluorescence spectroscopy', *J Food Sci*, **76**(6) 2397–2404.

90 ANDERSON M L and SKIBSTED L H (2002), 'Detection of early events in lipid oxidation by electron spin resonance spectroscopy', *European J Lipid Sci Technol*, **104**(1) 65–68.

91 SHAHIDI F and SPURVEY S A (1996), 'Oxidative stability of fresh and heat-processed dark and light muscles of mackerel', *J Food Lipids*, **3**(1) 13–26.

92 NINFALI P, BACCHIOCCA M, BIAGIOTTI E, SERVILI M and MONTEDORO G (2002), 'Validation of the oxygen radical absorbance capacity (ORAC) parameter as a new index of quality and stability of virgin olive oil', *J Am Oil Chem Soc*, **79**(10) 977–982.

93 ZIELINSKI T L, SMITH S A, PESTKA J J, GRAY J I and SMITH D M (2001), 'ELISA to quantify hexanal-protein adducts in a meat model system', *J Agric Food Chem*, **49**(6) 3017–3023.

94 KRESS-ROGERS E and BRIMELOW C J B (2001), *Instrumentation and Sensors for the Food Industry*, Cambridge, Woodhead Publishing.

14

Accelerated shelf-life tests

S. Mizrahi, Technion-Israel Institute of Technology

14.1 Introduction

The food industry has a great need to obtain, in a relatively short time, the necessary information for determining the shelf-life of its products. It has a very important impact on handling of the products' storage, distribution and shelf-life dating.[1] Moreover, it provides an essential tool to probe the possibilities of extending shelf-life through proper product formulation and processing techniques. For practical reasons, especially when the actual storage time is long, the industry resorts to accelerated test techniques that considerably shorten the process of obtaining the necessary experimental data. In the context of this chapter, therefore, accelerated shelf-life testing (ASLT) will refer to any method that is capable of evaluating product stability, based on data that is obtained in a significantly shorter period than the actual shelf-life of the product.

This chapter will discuss first the scientific basis of accelerated shelf-life testing. It will indicate what tools are available for carrying out the tests and explain the problems encountered when using them. At the end, an attempt is made to suggest where this important area of accelerated shelf-life testing is heading and what expectations one should have with regard to developing novel practical and reliable tools that the industry will find convenient to use.

14.2 Basic principles

ASLT is applicable to any deterioration process that has a valid kinetic model. That process may be chemical, physical, biochemical or microbial. The principles of the ASLT will be the same in all cases. However, most of the studies on ASLT have been done on chemical deterioration of foods and therefore the examples in this chapter will be based on them.

There are a number of approaches to ASLT but all are concerned with how to get reliable deterioration data in a short period, what model to use and how eventually to predict the actual shelf-life of the product. All these questions will be dealt with in relation to the different ASLT methods that are discussed in the following sections.

14.3 Initial rate approach

Conceptually, one of the simplest techniques for accelerating the shelf-life testing is the 'initial rate approach'.[2] It may be applicable to cases where the deterioration process can be monitored by an extremely accurate and sensitive analytical method. This method should be capable of measuring minute changes in the extent of deterioration after a relatively short storage time at actual conditions. In such a case, it is possible to get the kinetic data of the initial rate of the deterioration process at a very early stage of the process. To predict the actual shelf-life, one needs only to know or to evaluate how the deterioration process behaves as a function of time. In chemical reactions that information is provided by the order of reaction (n). In the case of monitoring the change in concentration c of a component of interest, the kinetic equation may be expressed as:

$$\frac{dC}{dt} = KC^n \qquad [14.1]$$

where K is the kinetic constant and t is time. For sake of simplicity, let us define an index of deterioration (D) that has the form:

$$dD = \frac{dC}{C^n} = K dt \qquad [14.2]$$

By doing that, the index of deterioration will be always linear with time and will have the following form:

$$D - D_0 = Kt \qquad [14.3]$$

where D_0 is the initial level of the index of deterioration. Equation 14.3 is the only kinetic model that is required to employ this approach to ASLT and the extrapolation process, after evaluating the value of K from the initial rate, is obviously very simple. The product shelf-life (t_s) is therefore:

$$t_s = \frac{D - D_0}{K} \qquad [14.4]$$

Fortunately, information about the order of reactions in many food systems is available in the literature. Most of the chemical deterioration reactions in foods follow either a zero or a first order kinetics. The value of the index of deterioration will be in these cases:

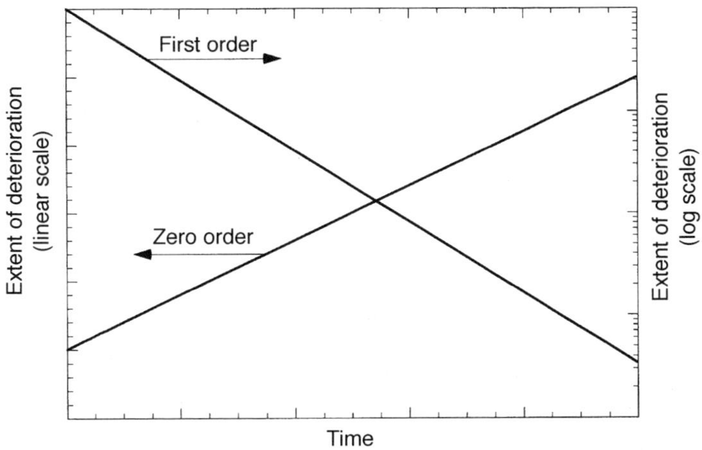

Fig. 14.1 Extent of deterioration as a function of time for zero and first order kinetics.

(a) Zero order ($n = 0$) $D = C$ [14.5]

(b) First order ($n = 1$) $D = C$ [14.6]

On a time scale it is translated to a linear or semi-logarithmic relationship, respectively (Fig. 14.1).

When the order of reaction is unknown, a simple accelerated test procedure may be used to evaluate it empirically. In that case the simplest version of the kinetic model approach, which is discussed in the following sections, may be used. Such a method uses any convenient kinetically active factor to accelerate the deterioration process.

The initial rate method, when applicable, can provide an ideal accelerated shelf-life testing technique. It has the advantage of obtaining, in a relatively short time, the kinetic data at the actual storage conditions and yet requires only the simplest kinetic model that relates solely to the order of reaction.

An example of using a relatively sensitive analytical method was attempted by Teixeira Neto *et al.* to determine the rate of oxygen uptake during oxidation of dehydrated foods.[3] The commonly used manometric techniques are notorious for being insensitive to minute changes in the relatively large mass of oxygen in the headspace.[4] Instead of using this method, Teixeira Neto *et al.*[3] determined the rate of oxygen uptake by analyzing the changes in the mass of the oxygen, which was adsorbed or entrapped in the product.[5] Since that mass is relatively much smaller than that of the manometric method, the data of the rate of oxygen uptake by the product was obtained in only a few days.

The discussion of the initial rate approach may serve also as an appropriate reminder to why there is a need to have other accelerated shelf-life testing methods. In the absence of a very sensitive and accurate analytical technique, the deterioration process should be allowed to progress for longer to enable the available method to detect the changes in a statistically significant way. The minimal time required to obtain significant data is therefore dependent on the accuracy and sensitivity of the analytical method; the worse they are the longer, the time needed to obtain the data. In a way, accelerated shelf-life testing is required to overcome the shortcoming of the analytical methods that are used by the industry. Therefore, the selection of the proper analytical techniques for monitoring the deterioration process is of great importance to shorten the period of the accelerated shelf-life testing.

14.4 Kinetic model approach

The kinetic model approach is the most common method for accelerated shelf-life testing. The basic process involves the following steps:

- Selection of the desired kinetically active factors for acceleration of the deterioration process.
- Running a kinetic study of the deterioration process at such levels of the accelerating factors that the rate of deterioration is fast enough.
- By evaluating the parameters of the kinetic model, extrapolating the data to normal storage conditions (Fig. 14.2).
- Use the extrapolated data or the kinetic model to predict shelf-life at actual storage conditions.

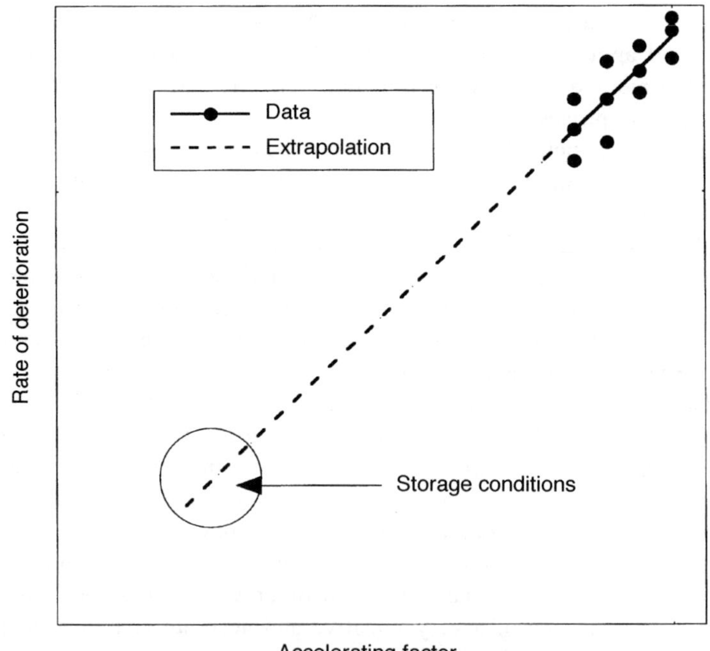

Fig. 14.2 Schematic diagram of data extrapolation in accelerated shelf-life testing.

The absolute requirement for using this procedure is to have a valid kinetic model for the deterioration process. The general and most comprehensive kinetic model for chemical reactions in foods includes all the factors that may affect their rate. These factors may be divided into two main groups, namely compositional (CF_i) and environmental factors (EF_j).[6] The model may be generally expressed as follows:

$$\frac{dD}{dt} = K(CF_i, EF_j) \qquad [14.7]$$

This equation indicates that the kinetic constant K is a function of these factors. In practice, however, one does not need a comprehensive kinetic model. For prediction of shelf-life at actual storage conditions, the model should include only those factors that change during storage (SF_i). Therefore, the required model should look as follows:

$$\frac{dD}{dt} = K(SF_i) \qquad [14.8]$$

The list of SF_i should include factors such as temperature, moisture content, light intensity, composition and others, but only if they change during storage. Obviously, when one is interested in predicting the shelf-life of a product at constant temperature, it is of no interest to have a kinetic model that includes this factor. Yet temperature can be used very effectively to accelerate the rate of the deterioration process. Therefore, the demands from a kinetic model for ASLT may be different from one that is used only to predict shelf-life. The model for accelerated shelf-life testing should contain two groups of factors. The first comprises those that are changing during storage (SF_i), as is in equation 15.8, and the second those that are used to accelerate the rate of reaction (AF_j). The kinetic model for ASLT therefore has the form:

$$\frac{dD}{dt} = K(SF_i, AF_j) \qquad [14.9]$$

The kinetic model for accelerated shelf-life testing may therefore be different from the one usually used to predict product stability at normal storage conditions. Obviously, any of the factors that are changing during storage may be used to accelerate the rate of reaction.

Equation 14.9 expresses a concept of great practical importance for ASLT. It indicates that it is possible to use any desired factor to accelerate the process of deterioration regardless of whether it is active during normal storage conditions. Weissman *et al.*[7] have suggested that one might even use compositional factors to accelerate the rate of deterioration. This implies that the composition of a product may be altered just for the benefit of accelerating the deterioration rate. Clearly, the information obtained is useful only if a valid kinetic model is available for these compositional factors. Such a concept can open a large number of creative avenues for conducting accelerated shelf-life testing.

14.4.1 Single accelerating factor

When applying the kinetic model approach, the first question that has to be considered is whether to use a single or multiple factors for accelerating the deterioration reaction, as well as which ones to choose. The simplest and most commonly used method of ASLT is based on employing only a single factor to expedite the deterioration process. The simplicity of such a method is related to both the experimental procedure and the extrapolation of data. As already stated, such tests requires a valid kinetic model. It should be emphasized that in ASLT, the validity of the kinetic model is crucial to obtaining accurate prediction of the shelf-life. Unfortunately, the validity of the model cannot be fully verified by the ASLT procedure, because the levels used for the accelerating factor do not include those of actual storage conditions. This is in contrast to the situation where the kinetic model is established and verified for actual storage conditions. Therefore, the selection of a model for ASLT must be based on prior knowledge of its validity. The latter may rely either on available empirical data or on a sound physical and/or chemical theory, which has been extensively tested in a large number of similar cases. The Arrhenius model that relates the rate of a chemical reaction to the changes in temperature is the best example of such a validated model. This model is represented by:

$$K = K_0 \exp\left(-\frac{E_a}{RT}\right) \qquad [14.10]$$

where K_0 is a constant, E_a the energy of activation, R the gas constant and T the absolute temperature. Since this model has been used in many cases, a large database is available, mainly of the energy of activation of different reactions. One may conveniently use this information to get a reasonable estimate of the extent a change in temperature may affect the rate of reaction. To simplify the process further, one may get over the need to evaluate K_0 by using a ratio between the rates of reaction when the temperature is changed by any arbitrary value. The most commonly used value is 10 °C and therefore the ratio between the rate of reactions is known as Q_{10}. The value of Q_{10} may be calculated using equation 15.8 to express the rate of reaction first for a temperature of $(T+10)$ and then for T and divide the two, namely:

$$Q_{10} = \frac{\frac{dD_2}{dt}}{\frac{dD_1}{dt}} = \frac{\exp\left(\frac{E_a}{R(T+10)}\right)}{\exp\left(\frac{E_a}{RT}\right)} = \exp\left(\frac{10E_a}{RT(T+10)}\right) \qquad [14.11]$$

The simplicity of using Q_{10} has made it a very popular method for estimating shelf-life. If prior knowledge or estimates of the value of the energy of activation are relied on, the accelerated tests must be run only at one elevated temperature. When choosing the maximal possible temperature, for which the Arrhenius model is still valid, the data are obtained in the shortest possible time by minimal experimental efforts. To improve the accuracy of this version of tests further, the energy of activation may also be evaluated. In that case, the rate of reaction must be obtained at a number of different temperatures below the maximal one in order to be within the range where the model is valid. Obviously, such a procedure takes a much longer time to run. The rule in accelerated stability tests is that to get more accurate data requires a longer experimental time.

The popularity of using the Arrhenius model has made it synonymous with ASLT. Most of the reported ASLTs are based on this model.[8–13] Owing to its popularity, the use of the Arrhenius model has received a lot of attention, especially with regard to two subjects. The most important one has to do with the validity of this model, especially when changes in the mechanism of reaction might take place due to phase transition, competitive reactions, glass transition, etc.[14] The second one is related to the evaluation of the statistical methods used to fit the model with the empirical data.[15, 16]

The most common way of accelerating the rate of reaction is by placing the product at elevated constant temperatures. However, non-isothermal procedures, using programmed changes in conditions, were also tested.[1, 17–19] This is an example of a dynamic test approach, which will be discussed later, where the accelerating factor changes with time. In a procedure where samples are withdrawn from the test for analysis, this type of approach has no advantage over the isothermal method. Moreover, it might have a severe drawback when the samples are not allowed to stay at the lower range of temperature for long enough. In such a case the data obtained for that range are not as accurate as the isothermal method where the sample is kept as long as necessary. The only possible practical advantage that the non-isothermal may have over the isothermal one is where the number of samples is very small and their deterioration process can be monitored continuously.

The use of the Arrhenius model is questionable if it has to deal with changes in the reaction mechanism mentioned above. However, even if it is valid, its use, or rather any approach that is based on a single accelerating factor, may be problematic with regard to the accuracy of the extrapolated data. To demonstrate that problem, let us consider first a simple case where the kinetic constant of the reaction is linearly related to the accelerating factor (Fig. 14.3). In this figure, the solid line represents the true relationship between the kinetic constant (y) and the accelerating factor (x). The point at the top end of the line represents the true kinetic constant (Y_e) at the level (X_e), which may be estimated from the experimental data. To extrapolate the data, the slope (a) of the line must be evaluated by curve fitting of the accelerated test's kinetic data. That value of the slope is used to extrapolate the line to actual storage conditions (X_s) where the true rate of reaction is supposed to be (Y_s). However, the error in the slope (Δa) may cause the extrapolated line to produce a predicted kinetic constant (Y_p (high) or Y_p (low)) which deviates from that true value (Y_s) by ΔY (Fig. 14.3). For the line that has a slope of ($a-\Delta a$), which is symmetrical to the one with a slope of ($a+\Delta a$), the following expression should hold:

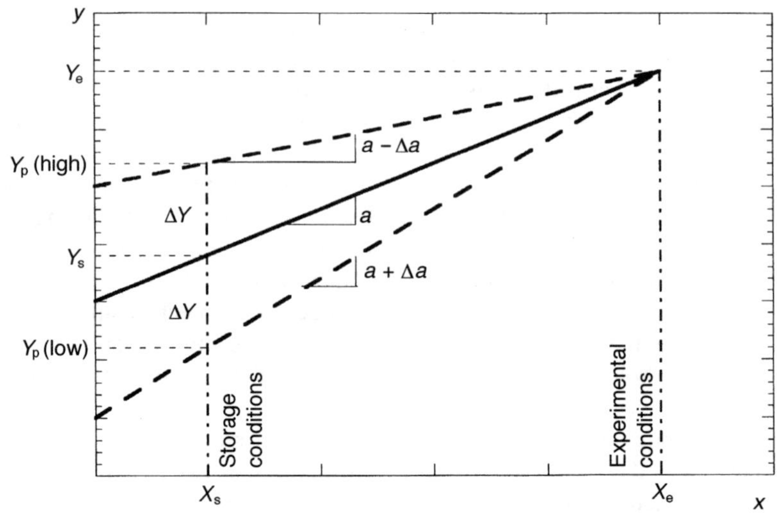

Fig. 14.3 Analysis of extrapolation error in linear plot.

$$\frac{Y_e - Y_p\,(\text{high})}{X_e - X_s} = \frac{Y_e - (Y_s + \Delta Y)}{X_e - X_s} = a - \Delta a \qquad [14.12]$$

For the true line:

$$\frac{Y_e - Y_s}{X_e - X_s} = a \qquad [14.13]$$

Subtracting equation 14.13 from equation 14.12 one obtains:

$$\frac{\Delta Y}{X_e - X_s} = \Delta a \qquad [14.14]$$

To find how the error in evaluating slope (a) affects the accuracy of the extrapolated value, equation 14.14 should be divided by equation 14.13, resulting in the following expression:

$$\frac{\Delta a}{a} = \frac{\Delta Y}{Y_e - Y_s} = \frac{\Delta Y}{Y_s\left(\frac{Y_e}{Y_s} - 1\right)} \qquad [14.15]$$

Therefore, the error in the extrapolated value is:

$$\frac{\Delta Y}{Y_s} = \frac{\Delta a}{a}\left(\frac{Y_e}{Y_s} - 1\right) \qquad [14.16]$$

Let us define the acceleration ratio (AR) as the rate of the accelerated reaction in reference to that at normal storage conditions. In case of the linear relationship between the kinetic constant and the accelerating factor, the value of that acceleration ratio is expressed as:

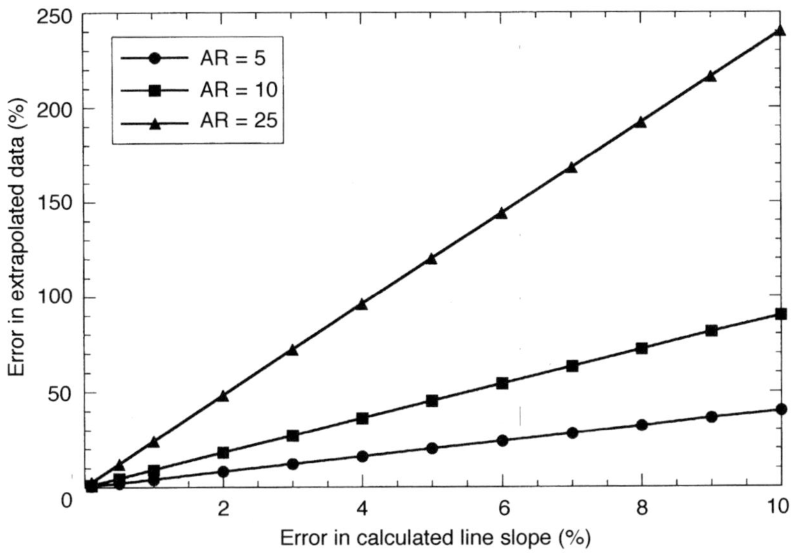

Fig. 14.4 Error in linear extrapolation.

$$\mathrm{AR} = \frac{Y_e}{Y_s} \qquad [14.17]$$

Therefore, the relative error of the predicted value of the kinetic constant is:

$$\frac{\Delta Y}{Y_s} = \frac{\Delta a}{a}(\mathrm{AR} - 1) \qquad [14.18]$$

The extrapolation process multiplies the experimental error of evaluating the slope of the line by the acceleration ratio minus one. The error of the predicted kinetic constant may be extremely high, especially when a very high acceleration ratio is used and if special care is not taken to reduce the experimental error to a very low value (Fig. 14.4). The magnitude of the error changes when the relationship between the kinetic constant and the accelerating factor is no longer linear. In the case, for example, when that relationship is exponential (Arrhenius model) or a power law, the extrapolation error may be different and it can be estimated by turning these models into their linear form and then using the above equations. The only step needed is to assign the y-axis the value of $\ln K$. In such a case, equation 14.18 will read:

$$\frac{\Delta \ln K}{\ln K_s} = \frac{\ln K_p - \ln K_s}{\ln K_s} = \frac{\ln \dfrac{K_p}{K_s}}{\ln K_s} = \frac{\Delta a}{a}\left(\frac{\ln K_e}{\ln K_s} - 1\right) \qquad [14.19]$$

Therefore:

$$\ln \frac{K_p}{K_s} = \frac{\Delta a}{a}(\ln K_e - \ln K_s) = \frac{\Delta a}{a}\left(\ln \frac{K_e}{K_s}\right) = \frac{\Delta a}{a}\ln \mathrm{AR} \qquad [14.20]$$

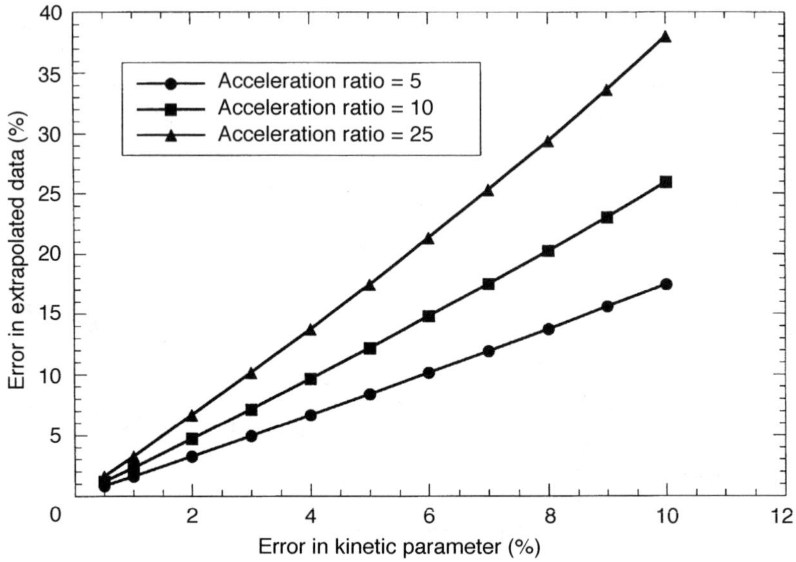

Fig. 14.5 Error in exponential or power law extrapolation.

That results in:

$$\frac{K_p}{K_s} = (\text{AR})^{\Delta a/a} \tag{14.21}$$

The error in the extrapolated data is:

$$\frac{K_p - K_s}{K_s} = \frac{K_s(\text{AR}^{\Delta a/a} - 1)}{K_s} = \text{AR}^{\Delta a/a} - 1 \tag{14.22}$$

It appears, therefore, that using a model like the Arrhenius equation involves a lower error in extrapolating data (Fig. 14.5) than in the case of a simple linear model (Fig. 14.4).

14.4.2 Glass transition models

One of the most interesting approaches to kinetic studies and their use for ASLT is based on glass transition models, which were borrowed from polymer science. Clearly, this approach may be applicable only to products that are in the physical state for which such models are valid. These models, such as the Williams, Landel and Ferry (WLF) model, relate changes in the system properties, which are related to the polymer molecular mobility, to the temperature within the range of the transition of the product from its glassy to rubbery state.[20] Based on the assumption that the rate of the deterioration reactions should relate to molecular mobility in much the same way, this approach yielded valuable information about processes of recrystallization, and losses of flavor and desired textural attributes caused by such structural changes.[21] When applicable, the glass transition models offer a number of very attractive features with regard to kinetic studies and ASLT. The first one is the fact that it combines both the effects of the temperature and the moisture content into one relatively simple equation.[22] The second one, which is even more interesting, is that the rate of the deterioration is related only to the physical state of the system, which can be independently determined in a very short time by readily available physical techniques. That considerably simplifies the experimental work since one needs only the kinetic data, at one high level of temperature or moisture content, and the physical characterization of the system. Unfortunately, that kind of interesting approach to ASLT has, so far, found very limited use. In general, the glass transition model was found to correspond closely to a stability limit with respect to physical processes, such as the ones mentioned above.[23] On the other hand, the glass transition model proved inadequate to account for different deterioration kinetics. [21,24–28] In general, the glass transition model failed to account for diffusion of some small molecules, especially water. However, it has been proposed that the glass transition model may be applicable to predict changes in the rate of chemical reactions in food deterioration but only if proven to be diffusion limited.

14.4.3 Multiple accelerating factors

The use of multiple accelerating factors presents an effective approach to obtain a high acceleration ratio of the deterioration reaction at a minimal cost of prediction error. To demonstrate this fact, let us consider a simple theoretical case of a kinetic model that has the following form:

$$K = (c_1F_1)(c_2F_2) = c_1c_2F_1F_2 \qquad [14.23]$$

where c_1 and c_2 are the estimated parameters of the accelerating factors F_1 and F_2, respectively. In order to evaluate the error in the kinetic constant due to that of the estimated parameters, equation 14.23 is differentiated with regard to these parameters, resulting in:

$$\mathrm{d}K = c_2F_1F_2\mathrm{d}c_1 + c_1F_1F_2\mathrm{d}c_2 \qquad [14.24]$$

When dividing equation 14.24 by equation 14.23 and combining it with equation 14.18, the estimated error is found from the following expression:

$$\frac{\Delta K}{K} = \frac{\Delta c_1}{c_1} + \frac{\Delta c_2}{c_2} = (\mathrm{AR}_1 - 1)\mathrm{RE}_1 + (\mathrm{AR}_2 - 1)\mathrm{RE}_2 \qquad [14.25]$$

where RE_1 and RE_2 are the experimental relative errors for the factors F_1 and F_2, respectively. By using multiple factors, a 100-fold acceleration of the deterioration reaction, e.g. a single one may be replaced by two factors each having an acceleration ratio of only 10. This one order of magnitude reduction in the acceleration ratio decreases considerably the extrapolation error. If, for example, the error in estimating the model parameter for each of these factors is only 1%, the extrapolated data might deviate from the real value by 99% (equation 14.18) for a single as compared to 18% (equation 14.25) for two accelerating factors. While the total acceleration effect of using two or more factors is a multiplication of their effect, the error is only the summation. Moreover, the required relatively low acceleration ratio is achieved by a much smaller change in the level of the kinetic factors and thus the system stays much closer to the actual storage conditions. Furthermore, when a narrower range of the accelerating factor is used, not only is the validity of the kinetic model better maintained but also the kinetic model may have a simpler form. The advantages of the multiple factors approach are obtained at a cost of running a more complicated experimental procedure. That is the result of the need to evaluate not only the effect of each factor on the reaction kinetics but also a possible interaction between them. The procedure, therefore, lacks the simplicity that makes such a technique more practical for the food industry.

A multiple factor acceleration of the deterioration reaction was carried out by Mizrahi *et al.* by combining the effect of temperature and moisture content (m).[8] It enabled a shelf-life that lasts for over one year to be predicted based on an experimental study that required only 10 days. The basic kinetic equation had the following general form:

$$K(m, T) = f(m)_{T_r} \exp[E_a/R\left(\frac{1}{T_r} - \frac{1}{T}\right)] \qquad [14.26]$$

where T_r is a reference temperature. Since moisture content in a food product is related to the water activity (a_w) by the sorption isotherm, the kinetic function at the constant reference temperature ($f(m)_{T_r}$) could be expressed also in terms of that water activity.

One form of such a function for non-enzymatic browning of cabbage is:[8]

$$K = K_0(a_w)^s \qquad [14.27]$$

The kinetic model shown in equation 14.26 indicates that the evaluation of the kinetic effect of moisture content is performed for a constant reference temperature (T_r). Theoretically, therefore, the evaluation of the kinetic model may be as simple as first running an experiment at an elevated constant temperature and changing only the moisture content and then keeping the latter constant at any desired level and varying the temperature. In many cases, especially when the range of temperature and moisture content changes are kept within a relatively narrow range, that procedure may be adequate. However, when that range is relatively large, a possible interaction between the two factors might play an important role in determining the accuracy of the shelf-life prediction. Such was the case in the study of the non-enzymatic browning of cabbage where the energy of activation happened to be affected by the moisture content.[8] The empirical expression that was used to describe the effect of the moisture content on the energy of activation was:

$$E_a = c_1 \exp(-c_2 m) \qquad [14.28]$$

where c_1 and c_2 are constants. That interaction between the factors greatly complicates the experimental procedure since the effect of the moisture content on the energy of activation should be tested by changing both factors at the same time. That requires much longer time and more experimental work, which may make this method very unattractive for practical use. However, as stated before, when a narrower range of the accelerating factors is used, that elaborate and cumbersome procedure may not be necessary.

14.4.4 Accelerated methods for establishing a kinetic model

The lack of well-proven general kinetic models often makes it necessary to establish or to validate a model for ASLT. Since the commonly used procedure to establish a reliable kinetic model may take a longer time than the actual shelf-life of the product, an accelerated method was developed to do it. Such a method is based on a dynamic testing procedure.[29–31] The product is subjected to conditions where the kinetically active factor is programmed to change with time in any desired way. That creates a situation where both the extent of deterioration and the value of the kinetic factor are changing with time (Fig. 14.6). At any given time, namely at a given level of the kinetic factor, the rate of reaction can be obtained by a numerical or graphical derivative of the deterioration curve. In that way one obtains the relationship between the value of the kinetic factor and the rate of reaction. The reason such a method requires a relatively short time is because most of the deterioration is taking place at the levels of the kinetic factor where the rates are very high. This casts a serious question on the accuracy of using the obtained data to establish or validate a model that should apply to those levels of the kinetic factor where the rate of

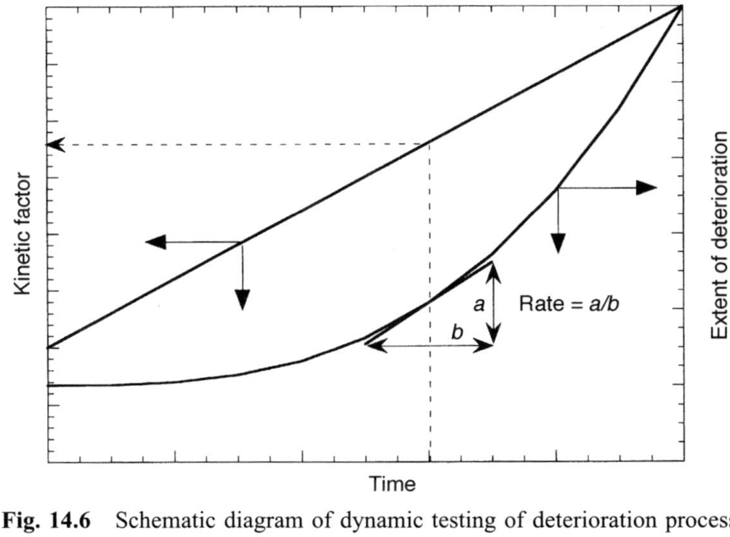

Fig. 14.6 Schematic diagram of dynamic testing of deterioration processes.

reaction is very low. Moreover, since the kinetic factor is programmed to change continuously, the system usually stays at the condition where the rate of reaction is very low for too short a time to develop any significant change in the extent of deterioration. The use of dynamic testing for accelerating the time for establishing a kinetic model is therefore not as accurate as the conventional process that takes a much longer time.

14.4.5 The 'no model' approach

The 'no model' approach is a term used for the accelerated shelf-life testing method that assumes that a valid kinetic model exists but does not require experiments to evaluate it. This approach may apply only to cases where the kinetically active factor (F) is changing during storage in a monotonically and continuous way. The ASLT technique is based on monitoring the extent of deterioration in the same product in which that factor is programmed to change in such a way that it goes through the 'storage' cycle in a shorter period. The obtained data are then converted into real storage conditions by a calculation that is based only on knowing how the kinetically active factor (F) is changing with time (t), namely on having the following function (g):

$$F = g(t) \tag{14.29}$$

The inverse of that equation yields the function (f) of how time relates to the changing factor:

$$t = f(F) \tag{14.30}$$

It should be noted that this equation might have an analytical expression, but may as well represent a numerical or graphical datum. Assuming that a valid kinetic model exists for the deterioration reaction, it will have the following form:

$$\mathrm{d}D = K(F)\mathrm{d}t \tag{14.31}$$

The value of dt may be replaced in this equation by using the derivative of equation 14.30, namely:

$$\mathrm{d}t = f'(F)\mathrm{d}F \tag{14.32}$$

Thus equation 14.31 changes into:

$$\mathrm{d}D = K(F)f'(K)\mathrm{d}F \tag{14.33}$$

When we have two samples of the same product, one at actual storage conditions and the other at accelerated test conditions (denoted by subscript s and a, respectively), the ratio between their rate of deterioration is:

$$\frac{(\mathrm{d}D)_\mathrm{s}}{(\mathrm{d}D)_\mathrm{a}} = \frac{[K(F)f'(F)\mathrm{d}F]_\mathrm{s}}{[K(F)f'(F)\mathrm{d}F]_\mathrm{a}} \tag{14.34}$$

Thus, the rate of deterioration at actual storage conditions is related to that at accelerated ones by:

$$(\mathrm{d}D)_s = \frac{[K(F)f'(F)\mathrm{d}F]_s}{[K(F)f'(F)\mathrm{d}F]_a}(\mathrm{d}D)_a \qquad [14.35]$$

Let us consider first a situation where the kinetic factor (F) is changing linearly with time both in storage and accelerated test conditions, thus having the following respective expressions:

$$F = F_0 + b_s t \qquad [14.36]$$

$$F = F_0 + b_a t \qquad [14.37]$$

where b is a constant. Using the inverse form of these equations, the ratio of their derivative is:

$$\frac{f_s'(F)}{f_a'(F)} = \frac{b_a}{b_s} \qquad [14.38]$$

Therefore, the ratio between the extent of deterioration in this case is:

$$(D - D_0)_s = \frac{b_a}{b_s}\frac{\left[\int_{F_0}^{F} K(F)\mathrm{d}F\right]_s}{\left[\int_{F_0}^{F} K(F)\mathrm{d}F\right]_a}(D - D_0) = \frac{b_a}{b_s}(D - D_0)_a \qquad [14.39]$$

Since both integrals in this equation are only functions of the factor F, they have the same value and therefore cancel out. The extent of deterioration at storage conditions is therefore obtained by accelerating the change in the kinetically active factor with time and multiplying the obtained data by the ratio of the rates of change.

So far, this method is applicable only to cases where the kinetic factor is changing linearly with time. The application of this approach may be extended also to the general situation, which is expressed by equation 14.35. In that case, it is possible to divide the whole range of these equations to n sections, each of which may be approximated by a straight line with a slope, which can be calculated from the derivative of this equation. The basic equation in this case will be:

$$(\Delta D_j)_s = \frac{f_s'(F_j)\mathrm{d}F}{f_a'(F_j)\mathrm{d}F}(\Delta D_j)_a = \frac{(b_j)_a}{(b_j)_s}(\Delta D_j)_a \qquad [14.40]$$

The extent of deterioration is therefore:

$$(D - D_0)_s = \sum_{j=1}^{n}(\Delta D_j)_s = \sum_{j=1}^{n}\frac{f_s'(F_j)}{f_a'(F_j)}(\Delta D_j)_a \qquad [14.41]$$

This 'no model' approach was developed and successfully tested for a moisture-sensitive dry product.[32] The product was packaged in a water vapour permeable plastic film. Since the water activity in common storage conditions of such a product is higher than that of the packaged foods, the product will continuously absorb moisture through the film. The accelerated shelf-life testing in this case was carried out by packing the same product in a film that has significantly higher water vapour permeability than the original one. In both the actual storage and the accelerated test conditions, the change in moisture content with time is not linear. In fact, the derivative of the relationship between time and moisture content, for the samples that were kept at external constant water activity (a_e) can be expressed:[32]

$$f'(m) = [\mathrm{k}P(a_e - h)(m)]^{-1} \tag{14.42}$$

where h denotes a function of moisture content (m), k is a constant and P is the packaging film permeability to water vapour. If different films are used for storage and for accelerated tests having a permeability of P_s and P_a, respectively, then:

$$\frac{f_s'(\mathrm{m})}{f_a'(\mathrm{m})} = \frac{P_a}{P_s} \tag{14.43}$$

In that case the extent of deterioration is given by:

$$(D - D_0)_s = \frac{P_a}{P_s}(D - D_0)_a \tag{14.44}$$

This is the same solution as the linear case owing to the fact that the external water activity is the same for storage and accelerated tests. Such an accelerated shelf-life testing method is simple to perform, especially since it does not require the evaluation of the kinetic model. However, there is one important problem that should be considered. It has to do with the fact that the higher the rate that one programs the change of the kinetic factor, namely the moisture content in this example, the lower the extent of deterioration. That is simply the result of the fact that the deterioration reaction is given less time to develop. This approach is therefore more effective the better the accuracy and sensitivity of the analytical method used to monitor the deterioration process. In any case, the acceleration ratio in this approach is very dependent on how small a fraction of the total acceptable extent of deterioration may be significantly determined.

14.4.6 Combination of approaches

The application of a combination of methods to accelerated shelf-life testing has the same advantages as using multiple accelerating factors. Such a combination may provide an effective approach in obtaining a high acceleration ratio of the deterioration reaction at a minimal cost of prediction error by staying closer to actual storage conditions. Moreover, this approach provides potentially the largest number of avenues to ASLT. One may use a combination of multiple factors together with initial rate and 'no model' approaches. Mizrahi and Karel have used a combination of the 'no model' approach together with elevated temperature for accelerated stability tests of moisture-sensitive products.[33] This combination presents an interesting case of how to link the effect of two methods where one requires evaluation of the kinetic model and the other one does not. The assumption was that the Arrhenius equation is a valid kinetic model for the rate of deterioration at different temperatures when the moisture content is kept constant. The procedure is based on packing the product in films of different permeability and placing them in an environment of the same, or different, water activity and elevated temperatures. The temperature changes not only the rate of reaction but also the moisture gain. Therefore, in order to evaluate the parameters of the Arrhenius equation one has to separate the two processes. The technique is based on the following steps:[33]

- Arbitrarily select a reference moisture gain curve. It may be, for example, the moisture gain of the product at actual storage conditions. For some cases, one may conveniently select a straight line.
- At each temperature, transform the extent of deterioration to the reference moisture gain line by using the procedure outlined in the 'no model' approach, namely by using equation 14.35 or 14.44 for the simple case where the ratio of the moisture gain is constant.
- Use the transformed data, which are now normalized to the same reference line, to obtain the parameters of the Arrhenius equation.
- Use the combination of the reference data and Arrhenius equation to extrapolate the data to actual storage conditions.

14.5 Problems in accelerated shelf-life tests

The problems that are related to ASLT may be classified into three main groups. The first has to do with those cases where no valid kinetic model is believed to exist for any accelerating kinetic factor. No accelerated test procedure is available for such a case. The second kind of problem is encountered when a model does exist but it is very complicated and requires the evaluation of too large a number of parameters. The experimental procedure in such a case may prove very cumbersome to a point where the ASLT procedure may not be practical. The third group of problems relates to the application of valid ASLT methods. These problems are discussed in the following section.

14.5.1 Absence of deterioration index

Food products may be judged on a basis of sensory evaluation that is influenced by the combined effect of a multitude of different reactions. In many cases, a measurable deterioration index, which correlates well with the sensory evaluation, is unavailable. The product may therefore be judged only on the basis of acceptable or unacceptable and not by a continuous scale, thus eliminating the possibility of using the 'initial rate' or the 'no model' approaches to accelerated stability tests. However, the kinetic model approach may be used in such cases simply by assigning the kinetic constant (K), at constant conditions, a value of:

$$K = 1/t_c \qquad [14.45]$$

where t_c is the critical time that marks the end of the shelf-life of the product. This approach arbitrarily assigns the point of product failure a value of one. As in any other kinetic study, this kinetic constant is evaluated by an experimental procedure that is carried out at different constant storage conditions. The obtained data of the values of the kinetic constant as a function of these conditions provide the basis for evaluating the kinetic model and its parameters. That model can be used for predicting shelf-life by integrating the kinetic equation and finding the time it takes to reach a degree of deterioration of one. This approach is exactly the same as the time–temperature tolerance (TTT) that has been extensively used to predict shelf-life mainly in frozen products.[34, 35]

14.5.2 Time-dependent effects

All available methods for accelerating the product stability tests are based on the ability to predict the progress of the deterioration process based on the order of reaction. This order of reaction can be evaluated by the ASLT procedure. However, the situation becomes much more complicated when other time-dependent effects play a major role in the deterioration process, namely when the deterioration rate is affected by the history of the process.[4, 36, 37] The effect of any specific storage history may be evaluated by carrying out kinetic studies only at actual storage conditions. So far, there is no way to simulate a given storage history by accelerated tests.

14.5.3 Statistical problems

Statistics is an essential part of designing the experimental procedures and analyzing the data both in common kinetic studies as well as in ASLT. It is essential that the proper statistical methods be used in ASLT. One particular subject in that respect, which relates to the validation of kinetic models, should be especially noted. The validity of the model is best established when kinetic data are available for both the actual storage and the accelerated tests conditions. Obviously, the ASLT technique by itself lacks the capability of verifying the validity of the model, especially an empirical one, for actual storage. Moreover, when any model is used its parameters are evaluated only by using the data of the very high rate of reaction. That may produce a large deviation of the extrapolated data to normal conditions. One should therefore use statistical methods that test the sensitivity of the model by a cross-validation method. In principle, these methods are using part of the data to verify the validity of the model. This requires a wider range of accelerated storage conditions. The closer they are to the actual storage conditions the better. Such an approach costs more both in time and in experimental efforts.

14.6 Future trends

Kinetic models of deterioration processes have been so far the main basis for the prediction of foods' shelf-life by accelerated storage tests. The common approach has been based on the assumption that the deterioration process progresses according to a certain model, that other parallel reactions do not affect the outcome of the main process and that the product's 'failure point' is well defined and independent of its handling, storage and distribution history. In reality, the deterioration processes in foods may be too complex to be characterized by a single well-defined chemical reaction. Therefore, in such cases, it is convenient to judge the product's overall quality by using trained or untrained consumer panels. The panelists' evaluations are usually analyzed by statistical methods that are based on the distribution of their responses in terms of the product's 'failure' or 'survival'.[38] One common approach is based on the Weibull distribution, which is often used to characterize the times to failure of many very different non-food products or items. It is therefore no wonder that the same approach and the use of the Weibull distribution as a model have also been adapted to the analysis of food products' shelf-life as perceived by consumers.[39]

According to this model, the product's 'level of survival' after a time t, $S(t)$, is expressed as a fraction of the total and is given by:

$$S(t) = \exp[-(t/\beta)^{\alpha}] \qquad [14.46]$$

where α and β are constants. α is commonly referred to as the shape parameter and β as the scale parameter. $Z = 1/\beta$ may be regarded as a 'rate parameter'.[40] When $\alpha = 1$, the above survival model becomes the familiar first-order kinetics equation, where $\alpha = 1$ is replaced by the exponential decay rate constant, k. The main advantage of the Weibull model is its versatility, which enables it to represent a wide range of failure phenomena. Moreover, recent results indicate that the Weibull distribution may also be a useful tool in the analysis of deterioration reactions kinetics.[40] This suggests that the 'Weibull approach' might offer a new way to design and interpret accelerated shelf-life tests. Such an application of the concept has not yet been tested but its feasibility deserves some consideration. In order to exploit this approach for the analysis of accelerated storage data and the results of shelf-life testing of a food product, it is necessary to assume that its failure pattern indeed follows the Weibull distribution. Because of the mathematical versatility of the model, this requirement is expected to be satisfied by many stored foods. A possible exception might be the cases where the product's deterioration is governed by zero-order kinetics, which the Weibull model does not account for. The same can be said about cases where the deterioration or failure pattern follows a different kind of distribution, in which case the model itself has to be reformulated.

Another consideration is the need to know how the Weibull model's two parameters depend on the contemplated acceleration factors, e.g., temperature, oxygen pressure, pH, moisture contents, etc. Fortunately, the data available so far suggests that the Weibull distribution's shape factor hardly changes with temperature and hence can be treated as being practically constant.[40, 41] If this observation could be proven to be generally true, then one would be able to determine the magnitude of the shape parameter in a relatively short time by running tests at elevated temperatures. In several cases, the rate parameter's temperature dependence has been found to follow the log–logistic relationship:[42]

$$Z(T) = \ln\{1 + \exp[c(T - T_c)]\} \qquad [14.47]$$

where c and T_c are constants. These constants can be easily determined from experimental $Z(T)$ vs. temperature data by non-linear regression. This is similar to the determination of the parameters of the Arrhenius equation, which has been traditionally employed to characterize the temperature dependence of the reaction's rate constant. However, the above model will not be applicable in cases where two or more influential parameters change simultaneously or when the rate parameter's increase with temperature at $T >> T_c$ is non-linear. For the latter case, the model can be modified:[42]

$$Z(T) = \ln\{1 + \exp[c(T - T_c)]\}^m \qquad [14.48]$$

where m is a power that is expected to be close to or higher than one. This model, if used, would require the determination of three adjustable parameters, which would make it less attractive. A shelf-life evaluation procedure based on such a model might not be practical, especially in cases where the conventional methods based on traditional models give satisfactory results. However, it should be noted that since the Weibull model is empirical, its usage does not require any assumption regarding the nature of the chemical mechanisms which control the deterioration process kinetics. Therefore, it will not be surprising if future research will reveal the existence of certain food products where the Weibull approach might have an advantage over the traditional methods to estimate shelf-life from high temperature storage data.

At least in principle, the concept can be extended to other factors that can accelerate the deterioration of foods, notably the mentioned moisture contents or the oxygen concentration, in which case $Z(T)$ will be expressed in terms of the corresponding parameters. i.e., $Z(M)$, $Z(O_2$ conc.), etc. As yet, there are no published reports about relationships between the Weibull distribution model's parameters and these factors. The only information that is available so far is that both the shape and the scale parameters are affected by moisture contents but exactly how is yet to be established.[42] Obviously, much more experimental data will be required to evaluate the potential utility of the concept of replacing or augmenting the effect of temperature as the accelerating factor by other agents.

Also worth mentioning is that the Weibull distribution can serve as the basis of model of *non-isothermal degradation* of nutrients, pigments and enzymes.[40] Such a model allows one not only to calculate the extent of deterioration when a product is exposed to any given thermal history but also to determine its parameters directly from non-isothermal data. Non-isothermal testing is a dynamic technique and hence, as indicated in a previous section, may considerably shorten the testing procedure.

The application of the 'Weibull distribution model' to shelf-life analysis is an example of how one can handle the complexity of deterioration reactions by using a statistical approach. The availability of a versatile mathematical model and the simplicity of its usage make this approach especially attractive and hence worth exploration. One should, therefore, expect that it will be tested more frequently in the future and that it will find a growing number of applications in shelf-life prediction and eventually in providing novel methods for accelerated shelf-life tests.

14.7 References

1. LABUZA TP, *Shelf-life Dating of Foods*, Westport, CT, Food & Nutrition Press, 1982.
2. MIZRAHI S, Novel approaches to accelerated storage tests. In *Engineering and Food*, Vol 2: *Preservation Processes and Related Techniques*, Spiess W E L and Schubert H, eds pp. 794–803, New York, Elsevier Applied Science, 1990.
3. TEIXEIRA NETO R O, KAREL M, SAGUY I and MIZRAHI S, 'Oxygen uptake and β-carotene decoloration in a dehydrated food model', *J of Food Science*, 1981 **46** 665–76.
4. QUAST D G and KAREL M, 'Effects of environmental factors on the oxidation of potato chips', *J of Food Science*, 1972 **37** 584.
5. SAGUY I, GOLDMAN M, HOREV B and KAREL M, 'An improved method for the determination of adsorbed/entrapped gases in dehydrated foodstuffs', *Z Lebensm Untess Forsch*, 1983 **177** 359–63.
6. SAGUY I and KAREL M, 'Modeling of quality deterioration during food processing and storage', *Food Technology*, 1980 **34** (2) 78–85.
7. WEISSMAN I, RAMON O, KOPELMAN I J and MIZRAHI S, 'A kinetic model for accelerated tests of Maillard browning in a liquid model system', *J of Food Processing and Preservation*, 1993 **17** 455–70.
8. MIZRAHI S, LABUZA T P and KAREL M, 'Feasibility of accelerated tests for browning in dehydrated cabbage', *J of Food Science*, 1970 **35** 804–7.
9. MIZRAHI S and KAREL M, 'Accelerated stability tests of moisture sensitive products in permeable packages at high rates of moisture gain and elevated temperatures', *J of Food Science*, 1977 **42** 1575–9.
10. LABUZA T P and RIBOH D, 'Theory and application of Arrhenius kinetics to the prediction of nutrient losses in food', *Food Technology*, 1982 **36** 66–74.
11. LABUZA T P and KAMMAN J, Reaction kinetics and accelerated tests simulation as a function of temperature. In *Application of Computers in Food Research*, Saguy I, ed., New York, Marcel Dekker, 1983.
12. LABUZA T P and SCHMIDL M K, 'Accelerated shelf-life testing of foods', *Food Technology*, 1988 **39** (9) 57–62, 64.
13. TSOUMBELI M N and LABUZA T P, 'Accelerated kinetic study of aspartame degradation in the neutral pH range', *J of Food Science*, 1991 **56** 1671–5.
14. NELSON K and LABUZA T P, 'Water activity and food polymer science: implications

of state on Arrhenius and WLF models in predicting shelf-life'. *J of Food Engineering*, 1994 **22** 271–90.

15. COHEN E and SAGUY I, 'Statistical evaluation of Arrhenius model and its applicability in prediction of food quality losses', *J of Food Processing and Preservation*, 1985 **9** 273–90.
16. HARALAMPU S G, SAGUY I and KAREL M, 'Estimating of Arrhenius model parameters using three least square method', *J of Food Processing and Preservation*, 1985 **9** 129–43.
17. LABUZA T P, 'A theoretical comparison of losses in foods under fluctuating temperature sequences', *J of Food Science*, 1979 **44** 1162.
18. LABUZA T P, BOHNSACK K and KIM M N, 'Kinetic of protein quality loss stored under constant and square wave temperature distributions', *Cereal Chemistry*, 1982 **59** 142.
19. RIBOH D K and LABUZA T P, 'Kinetics of thiamine loss in pasta stored in a sine wave temperature condition', *J of Food Processing and Preservation*, 1982 **6** (4) 253.
20. WILLIAMS M L, LANDEL R F and FERRY J D, 'The temperature dependence of relaxation mechanisms in amorphous polymers and other glass-forming liquids', *J of Chemical Engineering*, 1955 **77** 3701–7.
21. KAREL M, ANGLEA S, BUERA P, KARMAS R and LEVI G, 'Stability related transitions of amorphous foods', *Thermochimica ACTA*, 1994 **246** (2) 249–69.
22. BUERA P and KAREL M, 'Application of the WLF equation to describe the combined effect of moisture, temperature and physical changes on non-enzymatic browning rates in food systems', *J of Food Processing and Preservation*, 1993 **17** 31–47.
23. KAREL M, Food research tasks at the beginning of the new millennium – a personal vision. In *Water Management in the Design and Distribution of Quality Foods*, Roos Y H, Leslie R B and Lillford P J, eds, Lancaster PA, Technomic, pp. 535–59, 1999.
24. CHIRIFE J and BUERA P, 'Water activity, glass transition and microbial stability in concentrated semimoist food systems', *J of Food Science*, 1994 **59** 921–7.
25. BUERA M D P, CHIRIFE J and KAREL M, 'A study of acid-catalyzed sucrose hydrolysis in an amorphous polymeric matrix at reduced moisture contents', *Food Research International*, 1995 **28** (4) 359–65.
26. KARMAS R and KAREL M, 'Modeling Maillard browning in dehydrated food systems as a function of temperature, moisture content and glass transition temperature', *ACS Symp. Ser.*, 1995 **610** 64–73.
27. CARDONA S, SCHEBOR C, BUERA M P, KAREL M and CHIRIFE J, 'Thermal stability of invertase in reduced-moisture amorphous matrices in relation to glassy state and trehalose crystallization', *J of Food Science*, 1997 **62** (1) 105–12.
28. SCHEBOR C, BUERA M P, KAREL M and CHIRIFE J, 'Color formation due to non-enzymatic browning in amorphous glassy anhydrous model systems', *Food Chemistry*, 1999 **65** (4) 427–32.
29. SAGUY I, MIZRAHI S, VILLOTA R and KAREL M, 'Accelerated method for determining the kinetic model of ascorbic acid loss during dehydration', *J of Food Science*, 1978 **43** 1861–4.
30. HARALAMPU S G, SAGUY I and KAREL M, 'Identification of moisture sensitivity models of packaged materials under simulated storage conditions', *Mathematical Modelling*, 1986 **7** 1–13.
31. HARALAMPU S G, SAGUY I and KAREL M, 'The performance of a dynamic stability test for moisture sensitivity', *Mathematical Modelling*, 1986 **7** 15–25.

32. MIZRAHI S and KAREL M, 'Accelerated stability tests of moisture-sensitive products in permeable packages by programming rate of moisture content increase', *J of Food Science*, 1977 **42** 958–63.
33. MIZRAHI S and KAREL M, 'Accelerated stability tests of moisture sensitive products in permeable packages at high rates of moisture gain and elevated temperatures', *J of Food Science*, 1977 **42** 1575–9.
34. VAN ARSDEL W B, COPLEY M J and OLSON R L, *Quality and Stability of Frozen Foods Time–Temperature Tolerance*, New York, Wiley-Interscience, 1969.
35. JUL M, *The Quality of Frozen Foods*, London, Academic Press, 1984.
36. QUAST D G and KAREL M, 'Computer simulation of storage life of foods undergoing spoilage by two interactive mechanisms', *J of Food Science*, 1972 **37** 679.
37. LABUZA T P and RAGNARSSON J O, 'Kinetic history effect on lipid oxidation of methyl linoleate in model system', *J of Food Science*, 1985 **50** (1) 145.
38. GACULA M C JR and SINGH J, *Statistical Methods in Food and Consumer Research*, New York, Academic Press, 1984.
39. CARDELLI C and LABUZA T P, 'Application of Weibull hazard analysis to the determination of the shelf life of roasted and ground coffee', *Lebensmittel-Wissenschaft und-Technologie*, 2001 **34** 273–8.
40. CORRADINI M G and PELEG M, 'A model of non-isothermal degradation of nutrients, pigments and enzymes', *J of the Science of Food and Agriculture* (submitted).
41. VAN BOEKEL M A J S, 'On the use of the Weibull model to describe thermal inactivation of microbial vegetative cells', *International Journal of Food Microbiology*, 2002 **7** 139–59.
42. PELEG M, ENGEL R, GONZALEZ-MARTINEZ C and CORRADINI M G, 'Non-Arrhenius and non-WLF kinetics in food systems', *J of the Science of Food and Agriculture,* 2002 **82** 1346–55.

15

Shelf-life testing

C. M. D. Man, London South Bank University, UK

15.1 Introduction

Shelf-life is both an important concept and an important property of today's food products. Indeed, when expressed in terms of an appropriate indication of minimum durability as is legally required within the European Union (EU), it conveys to the consumer a vital piece of technical information about a food product essential for food safety and enjoyment of the food as it is intended. While minimum durability indication and shelf-life are not synonymous, the length of a shelf-life usually reflects the length of its minimum durability indication; the former provides a scientific basis for the setting of the latter. In the absence of a universal or legal definition, the one adopted by the UK Institute of Food Science and Technology (IFST) is used here, which defines shelf-life as (IFST, 1993): 'the period of time during which the food product will

1. remain safe;
2. be certain to retain its desired sensory, chemical, physical, microbiological and functional characteristics;
3. where appropriate, comply with any label declaration of nutrition data, when stored under the recommended conditions'.

Shelf-life is therefore a multifaceted property that is enormously important to food manufacturers and processors as well as consumers. It is clear from the definition that food safety and desired quality are the two main aspects of an acceptable shelf-life. In reality, it is impossible to have a food product that has quality but is unsafe. Food safety and quality are therefore inextricably linked. For a food product the safety of which has been called into question, for instance as suggested by a product recall, there is no useful shelf-life. Furthermore, the controlling factors for microbiological safety and quality in food are often identical and separate consideration of food safety and quality in the context of shelf-life evaluation is artificial.

What quality characteristics are desired, however, will depend on the food product in question as well as on its manufacturer. Different products naturally have different quality characteristics but similar or even competing products do not necessarily have the same characteristics and identical shelf-lives. Food manufacturers and processors must be able to define the quality characteristics of their products, which they intend their customers to enjoy or which are known to them to be important in determining consumer acceptance. This information is essential if consumer requirements and expectations regarding the product are to be fully and consistently met. Having defined the selling characteristics of a food product, the primary aim of a shelf-life study is to understand how these characteristics change during storage and to identify those changes that eventually cause the product to be unacceptable to the target consumers when properly stored. In practice, however, a manufacturer may well assign to its product a significantly shorter shelf-life than is revealed by the study for marketing and commercial reasons.

As food safety is both a legal and an unequivocal requirement, all shelf-life studies will necessarily include an evaluation of the product in respect of food safety. Once confidence about food safety has been established, the main task is to determine how and when the product reaches the end of its shelf-life, this being informed by prior knowledge of the quality characteristics critical to consumer acceptance as explained earlier. As these characteristics can be chemical, functional, microbiological or physical, and are product specific, manufacturers will need to carry out and repeat if necessary experiments designed to study the progress of shelf-life and its endpoint. It follows that it is unwise for a food business to assign shelf-lives to its products without having first conducted shelf-life experiments that are properly designed, based on sound scientific principles and up-to-date product information.

Despite the vast range and variety of food products available nowadays, much knowledge about food deterioration and spoilage has been accumulated. Consequently the following mechanisms can be used to explain the deterioration and spoilage of many foods (IFST, 1993):

- Moisture and/or water vapour transfer leading to gain or loss
- Physical transfer of substances other than moisture and/or water vapour, e.g. oxygen, odours or flavours
- Light-induced changes, i.e. changes caused and/or initiated by exposure to daylight or artificial light
- Chemical and/or biochemical changes
- Microbiological changes
- Other mechanisms or changes that cause the food to deteriorate through one or more of the above mechanisms, e.g. damage to pack caused by insect infestation.

In a sense, these are the main changes that can develop into major product quality hazards, which, if uncontrolled, will eventually limit the shelf-lives of many food products. Detailed treatment of these mechanisms can be found in the relevant chapters of this book.

15.2 Assuring food safety: the HACCP system

A test that can be used to establish the global safety of food products does not exist. In fact, it is now widely accepted by many enforcement agencies and the industry alike that end product testing alone is both an expensive and an ineffective way of assuring safety. For instance, microbiological examinations of food products for specific pathogens at best confirm the non-detection of these organisms in a specified sample of food and do not guarantee the complete absence of them in the batch or consignment of products from which the sample is taken. If a pathogen is indeed detected, the product concerned may have been despatched or even consumed in the case of short shelf-life products. Similar arguments apply to chemical contaminants in foods that can cause harm to consumers. Clearly, unless every product is tested, which itself cannot be a practicable proposition, there is little prospect of assuring food safety by means of product testing.

Today the most effective way to ensure the safety of food is to use the internationally recognised Hazard Analysis and Critical Control Point (HACCP) system as adopted and amended by the Codex Alimentarius Commission in 1997 and 1999 respectively. The HACCP system is science based and systematic and, when implemented correctly, provides preventative control of major food safety hazards, reducing reliance on end product testing. The system consists of the following seven principles (Codex, 1997):

Principle 1: Conduct a hazard analysis
Principle 2: Determine the critical control points (CCPs)
Principle 3: Establish critical limit(s)
Principle 4: Establish a system to monitor control of the CCP
Principle 5: Establish the corrective action to be taken when monitoring indicates that a particular CCP is not under control
Principle 6: Establish procedures for verification to confirm that the HACCP system is working effectively
Principle 7: Establish documentation concerning all procedures and records appropriate to these principles and their application.

Guidance for the application of these principles is given in the same document, and in many other useful publications on HACCP, notably those by Sara Mortimore and Carol Wallace (Mortimore and Wallace, 1998, 2001). A number of well-known factors are essential for an effective HACCP system. These include genuine commitment and involvement of management, use of a multidisciplinary approach, and training and involvement of the workforce.

When set up and operated correctly, a HACCP system can be profitably used by all sizes of food business and successfully applied throughout the food chain. While a HACCP study should always be carried out by a multidisciplinary team that comprises all the required knowledge and expertise, computer-based aids to HACCP have been a reality for some time (Smedley, 1997). In the UK, one of the better-known commercial software tools is the Safefood HACCP Documentation Software developed and produced by Campden and Chorleywood Food Research Association (CCFRA). The software is basically a Windows™ based documentation tool, designed to assist with HACCP record keeping (CCFRA, 2003). It uses a structured and organised approach to HACCP documentation to support audit trail requirements and help to demonstrate due diligence. It includes a flowcharting capability, allowing mapping of process flow diagrams. The software, however, does not provide detailed information and advice on specific hazards and control measures.

Other recent developments that are complementary to HACCP and will undoubtedly enhance its further development and implementation include predictive food microbiology and quantitative microbiological risk assessment (QMRA). Predictive models for food microbiology, which in some applications are now used routinely, can help to indicate whether a hazard exists at a particular process step (principle 1) (Walker, 2000). Models can also be used to assist in the setting of optimal target value at a CCP and the critical limit(s) that may be permitted (principle 3). Likewise, development of QMRA is expected to lead to better defined food safety objectives (FSOs) – statements of the maximum frequencies and/or concentrations of major microbiological hazards in food at the time of consumption that are considered to give an acceptable level of consumer protection. These FSOs will effectively become the targets that the industry should aim to achieve through the effective implementation of HACCP (Stringer, 2003). Further details about these developments can be found in the relevant chapters of this book.

15.3 Determining the shelf-life of food

15.3.1 Direct determination of shelf-life

While it is vital to establish as soon as possible whether or not a product being evaluated is safe, its HACCP study is usually conducted in parallel with its shelf-life determination. As pointed out before, the shelf-life investigator must know what the quality characteristics of the product are and, ideally, have some idea about the deterioration mechanism(s) that will spoil these characteristics, resulting in a loss of product acceptability. A shelf-life determination therefore involves an experimental study of the deterioration of the food, culminating in the identification of a point in time that marks the end of its shelf-life. In the UK and the EU hazard analysis is a legal requirement as contained in the European Food Hygiene Directive (93/43/EEC). When carried out alongside a shelf-life study, a HACCP study, i.e. a hazard analysis, will ensure that any changes in formulations, processing conditions, packaging and so on, particularly during product development, do not compromise food safety.

The most common and direct way of determining shelf-life is to carry out storage trials of the product under controlled conditions that mimic those it is likely to encounter during storage, distribution, retail display and consumer use. A fully comprehensive shelf-life determination, however, is rarely possible as conditions during distribution and retail display, for instance, are difficult to replicate. Consumer handling, too, is often highly unpredictable and the manufacturer has very little control over it. For example, it has been found that for chilled foods in the UK, many of which are particularly perishable, consumer handling is perhaps the most variable part of the chill chain, making it very difficult, if not impossible, to simulate consumer handling in a shelf-life study of chilled foods (Evans, 1998). For these and other reasons, most food manufacturers can only realistically carry out shelf-life studies under fixed storage conditions, which do not fully and precisely simulate the exact distribution of their products. It is debatable whether or not extreme handling conditions and consumer abuse should be included as they are naturally unpredictable and, in theory, can be of endless combinations and permutations. After all, a shelf-life study is no more than a controlled experiment and cannot be designed to cover all eventualities. Besides, any food, because of its perishable nature, is unlikely to survive its shelf-life when subjected to significant consumer abuse, be it deliberate or unintentional.

Fixed storage conditions that are commonly employed to determine shelf-life include (Man, 2002):

- Frozen: −18°C or lower (relative humidity is usually near 100%)
- Chilled: 0 to +5°C, with a maximum of +8°C (relative humidity is usually very high)
- Temperate: 25°C, 75% relative humidity
- Tropical: 38°C, 90% relative humidity.

Ideally, for a given set of storage conditions, the following variations should be available, which obviously put extra demands on resources:

- Optimum conditions – the most desirable conditions of storage. Storage under these conditions is intended to provide the most optimistic shelf-life data that may be used to support the longest shelf-life feasible.
- Typical or average conditions – the conditions most likely to be experienced by the product. Storage under these conditions is intended to provide shelf-life data that will apply to actual production, resulting in a shelf-life which is acceptable to both the manufacturer and the consumer.
- Worst-case conditions – the most harsh conditions that the product is expected to encounter. Storage under these conditions is intended to provide the most conservative shelf-life data, which if used to assign a shelf-life should give it a margin of safety ensuring a fail-safe but perhaps under-estimated shelf-life.

Whatever the storage conditions, the objective of the trial is of prime importance and must be clearly and accurately defined. For instance, a chilled ready-to-eat pasta with ham salad can be sold as such in a retail single-use consumer pack or through a refrigerated delicatessen counter where it is displayed in a larger open pack over a number of days. Given the same product, different destinations and/or different target consumers may well require very different experimental designs that cater for the different shelf-life requirements.

15.3.2 Experimental protocols for direct determination of shelf-life

There is no universal protocol for the direct determination of shelf-life. A number of designs have been put forward (Kilcast and Subramaniam, 2000a), including some based on a statistical approach (Gacula, 1975). All have advantages and disadvantages, as well as varying implications on resources. In general, the following are some possible protocols (Man, 2002):

- Short shelf-life products. For chilled foods with shelf-life of up to one week (e.g. ready meals), samples can be taken off daily for evaluation.
- Medium shelf-life products. For products with a shelf-life of up to three weeks (e.g. some ambient cakes and pastry), samples can be taken off on days 0, 7, 14, 19, 21 and 25.
- Long shelf-life products. For products with a shelf-life of up to one year (e.g. some breakfast cereals and heat-processed shelf-stable foods), samples can be taken off at monthly intervals or at months 0, 1, 2, 3, 6, 12 and (perhaps) 18. The exact frequency will depend on the product and on how much is already known of its storage behaviour.

The issue of control samples is an important one. Many food products can be deep frozen and thawed, with little effect on their organoleptic quality. When this is the case, frozen storage can be used as a means of keeping control samples for use in shelf-life studies. If it cannot be used, facilities must be available for the preparation of fresh control samples that are identical to the stored samples in every way, at any time during the trial. The importance of ensuring that sufficient control samples representative of what the company wants to sell are available cannot be overemphasised, as they are the references against which stored samples are to be assessed.

15.3.3 Shelf-life tests

An acceptable shelf-life is expected to retain desired sensory, chemical, functional, microbiological and physical characteristics of the product. Therefore, tests employed to measure shelf-life tend to be product-specific, reflecting the quality characteristics of the product in question. Figure 15.1 provides a schematic representation of the direct determination of shelf-life. In practice, it is important to recognise that more than one deterioration mechanism may be at work, either in sequence or in parallel. The challenge always is to

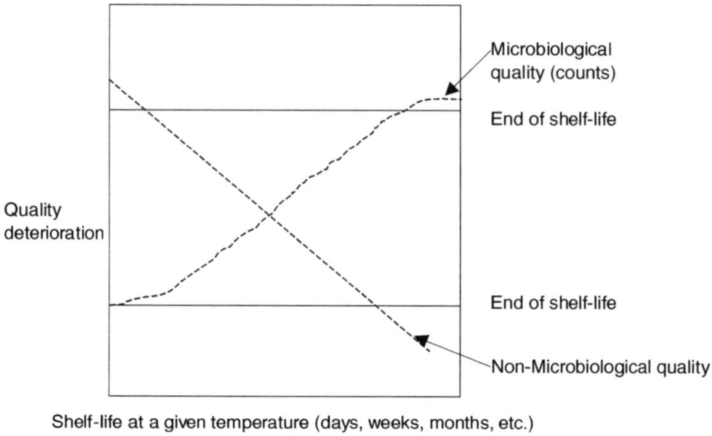

Fig. 15.1 Shelf-life of food – a schematic representation.

identify all of them as well as the one that is shelf-life limiting. A systematic and structured approach based on the HACCP principles has been used to put in place a control system designed to prevent rancidity in confectionery and biscuits (Frampton, 1994). In essence, the approach follows the same principles of HACCP in which a hazard is taken to mean a biological, chemical or physical agent in, or condition of, food with the potential to cause the food to deteriorate and spoil, terminating its shelf-life. Applying the principles systematically leads to the determination of the critical control points at which control can be applied and which are essential to eliminate or delay the shelf-life limiting hazard, preventing it from ending the required shelf-life prematurely. So, depending on the product and its deterioration mechanism(s), different types of tests can be used individually or in combination to measure the progress of shelf-life:

- Microbiological examination, including challenge testing
- Chemical analysis
- Physical testing, measurement and analysis such as rheological measurements, microscopical examination and so on
- Sensory evaluation.

Many shelf-life studies together with the tests employed have been published in both the primary and secondary literature. Table 15.1 contains some examples that illustrate the specific tests used to measure shelf-lives in the light of underlying mechanisms of deterioration. Naturally not all the tests need to be used regularly and permanently; without exception, once product safety has been established, sensory evaluation that includes taste is the test to be used in all studies as well as in all routine determinations. Guidance in sensory evaluation and its methodology can be found in various British and international standards, reference texts on the subject and elsewhere in this book. Table 15.2 lists the relevant standards in sensory analysis that may be of interest to readers.

Table 15.1 Examples of published shelf-life studies and their tests

Product	Storage conditions	Shelf-life tests	Approximate shelf-life	Reference
Marinated sardine in sunflower oil packaged in glass jars	4°C	• Total volatile basic nitrogen • Trimethylamine content • pH • Sensory evaluation (a 9-point descriptive scale)	120 days at 4°C	Gökoğlu *et al.*, 2004
Cooked shrimps in preservative-containing brine packed in modified atmosphere	0, 5, 8, 15 and 25°C	• Challenge tests with *Listeria monocytogenes* and spores of *Clostridium botulinum* • Sensory evaluation (odour, taste and texture and overall appearance using a 3-class scale) • pH • Total volatile nitrogen • Percentage drip loss	7 months at 0°C, 4–5 days at 25°C	Dalgaard and Jørgensen, 2000

Table 15.1 Continued

Product	Storage conditions	Shelf-life tests	Approximate shelf-life	Reference
Vacuum-packaged Chinese-style sausages preserved by different preservatives: sorbitol, sodium lactate, nisin	20°C	• Microbiological examinations (APC, LAB, etc.) • a_w • Volatile basic nitrogen • pH • Determination of dextran formation • Sensory evaluation using a 7-point scale	25 days at 20°C for sodium lactate sausages	Wang, 2000
Ambient sauces and dressings	25°C, 75% RH, i.e. temperate conditions	• Microbiological examinations including challenge tests • Sensory evaluation (colour, odour, taste, texture, emulsion appearance, etc.) • pH	Up to a year at ambient, dependent on product	Pourkomailian, 2000
Potato chips and savoury snacks in printed coextruded oriented polypropylene bags	25°C, 75% RH, i.e. temperate conditions	• Determination of peroxide value of extracted oil (fatty snacks) • Weighing of packs (to measure moisture uptake) • Sensory evaluation using a 7-point scale	Up to 3 months at ambient, dependent on product	Man, 2000

Table 15.2 Relevant standards in sensory analysis

BS	ISO	Title
5098:1992	5492	Glossary of terms relating to sensory analysis
5929-1:1986	6658	Methods for sensory analysis of food. General guide to methodology
5929-2:1982	5495	Methods for sensory analysis of food. Paired comparison test
5929-3:1984	4120	Methods for sensory analysis of food. Triangular test
5929-4:1986	6564	Methods for sensory analysis of food. Flavour profile methods
5929-5:1988	8588	Methods for sensory analysis of food. 'A'–'Not A' test
5929-6:1989	8587	Methods for sensory analysis of food. Ranking
5929-7:1992	3972	Methods for sensory analysis of food. Investigating sensitivity of taste
5929-8:1992	10399	Methods for sensory analysis of food. Duo–trio test
5929-9:1992	5496	Methods for sensory analysis of food. Initiation and training of assessors in the detection and recognition of odours
5929-10:1999	11037	Methods for sensory analysis of food. General guidance and test method for assessment of the colour of foods
7183:1989	8589	Guide to design of test rooms for sensory analysis of food
7667-1:1993	8586-1	Assessors for sensory analysis. Guide to the selection, training and monitoring of selected assessors
7667-2:1994	8586-2	Assessors for sensory analysis. Guide to the selection, training and monitoring of experts
BS ISO 4121:2003		Sensory analysis. Guidelines for the use of quantitative response scales
BS ISO 11056:1999		Sensory analysis. Methodology. Magnitude estimation method
BS ISO 13299:2003		Sensory analysis. Methodology. General guidance for establishing a sensory profile

15.3.4 Using shelf-life data to assign shelf-life

Assigning shelf-lives to products where relevant legal standards are available is relatively straightforward. For instance, in the UK, the Dairy Products (Hygiene) Regulations 1995 (HMSO, 1995) specify microbiological criteria for milk-based products, which may be used to fix shelf-lives. Likewise, the Coffee and Coffee Products Regulations 1978 (HMSO, 1978) require 'instant coffee' to contain not less than 95% coffee-based dry matter, i.e. not more than 5% moisture. Given a particular coffee powder and packaging format, an acceptable shelf-life can be found that will ensure this statutory limit is not exceeded at the point of sale. The level of tin in food, especially food packed in plain processed cans, can be a serious shelf-life limiting hazard, as the Tin in Food Regulations 1992 (HMSO, 1992) prohibit the sale of any food containing tin at more than 200 mg/kg. In the UK, between 1998 and 2002, no fewer than 10 product recalls appeared in various national newspapers involving canned chopped tomatoes, spaghetti in tomato sauce, and tomato soup. Thousands of products had to be withdrawn due to tin levels found to be in excess of the statutory limit.

Besides legal standards, voluntary guidelines can also be very useful in assisting companies to define shelf-life endpoints for their products. The UK

Health Protection Agency (incorporating the former Public Health Laboratory Service) has provided guidelines for the microbiological quality of some ready-to-eat foods (Gilbert *et al.*, 2000) and the IFST (UK) has published a useful guide on the 'Development and use of microbiological criteria for foods' (IFST, 1999). For foods such as breakfast cereals, many of which contain added vitamins, the declared levels of vitamins on the packs have become an important criterion in setting shelf-life.

In the absence of any standards, either legal or voluntary, manufacturers naturally have to set their own endpoints, be they based on microbiological, chemical or sensory criteria. In all cases, in order to ensure accurate and reproducible shelf-lives, manufacturers are likely to have to repeat their shelf-life experiments and take into account factors such as the following when interpreting shelf-life data:

- Age and quality of raw materials used in the experiments
- Scale effects that impact on shelf-life arising from scaling up to full production
- Anticipated batch-to-batch variations in product quality when full production commences
- Standard of hygiene in the production environment
- Need to use rework materials in on-going production.

15.4 Predicting the shelf-life of food

As the required or anticipated shelf-life gets longer, it becomes increasingly unrealistic for companies to rely solely on storage trials to determine shelf-life directly. The general desire to obtain an early answer to the shelf-life question coupled with competitive and commercial pressures has meant that an indirect and invariably quicker way of determining shelf-life is necessary. Consequently, procedures have been developed to predict or estimate shelf-lives quickly. While they are very attractive and useful, some based on accelerated procedures and others on computer-based models, all have to be used with caution. Accelerated procedures can only be used if the relationship exists between the storage behaviour under an ambient storage regime and that under an accelerated condition. Likewise predictions from models can only be used to assist in the setting of shelf-lives if the models have been validated in foods. The following sections give an outline of some of these indirect procedures; more detailed treatments can be found elsewhere in this book.

15.4.1 Accelerated shelf-life tests

In theory, accelerated shelf-life testing is applicable to any deterioration process that has a valid kinetic model. That process may be biochemical, chemical, microbiological or physical. In practice, most accelerated tests have been done on deterioration processes that are chemical in nature. The basic idea is that the rate of a shelf-life limiting chemical reaction is increased at an elevated storage temperature. The end of shelf-life is thereby reached much quicker and the data obtained can be extrapolated to provide an estimate of the shelf-life at normal or ambient storage conditions, usually by using the Arrhenius relationship. More detailed treatment of accelerated testing and its limitations is available (Labuza and Schmidl, 1985; Mizrahi, 2000). Despite its limitations and the fact that food deterioration and spoilage are generally far more complex than can be described accurately by a simple empirical relationship, a number of accelerated shelf-life studies based on the Arrhenius model have been published:

- Shelf-life and safety of minimally processed CAP/MAP chilled foods over a limited temperature range (Labuza *et al.*, 1992)
- Aspartame stability in commercially sterilised flavoured dairy beverages (Bell and Labuza, 1994)
- Accelerated shelf-life testing of whey-protein-coated peanuts (Lee *et al.*, 2003).

15.4.2 Other models

Because of the universal need to assure microbiological safety in foods, initial interest in predictive food microbiology has concentrated on the development of predictive models for foodborne pathogens. In the UK, a government-funded project during 1989–1994 led to the production of Food MicroModel, a software package that will predict the growth, survival and thermal death of major food pathogens and a number of food-spoilage organisms in a wide range of foods. Similar efforts in the USA led to the production of a predictive pathogen modelling program (PMP), which is freely available on the Internet (http://www.arserrc.gov/mfs/pathogen.htm). Since then, an Internet-based, publicly and freely available database of food microbiology data – ComBase – has been launched in 2003 (http://www.ifr.ac.uk/combase/), which is developed and supported by a consortium comprising of, in the UK, the Food Standards Agency and the Institute of Food Research at Norwich, and in the USA, the US Department of Agriculture, the Agricultural Research Service and the Eastern Regional Research Center at Wyndmoor. ComBase is recommended for research and training/education purposes. The consortium also intends to produce a package of unified predictive models known as ComBase Combined Database and Predictive Microbiology Program in the future.

In recent years, computer-based models for predicting microbiological spoilage have also become available:

- Forecast – a collection of predictive models developed by CCFRA, which can be used to assess the microbial spoilage rates or likely stability of foods. It is offered as a paid service by the Research Association.
- *Pseudomonas* Predictor – temperature function integration software developed at the Department of Agricultural Science, University of

Tasmania, Australia. The software has been commercialised and is marketed in Australia under the name Food Spoilage Predictor (Blackburn, 2000).
- Seafood Spoilage Predictor (SSP) – developed at the Danish Institute for Fisheries Research (DIFRES) in Lyngby. The software contains two types of model, namely the relative rates of spoilage (RRS) model and the microbial spoilage (SSO) model. It is available free of charge and can be downloaded from the Institute's website (http://www.dfu.min.dk/micro/ssp/).
- ERH CALC™ – part of a computer-based 'Cake Expert System' for the baking industry originally developed by the UK Flour Milling and Baking Research Association, now part of CCFRA. The program allows users to run simulations on flour confectionery formulations and rapidly calculate their theoretical equilibrium relative humidities (ERHs) and hence estimate their mould-free shelf-lives (MFSLs). The complete system is available for purchase from CCFRA. It has been suggested that the MFSLs of some 80% of ambient packaged cakes manufactured in the UK are now estimated by using this program.

15.5 Conclusions

In summary, success in determining the shelf-life of a food product depends on the following factors:

- Confidence in assurance of food safety
- Ability to define the critical quality characteristics that determine consumer acceptance
- Knowledge and understanding of the pertinent mechanisms of deterioration and spoilage
- Adequate capability, in terms of both technical know-how and appropriate facility, to measure shelf-life through either direct determination or indirect prediction and estimation, or both.

Few food companies can afford to have serious and persistent shelf-life problems. In respect of providing an adequate and reproducible shelf-life to consumers, both safety and quality are important; but the former must take precedence over the latter. A safe product of acceptable quality that consistently pleases its consumers has its origin in good product design that must include carefully planned and conducted shelf-life testing, evidence of which may be used to prove that the manufacturer has taken the trouble to consider his products and problems and that he does care about what he produces.

15.6 Sources of further information and advice

The huge amount of published information on the subject of shelf-life of foods, in both the primary as well as the secondary literature, has served to confirm the continuing interest in this field of food science and technology. While not intended to be exhaustive or even comprehensive, the following, in chronological order, are the better-known references on the subject.

Shelf-life Dating of Foods (Labuza, 1982)
Published in 1982, this is arguably the first major volume on the subject of shelf-life. It contains 22 chapters, covering 'Basic food preservation and degradation modes', 'Scientific investigation of shelf-life', 'Shelf-life of breakfast cereals', 'Shelf-life of bakery products', 'Shelf-life of dehydrated products', 'Shelf-life of frozen convenience products' and so on.

Evaluation of Shelf Life for Chilled Foods (CCFRA, 1990)
This is Technical Manual No. 28 published by CCFRA, UK. The document, first produced in 1990 by a Shelf Life Working Party of the Chilled Foods Panel of the Association, was part-revised in 1997. The intention is for it to be used as an outline structure for the evaluation of shelf-life of chilled foods including ingredients and products for retail sale. Throughout the document extensive reference is made to the HACCP approach to process assessment and control.

Shelf Life of Foods – Guidelines for its Determination and Prediction (IFST, 1993)
This is a publication of the IFST (UK). It was written on behalf of the Institute by an *ad hoc* working group consisting of members of the IFST. The aim of this publication is to provide concise advice to food business managers on the principles of shelf-life determination and prediction, at every point in the food chain. It also explains the factors influencing the shelf-life of foods and the various mechanisms of deterioration in foods, which form the basis of the scientific principles essential to all evaluation of shelf-life of foods. The booklet ends with a list of references.

Shelf Life Studies of Foods and Beverages – Chemical, Biological, Physical and Nutritional Aspects (Charalambous, 1993)
This is a substantial reference (1204 pages) consisting of 40 chapters by 89 contributors. The coverage ranges from shelf-lives of food commodities (e.g. meat, fish, fruit and vegetables) to those of manufactured foods (e.g. confectionery, bakery and extruded products) and drinks (e.g. tea, coffee, wines). It is an update of a similar title, published seven years earlier, from the same editor and publisher. Understandably, the bibliography is extensive.

Shelf-life Evaluation of Foods (Man and Jones, 2000)
This multi-authored reference, now in its second edition, concentrates primarily on the shelf-life of foods, although much of what is covered is equally applicable to drinks. The book, divided into two parts, begins with six chapters reviewing the principles of shelf-life evaluation. The remaining 10 chapters illustrate the practice of shelf-life evaluation using a number of selected product groups:

chilled yoghurt and other dairy desserts, fresh and lightly preserved seafood, ambient packaged cakes, potato crisps and savoury snacks, chocolate confectionery, ready-to-eat breakfast cereals, processed foods in containers other than cans, ambient-stable sauces and pickles, frozen foods and minimally processed, ready-to-eat, and ambient-stable meat products.

The Stability and Shelf-life of Food (Kilcast and Subramaniam, 2000b)
This is a multi-authored reference edited by two scientists at the UK Leatherhead Food Research Association. After an introduction, the book divides into two parts. Part 1 consists of five chapters, looking at ways of analysing and predicting shelf-life. Part 2 consists of seven chapters of case studies, covering 'Predicting packaging characteristics to improve shelf-life', 'Sous vide products', 'Milk and milk products', 'Confectionery products', 'Fruits and vegetables', 'Fats and oils' and 'Sauces and dressings'.

Shelf Life (Blackwell's Food Industry Briefing Series) (Man, 2002)
This is a small book that gives a concise overview of the subject at introductory level. In keeping with the primary objective of the series, the book aims to provide a concise and quick reference for busy food industry professionals, particularly those working in small and medium-sized enterprises. It is also suitable for senior undergraduate students and fresh graduates of food science and technology or a related discipline. The book is divided into three sections, namely, 'Introduction to shelf life of foods – frequently asked questions', 'The ways food deteriorates and spoils', and 'Determining shelf life in practice'.

Freshness and Shelf Life of Foods (Cadwallader and Weenen, 2003)
A publication of the American Chemical Society (ACS), this serves as a printed record of the ACS symposium entitled 'Freshness and Shelf Life of Foods' held in Chicago, Illinois, in 2001. The book is divided into four sections: 'Introduction and general issues', 'Flavor aspects', 'Texture' and 'Methods to improve freshness and shelf life'. All 21 chapters are either original review or research papers on the subject.

Indeed, the above books together with related book chapters and original research papers have combined to provide food scientists and technologists interested in shelf-life worldwide with a rich and diverse range of information sources based on which further developments will undoubtedly abound.

15.7 References

BELL, L N and LABUZA, T P (1994) Aspartame stability in commercially sterilised flavoured dairy beverages. *Journal of Dairy Science*, **77**, 34–38.

BLACKBURN, C DE W (2000) Modelling shelf-life. In: *The Stability and Shelf-life of Food*, Kilcast, D and Subramaniam, P (eds), pp 55–78, Woodhead Publishing, Cambridge.

CADWALLADER, K and WEENEN, H (eds) (2003) *Freshness and Shelf Life of Foods*. American Chemical Society, Washington DC.

CCFRA (1990) *Evaluation of Shelf Life for Chilled Foods*. Technical Manual No. 28, Campden and Chorleywood Food Research Association, Chipping Campden, UK.

CCFRA (2003) *HACCP Documentation Software Version 4.0*, Campden and Chorleywood Food Research Association Group, Chipping Campden, UK.

CHARALAMBOUS, G (ed) (1993) *Shelf Life Studies of Foods and Beverages Chemical, Biological, Physical and Nutritional Aspects*. Elsevier Science, Amsterdam.

CODEX COMMITTEE ON FOOD HYGIENE (1997) *HACCP System and Guidelines for its Application*, Annexe to CAC/RCP 1–1969, Rev 3, in Codex Alimentarius Commission Recommended International Code of Practice – General Principles of Food Hygiene, Food and Agriculture Organisation of the United Nations, World Health Organisation, Rome.

DALGAARD, P and JØRGENSEN, L V (2000) Cooked and brined shrimps packed in a modified atmosphere have a shelf life of >7 months at 0°C, but spoil in 4–6 days at 25°C. *International Journal of Food Science and Technology*, **35**, 431–442.

EVANS, J A (1998) Consumer perceptions and practice in the handling of chilled foods. In: *Sous Vide and Cook-chill Processing for the Food Industry*, Ghazala, S (ed.), Aspen Publishers, Gaithersburg, MD.

FRAMPTON, A (1994) Prevention of rancidity in confectionery and biscuits a Hazard Analysis Critical Control Point (HACCP) approach. In: *Rancidity in Foods*, 3rd edition, Allen, J C and Hamilton, R J (eds), pp 161–178, Blackie Academic & Professional, London.

GACULA, M C (1975) The design of experiments for shelf-life study. *Journal of Food Science*, **40**, 399–403.

GILBERT, R J, DE LOUVOIS, J, DONOVAN, T, LITTLE, C, NYE, K, RIBEIRO, C D, RICHARDS, J, ROBERTS, D and BOLTON, F J (2000) Guidelines for the microbiological quality of some ready-to-eat foods sampled at the point of sale. *Communicable Disease and Public Health*, **3** (3), 163–167.

GÖKOĞLU, N, CENGIZ, E and YERLIKAYA, P (2004) Determination of the shelf life of marinated sardine (*Sardina pilchardus*) stored at 4°C. *Food Control*, **15**, 1–4.

HMSO (1978) *The Coffee and Coffee Products Regulations 1978* (SI 1978/3118). Her Majesty's Stationery Office, London.

HMSO (1992) *The Tin in Food Regulations 1992* (SI 1992/496). Her Majesty's Stationery Office, London.

HMSO (1995) *The Dairy Products (Hygiene) Regulations 1995* (SI 1995/1086). Her Majesty's Stationery Office, London.

IFST (1993) *Shelf Life of Foods Guidelines for its Determination and Prediction*, Institute of Food Science and Technology (UK), London.

IFST (1999) *Development and Use of Microbiological Criteria for Foods*. Institute of Food Science and Technology (UK), London.

KILCAST, D and SUBRAMANIAM, P (2000a) Introduction. In: *The Stability and Shelf-life of Food*, Kilcast, D and Subramaniam, P (eds), pp 1–19, Woodhead Publishing, Cambridge.

KILCAST, D and SUBRAMANIAM, P (eds) (2000b) *The Stability and Shelf-life of Food*, Woodhead Publishing, Cambridge.

LABUZA, T P (1982) *Shelf-life Dating of Foods*. Food and Nutrition Press, Westport, CT.

LABUZA, T P and SCHMIDL, M K (1985) Accelerated shelf-life testing of foods. *Food Technology*, **9**, 57–62, 64, 134.

LABUZA, T P, FU, B and TAOUKIS, P S (1992) Prediction for shelf life and safety of minimally processed CAP/MAP chilled foods. *Journal of Food Protection*, **55**, 741–750.

LEE, S-Y, GUINARD, J-X and KROCHTA, J M (2003) Relating sensory and instrumental data to conduct an accelerated shelf-life testing of whey-protein-coated peanuts. In: *Freshness and Shelf Life of Foods*. Cadwallader, K and Weenen, H (eds), pp 175–187, American Chemical Society, Washington DC.

MAN, D (2000) Potato chips and savory snacks. In: *Shelf-life Evaluation of Foods*, 2nd edition, Man, D and Jones, A (eds), pp. 157–167, Aspen Publishers, Gaithersburg, MD.

MAN, D (2002) *Shelf Life*. Food Industry Briefing Series, Blackwell Science, Oxford.

MAN, D and JONES, A (EDS) (2000) *Shelf-life Evaluation of Foods*, 2nd edition, Aspen Publishers, Gaithersburg, MD.

MIZRAHI, S (2000) Accelerated shelf-life tests. In: *The Stability and Shelf-life of Food*, Kilcast, D and Subramaniam, P (eds), pp. 107–128, Woodhead Publishing, Cambridge.

MORTIMORE, S E and WALLACE, C A (1998) *HACCP: A Practical Approach*, 2nd edition, Aspen Publishers, Gaithersburg, MD.

MORTIMORE, S E and WALLACE, C A (2001) *HACCP*, Food Industry Briefing Series, Blackwell Science, Oxford.

POURKOMAILIAN, B (2000) Sauces and dressings. In: *The Stability and Shelf-life of Food*, Kilcast, D and Subramaniam, P (eds), pp. 311–331, Woodhead Publishing, Cambridge.

SMEDLEY, P (1997) *Development of Computer Based Aids to Hazard Analysis Critical Control Point (HACCP)*. PhD Thesis, South Bank University, London.

STRINGER, M (2003) Microbiological risk assessment. *Food Science and Technology*, **17**(4), 3134.

WALKER, S J (2000) The principles and practice of shelf-life prediction for microorganisms. In: *Shelf-life Evaluation of Foods*, 2nd edition, Man, D and Jones, A (eds), pp. 34–41. Aspen Publishers, Gaithersburg, MD.

WANG, F-S (2000) Effects of three preservative agents on the shelf life of vacuum packaged Chinese-style sausage stored at 20ºC. *Meat Science*, **56**, 67–71.

16

Lipid oxidation and the shelf-life of muscle foods

P. A. Morrissey and J. P. Kerry, University College Cork, Ireland

16.1 Introduction

Meat has traditionally held a special place in the human diet because of its appealing flavour and texture and its high nutritional value. Meat plays a very important role in the diet, by contributing quality protein, essential minerals and trace elements such as iron, zinc and copper, and a range of B vitamins and vitamins A and D. However, despite these important and highly acceptable attributes, meat consumption has come under close scrutiny in recent years. Emphasis on the role of fat and especially saturated fat in diseases associated with modern life, especially in developed countries, has resulted in a swing in attitude away from meat as a central part of a healthy diet associated with strength and virility in our culture. The concerns of people regarding food safety, animal welfare, and environmental issues have had an influence on animal production, animal nutrition, food processing and distribution, and food consumption patterns. In general, there is now a greater demand than ever by consumers for foods perceived as natural, fresh tasting, healthful and more nutritious (Morrissey *et al.*, 1998). In addition, changes in lifestyles, convenience, demographic characteristics, income, price, age and ethnicity impact on food purchasing, food preparation and consumption (Resurreccion, 2003). Katz (1999) noted that the number one requirement for foods in Europe is that they taste, smell and feel authentic.

One of the main factors limiting the quality and acceptability of meat and meat products is lipid oxidation. In raw meats, this results in the formation of brown pigments (especially in beef), increased drip losses and the development of unacceptable odour, while in cooked and stored meats it causes off-flavours, such as 'warmed-over' flavour (WOF) (Morrissey *et al.*, 1994a; Gray *et al.*,

1996) or 'meat flavour deterioration' (MFD) (St Angelo, 1996). In addition to having a negative impact on the acceptability of meat and meat products, lipid oxidation may also have food safety implications. Concerns have been expressed over the possible atherogenic effects of lipid oxidation products (e.g. malondialdehyde and cholesterol oxides). Preventing lipid oxidation during storage and retail display is therefore essential in order to maintain quality, wholesomeness and safety of meats, and to ensure that customers will make repeat purchases. However, the oxidative stability of meat is low and difficult for retailers to predict accurately (especially in the case of poultry or fish, which are rich in polyunsaturated fatty acids), and oxidation often results in meat (in particular beef) being discounted in price or removed from retail display before the end of its normal bacteriological shelf-life. The low oxidative stability of meat and meat products is a problem for all those involved in the meat production chain, including the primary producers, processors, distributors and retailers, and understanding and controlling the processes that lead to lipid oxidation is a major challenge for meat scientists.

The present review will focus on three critical phases of lipid oxidation that impact on shelf-life stability and muscle food quality. In phase one, we consider the production of reactive oxygen species, lipid peroxidation mechanisms, and antioxidant defence systems in the living animal. Phase two of oxidative damage occurs in the immediate post-slaughter period, while phase three occurs during post-slaughter handling, processing, storage and cooking. Here, the mechanisms are probably the same as those occurring in stressed tissue *in vivo*. Dietary factors capable of influencing lipid oxidation in muscle foods will be highlighted.

16.2 Lipid oxidation *in vivo*

16.2.1 Oxidative stress

The production of free radicals *in vivo* is a critical determinant of animal health, and consequently food quality, wholesomeness and acceptability of muscle foods by the consumer (Morrissey *et al.*, 2002). The most important potential stressors in aerobic organisms are reactive oxygen species (ROS). Free radical production in animal cells can be either accidental or deliberate. Four endogenous sources of reactive ROS appear to account for most of the oxidants produced in cells (Kehrer and Smith, 1994). Small amounts of ROS, including hydroxyl radical ($HO^{\bullet}$), superoxide anion ($O_2^{\bullet-}$) and H_2O_2, are produced during normal aerobic mitochondrial metabolism. The superoxide anion ($O_2^{\bullet-}$) appears to play a central role in the generation of ROS, since other reactive intermediates are formed in reaction sequence starting with $O_2^{\bullet-}$. Peroxisomes, which include fatty acyl CoA oxidase, dopamine-β-hydroxylase, urate oxidase and others, produce H_2O_2 as a by-product, which is then degraded by catalase. Some H_2O_2 escapes degradation and leaks into other compartments and increases oxidative damage. The cytochrome P-450 mixed-function oxidase system in animals constitutes one of the primary defence systems against natural toxic chemicals from plants, the major source of dietary toxins. The induction of these enzymes enhances production of free radicals and prevents acute toxic effects from foreign chemicals. On the other hand, some ROS are produced deliberately. For example, phagocytic cells defend against foreign organisms by deliberately synthesizing large amounts of $O_2^{\bullet -}$, H_2O_2, hypochlorite (^-OCl) and nitric oxide radical ($NO^\bullet$). Chronic infection by viruses, bacteria, parasitic worms and liver fluke results in chronic phagocytic activity and consequent chronic inflammation and failure to thrive.

A number of exogenous systems may also increase the endogenous free radical load. High dietary intakes of iron and copper or 'misplaced' iron as a result of tissue breakdown promote the generation of oxidizing radicals from peroxides (Halliwell, 1987). Normal plant foods with large amounts of natural phenolic compounds can generate oxidants by redox cycling. In addition, animal feeds containing spent commercial frying oils may also contribute to the radical load (Morrissey *et al.*, 1998).

The relevant ROS include: $O_2^{\bullet -}$, $HO^\bullet$ and other oxygen-centred radicals of organic compounds (peroxyl, $ROO^\bullet$ and alkoxyl, $RO^\bullet$) together with non-radical reactive compounds such as H_2O_2, singlet oxygen ($^1O_2$), hypochlorous acid (HOCl), and hydroperoxides and epoxide metabolites of endogenous lipids. The non-radical groups contain chemically reactive oxygen-containing groups (Kehrer, 1993). In addition, reactive nitrogen species (RNS) such as nitric oxide ($NO^\bullet$), nitrogen dioxide ($NO_2^\bullet$) and peroxynitrite ($ONOO^-$) are produced (Sies, 1991).

ROS can oxidize lipids, proteins, nucleic acids and other macromolecules, leading to cell death and tissue injury (Pompella, 1997). Although frequently occurring as a late event accompanying rather than causing cell damage or death (Halliwell and Chirico, 1993), lipid oxidation is probably still the most widely used measure of oxidative stress in animals.

16.2.2 Lipid oxidation

The fundamental principles of lipid oxidation were elucidated by the work of Farmer *et al.* (1942), Bolland and Gee (1946), Ingold (1961), Yagi *et al.* (1992) and Frankel (1985). Numerous lines of evidence indicate that lipid oxidation occurs *in vivo* (Frei, 1991; Frei *et al.* 1990, 1992; Morrow and Roberts 1991; Halliwell and Gutteridge, 1999). Lipid oxidation *in vivo*, and also in muscle food systems, is a free-radical-mediated mechanism and is initiated in the highly unsaturated phospholipid fraction in subcellular membranes. The process of initiation is the abstraction of a *bis*-allylic hydrogen from a polyunsaturated fatty acid (LH), by the action of the highly reactive radical ($HO^\bullet$), or certain Fe–O complexes, such as ferryl or perferryl radicals (Halliwell and Chirico, 1993). Buettner (1993) reviewed the thermodynamics of free-radical reactions. Using standard one-electron reduction potentials ($E^{o'}$), Buettner predicted a 'hierarchy, or pecking order' for the course of free-radical processes. Each oxidized (higher

Table 16.1 Electron reduction potential for free radicals and antioxidants

Free radicals or antioxidants	Reduction potential ($E^{o'}$, mV)
$HO^\bullet$, H^+/H_2O	2310
$H_3CH_2C^\bullet$, H^+/CH_3CH_3	1900
$LO^\bullet$, H^+/LOH (alkoxyl radical)	1600
$LOO^\bullet$, $H^+(LOOH)$ (peroxyl radical)	1000
$O_2^{\bullet -}$, $2H^+/H_2O_2$	940
$PUFA^\bullet$, $H^+/PUFA\text{-}H$ ($L^\bullet$, H^+/LH)	600
$Urate^\bullet$, $H^+/Urate\text{-}H$	590
Catechol-$O^\bullet$, H^+/Catechol-OH	530
α-Tocopheroxyl$^\bullet$, H^+/α-tocopherol	500
$Ascorbate^{\bullet -}$, H^+/ascorbate	282
Paraquat/$Paraquat^{\bullet -}$	−448

Source: adapted from Buettner (1993).

$E^{o'}$) species is capable of abstracting an electron (or hydrogen atom) from any reduced (lower $E^{o'}$) species listed below it if the reaction is kinetically feasible (Buettner, 1993). Thus, $HO^\bullet$ which is at the top of the pecking order ($E^{o'} = +2310$ mV), and is considered to be the most potent oxidant encountered in biological systems, can oxidize all the reduced species below it (Table 16.1), and it is remarkably effective in bringing about initiation of lipid peroxidation (Buettner, 1993; Morrissey *et al.*, 1994b). The driving force for the reaction is the difference in potential ($\Delta E^{o'}$) between the two reactions. For example, the $HO^\bullet$, H^+/H_2O couple has $E^{o'} = +2310$ mV, and the $L^\bullet$, H^+/LH has $E^{o'} = +600$ mV; $\Delta E^{o'} = +1700$. Thus, $HO^\bullet$ can, in principle, be remarkably effective in bringing about initiation of lipid peroxidation:

$$LH + HO^\bullet \xrightarrow{\Delta E^{o'} = +1700\,\text{mV}} L^\bullet + H_2O \tag{16.1}$$

The resulting fatty acyl radical ($L^\bullet$) reacts rapidly with O_2 to form a peroxyl radical ($LOO^\bullet$):

$$L^\bullet + O_2 \longrightarrow LOO^\bullet \tag{16.2}$$

The rate constant (k_1) for this reaction is 3×10^8 $mol^{-1}s^{-1}$ (Buettner, 1993). Because $LOO^\bullet$ is more highly oxidized (higher in the pecking order) than $L^\bullet$ or LH, it will preferentially oxidize further unsaturated fatty acids and propagate the chain reaction:

$$LOO^\bullet + LH \xrightarrow{\Delta E^{o'} = +400\,\text{mV}} LOOH + L^\bullet \tag{16.3}$$

The rate constant (k_2) for this step is relatively low (10^1–10^2 $mol^{-1}\,s^{-1}$) (Buettner, 1993). Lipid hydroperoxides (LOOH) formed in the propagation reaction are both products of the peroxidative process and substrates for further reactions with Fe^{2+} and Cu^{+} to yield $LOO^{\bullet}$ and alkoxyl radical ($LO^{\bullet}$) (Morrissey *et al.*, 1998). Fe^{2+} reductively cleaves LOOH as follows:

$$Fe^{2+} + LOOH \longrightarrow Fe^{3+} + LO^{\bullet} + OH^{-} \tag{16.4}$$

Both $LOO^{\bullet}$ and $LO^{\bullet}$ can initiate further reactions (e.g. (16.3) and the following):

$$LO^{\bullet} + LH \xrightarrow{\Delta E^{0'} = +1000\,\text{mV}} LOH + L^{\bullet} \tag{16.5}$$

The $LO^{\bullet}$ can also undergo β-scission and degrade to alkyl radicals ($R^1CH_2^{\bullet}$) and a range of aldehydes, depending on the particular hydroperoxide present (Morrissey *et al.*, 1994b). The aldehydes, including hexanal, malondialdehyde and 4-hydroxynonenal, can react with ϵ-amino groups of proteins to yield Maillard-type complexes. The alkyl radical is also higher on the pecking order than LH or LOOH and could also initiate additional chain reactions resulting in the formation of ethane or pentane, end products observed in the peroxidation of PUFAs.

Increasing the degree of unsaturation of lipid biomembranes or lipoproteins by dietary manipulation is likely to increase the concentration of labile *bis*-allylic hydrogen atoms, making it more likely that one of these hydrogen atoms may be abstracted by $LOO^{\bullet}$ or other radicals (Horwitt, 1986; Cosgrove *et al.*, 1987; Pryor, 1994; Frankel, 1998). The rate of oxidation of PUFA is proportional to the number of doubly allylic hydrogen atoms that a given PUFA molecule possesses (Cosgrove *et al.*, 1987; Pryor, 1994). On a scale where oleate (18:1) undergoes autoxidation too slowly to measure, linoleate (18:2), with two allylic hydrogens, undergoes oxidation half as fast as does linolenate (18:3) with four allylic hydrogens. The kinetic expression of the rate of oxidation of an olefin is presented as the oxidizable factor (Δ) (Pryor, 1994). The Δ value is equal to $k_p/(2k_t)^{0.5}$, where k_p is the rate constant for the propagation reaction ($LOO^{\bullet} + LH \longrightarrow LOOH + L^{\bullet}$) and k_t is the rate constant for the termination reaction ($LOO^{\bullet} + LOO^{\bullet} \longrightarrow$ non-radical products). The data from Cosgrove *et al.* (1987) and Pryor (1994) are presented in Table 16.2. Thus, the oxidizability of 22:6 (10 doubly allylic hydrogens atoms) is five times greater than that of 18:2 (two doubly allylic hydrogen atoms).

Table 16.2 Values of the oxidizability of PUFA with different numbers of doubly allylic hydrogens

PUFA	Symbol	H[1]	$\Delta^2 = k_p/(2k_t)^{0.5}$
Linoleate	18:2	2	20 ± 0.2
Linolenate	18:3	4	41 ± 2
Arachidonate	18:4	6	55 ± 2
Docosahexaenoate	18:6	10	102 ± 2

[1] The number of doubly allylic hydrogen atoms.
[2] Δ = oxidizability value.
Sources: Cosgrove *et al.*, 1987; Pryor, 1994.

16.3 Antioxidant defence systems

Cellular systems have evolved highly sophisticated mechanisms in order to maintain redox homeostasis and/or cope with the excess ROS produced during oxidative stress (Yu, 1994). Living organisms have an array of enzymes which serve to minimize production of $HO^{\bullet}$. Superoxide dismutase (SOD) enzymes serve to remove $O_2^{\bullet -}$ by accelerating the formation of H_2O_2. Mammalian cells have an SOD enzyme containing active site manganese (MnSOD) in mitochondria. SOD with active site copper and zinc (CuZn SOD) are present largely in the cytosol (Morrissey and O'Brien, 1998). SOD enzymes work in collaboration with H_2O_2- removing enzymes (catalase and glutathione peroxidase). The latter enzyme contains active site selenium.

An additional important primary antioxidant defence is the presence of metal ion storage and transport proteins such as transferrin, lactoferrin, haptoglobin, caeruloplasmin and metallothionein sequester transition metals in forms that inhibit catalyses of $O_2^{\bullet -}$ and H_2O_2 to the more damaging $HO^{\bullet}$ (Thurnham, 1990; Halliwell *et al.*, 1995). In addition, retinol may also make an important contribution to the antioxidant defence by limiting the decompartmentalization of highly catalytic iron through the maintenance of cellular integrity (Morrissey and O'Brien, 1998).

A significant amount of the work on antioxidant defence mechanisms has been confined to studies on the chain-breaking antioxidants vitamins C and E and the carotenoids (especially β-carotene). Lutein and other carotenoids as well as ubiquinol-10, thiols and uric acid are also capable of interrupting free-radical chain reactions (Stocker *et al.*, 1991; Halliwell and Gutteridge, 1999). Vitamin E (as α-tocopherol) (α-TOH) is the most important of these compounds in plasma lipids because it is present in concentrations at least 15-times higher than any of the others. The chemistry of vitamin E is rather complex because there are eight structurally related forms, four tocopherols (α-, β-, γ-, δ-) and four tocotrienols (α-, β-, γ-, δ-), which are synthesized from homogentisic acid and isopentenyl diphosphate in the plastid envelope of plants. The structures of α-, β-, γ-, δ-tocopherols are shown in Fig. 16.1. Vitamin E is an indispensable component of biological membranes and has membrane-stabilizing properties. The molecule is anchored in the highly hydrophobic hydrocarbon part of the membrane layer by the isoprenoid chain (Fig. 16.2) (Morrissey *et al.*, 2000). The chromanol nucleus lies at the surface of lipoprotein or at the membrane–water interface, and it is the phenolic hydroxyl group that quenches free radicals (Fig. 16.2). In this position, the chromanol ring has considerable mobility; it is able to quench peroxyl radicals that partition to the water–membrane interface and can be regenerated by harvesting the antioxidant capacity of other lipid-soluble antioxidants (e.g. ubiquinols) and water-soluble reductants, such as ascorbate and glutathione (Kagan *et al.*, 1992) and dihydrolipoate and thioredoxin (Packer *et al.*, 2001). The overall mechanisms of lipid peroxidation and antioxidant protection have been extensively reviewed (Buettner, 1993; Packer *et al.*, 2001; Morrissey and Kiely, 2002).

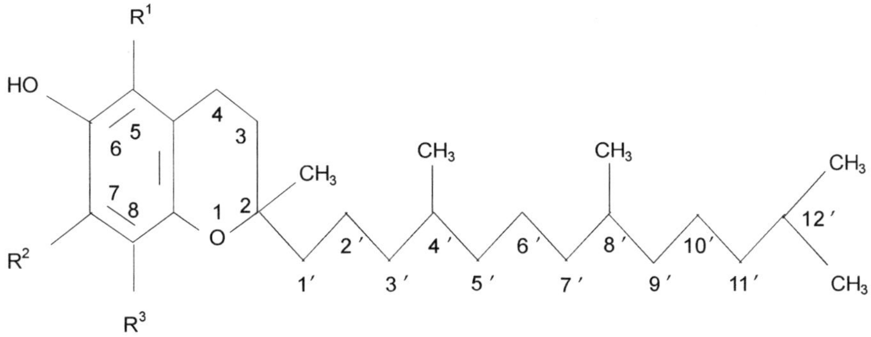

Compound	R^1	R^2	R^3
α-Tocopherol	CH_3	CH_3	CH_3
β-Tocopherol	CH_3	H	CH_3
γ-Tocopherol	H	CH_3	CH_3
δ-Tocopherol	H	H	CH_3

Fig. 16.1 The four major forms of vitamin E (α-, β-, γ- and δ-tocopherol) differ by the number and position of methyl groups on the chromanol ring. In α-tocopherol, the most biologically active form, the chromanol ring is fully methylated. In β- and γ-tocopherol, the ring contains two methyl groups, while in δ-tocopherol it is methylated at one position.

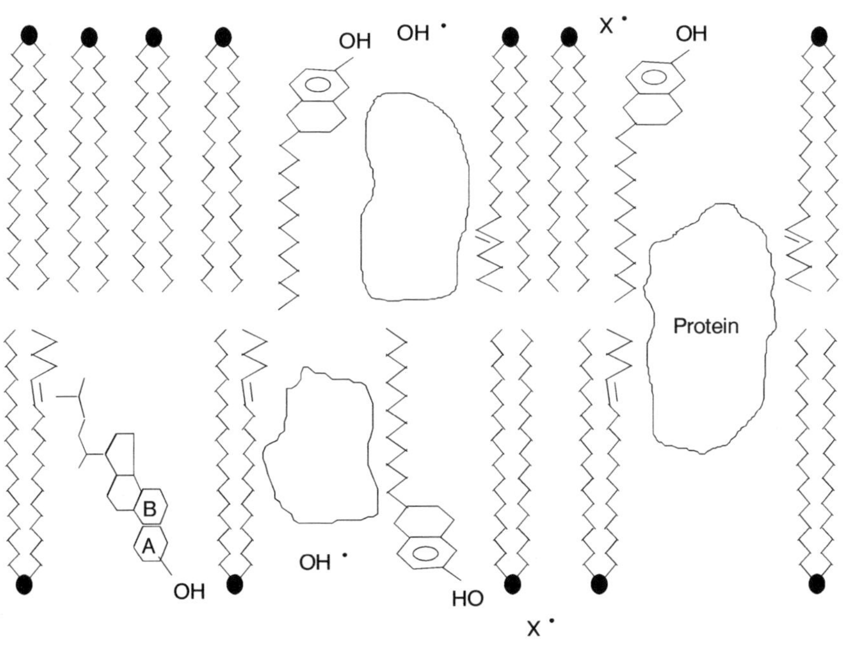

Fig. 16.2 Schematic representation of the lipid bilayer of a cell membrane, showing the possible positions of the tocopherol and cholesterol molecules. HO•, hydroxyl radical; X•, free radical (Morrissey *et al.*, 2000; reproduced with permission).

16.3.1 Antioxidant activity of vitamin E

The antioxidant activities of chain-breaking antioxidants are determined by how rapidly they scavenge free radicals, the ease of hydrogen transfer from an antioxidant to a free radical and the difference in the standard one-electron reduction potentials (Buettner, 1993). A compound whose reduction potential is lower than the reduction potential of a free radical or oxidized species can donate a hydrogen to that free radical if the reaction is kinetically feasible (Table 16.1). Thus, peroxyl radicals ($E^{o'} = +1000$ mV) have reduction potentials greater than α-TOH ($E^{o'} = +500$ mV), urate ($E^{o'} = +590$ mV), and ascorbate ($E^{o'} = +282$ mV), meaning that hydrogen transfer between any of these antioxidants and LOO$^\bullet$ is energetically favourable (Buettner, 1993; Morrissey *et al.*, 1994b). When the chromanol phenolic group of α-TOH encounters an LOO$^\bullet$ a hydroperoxide is formed, and in the process an α-tocopheroxyl radical (α-TO$^\bullet$) is generated:

$$\alpha\text{-TOH} + \text{LOO}^\bullet \xrightarrow{\Delta E^{o'} = +500\,\text{mV}} \text{LOOH} + \alpha\text{-TO}^\bullet \tag{16.6}$$

The rate constant (k_3) for this chain-inhibition reaction is 8×10^4 mol^{-1}s^{-1} (Buettner, 1993) or 2.35×10^6 mol^{-1}s^{-1} (Azzi and Stocker, 2000), which is considerably higher than that for the other tocopherols and related phenols. As the rate constant (k_2, reaction (16.3)) for the chain propagation reaction between LOO$^\bullet$ and LH is much lower than k_3, at approximately 10^2 mol^{-1}s^{-1}, α-tocopherol outcompetes the propagation reaction and scavenges the LOO$^\bullet$ ~10^4 times faster than LH reacts with LOO$^\bullet$. The concentration of α-TOH in biological membranes is approximately 1 mol per 1000–2000 mol phospholipids (i.e. ~1:10^3) (Burton *et al.*, 1983). This effectively means that about 90% of the LOO$^\bullet$ are scavenged by α-TOH before they can attack another LH (Buettner, 1993; Niki and Matsuo, 1993; Morrissey *et al.*, 1994b). Thus, the kinetic properties of antioxidants, and in particular α-TOH, require that only relatively small amounts be present in biological systems for them to act as effective antioxidants (Buettner, 1993; Morrissey *et al.*, 1994b). The overall antioxidant activity of α-TOH depends also on the fate of the α-TO$^\bullet$ generated when α-TOH scavenges lipid LOO$^\bullet$. The α-TO$^\bullet$ is relatively stable due to resonance delocalization (Nawar, 1996), the radical energy is low ($E^{o'} = +500$ mV) and it is less likely to promote the oxidation of other biomolecules. In biological systems, ascorbate can reduce α-TO$^\bullet$ back to α-TOH with a fairly high rate constant (1.5×10^6 mol^{-1}s^{-1}).

γ-Tocopherol is the most common isomer in plant foods such as corn, soy bean and palm oil, and nuts such as walnuts and peanuts. In contrast, α-tocopherol is the predominant form of vitamin E in most human and animal tissues (Clement and Bourre, 1997; Jiang *et al.*, 2001). γ-Tocopherol supplementation is known to spare α-tocopherol from being degraded, which would explain why γ-tocopherol supplementation results in an increase in tissue levels of α-tocopherol (Clement and Bourre, 1997). γ-Tocopherol has been reported to have greater scavenging activity than α-tocopherol in the presence of

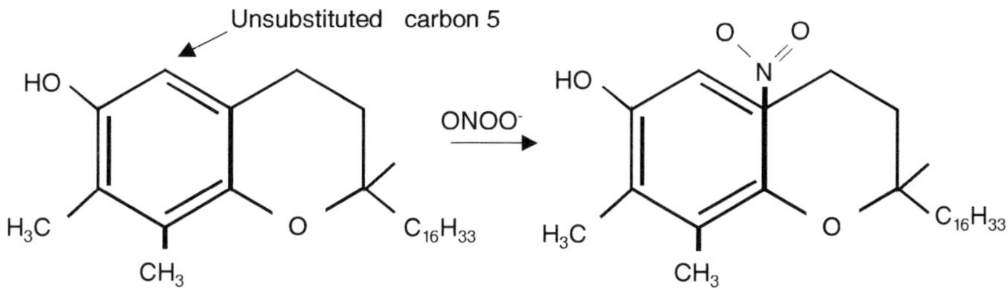

Fig. 16.3 Formation of 5-nitro-γ-tocopherol from the reaction of γ-tocopherol with peroxynitrite ($ONOO^-$).

peroxynitrite ($ONOO^-$) radicals (Wolf, 1997). γ-Tocopherol can act as a trap for membrane-soluble electrophiles such as reactive RNS, forming stable carbon-centred adducts through its nucleophilic 5-position, which is blocked in α-tocopherol (Christen *et al.*, 1997) (Fig. 16.3). Excess generation of RNS ($ONOO^-$, $NO^\bullet$, $NO_2^\bullet$) is associated with chronic inflammation-related diseases in humans and animals. γ-Tocopherol may contribute significantly to animal health and muscle food quality. It may also control peroxynitrite production during food processing and help improve food quality parameters such as flavour, colour and functional properties (Brannan *et al.*, 2001). However, further studies are required to elucidate its mechanism of action in foods and its potential to inhibit lipid oxidation and improve colour stability in meats.

16.3.2 Antioxidant activity of ascorbic acid

Ascorbate is a reversible biological reductant, and as such, provides reducing equivalents for a variety of biochemical reactions and is considered the most important antioxidant in extracellular fluid (Sies *et al.*, 1992). Ascorbate is thermodynamically close to the bottom of the list of one electron reducing potentials of oxidizing free radicals ($E^{o'} = +282\,mV$, Table 16.1). For this reason, ascorbate is the first line of defence against ROS and RNS in plasma (Briviba and Sies, 1994). It efficiently scavenges all oxidizing species with a greater one-electron potential (higher $E^{o'}$ values), which includes $O_2^{\bullet-}$ (rate constant $2.7 \times 10^5\,mol^{-1}s^{-1}$), $HO^\bullet$ (rate constant ~$10^9\,mol^{-1}s^{-1}$), $^1O_2$ (rate constant $10^7\,mol^{-1}s^{-1}$), water soluble $LOO^\bullet$ (rate constant $= 2 \times 10^8\,mol^{-1}s^{-1}$), ^-OCl and thiyl radical (Briviba and Sies, 1994). Overall, ascorbate is reactive enough to effectively interrupt oxidants in the aqueous phase before they can attack and cause detectable oxidative damage to DNA and lipids.

Ascorbate may also restore α-TOH by re-reducing $TO^\bullet$ to its native state. Ascorbate has a lower redox potential ($E^{o'} = +282$ mV) than α-TOH ($E^{o'} = +500$ mV) and, in addition, the α-$TO^\bullet$ is at the membrane–water interface, thereby allowing water-soluble ascorbate access to membrane-bound α-$TO^\bullet$ for the repair reaction (Reaction (16.7)), and recycling of the α-tocopherol (Frei *et al.*, 1989; Buettner, 1993; May *et al.*, 1999):

$$\text{Asc H}^- + \alpha\text{TO}^\bullet \xrightarrow{\Delta E^{o'} = +220\,\text{mV}} \text{Asc}^{\bullet -} + \alpha\text{-TOH} \tag{16.7}$$

The rate constant (k_5) for the reaction is $1.5 \times 10^6\,\text{mol}^{-1}\text{s}^{-1}$.

The resulting semi-dehydroascorbyl radical ($\text{Asc}^{\bullet -}$) dismutates to dehydroascorbate (Buettner, 1993) and is then regenerated back to ascorbate at the expense of glutathione, dihydrolipoate, thioredoxin and other enzyme systems. This process allows for the transporting of a radical load from a lipophilic compartment to an aqueous compartment where it is taken care of by efficient enzymatic defences (Diplock *et al.*, 1998).

It should be noted that as a reducing agent ascorbate has the ability to reduce Fe^{3+} to Fe^{2+} and Cu^{2+} to Cu^+, thereby increasing the prooxidant activity of the metals and generating $HO^\bullet$, $O_2^{\bullet -}$and H_2O_2 that initiate lipid peroxidation in biological systems (Buettner and Jurkiewicz, 1996). It is considered unlikely that ascorbate shows prooxidant properties *in vivo* since the concentrations of 'free' transition metals in healthy animals are very small because they are effectively bound by metals ion storage and transport proteins.

16.3.3 Other antioxidants

The antioxidant nature of carotenoids has been attributed to their unique structure, an extended system of double bonds (Stahl and Sies, 1996). The radical quenching and antioxidant activity of carotenoids have been extensively reviewed (Sies and Stahl, 1995; Palace *et al.*, 1999). Carotenoids are the most efficient naturally occurring quenchers of singlet oxygen. The rate constants for singlet oxygen scavenging by carotenoids are in the order of $10^{10}\,\text{mol}^{-1}\text{s}^{-1}$, and are even greater than that of α-TOH ($10^8\,\text{mol}^{-1}\text{s}^{-1}$) (Ojima *et al.*, 1993). It has been suggested that β-carotene scavenges $LOO^\bullet$ and forms a complex between the β-carotene and $LOO^\bullet$ yielding a resonance-stabilized structure or carotenoid radical that effectively terminates the lipid peroxidation process (Burton and Ingold, 1984). There is increasing evidence that both antioxidant and/or prooxidant mechanisms are exerted by β-carotene in biological systems. The balance between prooxidant and antioxidant behaviour is very delicate, and the antioxidant potential is more pronounced at low oxygen partial pressure (Burton, 1989; Jorgensen and Skibsted, 1993). On the other hand, at high, non-physiologic oxygen tension and relatively high concentrations (>500 μM), β-carotene loses its antioxidant potency and exhibits prooxidant activity (Zhang and Omaya, 2000, 2001). The antioxidant/prooxidant phenomenon of β-carotene has been reviewed by Palozza (1998) and Krinsky (2001). These authors conclude that the antioxidant activity of β-carotene and other carotenoids may shift into a prooxidant mode depending on the redox potential of the carotenoid and on the biologic environment in which it acts. A number of studies have demonstrated a synergestic effect between carotenoids and tocopherols in model systems (Mortensen and Skibsted, 2000). However, studies have shown that carotenoids adversely affect the absoption and concentration of α-tocopherol in plasma and liver (Pellet *et al.*, 1994).

Table 16.3 Dietary factors contributing to antioxidant defences *in vivo*

Vitamin A	Iron	Carotenoids
Vitamin C	Copper	Flavonoids and related compounds
Vitamin E	Zinc	Phytic acid
Riboflavin	Selenium	
Niacin	Manganese	
Folate	Magnesium	
Vitamin B_{12}		

The flavonoids are versatile antioxidants and can prevent lipid peroxidation by scavenging ROS, lipid peroxyl and alkoxyl radicals, chelating transition metal ions and/or regenerating α-TOH through reduction of α-TO$^{\bullet}$ (Rice-Evans *et al.*, 1996). However, their role as antioxidants in muscle foods has not been elucidated.

The production of ROS and the animal's antioxidant defences are approximately balanced *in vivo*. However, it is easy to tip the prooxidant–antioxidant balance in favour of ROS and create the situation of oxidative stress, which may cause tissue damage. Examples relating to diet include a high intake of PUFA or highly oxidized fat, or an inadequate intake of nutrients that contribute to the defence system (Morrissey *et al.*, 1994b, 1998). Packer (1993) noted that vitamin E requirements vary up to five-fold depending on tissue fatty acid composition due to previous dietary intake. Dietary factors that contribute to the antioxidant system and that may affect the optimal balance are listed in Table 16.3 (Morrissey *et al.*, 1998).

16.4 Lipid oxidation in muscle and muscle foods

Phase two of oxidative destabilization probably commences immediately pre-slaughter due to stress and certainly during the early post-slaughter phase (Morrissey *et al.*, 1994b, 1998). The biochemical changes that accompany post-slaughter metabolism and post-mortem ageing in the conversion of muscle to meat give rise to conditions where the process of lipid peroxidation in the highly unsaturated phospholipid fraction in sub-cellular membranes is no longer tightly controlled and the balance of prooxidant factors–antioxidant capacity favours oxidation (Morrissey *et al.*, 1994b). The conversion of muscle to meat is a direct result of the cessation of blood flow and the triggering or arresting of many metabolic processes. Orderly metabolic activity continues during the early post-slaughter period, but because of the cessation of blood flow, the product of glycogen breakdown becomes lactic acid. This accumulates in the tissue, gradually lowering the pH from near neutrality to a more or less mildly acid (pH ~ 5.5) value. In the post-slaughter phase, it is highly unlikely that the armoury of antioxidant defensive systems available to the cell in the live animal still function because of quantitative changes in pH, in several metabolites and physical properties. In many instances, the defensive systems may be weakened by nutritional deficiency. The potential for peroxidation in muscle is also increased by the rate of pH decline, the rate of carcass chilling and tenderizing techniques such as electrical stimulation, high pressure treatment, ultrasound, hydrodyne (explosive shock) process, and injection of organic acids. However, because of the anaerobic environment in post-slaughter muscle, lipid oxidation is likely to be limited during this phase.

The third phase, and in most cases the highly significant phase, of lipid oxidation occurs during post-slaughter handling, processing, storage, cooking and subsequent refrigerated storage. Disruption of the integrity of muscle membranes by mechanical deboning, mincing, restructuring or cooking alters cellular compartmentalization. During these processes iron is released from high molecular weight sources (e.g. haemoglobin, myoglobin, ferritin, haemosiderin) and made available to low molecular weight compounds such as amino acids, nucleotides and phosphates with which it is believed to form chelates (Decker *et al.*, 1993). The importance of these chelates or 'free iron' in the catalysis of lipid peroxidation in biological tissues and in foods has been defined by Halliwell and Gutteridge (1986) and Kanner *et al.* (1988). Depending on the severity of the above processing operations, the prooxidant–antioxidant balance may be substantially altered. The balance is also highly dependent on the concentrations of PUFA and low-molecular-weight antioxidants.

Overall, the interaction of prooxidants with PUFA is facilitated, resulting in the generation of free radicals and propagation of the oxidative reaction. The mechanisms of oxidation are likely to be similar to those occurring in stressed tissue *in vivo* (Morrissey *et al.*, 2002).

16.5 Factors influencing lipid stability in meat and meat products

16.5.1 Fatty acid composition

It is accepted that lipid oxidation in muscle foods is initiated in the highly unsaturated phospholipid fraction in subcellular membranes. The susceptibility to oxidation depends on animal species, metabolic type of muscle and cooking methods (Gandemer and Meynier, 1995). The ability of unsaturated fatty acids, essentially those with two or more double bonds, to oxidize leads to the development of rancidity and colour deterioration during storage. On the other hand, the propensity to oxidize is important in meat flavour development during cooking. In general, the susceptibility of muscle systems to lipid peroxidation is in the order fish > turkey > chicken > pork > beef > lamb (Allen and Fogeding, 1981; Tichivangana and Morrissey, 1985) which reflects the degree of unsaturation in the subcellular membranes (Wood *et al.*, 2003). In addition, the type of muscle may also influence the oxidizability of muscle. Oxidative ('red') muscles have a higher percentage of phospholipid than glycolytic ('white') ones and therefore a higher percentage of PUFA (Enser *et al.*, 1998).

The thigh muscles (oxidative) of poultry are more susceptible to peroxidation than breast muscle (glycolytic) (Gandemer and Meynier, 1995; Rhee *et al.*, 1996). It is generally accepted that thigh muscle is inherently more susceptible to lipid oxidation than breast muscle because it has a greater phospholipid content, is more oxygenated and has a higher haem content (Galvin *et al.*, 1997). Tang *et al.* (2001a) showed that the susceptibility of raw minced muscle to lipid oxidation during six days of refrigerated display was in the order mackerel > beef > duck > ostrich > pork ≥ chicken breast > whiting. On the other hand, the susceptibility of cooked samples during subsequent storage at 4°C was in the order mackerel ≫ duck > ostrich > beef > pork > chicken breast ≥ whiting. The susceptibility of cooked muscle was closely related to lipid content, concentration of PUFA and presence of iron in different species (Tang *et al.*, 2001b). The data suggest that the amount of substrate (PUFA) is the major determinant of inter-species differences in lipid oxidation when the integrity of muscle membranes is disrupted, and iron released by heating, etc.

In recent years, there has been considerable emphasis on modification of the fatty acid composition of animal tissues in an attempt to produce new 'designer' or nutritionally 'friendly' muscle foods with higher levels of *n*-3 PUFA and a more favourable (lower *n*-6:*n*-3 ratio). Larick *et al.* (1992) showed that pig muscle with elevated levels of 18:2 oxidized rapidly when heated. Muscles from pigs fed 5% soy oil had significantly higher 18:2 to 18:1 ratios in the neutral and polar lipid fractions of skeletal muscle and were significantly more susceptible to oxidation than muscles from pigs fed a tallow diet (Monahan *et al.*, 1992a). Enser *et al.* (2000) and Sheard *et al.* (2000) showed that the *n*-6:*n*-3 ratio in pork could be reduced by feeding crushed whole linseed, with no detectable adverse effects on meat quality. However, a recent study by Kouba *et al.* (2003) showed that the levels of *n*-3 PUFA in pigs fed a diet containing 6% whole crushed linseed produced higher TBARS values during retail display for seven days. They concluded that adverse effects on meat quality, defined in terms of shelf-life (lipid and myoglobin oxidation) and flavour, were only observed when linolenic acid (18:3) concentration approached 3% of neutral lipids or phospholipids and the processing conditions favoured oxidation.

Feeding broilers diets containing coconut oil, tallow, olive oil, linseed oil, or partially hydrogenate soya bean oil significantly altered the fatty acid composition of dark and white muscle (O'Neill *et al.*, 1998), neutral lipids and to a lesser extent the phospholipids of dark and white muscle (Lin *et al.*, 1989) and muscle membranal fractions (Lauridsen *et al.*, 1997). The fatty acid composition in turn influenced the oxidative stability of raw (Lin *et al.*, 1989) and cooked (O'Neill *et al.*, 1998) muscle during storage.

The presence of the rumen makes fatty acid composition in beef and sheep more difficult to manipulate by changing diet than in pigs and poultry (Wood *et al.*, 1999, 2003). Nevertheless, dietary lipids do affect tissue fatty acid composition (Scollen *et al.*, 2001; Vatansever *et al.*, 2000), TBARS (Vatansever *et al.*, 2000) and degradation products of lipid oxidation (saturated and unsaturated aldehydes, ketones and alcohol) in cattle (Elmore *et al.*, 1999).

Studies with beef and lamb showed that the concentrations of 18:3 and 20:5 in muscle phospholipids are higher when animals consume grass than when fed grain-based diets (Enser *et al.*, 1998; Fisher *et al.*, 2000; Warren *et al.*, 2002). These increases are due to the predominance of 18:3 in grass lipids and 18:2 in most other plants and seeds. Although grass-fed animals had higher concentrations of the more oxidizable *n*-3 PUFA, they were less susceptible to deterioration during storage (Warren *et al.*, 2002). The high content of vitamin E in grass results in higher tissue levels of α-tocopherol in tissues (Yang *et al.*, 2002) and greater stability of lipids to oxidation and oxymyoglobin retention.

16.5.2 Vitamin E supplementation

Numerous studies have shown that dietary vitamin E supplementation (in the form of *all-rac* α-tocopheryl acetate) consistently increases α-tocopherol levels in muscle and improves oxidative stability in meat from pigs (Monahan *et al.*, 1992a; Jensen *et al.* 1998), cattle (Liu *et al.*, 1995), lamb (Wulf *et al.,* 1995; Guidera *et al.*, 1997a,b), turkeys (Wen *et al.*, 1996, 1997a) and chickens (Sheehy *et al.*, 1993; Jensen *et al.*, 1995; Galvin *et al.*, 1997).

Dietary vitamin E supplementation (200 mg kg^{-1} fed) significantly increased the α-tocopherol levels in pig muscle (2.8-fold), when compared to pigs fed the basal level of α-tocopheryl acetate (10–50 mg kg^{-1}), and supplementation enhanced the oxidative stability of cooked ground pork during refrigerated storage for up to four days (Monahan *et al.*, 1992a). Supplemental vitamin E also improves oxidative stability in cooked chops packaged in modified atmospheres (Jensen *et al.*, 1998) and vacuum-packaged precooked chops and roasts (Cannon *et al.*, 1995; Kerry *et al.*, 1998). Other studies also reported that supra-nutritional vitamin E levels significantly increase concentrations of α-tocopherol in muscle and markedly decrease lipid oxidation in pork systems (Pfalgraf *et al.*, 1995; Jensen *et al.*, 1997; Hoving-Bolink *et al.*, 1998; O'Sullivan *et al.*, 1998). The effects of vitamin E are also apparent at the level of the subcellular membranes, the presumed site of lipid oxidation (Wen *et al.*, 1997b). Dietary supplementation of pigs with 1000 mg kg^{-1} α-tocopheryl acetate increased the concentrations of α-tocopherol in muscle, mitochondria and microsomes by 3.2-, 6.1- and 5.6-fold, respectively, compared to their counterparts from control animals. The increase in α-tocopherol was associated with a progressive decrease in susceptibility of the muscle and subcellular membranes to iron-ascorbate-induced lipid oxidation.

A significant protective effect of vitamin E supplementation against lipid oxidation has also been reported in poultry. Sheehy *et al.* (1993) observed that TBARS concentrations in cooked muscle from chicks fed on high α-tocopherol (65 or 180 mg α-tocopheryl acetate kg^{-1}) feeds were significantly lower during refrigerated or frozen storage than those in muscles in chicks fed low levels (5 or 25 mg) of α-tocopherol (Fig. 16.4). Supplementation of broiler diets with 200 mg α-tocopheryl acetate kg^{-1} feed for up to five weeks prior to slaughter

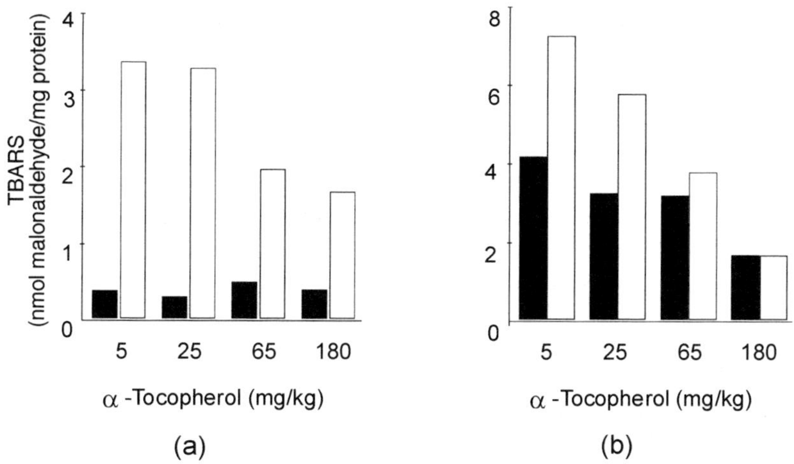

Fig. 16.4 Effect of feeding diets containing 5, 25, 65 and 180 mg α-tocopherol kg^{-1} on lipid oxidation (thiobarbituric acid-reactive substances, TBARS) in (a) raw and (b) cooked chicken muscle during 0 (■) to 2 (□) months frozen storage (Sheehy *et al.*, 1993; reproduced with permission).

significantly reduced the susceptibility of muscle homogenates to iron-ascorbate-induced lipid oxidation (Morrissey *et al.*, 1997), and also effectively reduced the detrimental effects of salt addition to ground poultry meat during refrigerated and frozen storage (Brandon *et al.*, 1993).

Poultry thigh meat has higher amounts of endogenous vitamin E deposited than breast meat when fed a basal (30 mg α-tocopheryl acetate kg^{-1} feed) or supplemented (200 mg kg^{-1} feed) diet for up to six weeks prior to slaughter (Galvin *et al.*, 1997; Morrissey *et al.*, 1997). However, thigh muscle was more susceptible than breast muscle to iron-ascorbate-induced oxidation (Morrissey *et al.*, 1997) and to oxidation during storage at 4°C for two weeks or at −20°C for three months (Brandon *et al.*, 1993).

The efficacy of dietary vitamin E supplementation as a protective mechanism against oxidation has been shown in whole turkey (Sklan *et al.*, 1982) and turkey meat composites (Sheldon *et al.*, 1997). Turkeys possess considerably lower levels of natural tocopherol in their tissues than chickens (Sklan *et al.*, 1982) and it takes longer for concentrations in muscle to reach saturation, because of their poor capacity to absorb the vitamin (Bartov, 1983) and greater faecal losses of tocopherol glucuronide conjugates (Sklan *et al.*, 1982).

The concentrations of α-tocopherol in raw burgers from turkeys fed 300 and 600 mg α-tocopheryl acetate kg^{-1} feed were 6.1- and 9.8-fold higher, respectively, compared with those from the control turkeys, and lipid oxidation was inhibited in both raw and cooked ground meat during refrigerated storage (Wen *et al.*, 1996). Vitamin E at dietary levels of 300 and 600 mg kg^{-1} were clearly protective against iron-ascorbate-induced oxidation (Wen *et al.*, 1997a). Regression analysis indicated that increasing breast muscle α-tocopherol to about 2.0–5.0 mg kg^{-1} and thigh muscle α-tocopherol to 4.0–8.0 mg kg^{-1}

reduced the susceptibility of the muscle to iron-ascorbate induced lipid oxidation. Supplementation of turkey poults with 600 mg α-tocopheryl acetate kg^{-1} feed for 21 weeks prior to slaughter inhibited lipid oxidation in vacuum and aerobic-packaged raw ground breast and thigh meat (Higgins *et al.*, 1998). However, a high concentration of α-tocopherol is of greater importance in aerobic-packaged than in vacuum-packaged turkey meat and, undoubtedly, is attributable to the greater oxidative stress encountered in the former (Higgins *et al.*, 1998). Supplementation with 300 or 600 mg kg^{-1} feed also reduced lipid oxidation in previously frozen turkey breast which was cooked, sliced and refrigerated in aerobic packaging (Higgins *et al.*, 1999a).

Several studies have demonstrated that dietary vitamin E supplementation is very effective in reducing oxidation of lipids and myoglobin in fresh, ground and frozen beef muscle, and these studies have been comprehensively reviewed by Liu *et al.* (1995, 1996a), Shaefer *et al.* (1995) and Kerry *et al.* (2000a). It is now well recognized that the improvement in lipid stability in beef is associated with a marked enhancement of colour (myoglobin) stability.

The strategy of supplementing meat-producing animals with supra-nutritional levels of vitamin E is to achieve sufficient muscle concentration of α-tocopherol to nearly maximize the antioxidant efficacy of the system. The objective is to achieve the minimum muscle α-tocopherol concentrations compatible with near-maximal suppression of triacylglycerol and cholesterol oxidation, WOF development, drip loss and metnyoglobin formation (in fresh beef). Higher supplementation may yield a small marginal benefit relative to a large marginal expense. The present evidence suggests that optimal dietary concentrations of vitamin E in poultry to prevent lipid oxidation are in the region of 200 mg α-tocopheryl acetate kg^{-1} feed for 35 days (Morrissey *et al.*, 1997). On the other hand, supplementation with 400 mg kg^{-1} feed may be necessary to limit cholesterol oxidation in cooked (Galvin *et al.*, 1998a) and cooked-irradiated (Galvin *et al.*, 1998b) poultry meat. Turkey diets should probably be supplemented with higher concentrations (~400 mg α-tocopheryl acetate kg^{-1} feed for 90 days) because of the slower uptake and deposition of α-tocopherol in turkeys (Wen *et al.*, 1996, 1997a). However, Sheldon *et al.* (1997) hypothesized that feeding turkeys high vitamin E supplements throughout the production cycle is not cost effective. Instead, they proposed feeding 200–300 mg α-tocopheryl acetate for 27 days immediately prior to slaughter. For pig production, the Meat and Livestock Commission in the UK is encouraging producers to include vitamin E at a minimum level of 100 mg α-tocopheryl acetate kg^{-1} feed (Warkup, 1994). Our studies (Morrissey *et al.*, 1996) suggest that pigs should be fed 200 mg α-tocopheryl acetate kg^{-1} feed for about 90 days in order to ensure optimal protection.

16.5.3 Ascorbic acid and carotenoid supplementation

One of the functions of ascorbic acid in living animals is to regenerate α-tocopherol from the α-tocopheroxyl radical (Packer and Kagan, 1993). This suggests that supplementation of animal diets with ascorbic acid might enhance the protection afforded by α-tocopherol against lipid oxidation in meats after slaughter. However, King *et al.* (1995) and Morrissey *et al.* (1998) did not observe any enhancement of oxidative stability in broiler muscle by dietary ascorbic acid over and above that provided by α-tocopheryl acetate. In beef cattle, jugular infusion of ascorbate (1.7 mol) 10 min prior to slaughter delayed oxymyoglobin oxidation extended the colour display life of *psoas*, *gluteus* and *longissimus* muscles (Schaefer *et al.*, 1995), but the authors concluded that dietary supplementation would probably not be as effective given the rapid rate at which ascorbate disappears from plasma. Data on the ability of ascorbate to regenerate α-tocopherol *in vivo* are conflicting, and Wenk *et al.* (2000) concluded that the 'sparing' action of ascorbate on α-tocopherol may be of negligible importance *in vivo* in animals that are not oxidatively stressed. As already outlined, ascorbate has the ability to promote lipid oxidation by reducing transition metal ions *in vitro* (Decker *et al.*, 1993). Therefore, the potential benefit of dietary ascorbate supplementation or its addition during processing of muscle foods as a means of improving oxidative stability is questionable.

King *et al.* (1995) reported that β-carotene added to the feed of chickens increased TBARS values and thus acted as a prooxidant. In another study, Ruiz *et al.* (1999) showed that β-carotene at low concentrations acted as an antioxidant in poultry meat, whereas it acted as a prooxidant at higher concentrations. In general, studies have mainly noted a prooxidative effect for β-carotene, but it appears that interaction with tocopherols is important for the balance between pro- and antioxidant activity (Mortensen and Skibsted, 2000). In addition, dietary supplementation with β-carotene adversely affects tissue levels of α-tocopherol.

16.5.4 Other antioxidants

Tea catechins are a predominant group of polyphenols present in green tea leaf (*Camellia sinensis* L). Tea catechins possess strong scavenging capacity for free radicals (Nanjo *et al.*, 1996; Unno *et al.*, 2000; Tang *et al.*, 2002; Jo *et al.*, 2003), have significantly higher scavenging activities compared with vitamin E and ascorbic acid (Tang *et al.*, 2002) and also possess some metal-chelating capacity (Record *et al.*, 1996; Tang *et al.*, 2002). Tea catechins are effectively incorporated into human blood plasma (Nakagawa *et al.*, 1997) and Tang *et al.* (2000) showed that the polyphenolic compounds in green tea are absorbed and enter the systemic circulatory system after ingestion and are distributed and retained in muscle and other tissues in chickens. In addition tea catechins, in particular epigallocatechin gallate, and epicatechin gallate, can permeate the lipid bilayer membrane structure (Hashimoto *et al.*, 1999). Supplementation of broiler diets with 200 and 300 mg tea catechins kg^{-1} feed for six weeks prior to slaughter significantly delayed lipid oxidation in raw breast and thigh meat that had been ground and stored at 4°C for up to 10 days (Tang *et al.*, 2001c). Tea catechins fed at a level of 200 mg kg^{-1} showed antioxidant activity equivalent to that of α-tocopheryl acetate fed at the same level during frozen storage (−20°C for three months). Tang *et al.* (2001a) showed that the antioxidant potential of tea catechins (300 mg kg^{-1}) added to uncooked minced pork, chicken, duck, ostrich, whiting and mackerel was 2- to 4-fold greater than α-tocopherol added at the same concentration.

Refined rosemary extract is a highly potent antioxidant in several foods, especially those containing animal fats and vegetable oils. Its antioxidant properties are related to the phenolic contents which scavenge lipid and hydroxy radicals (Cuvelier *et al.*, 1996; Mielnik *et al.*, 2003) and to their ability to chelate metal ions such as Fe^{2+} (Fang and Wada, 1993). Rosemary extracts inhibit lipid oxidation in restructured chicken (Lai *et al.*, 1991), pork fat (Chen *et al.*, 1992), fresh and precooked minced pork stored under refrigerated and frozen conditions (McCarthy *et al.*, 2001), and frozen, mechanically deboned turkey meat (Mielnik *et al.*, 2003). The optimum concentration of rosemary required to effectively inhibit oxidation is about 0.1% (McCarthy *et al.*, 2001). The combination of dietary α-tocopherol and rosemary extracts added during processing of beef exerted a stronger protective effect than either antioxidant alone (Formanek *et al.*, 2001). Wada and Fang (1992) proposed that rosemary extract acts as a synergist to regenerate α-tocopherol by providing the tocopheroxyl radicals with hydrogen atoms.

16.6 Cholesterol oxidation

Cholesterol oxidation has been studied extensively over the past decade or so because several of its oxidation products (oxysterols or cholesterol oxidation products, COPs) may be linked to a wide range of adverse biological effects in humans and animals, including atherogenesis, cytotoxicity, mutagenesis and carcinogenesis. Research on cholesterol oxidation has been extensively reviewed (Smith, 1981, 1996; Addis and Park, 1989; Addis and Warner, 1991; Guardiola *et al.*, 1996, 2002; Schroepfer, 2000). The production of COPs in meats has been recently reviewed (Kerry *et al.*, 2002).

Cholesterol, with its double bond between C-5 and C-6 of the B-ring (Fig. 16.5) readily undergoes autoxidation in the presence of molecular oxygen, light and metal ions through a free radical reaction to form some 66 COPs. Within the cell membrane, the hydrophobic cholesterol molecule is oriented parallel to fatty acids of adjacent phospholipids, the main site of oxidative attack on PUFAs (Fig. 16.2). The initial reaction of autoxidation process involves abstraction of a reactive allylic 7-hydrogen atom by peroxy or alkyl radicals of polyunsaturated fatty acids (Smith, 1992), yielding a carbon-centred radical, 3β-hydroxycholest-5-ene-7yl, which reacts with O_2 to yield the 3β-hydroxycholest-5-en-7-peroxyl radicals, stabilized in turn by hydrogen abstraction from an unsaturated fatty acid (LH) to provide 3β-hydroxycholest-5-ene-7-hydroperoxide (7-hydroperoxides) (Smith, 1992) (see Fig. 16.6). The 7-hydroperoxides undergo thermal degradation to yield a variety of products. The major products accumulating

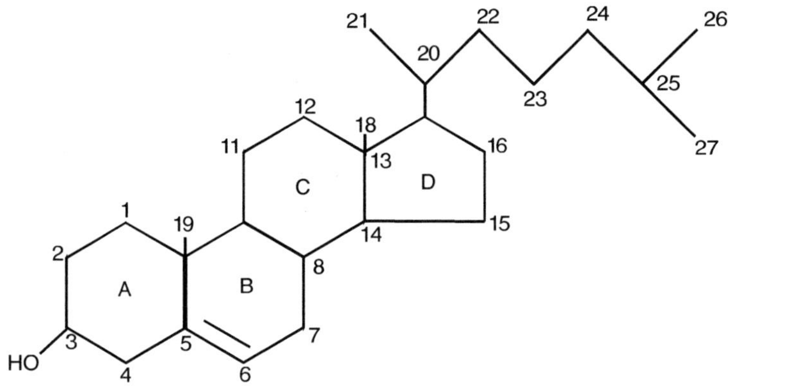

Fig. 16.5 Cholesterol molecule with ring labelling and carbon numbering.

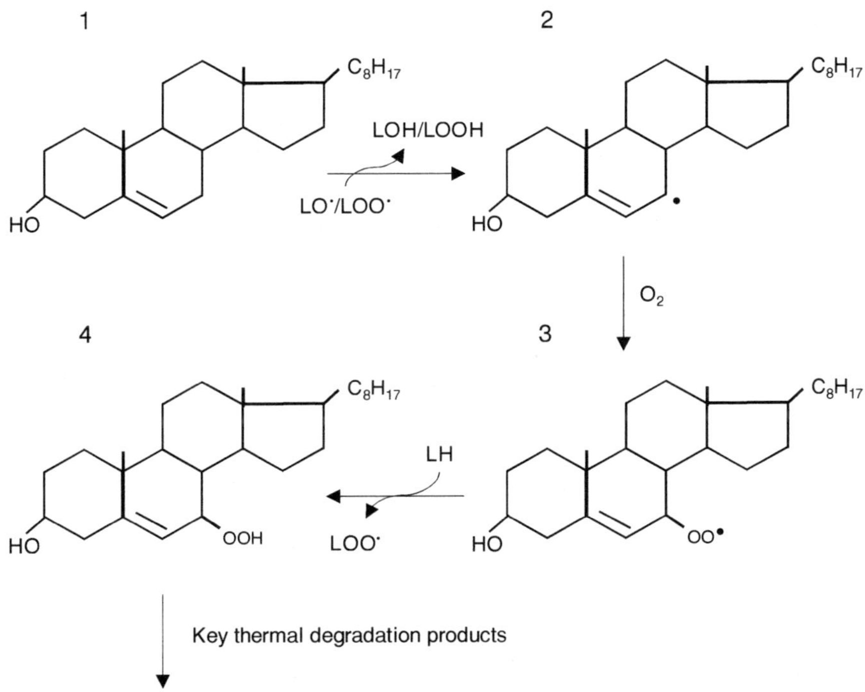

5-cholesten-3β-ol-7-one (7-ketocholesterol)
5-cholestene-3β, 7α-diol (7α-hydroxycholesterol)
5-cholestene-3β, 7β-diol (7β-hydroxycholesterol)
cholestan-5β, 6β-epoxy-3β-ol (cholesterol-5β, 6β-epoxide)

Fig. 16.6 Cholesterol autoxidation initiated by peroxy (LOO•) or alkoxy (LO•) radicals arising from peroxidation of polyunsaturated fatty acids (LH). Compounds are identified as follows: (1) cholesterol; (2) 3β-hydroxycholest-5-en-7-yl; (3) 3β-hydroxycholest-5-en-7-peroxy radical; (4) 3β-hydroxycholest-5-ene-7-hydroperoxide.

include 5-cholesten-3β-ol-7-one (7-ketocholesterol (7-keto)), 5-cholestene-3β, 7β-diol (7β-hydroxycholesterol (7β-OH)), 5-cholestene-3β, 7α-diol (7α-hydroxycholesterol) and cholestan-5β, 6β-epoxy-3β-ol (cholesterol-5β, 6β-epoxide (β-epoxide)) (Fig. 16.6). The 7-keto is the most abundant oxidation product. Oxidation of the cholesterol side-chain results in 20-, 24-, 25- and 26-hydroperoxides and their decomposition products. Side-chain oxidation usually takes place in solid cholesterol and not in aqueous dispersions.

In meats, processing conditions such as heating, freeze-drying, irradiation and storage conditions influence the formation of COPs in muscle foods, and the levels and types of COPs detected in a variety of muscle foods are extensively reviewed by Kerry *et al.* (2002) and will not be discussed in detail here. The present review will highlight methods used to control or prevent production of COPs in muscle foods.

Dietary α-tocopherol supplementation (200 mg α-tocopheryl acetate kg^{-1} feed) significantly lowered the levels of β-epoxide, 7β-OH and 7-keto and total COPs in refrigerated (for two and four days) cooked ground pork compared with pork from pigs fed the basal (10 mg kg^{-1}) diet (Monahan *et al.*, 1992b). Supplementation with 500 mg α-tocopherol per animal per day reduced total COPs by 65% in cooked veal following refrigerated storage for four days (Engeseth *et al.*, 1993). Galvin *et al.* (1998a) observed that dietary supplementation with 200 and 800 mg α-tocopheryl acetate kg^{-1} feed significantly reduced TBARS and formation of COPs in chicken during storage at 4°C. Total COPs formed after 12 days of storage were reduced by 42% and 75% in breast, and by 50% and 72% in thigh, at supplementation levels of 200 and 800 mg α-tocopheryl acetate kg^{-1} feed, respectively, compared with control samples. Similarly, supplementation with 200 mg kg^{-1} feed reduced total COPs by approximately 60% in cooked ground chicken refrigerated for four days (Lopez-Bote *et al.*, 1998). Vitamin E supplementation (200–225 mg kg^{-1} feed) delayed COP formation, regardless of dietary fat source (saturated or unsaturated), in cooked and refrigerated pork (Rey *et al.*, 2001) and in raw and cooked vacuum-packaged ground dark chicken meat following storage at 20% for seven months (Grau *et al.*, 2001). Monahan *et al.* (1992b) showed that total COPs and TBARS numbers were correlated in cooked pork. This correlation was later confirmed by Galvin *et al.* (1998a) (Fig. 16.7) and Maraschiello *et al.* (1998) for chicken meat. Galvin *et al.* (2000) also showed that dietary α-tocopheryl acetate supplementation in beef cattle (3000 mg α-tocopheryl acetate per head per day for 135 days prior to slaughter) reduced 7-keto formation in vacuum-packaged cooked M. *psoas major* steaks during refrigerated and frozen storage. Storage under aerobic conditions (4°C for 6 days) resulted in up to a 7-fold increase in COPs in cooked chicken samples (Conchillo *et al.*, 2003). On the other hand, vacuum packaging significantly inhibited COPs formation during storage.

Irradiation increases COPs in meat at doses permitted for food in beef, pork, veal, chicken and turkey (Hwang and Maerker, 1993; Galvin *et al.* 1998b; Ahn *et al.*, 2001; Du *et al.*, 2001; Nam *et al.*, 2001). In general, irradiation causes a significant increase in COPs formation in raw, cooked and stored meats under

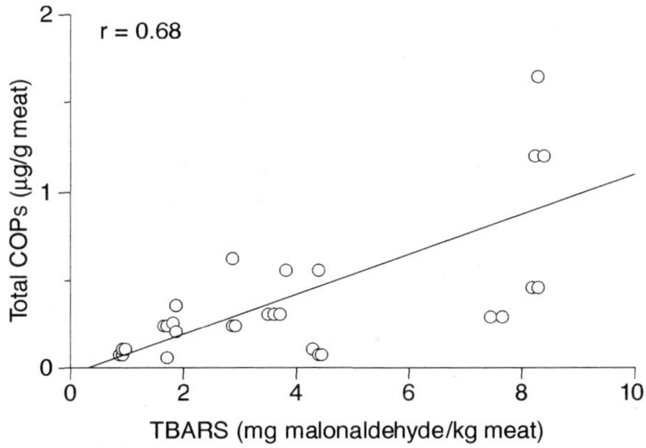

Fig. 16.7 Relationship between thiobarbituric acid-reactive substances (TBARS) and total cholesterol oxidation products (COPs) in cooked ground chicken breast stored at 4°C for up to 12 days (Galvin *et al.*, 1998a; reprinted from *Meat Science*, 48, Galvin *et al.* 'Cholesterol oxides in processed chicken muscle as influenced by dietary α-tocopherol supplementation', pp. 1–9, copyright 1998, with permission from Elsevier).

aerobic packaging conditions, whereas vacuum packaging was effective in preventing cholesterol oxidation (Du *et al.*, 2001; Nam *et al.*, 2001). The concentrations of COPs and lipid oxidation products (TBARS) were closely related to the content of PUFA present in the meat (Galvin *et al.*, 1998b; Ahn *et al.*, 2001). Recent studies conclude that packaging is far more important than irradiation in the formation of COPs and TBARS in cooked meats (Ahn *et al.*, 2001; Nam *et al.*, 2001). In general, these studies show that the sensitizing effects of irradiation on triacylglycerol and cholesterol oxidation can be overcome by storage under vacuum conditions. Galvin *et al.* (1998b) confirmed that dietary supplementation of α-tocopheryl acetate (up to 400 mg kg^{-1} feed) significantly reduced COPs levels in minced, cooked and irradiated breast and thigh meat.

16.7 Lipid oxidation and meat flavour, colour and drip loss

16.7.1 Meat flavour

The intrinsic flavour of meat is due to the production of volatile, odorous, lipid oxidation products during cooking and the involvement of these (mainly aliphatic aldehydes, ketones and alcohols) with Maillard reaction products to form other volatiles that contribute to odour and flavour (Wood *et al.*, 2003). The fat tissue and, in particular, the unsaturated phospholipid fatty acids are important in flavour development and consumer acceptability of meats (Mottram and Edwards, 1983; Mottram, 1996, 1998; Elmore *et al.*, 1999). On the other hand, uncontrolled oxidation results in the formation of off-odours and off-flavours and the phenomenon of 'warmed-over flavour' (WOF) or 'meat flavour deterioration' (MFD) (Pearson *et al.*, 1977; St Angelo, 1996; St Angelo *et al.*, 1987; Mottram, 1987). The terms WOF and MFD are used in flavour evaluation of pre-cooked, refrigerated ready-to-eat meat products to describe the increase in off-flavour notes and decrease in desirable flavour notes that occur with storage.

Aldehydes are quantitatively the most dominant class of volatiles in meat and are probably the most interesting contributors to flavour and off-flavour of cooked meats since they have low odour thresholds (Mottram, 1987; Elmore *et al.*, 1999). The latter observed that the levels of aldehydes and aliphatic alcohols in cooked beef reflected the levels of *n*-3 PUFAs in the meat. However, the alkanals, 2-alkenals and alkanols found in the aroma extracts contained alkyl chains with between four and nine saturated carbon atoms that are more likely to be derived from C18:1 *n*-9 and C18:2 *n*-6 fatty acids rather than *n*-3 PUFA. Elmore *et al.* (1999) suggested that in high PUFA meat, autoxidation is initiated much more readily by the presence of higher quantities of C18:3 *n*-3, C20:5 *n*-3, and C22:6 *n*-3, than from C18:1 and C18:2. Once the initial initiation and propagation reactions started, the subsequent chain reaction was less dependent on the nature of the unsaturated fatty acids; therefore the mass law of chemical reactions favoured the breakdown of the abundant oleic and linoleic acids.

Aldehydes such as hexanal, pentanal and 2,4-decadienal have been used to evaluate the oxidative state of meats from different animals (St Angelo *et al.*, 1987; Drumm and Spanier, 1991; Wen *et al.*, 1997c; Byrne *et al.*, 2002). The latter group, using GC-MS, observed large increases during storage in lipid-derived compounds such as hexanal, 2-heptenal, 2,4-alkadienals, 1-octen-3-ol, 2,3-octanedione and tridecane in cooked chicken. Descriptive sensory profiling indicated that WOF/MFD development involved an increase in 'rancid' and 'sulphur/rubber' sensory notes and a concurrent decrease in chicken 'meaty' flavour. In another study from the same laboratory, O'Sullivan *et al.* (2003a) observed that many compounds associated with lipid oxidation, e.g. pentanal, 2-pentylfuran, octanal, nonanal, 1-octen-3-ol and hexanal, correlated with oxidative sensory descriptors and WOF development.

Grass-fed beef and lamb have naturally high levels of 18:3 and long chain *n*-3 PUFA and the intensity of flavour is greater in comparison with grain-fed animals that consume and deposit relatively more linoleic acid (18:2) (Warren *et al.*, 2002). Larick *et al.* (1987) and Larick and Turner (1990) observed that lipid breakdown products such as aldehydes and ketones were more apparent in volatile compounds from beef produced on grass rather than on grain. Supplementation of pig diets with 100 or 200 mg α-tocopheryl acetate kg^{-1} feed reduced levels of saturated aldehydes (pentanal, hexanal, heptanal) in raw pork (Cava *et al.*, 2000), and resulted in fresher flavour in refrigerated cooked pork compared to meat from animals fed 60 mg kg^{-1} feed (Dirinck *et al.*, 1996). O'Sullivan *et al.* (2003a) observed that less WOF developed in cooked pork from pigs supplemented with vitamin E (200 mg α-tocopheryl acetate kg^{-1}feed) and more in cooked pork from pigs supplemented with iron (7 g iron(II) sulphate kg^{-1} feed). Vitamin E supplementation (200 mg kg^{-1} feed) inhibited the formation of saturated and unsaturated aldehydes in raw chicken breast and

Table 16.4 Effect of α-tocopherol supplementation on thiobarbituric acid-reactive substances (TBARS) and on warmed-over flavour (WOF) development in cooked chicken thigh meat following storage at 4°C for up to 5 days

α-Tocopheryl acetate (mg kg^{-1} diet)	TBARS[1] (days at 4°C)		WOF score[2] (days at 4°C)	
	3	5	3	5
30	3.16 ± 0.04	4.26 ± 0.02	6.88 ± 0.16	9.11 ± 0.10
200	0.81 ± 0.03	1.61 ± 0.06	1.84 ± 0.17	4.12 ± 0.16

1TBARS are expressed as mg malonaldehyde kg^{-1} meat.
2 Samples were scored on a 10 cm line scale, where 0 = no WOF and 10 = extreme WOF.
Adapted from O'Neill *et al.* (1998).

thigh meat and reduced off-flavour in cooked meat (De Winne and Dirinck, 1996), and also reduced WOF and TBARS development in cooked ground chicken thigh meat following refrigerated storage for up to five days (Table 16.4) (O'Neill *et al.*, 1998). In addition, a significant ($P < 0.001$) linear relationship was observed between TBARS and WOF scores ($r = 0.94$) (O'Neill *et al.*, 1998) (Fig. 16.8).

Supplementation of turkey diets with 600 mg α-tocopheryl acetate kg^{-1} feed reduced hexanal development (Wen *et al.*, 1997c) and WOF development in ground and cooked turkey meat (Higgins *et al.*, 1999b). Sheldon *et al.* (1997) observed that supplementation with up to 25 times the normal vitamin E requirement consistently produced higher turkey meat flavour and aftertaste scores and lower oxidized meat flavour and aftertaste scores in cooked ground meat during refrigerated storage for up to eight days. Supplementation with over

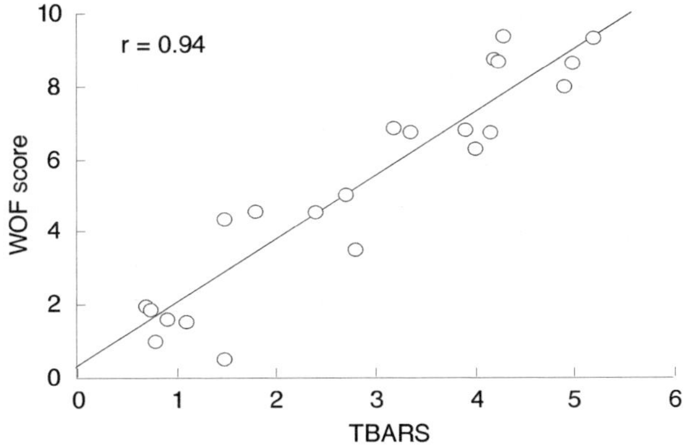

Fig. 16.8 Relationship between thiobarbituric acid-reactive substances (TBARS) and warmed-over flavour (WOF) in cooked, minced chicken thigh muscle (O'Neill *et al.*, 1998; reproduced with permission; http://www.tandf.co.uk/journals/tf/00071668.html).

10 times the normal requirement reduced total headspace aldehydes in raw samples refrigerated for seven days.

Irradiation is known to induce off-odours in meats. Patterson and Stevenson (1995) reported that dietary vitamin E supplementation (800 mg α-tocopheryl acetate kg^{-1} feed) resulted in a 39% and 44% reduction in total volatiles at doses of 2.5 and 10.0 kGy, respectively, in thigh meat.

16.7.2 Drip loss

Excessive drip loss from fresh meat signifies not only financial losses associated with such meats but losses in valuable vitamins, minerals, flavour compounds and water (Morrissey *et al.*, 2002). Loss of the latter component can affect overall eating quality, producing meat that can be described as tough and having poor mouthfeel characteristics. Vitamin E may have beneficial effects on drip loss, as it is involved in stabilizing lipid membranes (Zimmer *et al.*, 1993). The efficacy of dietary supplemental α-tocopheryl acetate in preventing drip loss during simulated display has been shown in pigs (Asghar *et al.*, 1991; Monahan *et al.*, 1994), beef (Mitsumoto *et al.*, 1995) and broilers (O'Neill *et al.*, 1998). Lawlor *et al.* (1999) observed that the initial vitamin E content was negatively correlated with drip loss after refrigerated storage, and the results further suggested that a substantial reduction in drip loss would occur if vitamin E concentrations of >6 mg kg^{-1} in M. *rhomboideus* and M. *seratus ventralis*, and >5 mg kg^{-1} in M. *semitendinosus* and M. *senimembranosus*, were achieved.

Although the precise mechanisms of how vitamin E decreases drip loss have not been defined, previous authors have suggested that the effect of vitamin E arises from its ability to react with free radicals thereby preventing oxidation of membranal lipids during storage (Asghar *et al.*, 1991; Monahan *et al.*, 1994; Mitsumoto *et al.*, 1995). It is recognized that lipid peroxidation can cause severe damage to membranal proteins (Halliwell and Gutteridge, 1999), decreases the relative proportion of PUFAs in biological membranes and increases the molecular order of membranal lipids (Nozawa *et al.*, 1980). These changes lead to a decrease in membrane fluidity and increase in 'leakiness' of the membrane bilayer (Richter, 1987). Vitamin E, because of its chain-breaking antioxidant activity, preserves the integrity of cell membranes and maintains membrane fluidity, thereby inhibiting the passage of sarcoplasmic fluid through muscle cell membranes during storage. However, the mechanism by which vitamin E modifies membrane fluidity and permeability of phospholipids is quite complex and depends upon the composition of the phospholipid and the tocopherol–phospholipid molar ratio (Halliwell, 1984).

16.7.3 Meat colour

At the point of sale, colour and colour stability are the most important attributes of fresh meat quality and various approaches have been used to meet consumer expectation that an attractive bright-red colour is compatible with long shelf-life and good eating quality (Hood and Mead, 1993). In red meats, particularly beef, a bright cherry-red colour (bloom) is perceived by consumers as being indicative of freshness, while they discriminate against beef which has turned brown (Buckley and Morrissey, 1992). Oxidation of the fresh muscle pigment oxymyoglobin (cherry red colour) to metmyoglobin (brown colour) leads to the discoloration of red meats. There is a general consensus that the processes of oxymyoglobin and lipid oxidation in muscle foods, while independent of each other, can be interrelated. However, the exact nature of this interrelationship has not been established. For instance, one hypothesis is that oxymyoglobin oxidation initiates the first step in a sequence of chemical reactions leading to the production of radicals (porphyrin cation radicals), which in turn bring about the initiation of lipid oxidation (Kanner and Harel, 1985). Conversely, another hypothesis is that muscle lipids and liposomes can catalyse oxymyoglobin oxidation (Renerre, 2000; Faustman and Wang, 2000). Lynch and Faustman (2000) reported that lipid oxidation products (aldehydes) alter myoglobin stability by increasing oxymyoglobin oxidation, decreasing the ability of metmyoglobin to be enzymatically reduced, thereby enhancing the proxidant activity of metmyoglobin. According to Faustman and Wang (2000) α-tocopherol delays the release of proxidative products of lipid oxidation from biomembranes, which in turn delays oxymyoglobin oxidation and the concomitant loss of desirable beef colour.

A comprehensive review of the oxidative stability of beef and the quality-enhancing role of vitamin E has been published by Kerry *et al.* (2000a). The majority of studies carried out on dietary supplementation with α-tocopheryl acetate to cattle have consistently shown improved colour stability on subsequent retail storage of all meat cuts (Faustman *et al.*, 1989a,b; Arnold *et al.*, 1992; Arnold *et al.*, 1993a,b; Liu *et al.*, 1996a,b; Sherbeck *et al.*, 1995; Sanders *et al.*, 1997). With the exception of the report by Strohecker *et al.* (1997), similar observations have been made for lamb meat colour (Wulf *et al.*, 1995; Guidera *et al.*, 1997a,b). Faustman *et al.* (1989b) and Arnold *et al.* (1993a) defined the relationship between muscle α-tocopherol concentration and metmyoglobin percentage and concluded that the target α-tocopherol level in fresh muscle for optimum protection against discoloration was in the region of 3–3.5 mg α-tocopherol kg^{-1}, depending on the muscle in question. Arnold *et al.* (1993b) stated that feeding cattle for a minimum of 44 days at a rate of 1300 mg *all-rac*-α-tocopheryl acetate per day was necessary to incorporate adequate α-tocopherol within M. *longissimus dorsi*. Lui *et al.* (1995) recommended a supplementation strategy of 500 mg *all-rac*-α-tocopheryl per steer per day for 126 days in order to attain the desirable levels in muscle.

16.8 Meat packaging and shelf-life quality

Walsh and Kerry (2002) recently reviewed the role of packaging in the meat industry. The overriding factor that will affect meat packaging performance and overall shelf-life stability is product composition. The two principal spoilage mechanisms affecting the shelf-life of meat are microbial growth and oxidation of oxymyoglobin and/or lipids (Kerry *et al.*, 2000a). As a result, two primary forms of packaging have emerged to control these primary meat spoilage processes: vacuum packaging and gas pack flushing technologies. The use of modified atmosphere packaging (MAP) for fresh red meats does not strictly control both spoilage processes in that high levels of oxygen (70–80%) are used to purposefully oxidize myoglobin to oxymyoglobin to produce acceptable consumer meat colour; however, this level of oxygen equally oxidizes meat lipids. With white meat systems, the use of high levels of oxygen is unnecessary and is often avoided as high oxygen levels do not have any significant beneficial effects on colour and in general drive oxidative and microbiological spoilage processes at a faster rate than observed for red meats. Therefore, white meats are often packaged using modified atmospheres of carbon dioxide (for bacteriostatic purposes) and nitrogen (for pack-filling purposes to prevent package collapse) mixes.

One approach to overcoming the problem of lipid oxidation in MAP meats (particularly red meats using high oxygen levels to enhance colour) is through the dietary supplementation of vitamin E or other potential antioxidants to red meat-producing animals (Kerry *et al.*, 2000a). Numerous reports have been published demonstrating the shelf-life extension of MAP beef in combination with dietary vitamin E supplementation (Taylor *et al.*, 1994; Lanari *et al.*, 1995; Kerry *et al.*, 1996; Formanek *et al.*, 1998; O'Grady *et al.*, 1998) but less so for sheep meat (Kerry *et al.*, 1998). The exogenous treatment of red meats with antioxidants in various packaging systems has also been reviewed (Mitsumoto, 2000) or investigated (McCarthy *et al.*, 2001; Formanek *et al.*, 2001, 2003; Tang *et al.*, 2001a).

The use of high oxygen MAP systems for red meat packaging not only promotes higher levels of lipid oxidation in these meats compared to vacuum or over-wrap packaging systems but often produces more definite quality trends in terms of shelf-life, in particular where animals have received different dietary components at either the macro- or micro-nutrient level. O'Sullivan *et al.* (2003b) reported that when steaks from cattle fed all-concentrate, mixed ratios of concentrate/forage or all-forage diets were over-wrapped and held in refrigerated display during shelf-life stability tests, no significant trends were observed for colour. However, when steaks from the same animals were packaged under MAP, significantly higher redness values were recorded for steaks taken from cattle fed the all-forage diets compared to those taken from cattle fed the all-concentrate diets. Therefore, the use of different packaging formats may in its own unique way unveil different and unforeseen meat characteristics which may be brought about principally through gross or minute differences in basic meat composition. It is essential that simplistic approaches towards meat packaging be replaced by a more scientific process that recognizes the synergy that exists between the muscle food in question and the packaging format, the synergy being constant only when the muscle composition and packaging format remain the same.

16.9 Future developments

Lipid stability in meat and meat products is influenced by many factors, including species, muscle type, the amount and type of fat in the diet, the nutritional status of the animal at slaughter, the presence or absence of disease or infection, and, increasingly, the type of post-slaughter conditioning and processing. There is now little doubt that altering the concentrations of substrate (PUFAs) and antioxidants in the live animal is the best way of maximizing oxidative stability. Dietary vitamin E supplementation significantly above physiological levels reduces lipid oxidation and cholesterol oxidation as well as myoglobin oxidation and drip loss.

However, the mechanism whereby vitamin E maintains myoglobin in meat and modifies membrane fluidity and permeability needs to be investigated. The situation is less clear-cut regarding other components of the diet. Dietary vitamin C supplementation appears to have few, if any, beneficial effects on meat stability. The practical effects of feeding higher levels of carotenoids to animals also requires further study, including their potentially negative effects on α-tocopherol status. Other dietary components that have attracted attention include α-lipoic acid and its reduced derivative dihydrolipoate. These seem to have several different kinds of antioxidant properties (e.g. metal chelating, radical scavenging and α-tocopherol-regenerating ability) in specific model systems. Some antioxidants can effectively function as prooxidants under certain conditions and it is important to evaluate interactions between antioxidants and the effectiveness of multicomponent, biphasic antioxidant systems in skeletal muscle during processing. Understanding how processing operations influence endogenous skeletal muscle antioxidants and the release of transition metals could lead to new technologies that increase the oxidative stability of muscle. Oxidative deterioration of muscle foods may be ameliorated by combining dietary antioxidant supplementation with optimization of the iron and copper contents of the feed, and development of new and innovative minimal processing technologies and packaging systems that help to maintain a favourable antioxidant–prooxidant balance. Research in these areas is likely to provide a safer, more consistent product that satisfies the increased consumer demands for safety, freshness and variety.

16.10 References

ADDIS PB and PARK SW (1989), 'Role of lipid oxidation products in atherosclerosis', in Taylor SL and Scanlan RA, *Food Toxicology: A Perspective on the Relative Risks*, New York, Marcel Dekker, 297–330.

ADDIS PB and WARNER GJ (1991), 'The potential health aspects of lipid oxidation products in foods', in Arouma OI and Halliwell B, *Free Radicals and Food Products*, London, Taylor and Francis, 77–119.

AHN DU, NAM KC, DU M and JO C (2001), 'Effect of irradiation and packaging conditions after cooking on the formation of cholesterol and lipid oxidation products in meats during storage', *Meat Sci*, 57, 413–418.

ALLEN CE and FOEGEDING EA (1981), 'Some lipid characteristics and interactions in muscle foods: a review', *Food Technol*, 35(5), 253–257.

ARNOLD RN, SCHELLER KK, ARP SC, WILLIAMS SN, BUEGE DR and SCHAEFER DM (1992), 'Effect of long- and short-term feeding of α-tocopheryl acetate to Holstein and crossbred beef steers on performance, carcass characteristics and beef color stability', *J Anim Sci*, 70, 3055–3065.

ARNOLD RN, ARP SC, SCHELLER KK, WILLIAMS SN and SCHAEFER DM (1993a), 'Tissue equilibrium and subcellular distribution of vitamin E relative to myoglobin and lipid oxidation in displayed beef', *J Anim Sci*, 71, 105–118.

ARNOLD RN, SCHELLER KK, ARP SC, WILLIAMS SN and SCHAEFER DM (1993b), 'Dietary α-tocopheryl acetate enhances beef quality in Holstein and beef breed steers', *J Food Sci*, 58, 28–33.

ASGHAR A, GRAY JI, BOOREN AM, GOMAA EA, ABOUZIED MM, MILLER ER and BUCKLEY DJ (1991), 'Effects of supranutritional dietary vitamin E levels on subcellular deposition of α-tocopherol in muscle and on pork quality', *J Sci Food Agric*, 57, 31–41.

AZZI A and STOCKER A (2000) 'Vitamin E: non-antioxidant roles', *Prog Lipid Res*, 39, 231–255.

BARTOV J (1983), 'Effect of various dietary factors and age on plasma α-tocopherol concentration of turkeys', *Poultry Sci*, 62, 635–641.

BOLLAND JL and GEE G (1946), 'Kinetic studies in the chemistry of rubber and related materials', *Trans Faraday Soc*, 42, 236–252.

BRANDON S, MORRISSEY PA, BUCKLEY DJ and FRIGG M (1993), 'Influence of dietary α-tocopheryl acetate on the oxidative stability of chicken tissues', in Corlin P, Culioli J and Richard FF, *Proceedings of the 11th European Symposium on the Quality of Poultry Meat*, France, World's Poultry Science Association, 397–403.

BRANNAN RG, CONNOLLY BJ and DECKER EA (2001), 'Peroxynitrite; a potential initiator of lipid oxidation in food', *Trends Food Sci Technol*, 12, 164–173.

BRIVIBA K and SIES H (1994), 'Nonenzymatic antioxidant defence systems', in Frei B, *Natural Antioxidants in Human Health and Disease*, London, Academic Press, 107–128.

BUCKLEY DJ and MORRISSEY PA (1992), 'Vitamin E and meat quality', F. Hoffman-La Roche Ltd, Vitamins and Fine Chemicals Division, Basel, Switzerland

BUETTNER GR (1993), 'The pecking order of free radicals and antioxidants: lipid peroxidation, α-tocopherol, and ascorbate', *Arch Biochem Biophys*, 300, 535–543.

BUETTNER GR and JURKIEWICZ BA (1996), 'Catalytic metals, ascorbate and free radicals: combinations to avoid', *Radiat Res*, 145, 532–541.

BURTON GW (1989), 'Antioxidant action of carotenoids', *J Nutr*, 119, 109–111.

BURTON GW and INGOLD KU (1984), 'β-Carotene: an unusual type of lipid antioxidant', *Science*, 224, 569–573.

BURTON GW, JOYCE A and INGOLD KU (1983), 'Is vitamin E the only chain-breaking antioxidant in human blood plasma and erythrocyte membranes?', *Arch Biochem Biophys*, 221, 281–290.

BYRNE DV, BREDIE WLP, MOTTRAM DS and MARTENS M (2002), 'Sensory and chemical investigations on the effect of oven cooking on warmed-over flavour development in chicken meat', *Meat Sci*, 61, 127–139.

CANNON JE, MORGAN JB, SCHMIDT GR, DELMORE RJ, SOFOS JM, SMITH GC and WILLIAMS SN (1995), 'Vacuum-packaged precooked pork from hogs fed supplemental vitamin E:

chemical shelf-life and sensory properties', *J Food Sci*, 60, 1179–1182.

CAVA R, VENTANAS J, TEJEDA JF, RUIZ J and ANTEQUERA T (2000), 'Effect of free-range rearing and α-tocopherol and copper supplementation on fatty acid profiles and susceptibility to lipid oxidation of fresh meat from Iberian pigs', *Food Chem*, 68, 51–59.

CHEN Q, SHI H and HO CT (1992), 'Effects of rosemary extracts and major constituents on lipid oxidation and soybean lipoxygenase activity', *J Amer Oil Chem Soc*, 69, 999–1002.

CHRISTEN S, WOODALL AA, SHIGENEGA MK, SOUTHWELL-KEELY PJ, DUNCAN MW and AMES BN (1997), 'γ-Tocopherol traps mutagenic electrophiles such as NO_x and complements α-tocopherol: physiological implications', *Proc Natl Acad Sci*, 94, 3217–3222.

CLEMENT M and BOURRE JM (1997), 'Graded dietary levels of RRR-gamma-tocopherol induces a marked increase in the concentration of alpha- and gamma-tocopherol in nervous tissue, heart, liver and muscle of vitamin E-deficient rats', *Biochim Biophys Acta*, 1334, 173–181.

CONCHILLO A, ANSORENA D and ASTIASARÁN I (2003), 'Combined effect of cooking (grilling and roasting) and chilling storage (with and without air) on lipid and cholesterol oxidation in chicken breast', *J Food Prot*, 66, 840–846.

COSGROVE JP, CHURCH DF and PRYOR WA (1987), 'The kinetics of the autoxidation of polyunsaturated fatty acids', *Lipids*, 22, 299–304.

CUVELIER ME, RICHARD H and BERSET C (1996), 'Antioxidant activity of phenolic composition of pilot-plant and commercial extracts of sage and rosemary', *J Amer Oil Chem Soc*, 73, 645–652.

DE WINNE A and DIRINCK P (1996), 'Studies on vitamin E and meat quality 2: Effect of feeding high vitamin E levels on chicken meat quality', *J Agric Food Chem*, 44, 1691–1696.

DECKER EA, CRUM AD, SHANDRA NC and MORRISSEY PA (1993), 'Catalysis of lipid oxidation by iron from an insoluble fraction of beef diaphragm muscle', *J Food Sci*, 58, 233–236, 238.

DIPLOCK AT, CHARLEUX J-L, CROZIER-WILLI G, KOK FJ, RICE-EVANS C, ROBERFROID M, STAHL W and VINA-RIBES J (1998), 'Functional food science and defence against reactive oxygen stress', *Br J Nutr*, 80, S77–S112.

DIRINCK P, DE WINNE A, CASTEELS M and FRIGG M (1996), 'Studies on vitamin E and meat quality 1. Effect of feeding high vitamin E levels on time-related pork quality', *J Agric Food Chem*, 44, 65–68.

DRUMM TD and SPANIER AM (1991), 'Changes in the lipid content of autoxidation and sulphur-containing compounds in cooked beef during storage', *J Agric Food Chem*, 49, 336–343.

DU M, NAM KC and AHN DU (2001), 'Cholesterol and lipid oxidation products in cooked meats as affected by raw meat packaging and irradiation and by cooked meat packaging and storage times', *J Food Sci*, 66, 1396–1401.

ELMORE JS, MOTTRAM DS, ENSER M and WOOD JD (1999), 'Effect of polyunsaturated fatty acid composition of beef muscle on the profile of aroma volatiles', *J Agric Food Chem*, 47, 1619–1625.

ENGESETH NJ, GRAY JI, BOOREN AM and ASHGAR A (1993), 'Improved oxidative stability of veal lipids and cholesterol through dietary vitamin E supplementation', *Meat Sci*, 35, 1–15.

ENSER M, HALLETT KG, HEWETT B, FURSEY GAJ and WOOD JD (1996), 'Fatty acid content and composition of English beef, lamb and pork at retail', *Meat Sci*, 44, 443–458.

ENSER M, HALLETT KG, HEWETT B, FURSEY GAJ, WOOD JD and HARRINGTON G (1998), 'Fatty acid content and composition of UK beef and lamb muscle in relation to production system and implications for human nutrition', *Meat Sci,* 49, 329–341.

ENSER M, RICHARDSON RI, WOOD JD, GILL BP and SHEARD PR (2000), 'Feeding linseed to increase the n-3 PUFA of pork: fatty acid composition of muscle, adipose tissue, liver and sausages', *Meat Sci*, 55, 201–212.

FANG X and WADA S (1993), 'Enhancing the antioxidant effect of α-tocopherol with rosemary in inhibiting catalysed oxidation caused by Fe^{2+} and hemoprotein', *Food Res Int*, 26, 405–411.

FARMER EH, BLOOMFIELD GF, SUNDRALMGIN A and SUTTON DA (1942), 'The course and mechanisms of auto-oxidation reactions in olefinic and polyolefinic substances, including rubber', *Trans Faraday Soc*, 38, 348–356.

FAUSTMAN C and WANG K-W (2000), 'Potential mechanisms by which vitamin E improves oxidative stability of myoglobin', in Faustman C, Decker E and Lopez-Bote CJ *Antioxidants in Muscle Foods: Nutritional Strategies to Improve Quality*, New York, John Wiley, 135–152.

FAUSTMAN C, CASSENS RG, SCHAEFER DM, BUEGE DR and SCHELLER KK (1989a), 'Vitamin E supplementation of Holstein steer diets improves sirloin steak color', *J Food Sci*, 54, 485–486.

FAUSTMAN C, CASSENS RG, SCHAEFER DM, BUEGE DR, WILLIAMS SN and SCHELLER KK (1989b), 'Improvement of pigment and lipid stability in Holstein steer beef by dietary supplementation with vitamin E', *J Food Sci*, 54, 858–862.

FISHER AV, ENSER M, RICHARDSON RI, WOOD JD, NUTE GR, KURT E, SINCLAIR LA and WILLIAMSON RG (2000), 'Fatty acid composition and eating quality of lamb types derived from four diverse breeds × production systems', *Meat Sci*, 55, 141–147.

FORMANEK Z, KERRY JP, BUCKLEY DJ, MORRISSEY PA and FARKAS J (1998), 'Effects of dietary vitamin E supplementation and packaging on the quality of minced beef', *Meat Sci*, 50, 203–210.

FORMANEK Z, KERRY JP, HIGGINS FM, BUCKLEY DJ, MORRISSEY PA and FARKAS J (2001), 'Addition of synthetic and natural antioxidants to α-tocopheryl acetate supplemented beef patties: effects of antioxidants and packaging on lipid oxidation', *Meat Sci*, 58, 337–341.

FORMANEK Z, LYNCH A, GALVIN K, FARKAS J and KERRY JP (2003), 'Combined effects of irradiation and the use of natural antioxidants on the shelf-life stability of overwrapped minced beef', *Meat Sci*, 63, 433–440.

FRANKEL EN (1985), 'Chemistry of free radical and singlet oxidation of lipids', *Prog Lipid Res*, 23, 197–221.

FRANKEL EN (1998), *Lipid Oxidation*, Dundee, UK, The Oily Press.

FREI B (1991), 'Ascorbic acid protects lipids in human plasma and low-density lipoprotein against oxidative damage', *Am J Clin Nutr*, 54, 1113S–1118S.

FREI B, ENGLAND L and AMES BN (1989), 'Ascorbate is an outstanding antioxidant in human blood plasma', *Proc Nat Acad Sci*, 86, 6371–6381.

FREI B, STOCKER R, ENGLAND L and AMES BN (1990), 'Ascorbate: the most effective antioxidant in human blood plasma', in Emerit I and Packer L, *Antioxidants in Therapy and Preventative Medicine*, New York, Plenum, 155–163.

FREI B, STOCKER R and AMES BN (1992), 'Small molecular antioxidant defenses in human extracellular fluids', in Scandalios J, *Molecular Biology of Free Radical Scavenging Systems*, Cold Spring Harbor, NY, Cold Spring Harbor Lab. Press, 23–45.

GALVIN K, MORRISSEY PA and BUCKLEY DJ (1997), 'Influence of dietary vitamin E and oxidized sunflower oil on the storage stability of cooked chicken muscle', *Br Poultry Sci*, 38, 499–504.

GALVIN K, MORRISSEY PA and BUCKLEY DJ (1998a), 'Cholesterol oxides in processed chicken muscle as influenced by dietary α-tocopherol supplementation', *Meat Sci*, 48, 1–9.

GALVIN K, MORRISSEY PA and BUCKLEY DJ (1998b), 'Effect of dietary α-tocopherol supplementation and gamma-irradiation on α-tocopherol retention and lipid oxidation in cooked minced meat', *Food Chem*, 62, 185–190.

GALVIN K, LYNCH A-M, KERRY JP, MORRISSEY PA and BUCKLEY DJ (2000), 'Effect of dietary vitamin E supplementation on cholesterol oxidation in vacuum packaged cooked beef steaks', *Meat Sci*, 55, 7–11.

GANDEMER G and MEYNIER A (1995), 'Lipid degradation and flavour compounds in meat products', in Chizzolini R, *Lipids and Quality Meat Products*: a *Comprehensive Analysis*, Parma, Italy, University of Parma, 31–43.

GRAU A, CODONY R, GRIMPA S, BAUCELLS MD and GUARDIOLA F (2001), 'Cholesterol oxidation in frozen dark chicken meat: influence of dietary fat source, and α-tocopherol and ascorbic acid supplementation', *Meat Sci*, 57, 197–208.

GRAY JI, GOMAA EA and BUCKLEY DJ (1996), 'Oxidative quality and shelf-life of meats', *Meat Sci*, 43, 5111–5123

GUARDIOLA F, CODONY R, ADDIS PB, RAFECAS M and BOATELLA J (1996), 'Biological effects of oxysterols: current status', *Food Chem Toxicol*, 34, 193–211.

GUARDIOLA F, DUTTA PC, CODONY R and SAVAGE GP (2002), *Cholesterol and Phytosterol Oxidation Products: Analysis, Occurrence, and Biological Effects*, Champaign, IL, AOCS Press.

GUIDERA J, KERRY JP, BUCKLEY DJ, LYNCH PB and MORRISSEY PA (1997a), 'The effect of dietary vitamin E supplementation on the quality of fresh and frozen lamb meat', *Meat Sci*, 45, 33–43.

GUIDERA J, KERRY JP, BUCKLEY DJ, LYNCH PB and MORRISSEY PA (1997b), 'The effect of dietary α-tocopheryl acetate supplementation on muscle α-tocopherol levels and lamb quality', *Irish J Agric Food Res*, 36, 241–247.

HALLIWELL B (1984), 'Oxygen radicals: a commonsense look at their nature and medical importance', *Med Biol*, 62, 71–77.

HALLIWELL B (1987), 'Oxidants and human disease: some new concepts', *FASEB J*, 1, 358–364.

HALLIWELL B (1989), 'Superoxide, iron, vascular endothelium and reperfusion injury', *Free Rad Res Comm*, 5, 315–318.

HALLIWELL B and CHIRICO S (1993), 'Lipid peroxidation: its mechanism, measurement and significance', *Amer J Clin Nutr*, 57, 715S-725S.

HALLIWELL B and GUTTERIDGE JMC (1986), 'Oxygen free radicals and iron in relation to biology and medicine. Some problems and concepts', *Arch Biochem Biophys*, 246, 501–514.

HALLIWELL B and GUTTERIDGE JMC (1999), *Free Radicals in Biology and Medicine*, 3rd edition, Oxford, Oxford University Press.

HALLIWELL B, MURCIA MA, CHIRICO S and ARUOMA OI (1995), 'Free radicals and antioxidants in food and *in vivo*: what they do and how they work', *Crit Rev Food Sci Nutr*, 35, 7–20.

HASHIMOTO T, KUMAZAWA S, NANJO F, HARN Y and NAKAYAMA T (1999), 'Interaction of tea catechins with lipid bilayers investigated with liposome systems', *Biosci Biotech*

Biochem, 63, 2252–2255.

HIGGINS FM, KERRY JP, BUCKLEY DJ and MORRISSEY PA (1998), 'Assessment of α-tocopheryl acetate supplementation, addition of salt and packaging on the oxidative stability of raw turkey meat', *Br Poultry Sci*, 39, 596–600.

HIGGINS FM, KERRY JP, BUCKLEY DJ and MORRISSEY PA (1999a), 'Effect of dietary α-tocopheryl acetate supplementation on α-tocopherol distribution in raw turkey muscles and its effect on the storage stability of cooked turkey meat', *Meat Sci*, 50, 373–383.

HIGGINS FM, KERRY JP, BUCKLEY DJ and MORRISSEY PA (1999b), 'Effects of α-tocopheryl acetate supplementation and salt addition on the oxidative stability (TBARS) and warmed-over flavour (WOF) of cooked turkey meat', *Br Poultry Sci*, 40, 59–64.

HOOD DE and MEAD GC (1993), 'Modified atmosphere storage of fresh meat and poultry', in Parry RT, *Principles and Applications of Modified Atmosphere Packaging of Food*, London, Blackie Academic and Professional, 269–298.

HORWITT MK (1986), 'Interpretation of requirements for thiamine, riboflavin, niacin-tryptophan and vitamin E plus comments on balance studies and vitamin B_6', *Am J Clin Nutr*, 44, 973–985.

HOVING-BOLINK AH, EIKELENBOOM G, VAN DIEPEN JM, JONGBLOED AW and HOUBEN JH (1998), 'Effect of dietary vitamin E supplementation on pork quality', *Meat Sci*, 49, 205–212.

HWANG KT and MAERKER G (1993), 'Quantification of cholesterol oxidation products in unirradiated and irradiated meats', *J Amer Oil Chem Soc*, 70, 371–375.

INGOLD KU (1961), 'Inhibition of the autoxidation of organic substances in the liquid phase', *Chem Rev*, 61, 563–589.

JENSEN C, SKIBSTED LH, JAKOBSEN K and BERTELSEN G (1995), 'Supplementation of broiler diets with all *rac*-α or a mixture of natural sources RRR-α-, β-, γ-, δ-tocopheryl acetate. 2: Effects on the oxidative stability of raw and precooked broiler meat products', *Poultry Sci*, 74, 2048–2056.

JENSEN C, GUIDERA JP, SKOVGAARD IM, STAUM H, SKIBSTED LH, JENSEN, SK, MOLLER AJ, BUCKLEY DJ and BERTELSEN G (1997), 'Effect of dietary α-tocopheryl acetate supplementation on α-tocopherol deposition in porcine M *psoas major* and M *longissimus dorsi* and on drip loss colour stability and oxidative stability of pork meat', *Meat Sci*, 45, 491–500.

JENSEN C, FLENSTED-JENSEN M, SKIBSTED LH and BERTLELSEN G (1998), 'Effect of dietary rape seed oil, copper(II) sulphate and vitamin E on drip loss, colour and lipid oxidation of chilled pork chops packed in atmospheric air or in a high oxygen atmosphere', *Meat Sci*, 50, 211–221.

JIANG Q, CHRISTEN S, SHIGENAGE MK and AMES BN (2001), 'γ-Tocopherol, the major form of vitamin E in the US diet, deserves more attention', *Am J Clin Nutr*, 74, 714–722.

JO C, SON JH, SON CB and BYUN MW (2003), 'Functional properties of raw and cooked pork patties with added irradiated, freeze-dried green tea leaf extract powder during storage at 4°C', *Meat Sci*, 64, 13–17.

JORGENSEN K and SKIPSTED LH (1993), 'Carotenoid scavenging of radicals. Effect of carotenoid structure and oxygen partial pressure an antioxidant activity', *Z Lebensm Unters Forsch*, 196, 423–429.

KAGAN VE, SERBINOVA EA, FORTE T, SCITA G and PACKER L (1992), 'Recycling of vitamin E in human low density lipoprotein', *J Lipid Res*, 33, 385–397.

KANNER J and HAREL S (1985), 'Initiation of membranal lipid peroxidation by activated metmyoglobin and methemoglobin', *Arch Biochem Biophys*, 237, 314–321.

KANNER J, HAZAN R and DOLL L (1988), 'Catalytic "free" iron ions in muscle foods', *J Agric Food Chem*, 36, 412–415.

KATZ F (1999), 'Top product development trends in Europe', *Food Tech*, 53(1), 38–42.

KEHRER JP (1993), 'Free radicals as mediators of tissue injury and disease', *Crit Rev Toxicol*, 23, 21–48.

KEHRER JP and SMITH CV (1994), 'Free radicals in biology: sources, reactivities and roles in the etiology of human diseases', in Frei B, *Natural Antioxidants in Health and Disease* London, Academic Press, 25–62.

KERRY JP, LYNCH M, BUCKLEY DJ, MORRISSEY PA, MONAHAN FJ and ALLEN P (1996), 'Effect of dietary vitamin E supplementation on the quality of fresh and vacuum-packaged beef', in *Proc 42nd Int Congr Meat Sci Technol*, Lillehammer, Norway, 96–97.

KERRY JP, BUCKLEY DJ, MORRISSEY PA, O'SULLIVAN K and LYNCH PB (1998), 'Endogenous and exogenous α-tocopherol supplementation: effects on lipid stability (TBARS) and warmed-over flavour (WOF) in porcine M *longissimus dorsi* roasts held in aerobic and vacuum packs', *Food Res Int*, 31, 211–216.

KERRY JP, BUCKLEY DJ and MORRISSEY PA (2000a), 'Improvement of oxidative stability of beef and lamb with vitamin E', in Faustman C, Decker E and Lopez-Bote CJ, *Antioxidants in Muscle Foods: Nutritional Strategies to Improve Quality*, New York, John Wiley, 229–261.

KERRY JP, O'SULLIVAN MG, BUCKLEY DJ, LYNCH PB and MORRISSEY PA (2000b), 'The effects of dietary α-tocopheryl acetate supplementation and modified atmosphere packaging (MAP) on the quality of lamb patties', *Meat Sci*, 56, 61–66.

KERRY JP, GILROY DA and O'BRIEN NM (2002), 'Formation and content of cholesterol oxidation products in meat and meat products', in Guardiola F, Dutta PC, Codony R and Savage GP, *Cholesterol and Phytosterol Oxidation Products: Analysis, Occurrence, and Biological Effects*, Champaign, IL, AOCS Press, 162–185.

KING AJ, UIJTTENBOOGAART TG and DE VRIES AW (1995), 'α-Tocopherol, β-carotene and ascorbic acid as antioxidants in stored poultry muscle', *J Food Sci*, 60, 1009–1012.

KOUBA M, ENSER M, WHITTINGTON FM, NUTE RG and WOOD JD (2003), 'Effect of a high linolenic acid diet on lipogenic enzyme activities, fatty acid composition and meat quality in the growing pig', *J Anim Sci*, 81, 1967–1979

KRINSKY NI (2001), 'Carotenoids as antioxidants', *Nutrition*, 17, 815–817.

LAI S-M, GRAY JI, SMITH DM, BOOREN AM, CRACKEL RL and BUCKLEY DJ (1991), 'Effects of oleoresin rosemary, tertiary butylhydroquinone, and sodium tripolyphosphate on the development of oxidative rancidity in restructured chicken nuggets', *J Food Sci*, 56, 616–620.

LANARI MC, SCHAEFER DM, CASSENS RG and SCHELLER KK (1995), 'Atmosphere and blooming time affect color and lipid stability of frozen beef from steers supplemented with vitamin E', *Meat Sci*, 40, 33–44.

LARICK, DK and TURNER BE (1990), 'Flavour characteristics of forage- and grain-fed beef as influenced by phospholipids and fatty acid compositional differences', *J Food Sci*, 55, 312–368.

LARICK DK, HEDRICK HB, BAILEY ME, WILLIAMS JE, HANCOCK DL, GARNER GB and MORROW RE (1987), 'Flavour constituents of beef as influenced by forage and grain feeding', *J Food Sci*, 52, 245–251.

LARICK DK, TURNER BE, SCHOENHERR WD, COFFEY MT and PILKINGTON DH (1992), 'Volatile compound contents and fatty acid composition of pork as influenced by linoleic acid content of the diet', *J Anim Sci*, 70, 1397–1403.

LAURIDSEN C, BUCKLEY DJ and MORRISSEY PA (1997), 'Influence of dietary fat and vitamin E supplementation on α-tocopherol levels and fatty acid profiles in chicken muscle membranal fractions and on susceptibility to lipid peroxidation', *Meat Sci*, 46, 9–22.

LAWLOR JBP, SHEEHY PJA , KERRY JP, BUCKLEY DJ and MORRISSEY PA (1999),'A note on the relationship between initial vitamin E content of four beef muscles and drip loss during refrigerated storage', *Irish J Agric Food Res*, 38, 261–266.

LIN CF, GRAY JI, ASGHAR A, BUCKLEY DJ, BOOREN AM and FLEGAL CJ (1989), 'Effect of dietary oils and α-tocopherol supplementation on lipid composition and stability of broiler meat', *J Food Sci*, 54, 1457–1460.

LIU Q, LANARI MC and SCHAEFER DM (1995), 'A review of dietary vitamin E supplementation for improving beef quality', *J Anim Sci*, 73, 3130–3140.

LIU Q, SCHELLER KK, ARP SC, SCHAEFER DM and WILLIAMS SN (1996a), 'Titration of fresh meat color stability and malondialdehyde development with Holstein steers fed vitamin E-supplemented diets', *J Anim Sci*, 74, 106–116.

LIU Q, SCHELLER KK, ARP SC, SCHAEFER DM and WILLIAMS SN (1996b), 'Color coordinates for assessment of dietary vitamin E effects on beef color stability', *J Anim Sci*, 74, 117–126.

LOPEZ-BOTE CJ, GRAY JI, GOMAA EA and FLEGAL CJ (1998), 'Effects of dietary oat administration on lipid stability in broiler meat', *Br Poultry Sci*, 39, 57–61.

LYNCH MP and FAUSTMAN C (2000), 'Effect of aldehydes lipid oxidation products on myoglobin', *J Agric Food Chem*, 48, 600–604

MARASCHIELLO C, ESTEVE E and GARCIA REGUEIRO JA (1998), 'Cholesterol oxidation in meat from chicken fed α-tocopherol and β-carotene-supplemented diets with different unsaturation grades', *Lipids*, 33, 705–713.

MAY JM, QU Z and MENDIRATTA S (1999), 'Protection and recycling of α-tocopherol in human erythrocytes by intracellular ascorbic acid', *Arch Biochem Biophys*, 349, 281–289.

MCCARTHY TL, KERRY JP, KERRY JF, LYNCH PB and BUCKLEY DJ (2001), 'Assessment of the antioxidant potential of natural food and plant extracts in fresh and previously frozen pork patties', *Meat Sci*, 57, 177–184.

MIELNIK MB, AABY K and SKREDE G (2003), 'Commercial antioxidants control lipid oxidation in mechanically deboned turkey meat', *Meat Sci*, 63, 1147–1155.

MITSUMOTO M (2000), 'Dietary delivery versus exogenous addition of antioxidants', in Faustman C, Decker E and Lopez-Bote CJ *Antioxidants in Muscle Foods: Nutritional Strategies to Improve Quality*, New York, John Wiley, 315–343.

MITSUMOTO M, ARNOLD RN, SCHAEFER DM and CASSEN RG (1995), 'Dietary vitamin E supplementation shifted weight loss from drip to cooking losses in fresh beef *longissumus* during display', *J Anim Sci* 73, 2289–2294.

MONAHAN FJ, BUCKLEY DJ, MORRISSEY PA, LYNCH PB and GRAY JI (1992a), 'Influence of dietary fat and α-tocopherol sypplementation on lipid oxidation in pork', *Meat Sci*, 31, 229–241.

MONAHAN FJ, GRAY JI, BOOREN AM, MILLER ER, BUCKLEY DJ, MORRISSEY PA and GOMAA EA (1992b), 'Influence of dietary treatment on lipid and cholesterol oxidation in pork', *J Agric Food Chem*, 40, 1310–1315.

MONAHAN FJ, GRAY JI, ASGHAR A, HAUG A, STRASBURG GM, BUCKLEY DJ and MORRISSEY PA (1994), 'Influence of diet on lipid oxidation and membrane structure in porcine muscle microsomes', *J Agric Food Chem*, 42, 59–63.

MORRISSEY PA and KIELY M (2002), 'Vitamin E, nutritional significance', in Roginski H, Fuguay JW and Fox PF, *Encyclopedia of Dairy Science*, London, Elsevier Sciences, 2670–2677.

MORRISSEY PA and O'BRIEN NM (1998), 'Dietary antioxidants in health and disease', *Internat Dairy J*, 8, 463–472.

MORRISSEY PA, BUCKLEY DJ, SHEEHY PJA and MONAHAN FJ (1994a), 'Vitamin E and meat quality', *Proc Nutr Soc*, 53, 289–295.

MORRISSEY PA, QUINN PB and SHEEHY PJA (1994b), 'Newer aspects of micronutrients in chronic disease: vitamin E', *Proc Nutr Soc*, 53, 571–582.

MORRISSEY PA, BUCKLEY DJ, SISK H, LYNCH PB and SHEEHY PJA (1996),'Uptake of α-tocopherol in porcine plasma and tissues', *Meat Sci*, 44, 275–283.

MORRISSEY PA, BRANDON S, BUCKLEY DJ, SHEEHY PJA and FRIGG M (1997), 'Tissue content of α-tocopherol and oxidative stability of broilers receiving dietary α-tocopheryl acetate supplement for various periods pre-slaughter', *Br Poultry Sci*, 38, 84–88.

MORRISSEY PA, SHEEHY PJA, GALVIN K, KERRY JP and BUCKLEY DJ (1998), 'Lipid stability in meat and meat products', *Meat Sci*, 49 (Suppl. 1), 573–586.

MORRISSEY PA, BUCKLEY DJ and GALVIN K (2000), 'Vitamin E and the oxidative stability of pork and chicken', in Decker EA, Faustman C and Lopez-Bote CJ, *Antioxidants in Muscle Foods: Nutritional Strategies to Improve Quality*, New York, John Wiley, 263–287.

MORRISSEY PA, KERRY JP and GALVIN K (2002), 'Lipid oxidation in muscle foods' in Kadwallader KR and Weener H, *Freshness and Shelf Life of Foods*, Washington, DC, American Chemical Society, 188–200.

MORROW JD and ROBERTS LJ II (1991), 'Quantification of noncyclooxygenase derived prostanoids as a member of oxidative stress', *Free Rad Biol Med*, 10, 195–200.

MORTENSEN A and SKIBSTED LH (2000), 'Antioxidant activity of carotenoids in muscle foods', in Decker EA, Faustman C and Lopez-Bote CJ, *Antioxidants in Muscle Foods: Nutritional Strategies to Improve Quality*, New York, John Wiley, 61–83.

MOTTRAM DS (1987), 'Lipid oxidation and flavour in meat and meat products', *Food Sci Technol Today*, 1, 159–162.

MOTTRAM DS (1996), 'The role of phospholipids in meat flavour: an overview', in Pickenhagen W, Ho C-T and Spanier AM, *Contribution of Low and Nonvolatile Material to the Flavor of Foods*, Allured Publishing, Carol Stream IL, 193–206.

MOTTRAM DS (1998), 'Flavour formation in meat and meat products: a review', *Food Chem*, 62, 415–424.

MOTTRAM DS and EDWARDS RA (1983), 'The role of triglycerides and phospholipids in the aroma of cooked beef', *J Sci Food Agric*, 34, 517–522.

NAKAGAWA K, OKUDA S and MIYAZAWA T (1997), 'Dose-dependent incorporation of tea catechin (−)-epigallocatechin-3-gallate and (−)-epigallocatechin into human plasma', *Biosci Biotech Biochem*, 61, 1981–1985.

NAM KC, DU M, JO C and AHN DU (2001), 'Cholesterol oxidation products in irradiated raw meat with different packaging and storage conditions', *Meat Sci*, 58, 431–435.

NANJO F, GOTO K, SETO R, SUZUKI M, SAKAI M and HARA Y (1996), 'Scavenging effects of tea catechins and their derivatives on 1,1-diphenyl-2-picryl-hydrazyl radical', *Free Rad Biol Med*, 21, 895–902.

NAWAR WW (1996), 'Lipids', in Fennema OR, *Food Chemistry*, 3rd edition, , New York, Marcel Dekker, 225–319.

NIKI E and MATSUO M (1993) 'Rates and products of reactions of vitamin E with oxygen radicals', in Packer L and Fuchs J, *Vitamin E in Health and Disease*, New York, Marcel Dekker, 121–130.

NOZAWA Y, KASAI R, KAMEYAMA Y and OHKI K (1980), 'Age-dependent modification in membrane lipids: lipid composition, fluidity and palmitolyl CoA desaturase in

Tetrahymena membranes', *Biochim Biophys Acta*, 599, 232–245.

O'GRADY MN, MONAHAN FJ, BAILY J, ALLEN P, BUCKLEY DJ and KEANE MG (1998), 'Colour-stabilising effect of muscle vitamin E in minced beef stored in high O_2 packs', *Meat Sci*, 50, 73–80.

O'NEILL LM, GALVIN K, MORRISSEY PA and BUCKLEY DJ (1998), 'Comparison of effects of dietary olive oil, tallow and vitamin E on the quality of broiler meat and meat products', *Br Poultry Sci*, 39, 365–371.

O'SULLIVAN MG, KERRY JP, BUCKLEY DJ, LYNCH PB and MORRISSEY PA (1998), 'The effect of dietary vitamin E supplementation on quality aspects of porcine muscles', *Ir J Agric Food Res*, 37, 227–235.

O'SULLIVAN MG, BYRNE DV, JENSEN MT, ANDERSEN HJ and VESTERGAARD J (2003a), 'A comparison of warmed-over flavour in pork, by sensory analysis, GC/MS and the electronic nose', *Meat Sci*, 63, 1125–1138

O'SULLIVAN A, GALVIN K, MOLONEY AP, TROY DJ and KERRY JP (2003b), 'Effect of pre-slaughter rations of forage and/or concentrates on the composition and quality of retail packaged beef', *Meat Sci*, 63, 279–286.

OJIMA F, SAKAMOTO H, ISHIGURO Y and TERAO J (1993), 'Consumption of carotenoids in photosensitized oxidation of human plasma and plasma low density lipoprotein', *Free Rad Biol Med*, 15, 377–384.

PACKER L (1993), 'Vitamin E: Biological activity and health benefits: Overview', in Packer L and Fuchs J, *Vitamin E in Health and Disease*, New York, Marcel Dekker, 977–984.

PACKER L and KAGAN NE (1993), 'The antioxidant harvesting centre of membranes and lipoproteins', in Packer L and Fuchs J, *Vitamin E in Health and Disease*, New York, Marcel Dekker, 172–192.

PACKER L, WEBER SU and RIMBACH G (2001), 'Molecular aspects of α-tocotrienol antioxidant action and cell signalling', *J Nutr*, 131, 369S-373S.

PALACE VP, KHAPER N, QIN Q and SINGAL PK (1999), 'Antioxidant potentials of vitamin A and carotenoids and their relevance to heart disease', *Free Rad Biol Med*, 26, 746–761.

PALOZZA P (1998), Prooxidant actions of carotenoids in biologic systems', *Nutr Rev*, 56, 257–265.

PATTERSON RLS and STEVENSON MA (1995), 'Irradiation-induced off-odours in chicken and its possible control', *Br Poultry Sci*, 36, 425–441.

PEARSON AM, LOVE JD and SHORLAND FB (1977), 'Warmed-over flavor in meats, poultry and fish', *Adv Food Res*, 23, 1–74.

PELLET LJ, ANDERSEN HJ, CHEN H and TAPPEL AL (1994), 'β-Carotene alters vitamin E protection against heme protein oxidation and lipid peroxidation in chicken liver slices', *J Nutr Biochem*, 5, 479–484.

PFALZGRAF A, FRIGG M and STEINHART H (1995), 'α-Tocopherol contents and lipid oxidation in pork muscle and adipose tissue during storage', *J Agric Food Chem*, 43, 1339–1342.

POMPELLA A (1997), 'Biochemistry and histochemistry of oxidant stress and lipid peroxidation', *Internat J Vit Nutr Res*, 67, 289–297.

PRYOR WA (1994), 'Free radicals and lipid peroxidation: what they are and how they got that way', in Frei B, *Natural Antioxidants in Human Health and Disease*, London, Academic Press, 1–24.

RECORD IR, MCINERNEY JK and DREOSTI IE (1996), 'Black tea, green tea, and tea polyphenols: effects on trace element status in weanling rats', *Biol Trace Element Res*, 53, 27–43.

RENERRE M (2000), 'Oxidative processes and myoglobin', in Faustman C, Decker E and Lopez-Bote CJ, *Antioxidants in Muscle Foods: Nutritional Strategies to Improve Quality*, New York, John Wiley, 113–133.

RESURRECCION AVA (2003), 'Sensory aspects of consumer choice for meat and meat products', *Meat Sci*, 66, 11–20.

REY AI, KERRY JP, LYNCH PB, LOPEZ-BOTE CJ, BUCKLEY DJ and MORRISSEY PA (2001), 'Effect of dietary oils and α-tocopheryl acetate supplementation on lipid (TBARS) and cholesterol oxidation in cooked pork', *J Anim Sci*, 79, 1201–1208.

RHEE KS, ANDERSON LM and SAMS AR (1996), 'Lipid oxidation potential of beef, chicken and pork', *J Food Sci*, 61, 8–12.

RICE-EVANS CA, MILLER NJ and PAGANGA G (1996), 'Structure–antioxidant activity relationship of flavonoids and phenolic acids', *Free Rad Biol Med*, 20, 933–956.

RICHTER C (1987), 'Biophysical consequences of lipid peroxidation in membranes', *Chem Phys Lipids*, 44, 175–189.

RUIZ JA, PEREZ-VENDRELL AM and ESTEVE-GARCIA E (1999), 'Effect of β-carotene and vitamin E on oxidative stability in leg meat of broilers fed different supplemental fats', *J Agric Food Chem*, 47, 448—454.

ST ANGELO AJ (1996), 'Lipid oxidation in foods', *Crit Rev Food Sci Nutr*, 36, 175–224.

ST ANGELO AJ, VERCELLOTTI JR, LEGENDRE MG, VINNETT CH, KUAN JW, JAMES C JR and DUPUY HP (1987), 'Chemical and instrumental analyses of warmed-over flavor in beef', *J Food Sci*, 52, 1163–1168.

SANDERS SK, MORGAN JB, WULF DM, TATUM JD, WILLIAMS SN and SMITH GC (1997), 'Vitamin E supplementation of cattle and shelf-life of beef for the Japanese market, *J Anim Sci*, 75, 2634–2640.

SCHAEFER DM, LIU Q, FAUSTMAN C and YIN MC (1995) 'Supranutritional administration of vitamin E and C improves oxidative stability of beef', *J Nutr*, 125, 1792S-1798S.

SCHROEPFER GJ JR (2000), 'Oxysterols: modulators of cholesterol metabolism and other processes', *Physiol Rev*, 80, 361–554.

SCOLLEN ND, CHOI N-J, KURT E, FISHER AV, ENSER M and WOOD JD (2001), 'Manipulating the fatty acid composition of muscle and adipose tissue in beef cattle', *Br J Nutr*, 85, 115–124.

SHEARD PR, ENSER M, WOOD JD, NUTE GR, GILL BP and RICHARDSON RI (2000), 'Shelf life and quality of pork and pork products with a raised n-3 PUFA', *Meat Sci*, 55, 213–221.

SHEEHY PJA, MORRISSEY PA and FLYNN A (1993), 'Increased storage stability of chicken muscle by dietary α-tocopherol supplementation', *Irish J Agric Food Res,* 32, 67–73.

SHELDON BW, CURTIS PA, DAWSON PL and FERKET PR (1997), 'Effect of dietary vitamin E on the oxidative stability, flavour, colour and volatile profiles of refrigerated and frozen turkey breast meat', *Poultry Sci*, 76, 634–641.

SHERBECK JA, WULF DM, MORGAN JB, TATUM JD, SMITH GC and WILLIAMS SN (1995), 'Dietary supplementation of vitamin E to feedlot cattle affects retail display properties', *J Food Sci*, 60, 250–252.

SIES H (1991), 'Oxidative stress: introduction', in Sies H, *Oxidative Stress: Oxidants and Antioxidants*. Academic Press, London, xv–xxii.

SIES H and STAHL W, (1995), 'Vitamins E and C, β-carotene and other carotenoids as antioxidants', *Amer J Clin Nutr*, 62, 1315S-1321S.

SIES H, STAHL WM and SANDQUIST AR (1992), 'Antioxidant functions of vitamins. Vitamins E and C, β-carotene and other carotenoids', *Ann New York Acad Sci*, 669, 7–20.

SKLAN D, BARTOV I and HURWITZ S (1982), 'Tocopherol absorption and metabolism in the chick and turkey', *J Nutr*, 112, 1394–1400.

SMITH LL (1981), *Cholesterol Autoxidation*, New York, Plenum Press.

SMITH LL (1992), 'The oxidation of cholesterol', in Peng SK and Morin RJ, *Biological Effects of Cholesterol Oxides*, Boca Raton, FL, CRC Press, 7–31.

SMITH LL (1996), 'Review of progress in sterol oxidation: 1987–1995', *Lipids*, 31, 453–487.

STAHL W and SIES H (1996), 'Lycopene: a biologically important carotenoid for humans?', *Arch Biochem Biophys*, 336, 1–9.

STOCKER R, BOWRY VW and FREI B (1991), 'Ubiquinol-10 protects human low density lipoprotein more effectively against lipid peroxidation than does α-tocopherol, *Proc Nat Acad Sci*, 80, 1645–1650.

STROHECKER MG, FAUSTMAN C, FURR H, HOAGLAND TA and WILLIAMS SN (1997), 'Vitamin E supplementation effects on lipid and colour stability of whole and ground lamb', *J Muscle Foods*, 8, 413–426.

TANG SZ, KERRY JP, SHEEHAN D, BUCKLEY DJ and MORRISSEY PA (2000), 'Dietary tea catechins and iron-induced lipid oxidation in chicken meat, liver and heart', *Meat Sci*, 56, 285–290.

TANG SZ, SHEEHAN D, BUCKLEY DJ, MORRISSEY PA and KERRY JP (2001a), 'Anti-oxidant activity of added tea catechins on lipid oxidation of raw minced red meat, poultry and fish muscle', *Int J Food Sci Technol*, 36, 685–692.

TANG SZ, KERRY JP, SHEEHAN D, BUCKLEY DJ and MORRISSEY PA (2001b), 'Antioxidative effect of added tea catechins on susceptibility of cooked red meat, poultry and fish patties to lipid oxidation', *Food Res Int*, 34, 651–657.

TANG SZ, KERRY JP, SHEEHAN D, BUCKLEY DJ and MORRISSEY PA (2001c), 'Antioxidative effect of dietary tea catechins on lipid oxidation of long-term frozen stored chicken meat', *Meat Sci*, 57, 331–336.

TANG SZ, KERRY JP, SHEEHAN D and BUCKLEY DJ (2002), 'Antioxidative mechanisms of tea catechins in chicken meat systems', *Food Chem*, 76, 45–51.

TAYLOR AA, VEGA L, WOOD JD and ANGOLD M (1994), 'Extending colour life of MA packed beef by supplementing feed with vitamin E', in *Proc 40th Int Congr Meat Sci Technol*, The Hague, The Netherlands, S-IVA.44.

THURNHAM DI (1990), 'Antioxidants and prooxidants in malnourished populations', *Proc Nutr Soc*, 49, 247–259.

TICHIVANGANA JZ and MORRISSEY PA (1985), 'Metmyoglobin and inorganic metals as prooxidants in raw and cooked muscle systems', *Meat Sci*, 15, 107–116.

UNNO T, SUGIMOTO A and KAKUDA T (2000), 'Scavenging effect of tea catechins and their epimers on superoxide anion radicals generated by a hypoxanthine and xanthine oxidase system', *J Sci Food Agric*, 80, 601–606.

VATANSEVER L, KURT E, ENSER M, NUTE GR, SCOLLAN ND, WOOD JD and RICHARDSON RI (2000), 'Shelf life and eating quality of beef from cattle of different breeds given diets differing in n-3 polyunsaturated fatty acid composition', *J Anim Sci*, 71, 471–482.

WADA S and FANG X (1992), 'The synergistic antioxidant effect of rosemary extract and α-tocopherol in sardine oil model system and frozen-crushed fish meat', *J Food Proc Pres*, 16, 263–274.

WALSH H and KERRY JP (2002), 'Meat packaging', in Kerry JP, Kerry JF and Ledward D *Meat Processing: Improving Quality*, Cambridge, UK, Woodhead Publishing, 417–451.

WARKUP C (1994), 'Vitamin E and pig meat quality', *Proc Symp Vitamin E and Meat Quality*, Bologna, Italy, Institute delle Vitamine, Milan, 1–16.

WARREN HE, SCOLLAN ND, HALLETT K, ENSER M, RICHARDSON RI, NUTE GR and WOOD JD (2002), 'The effects of breed and diet on the lipid composition and quality of bovine muscle', *Proc. 48th Congr Meat Sci Technol*, 1, 370–371.

WEN J, MORRISSEY PA, BUCKLEY DJ and SHEEHY PJA (1996), 'Oxidative stability and α-tocopherol retention in turkey burgers during refrigerated and frozen storage as influenced by dietary α-tocopheryl acetate', *Br Poultry Sci*, 37, 787–795.

WEN J, MCCARTHY SN, HIGGINS FMJ, MORRISSEY PA, BUCKLEY DJ and SHEEHY PJA (1997a), 'Effect of dietary α-tocopheryl acetate on the uptake and distribution of α-tocopherol in turkey tissues and lipid stability', *Ir J Agric Food Res*, 36, 65–74.

WEN J, MORRISSEY PA, BUCKLEY DJ and SHEEHY PJA (1997b), 'Supranutritional vitamin E supplementation in pigs: influence on subcellular deposition of α-tocopherol and on oxidative stability by conventional and derivative spectrophotometry', *Meat Sci*, 47, 301–310.

WEN J, MORRISSEY PA, WALTON J and SHEEHY PJA (1997c), 'Rapid and quantitative determination of hexanal in cooked muscle foods', *Irish J Agric Food Res*, 36, 75–84.

WENK C, LEONHARDT M and SCHEEDER ML (2000), 'Monogastric nutrition and potential for improving muscle quality', in Decker E, Faustman C and Lopez-Bote CL, *Antioxidants in Muscle Food: Nutritional Strategies to Improve Quality*, New York, John Wiley, 199–227.

WOLF G (1997), 'γ-Tocopherol: an efficient protector of lipids against nitric oxide-initiated peroxidative damage', *Nutr Rev*, 55, 376–378.

WOOD JD, ENSER M, FISHER AV, NUTE GR, RICHARDSON RI and SHEARD PR (1999), 'Manipulating meat quality and composition', *Proc Nutr Soc*, 58, 363–370.

WOOD JD, RICHARDSON RI, NUTE GR, FISHER AV, CAMPO MM, KASAPIDOU E, SHEARD PR and ENSER M (2003), 'Effects of fatty acids on meat quality: a review', *Meat Sci*, 63, 21–32.

WULF DM, MORGAN JB, SANDERS SK, TATUM JD, SMITH GC and WILLIAMS J (1995), 'Effects of dietary supplementation of vitamin E on storage and case-life properties of lamb retail cuts', *J Anim Sci*, 73, 399–405.

YAGI K, ISHIDA N, KOMURA S, OHISHI N, KUSAI M and KOHNO M (1992), 'Formation of hydroxyl radicals from lipid hydroperoxides', in Yagi K, Kondo M, Niki E and Yoshikowa T, *Oxygen Radicals*, New York, Elsevier, 223–226.

YANG A, BREWSTER MJ, LANARI MC and TUME RK (2002), 'Effect of vitamin E supplementation on α-tocopherol and β-carotene concentrations in tissues from pasture and grain-fed cattle', *Meat Sci*, 60, 35–40.

YU BP (1994), 'Cellular defences against damage from reactive oxygen specials', *Physiol Rev*, 74, 139–162.

ZHANG P and OMAYA ST (2000), 'β-Carotene and protein oxidation: effects of ascorbic acid and α-tocopherol', *Toxicology*, 146, 37–48.

ZHANG P and OMAYA ST (2001), 'Antioxidant and prooxidant roles for β-carotene, α-tocopherol and ascorbic acid in human lung cells', *Toxicol in vitro*, 15, 13–24.

ZIMMER G, THÜRICH T and SCHEER B (1993), 'Membrane fluidity and vitamin E', in Packer L and Fuchs J, *Vitamin E in Health and Disease*, New York, Marcel Dekker, 207–222.

Index